T0192866

DYNAMICS OF
COASTAL SYSTEMS

SECOND EDITION

ADVANCED SERIES ON OCEAN ENGINEERING

Series Editor-in-Chief
Philip L- F Liu (*Cornell University*)

*For the complete list of titles in this series, please write to the Publisher.

Advanced Series on Ocean Engineering — Volume 41

DYNAMICS OF COASTAL SYSTEMS

SECOND EDITION

Job Dronkers

Netherlands Centre of Coastal Research, The Netherlands

 World Scientific

NEW JERSEY · LONDON · SINGAPORE · BEIJING · SHANGHAI · HONG KONG · TAIPEI · CHENNAI · TOKYO

Published by

World Scientific Publishing Co. Pte. Ltd.
5 Toh Tuck Link, Singapore 596224
USA office: 27 Warren Street, Suite 401-402, Hackensack, NJ 07601
UK office: 57 Shelton Street, Covent Garden, London WC2H 9HE

Library of Congress Cataloging-in-Publication Data
Names: Dronkers, J. J., author.
Title: Dynamics of coastal systems / Job Dronkers, Netherlands Centre of
    Coastal Research, The Netherlands.
Description: Second edition. | New Jersey : World Scientific, 2016. | Series:
    Advanced series on ocean engineering ; v. 41 | Includes bibliographical references and index.
Identifiers: LCCN 2016015330 | ISBN 9789814725132 (hardcover)
Subjects: LCSH: Coast changes.
Classification: LCC GB451.2 .D76 2016 | DDC 551.45/7--dc23
LC record available at https://lccn.loc.gov/2016015330

British Library Cataloguing-in-Publication Data
A catalogue record for this book is available from the British Library.

Typeset by Stallion Press
Email: enquiries@stallionpress.com

Printed in Singapore

Preface to the Second Edition

The spring tides of March 2015, shortly after the solar eclipse, were extreme. In the same period, strong ocean swell rushed onto the coast of Senegal. The fishermens houses at Saint-Louis are built on the beach. I saw the seafront houses be torn by the waves and tumble into the sea. The local people tried to build a seawall of the debris, in an attempt to protect the next row of houses. At first they were indignant, when they saw a stranger taking pictures of their distress.

I was with a colleague in Senegal for advice on the breach of the Senegal river; our presence that day at Saint-Louis was a coincidence. A few days later we were at a meeting with the minister of Fisheries and Maritime Affairs in Dakar, where we informed him about the situation in Saint-Louis and discussed possible measures. But what measures? Are there other choices than retreating inland? And where inland? The Langue de Barbarie, a narrow strip of land between the ocean and the Senegal River, is densely built-up. One may expect that the beach starts recovering when the weather calms down. The fishermen are likely to rebuild their homes, waiting for the next high tide or the next storm....

Understanding the Dynamics of Coastal Systems does not help much in emergency situations like this. But it is a crucial condition for coastal zone planning, in order to avoid such situations. Another crucial condition, of course, is the institutional capacity to implement such plans.

Traditionally, coastal planning and protection measures were based on lessons from the past; often costly lessons in terms of property and human lives. Today we rely more on model prognoses. In many cases, modern numerical models can hindcast quite well the impacts of past events. However, this does not guarantee reliable predictions. Mother nature can still surprise us.

The movable storm surge barrier at the Eastern Scheldt inlet was planned to protect the south-western part of the Netherlands against inundation, while preserving a unique marine ecosystem. The engineering works reduced the tidal prism by about 30%. The channel system responded by eroding sand from the tidal flats, instead of importing sand from the North Sea. Scour holes at both sides of the barrier appeared to be efficient sediment traps. We now expect that at the long term, no tidal flats will be left. This was not foreseen when the construction of the storm surge barrier was decided. We did neither foresee that the outer-delta sandbanks in front of the closed inlets would migrate onshore and emerge above low water level — at some places even above high water. At present a new lagoon-type environment is developing in the nearshore zone, compensating in some way for the loss of intertidal area in the Eastern Scheldt. This illustrates that it remains a challenge to predict the evolution of coastal systems and their response to interventions, without overlooking any major impact.

When reviewing the first edition of this book, I realised that a considerable amount of new evidence on coastal morphodynamic processes has become available in the past ten years. I therefore felt that major parts of the book had to be rewritten. I made a selection of what I consider the most essential advances in our understanding of coastal morphodynamic processes and in our understanding of the way coastal systems respond to changes in external conditions.

I included a chapter on sediment transport, discussing, in particular, new insight that has been gained in the dynamics of the wave boundary layer. This insight is crucial for better understanding the onshore and offshore sand fluxes that shape the coastal profile. I have completely rewritten the chapters on tide-topography interaction and wave-topography interaction. Morphodynamic equilibrium theories have been reformulated in a more compact and consistent way. Channel meandering and tidal flat dynamics are dealt with in more detail, taking into account the role of biota. I have taken advantage of a large body of field evidence that has become available, for illustration of the theories.

Many aspects of wave-topography interaction are still not well understood today. Results of recent field and model studies show that temporal variability and morphodynamic feedbacks are crucial elements, for the short-term response to change as well as for the long-term evolution of coastal morphology. It has become clear that the reliability of models which do not consider these elements, is highly questionable.

The last section of the book deals with coastal erosion — the issue I talked about in the beginning and which is a major concern worldwide. Twenty-five years of experience with the Dutch coastal maintenance programme provides new insight in the effectiveness of 'working with-nature' measures to combat coastal erosion. Coastal nourishments are a major element of this programme and the experience provides a realistic picture of what can be achieved with this practice.

I found several mistakes in the first edition, which I corrected. I also added illustrations of coastal systems worldwide, with thanks to Google Earth.

I maintained the structure of the book, which is based on the principles of morphodynamic self-organisation. This approach allows for a coherent treatment of the numerous and diverse processes which determine the evolution of coastal systems. I tried to keep explanations as simple as possible, emphasising the range of validity of the underlying assumptions.

'Nothing is as practical as a good theory', is an often quoted statement of social psychologist Kurt Lewin. However, all coastal morphodynamics theories have their limitations, because of the high diversity and complexity of coastal systems. I would rather say: 'Nothing is as practical as good understanding'. This summarises my intention of this second edition of Dynamics of Coastal Systems.

Acknowledgements

Rewriting Dynamics of Coastal Systems took more time than I had anticipated. I had to absorb and digest a mountain of literature. I discovered a lot of new information and got on the track of several new insights. However, it was not easy to order this into a coherent treatise, that introduces the reader into the state-of-the-art of coastal morphodynamics, without the need to consult all the underlying literature. Fortunately, I could rely on a number of colleagues to help me. Greg Rozynski has gone through almost all the chapters and thanks to him many inaccuracies in the manuscript have disappeared. Giovanni Coco provided very enriching comments on the emergence of morphological patterns and enlightened me on several new concepts. Huib de Vriend took a very critical look at the chapters on tide- and wave-driven driven morphodynamics, pointing out several mistakes. I owe suggestions about rearranging some chapters to David Prandle and I could benefit of his insightful comments on tidal dynamics. Ashis Mehta, Jan Ribberink and Leo van Rijn focused on sediment transport and put right a number of inaccuracies. Gerben Ruessink explained me in detail the dynamics of bar formation in the coastal zone. Andrew Chadwick provided additional insight into gravel shores, we do not know in the Netherlands. Finally, Zeng Bing Wang and Co van de Kreeke did a critical review of the texts about tidal inlet systems. I am greatly indebted to these colleagues for sharing their vast knowledge and expertise, and for the time and effort they have spent to take this book to a higher level.

List of Symbols

a	tidal amplitude *or* half wave orbital excursion $(=U/\omega)$ [m]
A	total channel cross-sectional area [m^2] *or* coefficient
A_C	channel cross-section (flow conveying part) [m^2]
A_u	acceleration skewness (3.70)
b	instantaneous width [m] *or* time-averaged width
b_S	surface (storage) width [m]
b_C	channel (conveyance) width
c	wave propagation speed [m/s] *or* volumetric concentration of suspended sediment
c_D	friction (drag) coefficient $(=\tau_b/\rho\overline{u}^2)$
C	sediment load (volumetric) [m] *or* roughness coefficient $C = u(\delta_r)/u_*$ *or* coefficient
$\mathbb{C}$	channel curvature $(=1/R)$ [m^{-1}]
d	grain diameter of sediment particles [m]
d_*	dimensionless grain diameter $(=d(g\,\Delta\rho/\rho v^2)^{1/3})$
D	total instantaneous water depth $(=Z_s - Z_b)$ [m]
D_L	longitudinal dispersion coefficient [m^2/s]
D_S	propagation depth $(=A_C/b_S)$ [m]
De	deposition rate (volumetric) [m/s]
e	2.7183
E	wave energy density *or* tidal energy density [J/m^2]
Er	erosion rate (volumetric) [m/s]

f Coriolis parameter [s^{-1}] *or* a function (amplitude)

f_w wave friction coefficient

F Froude number ($=u/\sqrt{gh}$) *or* wave energy flux ($=c_g E$)

g gravitational acceleration ($=9.8$ ms^{-2})

G velocity shear rate (3.49) [s^{-1}]

h time-averaged water depth [m] *or* channel depth

h_r average ripple height [m]

h_S time-averaged propagation depth ($=hb_C/b_S$) [m]

h_{br} water depth at onset of wave breaking [m]

h_{cl} closure depth [m] (7.15)

H wave height ($=H_{rms} = \sqrt{8E/g\rho} = 2a$) [m]

H_{br} wave height at the breaker line [m]

HW   time of high water (also indicated by superscript $^{+}$)

HWS   time of high slack tide

i $\sqrt{-1}$ *or* index

j index

k wave number (in x-direction) [m^{-1}]

K diffusion coefficient *or* eddy diffusivity [m^2/s]

I time-averaged water surface slope

$\Im$ imaginary part

l length (in particular basin length) [m]

l_A length of a lagoon basin along the flow axis [m]

l_b convergence length of an estuary [m]

L wavelength of tide *or* wave [m]

L_d Kolmogorov microscale ($=\nu^{3/4}\epsilon^{-1/4}$) [m]

LW   time of low water (also indicated by superscript $^{-}$)

LWS   time of low slack tide

m   metre

n velocity exponent in sediment flux formula *or* cross-channel coordinate

     or ratio of wave-group and wave propagation speed

N turbulent viscosity (eddy-viscosity) [m^2/s]

$O..$ order of magnitude estimate of ...

p pressure [Nm^{-2}] *or* parameter

p_b, p	bed porosity ($\approx 0.3 - 0.4$)
P	tidal prism [m^3]
q	volumetric sediment flux per unit width [m^2/s] *or* total flux [m^3/s]
q_b	bed load transport [m^2/s]
q_s	suspended load transport [m^2/s]
q_{br}	littoral drift (longshore sediment transport) [m^3/s]
Q	water discharge per unit width [m^2/s] *or* total discharge [m^3/s]
Q_R	average river discharge [m^3/s]
r	coefficient for linearised bottom friction ($\approx 8 c_D u_{max}/3\pi$) [m/s]
R	friction/inertia ratio ($= r/h\omega$) *or* curvature radius of channel-bend [m]
Re	Reynolds number
$\Re$	real part
Ri	Richardson-number
s	second; s= along-channel coordinate
S	salinity [ppt=parts per thousand]
$S^{(xx)}$, $S^{(xy)}$	radiation stresses (D.33) [Nm^{-1}]
S_u	velocity skewness (3.70)
t	time [s]
T	wave-period *or* tidal period [s]; T_p= peak wave period
T_s	swash period [s]
T_{sed}	settling-resuspension time lag [s]
$\vec{u}$	flow velocity vector [m/s]
u	flow velocity in x-direction [m/s]
u_c	steady current velocity [m/s]
u_w	wave-orbital velocity [m/s]; $u_{cw} = u_c + u_w$
u_{cr}	critical flow velocity for incipient sediment motion [m/s]
u_s	streaming velocity [m/s]
u_*	shear velocity $\sqrt{\tau_b/\rho}$ [m/s]

U	amplitude of tidal velocity *or* amplitude of wave-orbital velocity [m/s]
U_A	velocity asymmetry (6.64) [m/s]
U_R	velocity related to river discharge ($=Q_R/A$) [m/s]
U_S	Stokes drift velocity [m/s]
v	flow velocity in lateral y-direction [m/s]
V	volume [m^3] *or* (longshore) current velocity [m/s]
V_{br}	breaker-induced longshore current velocity [m/s]
w	flow velocity in vertical z-direction [m/s]
w_s	settling velocity [m/s]
x	longitudinal coordinate [m]
x_{cl}	location of closure depth
$X(t)$	shoreline position *or* longitudinal position in a moving frame [m]
X_{bar}	bar location [m]
X_{br}	width breaker zone (surf zone), location where waves start breaking [m]
X_s	swash excursion (runup) [m]
y	lateral coordinate (x, y, z right-turning coordinate system) [m]
z	vertical coordinate (z-axis upward) [m]
z_0	intercept $u = 0$ of logarithmic velocity profile
z_b	seabed height relative to unperturbed seabed [m]
Z_b	seabed height in a fixed reference frame [m]
Z_s	water surface height in a fixed reference frame [m]
α	coefficient sediment flux [m^{2-n}s^{n-1}]
β	bed slope/beach slope/shoreface slope ($=h_x$)
β_S	inverse Schmidt number
γ	bed-slope coefficient in sediment transport formula
γ_{br}	criterion for wave breaking ($=H_{br}/h_{br}$)
δ ..	small modification of ..
δ_r	height roughness layer [m]
δ_w	height wave boundary layer; $\delta_l=$ height log layer

Δ — difference

Δ_S — average time delay between HW/LW and corresponding slack tides [s]

Δ_{EF} — difference in the duration of ebb and flood [s]

Δ_{FR} — difference in the duration of falling tide and rising tide [s]

$\Delta\rho$ — density difference between sediment and water [kg/m^3]

∂ — partial derivative

ϵ — infinitesimal quantity *or* energy dissipation rate per unit mass [m^{2}s^{-3}]

ε_b — efficiency coefficient of bed-load transport

ε_s — efficiency coefficient of suspended-load transport

ζ — vorticity ($=\bar{v}_x - \bar{u}_y$) [s^{-1}]

ζ_{pot} — potential vorticity ($= (\zeta + f)/D$)

η — departure from mean water level
or water level with respect to horizontal reference [m]

θ — angle of wave incidence *or* angle between flow and bathymetry [rad]

θ_{cw} — angle between steady current and wave propagation direction

ϑ — Shields parameter ($=\tau_b/gd\Delta\rho$)

ϑ_{cr} — critical Shields parameter for erosion

κ — complex wavenumber ($=k + i\mu$) [m^{-1}] *or* Von Karman constant (≈ 0.4)

λ — wavelength of rhythmic seabed pattern [m]

μ — imaginary part wave number *or* coefficient exponential damping [m^{-1}]
or erosion rate coefficient

ν — kinematic viscosity [m^2/s]

ξ — Irribarren number ($=\beta/\sqrt{H/L}$) *or* variable

π — 3.14

ρ — density of (sea)water [kg/m^3]; sediment density ρ_{sed}

σ — complex radial frequency of a rhythmic seabed perturbation [s^{-1}]

σ_r radial frequency of a rhythmic seabed perturbation $\Re\sigma$ $[\text{s}^{-1}]$

σ_i growth rate (positive or negative) of a seabed perturbation $\Im\sigma$ $[\text{s}^{-1}]$

τ shear stress $[\text{Nm}^{-2}]$; τ_b= bed shear stress

τ_{cr} critical bed shear stress for erosion $[\text{Nm}^{-2}]$

ϕ phase angle between bed perturbation and sediment flux *or* shoreline angle X_y

φ phase angle between tidal variation of velocity and water level (phase angle between HWS and HW or between LWS and LW)

φ_r angle of repose

Φ flow potential ($\Phi_x = u$, $\Phi_z = w$) $[\text{m}^2/\text{s}]$ *or* dimensionless sediment transport ($q/\sqrt{gd^3\Delta\rho/\rho}$)

Φ_b dimensionless bed load transport

Φ_s dimensionless suspended load transport

χ (complex) function *or* factor

ψ stream function ($\psi_x = w$, $\psi_z = -u$) $[\text{m}^2/\text{s}]$ *or* ($\psi_x = v$, $\psi_y = -u$)

Ψ mobility parameter ($=\rho U^2/gd\Delta\rho$) *or* dispersive transport

ω angular frequency waves *or* angular frequency (semi-diurnal) tide $[\text{rad/s}]$

Ω Dean parameter ($=H_{br}/(w_s T)$)

Subscripts

$i = 0; 1; 2; ..$ residual component; basic frequency component; first higher harmonic component; ..

or   unperturbed state; 1st; 2nd order perturbation; ... ,

$x; y; z; t$ partial derivative

$f_x = \partial f/\partial x$; $f_t = \partial f/\partial t$; $f_{xx} = \partial^2 f/\partial x^2$; etc.

eq	equilibrium value f_{eq}
sat	saturation value f_{sat}
b	refers to seabed f_b
br	value at the breaker line f_{br}
c	refers to current
C	refers to the tidal channel f_C
R	refers to river flow f_R
T	refers to tidal flow f_T
w	refers to wave
cw	refers to combined wave and current
cr	refers to critical value for seabed erosion

Superscripts

$(x; y; z)$	$x; y; z$-component, $\vec{u} = (u^{(x)}, u^{(y)}, u^{(z)})$
$+ ; -$	value at HW (f^+) ; LW (f^-)
$'$	perturbation from equilibrium f'
$*$	dimensionless variable f^*

Averaging

$\langle f \rangle$	time-average $\frac{1}{T} \int_t^{t+T} f\, dt$
$\overline{f}$	depth-average $\frac{1}{D} \int_{-h}^{\eta} f\, dz$
$\overline{\overline{f}}$	cross-sectional average $\frac{1}{A} \int \int_A f\, dy\, dz$

Contents

Chapter 1

Introduction

1.1. What is this Book About?

Why are coasts the way they are?

Why are coasts the way they are: Multiform, infinitely complex, always changing and unpredictable in many aspects. But also: Why have coasts so many features in common? Any answer to these questions should be based on the universal nature of fundamental physical laws. This basis is known as 'coastal morphodynamics'.

This book is about the physical processes that shape sedimentary coastal landscapes. It introduces the reader to the physical-mathematical concepts developed during the past decades to explain the underlying basic principles. Most of the coastal landscape is under water, so landscape is a confusing word; we will use the term morphology instead.

Coastal morphology is shaped essentially by waves and currents, with often an important role of the tide. This seems like an obvious statement, but it raises a puzzling question. Coastal morphology exhibits in general a great richness of structures at many scales, from very small to very large. This strongly contrasts with the comparatively gradual spatial variability of waves, tides and wind-driven currents. One of the central themes of this book is to suggest explanations for this apparent contradiction.

1

Sediment in motion

Coastal zones are among the most dynamic and energetic environments on earth. Waves, currents and tides are the very visible expression of this dynamic nature. Acting on the shore and the seabed, the motion of the sea causes erosion and transport of seabed material. In the coastal zone, large quantities of sedimentary particles are perpetually in motion. In the Dutch coastal waters, for instance, the average quantity of sediment in motion at any moment is comparable to the net annual volume of coastal erosion (a few million cubic metres). The magnitude and direction of this sediment transport depend on waves, currents and tides, which are driven by external forces such as incoming waves, incoming tides and wind.

The coastal zone is characterised by the interdependence of water motion and seabed morphology

A crucial notion is that waves, currents, tides and sediment transport do not depend only on external forces, but also on the local topography and composition of the seabed. Hence, the magnitude and direction of sediment transport is not the same at different places in the coastal zone. At some places there will be erosion and at other places there will be deposition of sediment. As a result, seabed and shoreline are continuously changing: Changing position, changing form and changing composition. This change, in turn, affects waves, currents and tides. In other words: Coastal morphology and water motion evolve in an interdependent way. Sedimentary coastal environments are characterised by the continuous mutual adaptation of coastal morphology and water motion. Coastal systems in this book are defined by:

- water motion is substantially influenced by oceanic conditions;
- water motion and coastal morphology are interdependent.

Typical coastal systems corresponding to this definition are sketched in Figs. 1.1 and 1.2.

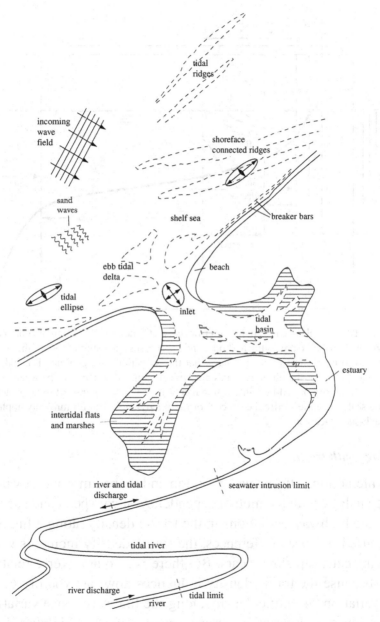

Fig. 1.1. The coastline may ingress far inland. The systems sketched in the figure are characteristic for low-lying coastal plains. Nonlinear interaction with water motion plays an important role in shaping the morphology of these coastal systems and the morphology of the near-shore seabed. The figure presents several features of the coastal environment and the corresponding terms used in this book.

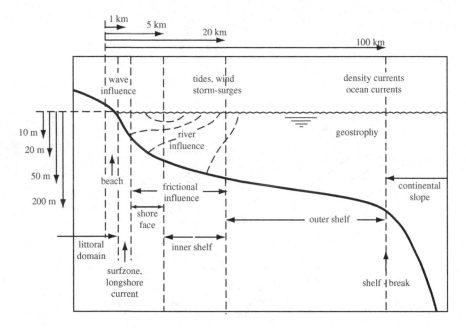

Fig. 1.2.   The coastal zone is a continuum from beach to shelf break. Different zones can be distinguished related to prevailing hydrodynamic processes and external forces; these zones correspond to characteristic depth ranges, but the transitions are gradual. The indicated depths and widths of the different zones are typical values that can vary between coastal zones. The intensity of bed-flow interaction increases from the outer shelf to the beach. At the time scale of sea-level rise the concept of morphodynamic equilibrium only applies to the beach-shoreface zone.

Analogy with traffic

The interaction between water motion and seabed morphology bears some analogy with the mutual dependency of car speed and car density on a highway. Variations in the traffic density along a highway are caused by speed differences; the traffic density increases where fast cars catch up slower cars. But there is also an inverse relationship, because the traffic density influences how fast you drive your car. Variations in traffic density along the highway cause variations in car speed and variations in car speed along the highway cause variations in traffic density. Hence, both evolve continuously in a mutually dependent way. If traffic density and speed differences are high, this mutual dependency leads to 'spontaneous' amplification of car density fluctuations. Amplification occurs when the 'inflow' of cars at the back of an initial fluctuation is greater than the 'outflow' at the front. Morphologic structures, like ripples or dunes, emerge when the sediment influx at the stoss side of a seabed perturbation is greater than the sediment outflow at the lee side.

Morphodynamics and morphodynamic equilibrium

In sedimentary coastal environments, the seabed is all the time subject to erosion and sedimentation processes. A perfect instantaneous balance between erosion and sedimentation does not exist; it makes no sense to talk of an instantaneous morphological equilibrium. In general, we are not interested in the instantaneous change of the seabed. What we want to know in practice, is the evolution of the seabed morphology averaged over a certain time interval. This time interval can be a wave period, a tidal period, a spring-neap tide cycle, a meteorological season or a number of years. We use the term 'morphodynamics' for the processes that steer morphological evolution of the seabed over different time intervals. At small spatial scales, seabed, morphology and water motion adapt to each other in a short delay, but at large spatial scales the adaptation period can be very long. If erosion and sedimentation are in balance, averaged over large temporal and spatial scales, an imbalance may exist at smaller scales or vice versa. In fact, the phenomena erosion, sedimentation and sediment transport always have to be defined in relation to particular temporal and spatial scales. If erosion and sedimentation are in balance over a certain time interval, we say that the seabed is in 'morphodynamic equilibrium'. A morphodynamic equilibrium defined with respect to a given temporal and spatial scale, does not necessarily imply a balance of erosion and sedimentation processes at smaller or larger temporal and spatial scales.

Stability of a morphodynamic equilibrium

Suppose a situation where erosion and sedimentation are in balance, averaged over a long-time interval and large spatial scale. What happens if this morphodynamic equilibrium is perturbed, for instance, by adding a small sedimentary deposit? There are several possibilities. The sediment may be dispersed and the deposit may disappear after some time; in that case the equilibrium is stable. The additional sedimentary deposit may also remain unaffected; in that case the equilibrium is called marginally stable. The third possibility is that the deposit starts to grow. In this last case the equilibrium is unstable.

Formation of morphologic features

The last possibility seems counter-intuitive if we agree that the added sediment does not possess any special attractive force. What makes the deposit grow? There is no unique answer to this question; it will be shown later that different processes may play a role. All these processes have in common that the flow perturbation produced by the sedimentary deposit affects the existing sediment motion in such a way that sediment converges at the deposit. This phenomenon is inherent to the nonlinearity of the equations describing water motion and sediment transport. However, the underlying principles are more general and also play a role in other systems with nonlinear feedback processes, such as the earlier mentioned occurrence of traffic jams on a highway. In Chapter 2 it will be shown that these basic principles can be captured in the concept of symmetry breaking, which is inherent to the nonlinear nature of the interaction between seabed morphology and water motion [631]. The concept of symmetry breaking applies to the emergence and evolution of most morphologic structures in sedimentary coastal environments, from ripples to sand banks, from creeks to tidal inlets.

Time scales of coastal morphodynamics

The physical processes that determine coastal morphology span a range of temporal scales covering more than ten orders of magnitude,

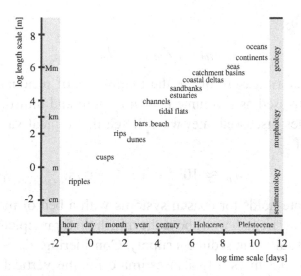

Fig. 1.3. Geomorphologic patterns in coastal systems span a very large range of space and time scales. Space and time scales are closely related. Adapted from [404].

see Fig. 1.3. At the lower end of this range we will not discuss processes at time scales smaller than the time scale of turbulence; this time scale is typically between one second and a few hours, depending on the type of flow. For these processes we will adopt empirical parametric descriptions. At the upper end we will limit the discussion to the time scale of substantial sea level change, which is in the order of ten thousand years. This excludes long-term geological processes, such as plate tectonics and glacial cycles, which are responsible for the global distribution of coastal environments. We will not try to answer the question why certain types of coastal systems are where they are; we accept inherited large-scale characteristics, such as the width of the continental shelf, the seabed composition and the external hydrodynamic forcing by waves, wind and tides.

Spatial scales of coastal morphodynamics

The restriction on the range of time scales is equivalent to a restriction on the range of morphologic scales. The temporal scale T_m and the spatial scales L_m (length), Z_m (scale of depth variation) are related by

$$q_m = Z_m \, L_m \, / \, T_m, \tag{1.1}$$

where q_m is a measure of the the magnitude of residual sediment fluxes (expressed as a volume per unit width and unit time). From the examples discussed later we will see that typical values are on the order of

$$q_m \approx 10^{-5} - 10^{-4} \ \mathrm{m^2/s}. \tag{1.2}$$

This estimate holds for coastal systems with a bed of mobile sediment, current strength well above the threshold for incipient sediment motion and sufficient sediment supply. Considering $T_m \sim 100$ years and $Z_m \sim 0.3$ m as a typical estimate for the vertical scale, we find $L_m \sim 100$–1000 km. According to (1.1), this is an estimate of the largest spatial scale at which coastal morphology can adapt to the present sea-level rise. Therefore we may expect that, under conditions where (1.2) holds, the large-scale morphology of coastal systems at spatial scales less than 100 km will not be far from equilibrium. This holds only for natural coastal systems, which have not been strongly modified by human interventions. Examples of such coastal systems are sketched in the Figs. 1.1 and 1.2.

1.2. Why this Book?

Breakthrough in understanding

Our understanding of coastal morphology has made spectacular progress during the past decades. This progress in understanding has benefited from technical innovation: Refined observation techniques and increased computing power. But a major key to this progress is conceptual innovation: Relating coastal morphology to symmetry-breaking properties of the nonlinear feedback with water motion. It appears that many patterns in seabed morphology can be related to the instability of an initially uniform equilibrium morphology under the action of currents. Other topographic characteristics can be related to the residual sediment fluxes generated by waves that lose their initial symmetry when propagating into the coastal zone. Symmetry breaking is the leading concept to which we will refer throughout this book as a unifying approach of the very diverse processes involved in coastal morphodynamics.

A *laboratory of stereotypical idealised coastal settings*

The basic mechanisms of symmetry breaking can best be illustrated for stereotypical idealised settings in which particular processes are singled out and most of real-life complexity is ignored. This artificial reduction of complexity makes the physics easier to understand; it cannot claim, however, to provide accurate predictions in real-life situations and entails the risk that the true complexity of the coastal environment is underexposed.

Stereotypical idealised settings that will be discussed include unbounded steady flow over a flat horizontal or sloping seabed, unbounded oscillating flow over a flat horizontal or sloping seabed, tidal flow in basins with uniform geometry or converging geometry.

Each time, we explain symmetry-breaking feedback processes in three complementary ways. First we introduce the process through a qualitative discussion illustrated with field observations, then we explain the process schematically in a figure and finally we provide a mathematical description. This last part can be skipped by the less mathematically inclined readers, although some features only appear through the mathematics. The mathematical analysis of spatial symmetry breaking is based on linear stability analysis. This imposes a major restriction: Only the initial emergence of morphologic patterns can be described in this way and not the evolution towards fully developed patterns.

Important *contributions to the discipline*

The great progress in physical coastal science of the past fifty years is related to the economic expansion after the second world war. A major trigger came from investments in coastal and harbour development.

It was soon recognised that such interventions interfere with the natural dynamics of coastal systems, which often respond in an unin-
tended way. In many countries research laboratories were established
or expanded in order to enable better control on coastal behaviour.
This book highlights essential discoveries made by these research
groups and attempts to bring them together in a coherent framework.
It focuses on understanding basic principles; it does not pretend
an in-depth review of all scientific developments. More complete
reviews on dedicated subjects can be found in other recent textbooks
[216,492,575,963]. I took inspiration for the approach based on self-
organisation principles in coastal morphodynamics from the work of
Nicolis and Prigogine [631] on complex-system dynamics and from
the symposium volume 'River, Coastal and Estuarine Morphodynamics' [766], edited by Seminara and Blondeaux (2001), with many
excellent review articles. The editors rightly mention R. A. Bagnold
[45] and F. Engelund [284] as leading pioneers of sediment transport
mechanics. Their work has led to important breakthroughs in understanding sedimentary processes in both the fluvial and the coastal
environment. Due to the many similarities between the two fields,
river research has strongly stimulated progress in coastal morphodynamics. Important generic insight has been gained from research
on the formation of ripples, dunes, bars and meanders in rivers, with
leading contributors such as L. B. Leopold [519], J. F. Kennedy [479],
M. S. Yalin [972], W. H. Graf [352], G. Parker [661] and S. Ikeda
[440]. Recent advances in this field have been made by the French
research group of B. Andreotti [310].

Sediment transport mechanics

The Danish school of river and coastal morphodynamics founded
by F. Engelund, with such leading scientists as J. Fredsøe and R.
Deigaard [318], has greatly contributed to the development of a theoretical framework for sediment transport mechanics and bed-flow
interaction. Important insight in coastal zone wave dynamics and
its influence on sediment transport is due to M. S. Longuet-Higgins

(1953) [537] and to later work of J. Battjes [59] on wave-breaking processes. Major contributions to the practical description of sediment transport in the coastal zone were made by E. Bijker (1967) [77], L. van Rijn [904], D. Huntley [433], R. Soulsby [796], P. Nielsen [634], J. S. Ribberink [717] and others. Our knowledge of the behaviour of cohesive sediments was greatly advanced by the pioneering experimental work of R. B. Krone [496] and A. J. Mehta [592] and the modelling work of P. LeHir [514] and J. Winterwerp [956].

Beach processes

The introduction of the equilibrium profile concept (Fig. 1.2) was a crucial step in coastal morphodynamics. It was first proposed a century ago [186, 187, 465], but in 1954 P. Bruun highlighted the implications of this concept for predicting long-term coastal evolution [113]; the theoretical foundation of this concept was further advanced by R. G. Dean (1973) [211] and A. J. Bowen (1980) [104]. The first comprehensive description of beach processes was made by L. D. Wright and A. D. Short (1984) [966], members of a very productive Australian coastal science community including other leading scientists such as R. W. G. Carter [138] and C. D. Woodroffe [962]. Morphodynamic feedback leading to the formation of shoreline patterns was demonstrated in cellular automata models by B. T. Werner [940], A. A. Ashton [33], A. B. Murray [623] and G. Coco [173]. After first exploratory work by M. Hino in 1974 [395], the Catalan research group of A. Falqués [293] with coworkers D. Calvete [128] and F. Ribas [713] managed to unravel the basic mechanisms responsible for pattern formation in the surf zone. The observational basis of this work was laid by R. Holman [404], who developed the ARGUS-camera technique and initiated the establishment of a worldwide network of observation sites. Major insight in bar dynamics in the surf zone is based on field studies by G. Masselink [576], G. Ruessink [739], A. Kroon [498], T. Aagaard [2] and B. Castelle [144].

Seabed structures

The coastal environment is characterised by a broad spectrum of seabed structures; sand ripples are at the small-scale end of this spectrum and tidal sand banks at the large-scale end. The mechanics of ripple formation has long been one of the most challenging coastal phenomena; the basic mechanisms were first described in detail by J. R. L. Allen [10] and J. F. A. Sleath [781] and later simulated with process-based analytical models developed by the Genua morphodynamics school of P. Blondeaux [88] and G. Vittori [922]. Important field investigations of tidal and nontidal sandbanks have been conducted by the research groups of T. Off [647], J. J. H. C. Houbolt [412], J. H. J. Terwindt [838], I. N. McCave [585], D. J. P. Swift [824], B. A. O'Connor [641] and V. R. M. Van Lancker [898]. The mechanics of tidal sand banks were first described by J. D. Smith [788], J. T. F. Zimmerman [989] and J. M. Huthnance [436,437]. Much of our understanding of bedform generation in the coastal zone is due to the work of H. De Swart [227] and S. Hulscher [424].

Tidal morphodynamics

The first morphodynamic description of tidal basins was made by J. van Veen (1950) [916]. This description was corroborated and extended with observational evidence on depositional processes in the estuarine environment by the research groups of G. P. Allen [13], R. W. Dalrymple [197], J. S. Pethick [670], C. L. Amos [20] and others. The essential role of tidal asymmetry and its morphological consequences was first recognised in 1954 by H. Postma [682]. Generation mechanisms of tidal asymmetry were first studied by P. H. LeBlond [511], D. Prandle [687] and D. G. Aubrey [38]; the relationship with estuarine morphology was further explored by C. T. Friedrichs [323], D. A. Jay [455], G. Seminara [768], S. Lanzoni [507] and others. Important steps towards process-based models of the full coastal system, including tidal basins and the adjacent shoreline, were realised within the Netherlands Centre

for Coastal Research, with major contributions by H. De Vriend,
M. J. F. Stive and D. J. A. Roelvink [805].

Bio-geomorphodynamics

Feedbacks between biological activity and morphology determine
to a large degree the dynamics of sheltered coastal areas, such as
the intertidal zones of lagoons and estuaries. The diversity of disciplines required for studying this mutual interaction has long been
an obstacle, but in recent years serious progress has been made in
this field. Pioneering work was done by the Dutch-Belgian research
group of P. Herman [876] and T. Bouma [101] and colleagues
S. Temmerman [833] and B. Borsjc [98]. Important insight in
tidal marsh dynamics has been gained from the modelling work of
S. Fagherazzi [290] and G. Mariotti [902].

A complementary approach

This short overview of important contributions to our present understanding of sea-land interaction is far from complete. No existing
textbook covers the entire field. The classical textbook on coastal
sedimentary processes is K. Dyer's 'Coastal and Estuarine Sediment
Dynamics' (1986) [258]. Other important textbooks are 'Beach Processes and Sedimentation' by P. D. Komar (1998) [492], 'Coasts' by
C. D. Woodroffe (2002) [963], 'Coastal processes' by R. G. Dean and
A. Dalrymple [216] and 'Introduction to coastal processes and geomorphology' by G. Masselink and M. Hughes (2002) [575]. Estuarine
dynamics are dealt with in the textbook by D. Prandle [691] and fine
sediment dynamics in the textbook by A. Mehta [594]. The present
book complements these recent overviews by its focus on the basic
physical principles underlying sea-land interaction.

1.3. Who is this Book Intended For?

Sustainable coastal management

Coastal zones directly support a growing part of the world population [785]. Highest densities of urbanisation are found in sedimentary

coastal plains, which have been shaped by land-sea interaction. This interaction has not ceased, and coastal morphology continuous to evolve. Coastal evolution can be frozen, locally and temporarily, by engineering interventions. But the large-scale evolution can hardly be stopped. Artificial constraints may even speed up coastal evolution instead of slowing it down; this will occur if the coastal system is brought further away from equilibrium. Sustainable coastal planning aims to avoid conflicts with natural coastal evolution; therefore it has to rely on a solid understanding of the mechanisms of land-sea interaction.

Continuing efforts to maintain the coastline

The ancient village of Egmond along the Dutch North Sea coast disappeared two centuries ago in the sea (Fig. 1.4). It could have been protected against the natural retreat of the coastline by the construction of sea walls, but for how long? At present, the coast around

Fig. 1.4.   The church of Egmond was swallowed by the sea in the 18th century. The picture was drawn shortly before collapse.

Egmond is nourished with sand taken from far offshore, in order to maintain the coastline and to improve the stability of this coastal stretch. It is expected that sand nourishment will have to continue for centuries, probably with increasing quantities to combat the effect of increasing sea levels. Coastline maintenance can only be sustained by using the North Sea bottom as a permanent sand source for coastal nourishment [494].

Cost-effectiveness of coastal policies

The example of Egmond illustrates the dilemmas related to coastal development. It implies striking a balance between benefits and costs; this balance is more easily evaluated for the short term than for the long term, but the latter is often more significant. Several policies for responding to coastline retreat are possible, ranging from hard protection structures to abandonment of settlements, from supratidal beach nourishment to subtidal shoreface nourishment. The effectiveness of different policies depends to a large degree on the long-term response of the coastal system to the intervention. A coast eroded by tidal currents may require other measures than a wave-dominated coast. Estimating the cost-effectiveness of measures is possible only if the natural dynamics of the coastline is well understood.

Coastal observation is essential

Any attempt to understand the coast should start with observing the coast. This is a major investment, because the time scale of coastal evolution is long and trends are masked by short-term fluctuations. In the Netherlands a coastal monitoring programme started since the 1960s. Each year, the shoreface-profile is measured over a cross-shore distance of 800 m along the entire coastline, with a long-shore spacing of 200 m. The coastal data set already covers almost 50 years; this information guides the annual coastal nourishment planning.

Since a few decades ARGUS video cameras have become a frequently used technique for coastal observation. The continuously sampled images give insight in the processes underlying coastal erosion and accretion at timescales ranging from hours to decades. A worldwide network of cameras is presently in operation.

Interpreting these data sets is essential to coastal zone planning. Each year large fluctuations occur in the coastline position, but many of these fluctuations do not correspond to long-term trends. Effective coastline management requires a thorough interpretation of observed coastline behaviour.

Management of river mouths, tidal basins and coastal wetlands

Sustainable management of river mouths, estuaries, tidal basins and coastal wetlands poses dilemmas similar to those in shoreline development and maintenance. Such dilemmas arise from the wish to protect existing features, or from new claims on the coastal environment, which are mutually conflicting or conflicting with natural coastal evolution. Decision-makers need to know whether an intervention may conflict with natural coastal evolution and at which scale, in space and time. How can such conflicts be avoided or mitigated? At which costs? Does the intervention influence coastal evolution? Will this affect existing uses and opportunities? How can competing claims and interests be reconciled? Sound policy choices make the the best possible use of the progress achieved during the past decades in understanding coastline dynamics; this understanding has increased substantially the capability to produce more reliable predictions of the natural evolution of the coastal environment and its response to intervention.

Models, rules and analogies

Present knowledge is not yet sufficiently advanced, however, for delivering fully reliable standard tools for predicting large-scale and long-term coastal evolution. Certain (semi-)empirical relationships are often successful, in situations that correspond to conditions for which the rules have been derived. We will discuss the 'Bruun rule' for coastal retreat (Sec. 7.6.2), the 'Dean rule' for the shoreface slope (Sec. 7.2.1), the 'O'Brien rule' for the cross-sectional area of tidal inlets (Sec. 6.2.1) and the 'Walton rule' for the ebb-delta volume (Sec. 6.2.2). In this book some additional relationships are derived from simple models, which are useful for predicting coastal evolution and the coastal response to intervention. It must be emphasised that these relationships have a restricted validity; they are applicable only under certain conditions. Understanding these conditions is even more important than knowing the relationship. It requires for each case knowledge of the specific field situation, which can only be obtained through observation programmes. The most important benefit of the morphologic relationships is the help they offer for interpreting the observations and for drawing the right conclusions. Models and rules can easily be misused. The emphasis in this book is therefore primarily on the processes governing land-sea interaction, rather than on specific field cases that may give rise to misleading analogies.

Coastal diversity

The land-sea interaction mechanisms discussed in this book are often illustrated with field evidence from the coastal zone of the Netherlands. This is a deliberate choice; many of these data were collected for special coastal management purposes by the Dutch Ministry of Public Works and Water Management and are not available in the open scientific literature. It must be realised that these data are not necessarily representative of other coastal environments. Major characteristics of the Dutch coast are the minor influence of ocean swell, medium tidal range (2–4 m), the modest fluvial sediment supply and the geological setting of a slowly but steadily subsiding basin. The basic land-sea interaction mechanisms are the same in most sedimentary coastal environments; however, the relative importance of these processes may differ greatly from one coastal zone to another, even over short distances.

Students and coastal professionals

This book was written in the first instance for learning purposes; it is based on a lecture course on coastal morphodynamics given at the Universities of Utrecht and Delft in the Netherlands for graduate students who are familiar with the basic concepts of coastal hydrodynamics. I have extended the material to provide an overview of the main physical principles of land-sea interaction. It enables coastal engineers to complete their background knowledge and to facilitate access to cutting-edge scientific literature on specific topics. The book may also serve to familiarise professionals in other coastal disciplines with modern concepts of land-sea interaction. The mathematical subsections, presented at the end of several sections under the heading 'A simple model', can often be skipped without affecting the understanding of the essential concepts.

1.4. How is this Book Organised?

Basic mechanisms

Throughout this book much emphasis is placed on explaining the mechanisms responsible for the generation of coastal morphology. A rough indication of the morphologic spatial and temporal scales can already be derived from these qualitative descriptions. These descriptions also provide an understanding of the conditions under which different topographies develop. For each interaction mechanism, a qualitative discussion is followed by an analytical model which is solved for a particular idealised stereotypical situation. The solution repeats the qualitative discussion, but sharpens the underlying assumptions. The predictive value of these models for real-life situations is limited; they should be considered primarily as tools for analysis and understanding.

Chapter 2: Principles of symmetry breaking

The concepts of spatial and temporal symmetry breaking provide a unifying framework for the multiple morphodynamic feedback processes discussed in the following chapters. The tidal bore in the Qiantang estuary (China) is discussed as an illustration of time symmetry breaking in wave propagation. Spatial symmetry breaking is illustrated by the instability of a system of parallel channels. The history of the Rhine delta (at a timescale of centuries) and the history of the Ameland reef (at a timescale of years) are discussed as examples.

Chapter 3: Boundary layers and sediment transport

Sediment transport depends crucially on the characteristics of flow near the seabed. Different cases are discussed: Steady currents, waves and a combination of both. The near-bottom flow depends on the fine structure of the seabed, but also influences this fine structure by reworking the seabed. The characteristics of sedimentary particles play an important role, as well as their interactions in the seabed matrix. These interactions are very complex, especially when fine sediments are present. They are influenced by chemical interactions and biotic activity. Up till present, sediment transport can only be described by empirical formulas. A thorough understanding of the underlying microscale processes is essential when these formulas are used in practice. This chapter provides major background notions on which the following chapters rely.

Chapter 4: Current-seabed interaction

Seabed instability inherent to the interaction of currents with topography results in spatial symmetry breaking and the generation of rhythmic seabed patterns. Spatial symmetry breaking occurs for steady currents as well as for oscillating currents (tidal flow and wave-orbital flow). The basic principle is the same in both cases, but the resulting patterns are different. Several symmetry-breaking processes are discussed, which produce seabed structures at very different scales, from ripples (wavelength of tens of centimetres) to tidal sandbanks (wavelength of kilometres). It will also be shown how fine and coarse sediments can selforganise into separate patterns.

Chapter 5: Current-channel interaction

Tidal flow in estuaries and tidal lagoons creates a topography of channels and tidal flats. This chapter deals with processes of spatial symmetry-breaking that underly the development of this topography. The primary focus is on the interrelated processes of channel meandering and tidal flat building. It is shown that wave action and biotic activity play an important role in the development to a mature channel-flat system.

Chapter 6: Tide-topography interaction

Tide-topography interaction in estuaries and tidal lagoons affects tidal propagation in such a way that differences arise in the strength of flood and ebb currents and in the duration of high-water slack tide and low-water slack tide. This asymmetry generates net sediment transport and determines to a large extent the large-scale morphological characteristics of these systems (for example, intertidal flat area, channel convergence length): Tidal asymmetry is therefore both cause and result of morphologic development. Stability criteria of estuaries and tidal lagoons are discussed with reference to field evidence of a large number of tidal systems documented in the literature. The results are relevant for estimating the response of these systems to human interventions (dredging, sand mining, tidal flat reclamation) and to sea level rise.

Small-sized particles are a major constituent of the sediments in many estuaries and tidal lagoons. Their transport dynamics differ notably from the medium-coarse sediment fraction. We discuss in particular transport related to settling- and resuspension-lag effects and transport by large-scale dispersive processes.

Chapter 7: Wave-topography interaction

Sediment transport in the nearshore zone is primarily caused by wave activity. Waves interact with coastal topography in several ways, involving morphological feedback to wave dissipation, wave refraction and amplification of wave asymmetry. Wave asymmetry in shallow water is at the root of net sediment fluxes and topographic adjustment, which shapes the coastal profile. The mutual interaction of wave dissipation and wave refraction with the seabed in the nearshore zone generates a broad spectrum of seabed patterns, such as transverse bars, rip cells and beach cusps. Dissipation of obliquely incident waves generates a longshore current in the nearshore zone; the interaction of this current with the longshore morphology can transform a straight shoreline in an undulating shoreline (and vice versa): The last sections of this chapter deal with coastal erosion processes and different options for dealing with coastal erosion. Particular attention is given to the question: How can we take advantage of the forces of nature for managing our coasts, instead of fighting these forces.

Appendices

The basic equations governing tidal flow and wave propagation are presented in appendices, together with solutions of these equations relevant for the models discussed in the different chapters. The appendices are intended for the reader familiar with basic mathematical techniques. They should provide the information needed to reproduce the results of the different analytical models discussed in this book, without the help of other sources.

Chapter 2

Morphodynamic Feedback

2.1. Pattern Generation

A remarkable coastal feature

Once I went for a walk with friends along the beach of northern Brittany (France), near Morlaix. The sight of this coastal landscape is fascinating, with its red granite rocks and pocket beaches, nested in between high cliffs protruding far into the sea. The lower part of the beaches is sandy and smooth, with a half-moon shaped shoreline. Higher up, the beach becomes steeper and it is covered with shingle. The upper part of the beach is made of nicely rounded and coloured boulders, a few tens of centimetres in size. It was low water, about 3 m below mean tidal level, and we could walk from one pocket-beach to the next by climbing the rocky promontories. One of the pocket-beaches looked different from the others, but it took some time before we realised the cause of this. The boulders at the top of the beach were arranged in a sequence of bows, terminating in sharp cusps of about one metre high, pointing seaward. The spacing between the cusps was the same everywhere; about 15 metres. We counted more than ten cusps along the upper rim of the beach, over a distance of a few hundred metres. This regular pattern looked quite artificial. My friends asked me whether the boulders could have been placed in cusps for coastal defence reasons. I answered that such coastal defence structures are most unusual, and that I could not see any good reason for coastal defence structures in this place. But if the boulder pattern had not been built by humans, what could be the explanation?

And why did the cusp pattern not appear on other pocket beaches? My friends proposed that it might be the work of extraterrestrials.

A *symmetry-breaking event*

Talking about extraterrestrials, let us imagine that we can travel back in time. We go back to the time that the boulders are still uniformly scattered on the beach; the beach now has a perfectly uniform appearance. But suddenly it gets dark; black clouds cover the sky and the wind starts blowing very hard. A heavy storm rises and huge ocean waves beat against the coast. Breaking waves lift the boulders, which are entrained by a violent swash flow and hurled higher up the beach with a loud roar. After a few terrible hours the storm calms down, the clouds move away and it gets clear enough to see the beach. The lower sandy beach strip has become much steeper, but the most striking is the regular pattern of the boulders, which are now placed in equally spaced cusps [938]. The beach has lost is original uniformity (see Fig. 2.1). The boulder pattern introduces a new spatial scale that

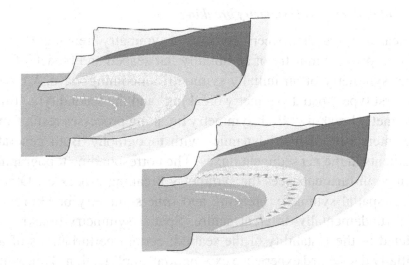

Fig. 2.1. Pocket-beach enclosed by rocky headlands on an exposed, macrotidal coast, at low water. The lower part of the beach is sandy, the higher part is covered with cobbles and boulders. Upper left: Longshore uniform beach. Below right: Rhythmic pattern of beach cusps on the higher part of the beach. See also the beach cusp pictures in Fig. 7.37.

did not exist before. But, even stranger, a similar scale did not appear in the waves hitting the coast, which had a much broader front and a far greater wavelength.

Is natural symmetry breaking possible?

In this story waves and boulders together played a game with a remarkable outcome. Could this really happen as described above? Could the transformation from a uniform to a non-uniform coast be achieved without human intervention? Would physical laws allow that a perfect symmetry is broken? How can the cusp pattern be so regular and what can determine the spatial scale? We will leave these questions till later as far as this particular example is concerned. We notice, however, that many other phenomena in nature raise the same type of questions [631]. For instance, how can a fluid change into crystals? How does a regular wave field arise from a flat ocean surface? Symmetry breaking and pattern formation are fundamental properties of the physical laws postulated by Newton.

Spatial and temporal symmetry breaking

We can distinguish between two kinds of symmetry breaking: Breach of the spatial symmetry of an initially flat seabed and breach of the time symmetry of an initially symmetric incoming wave. We call the first type 'spatial symmetry breaking' and the second type 'time symmetry breaking'. Both symmetry breaking processes result from the interaction of hydrodynamics with topography. Both generate gradients in the net sediment fluxes. The corresponding topographic change in turn enhances the symmetry breaking processes. Otherwise, spatial symmetry breaking and time symmetry breaking are of a fundamentally different nature. Spatial symmetry breaking is related to the instability of the seabed; certain perturbations of an initially flat seabed experience exponential amplification. Time symmetry breaking is not related to seabed instability, but to the influence of topography on the propagation of incoming waves — wind waves, swell or tidal waves. The initial symmetry of these waves, with mirrored forward and backward fluid motions, is broken; duration and strength of forward and backward fluid motions become different. Time symmetry breaking plays a major role in long-term coastal evolution, as discussed in Sec. 2.2. Spatial symmetry breaking is discussed in Sec. 2.3. It may explain, for instance, the formation of cusps on an initially uniform beach. Later we will provide evidence that many other morphologic patterns come about as a result of spatial symmetry breaking processes.

2.2. Time-Symmetry Breaking

2.2.1. *Wave Asymmetry*

Propagation of a disturbance

When we throw a stone in a pond, waves will radiate away from the location where the stone has hit the surface (see Fig. 2.2). More generally, when a water body is disturbed locally, the disturbance will propagate through the water body away from the initial location. With time, an increasing part of the water body will be affected, depending on the propagation speed of the disturbance and on the rate at which the disturbance dissipates. For water bodies with a free surface — the ocean, a shelf sea or a tidal river — the propagation speed depends mainly on water depth if the disturbance length scale (the wavelength) is much larger than the water depth. The water depth, however, is not an intrinsic property of the water body; it is influenced by the disturbance, at least to some degree. If the water surface disturbance is only a very small fraction of water depth, the

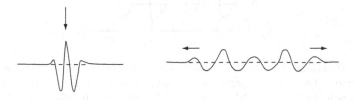

Fig. 2.2. A local impact on the water surface produces a disturbance which propagates away from the initial impact location.

influence of the disturbance on the propagation speed can be ignored. If, in addition, the disturbance height is very small compared to the wavelength, then the disturbance will propagate throughout the water body without being distorted. Swell waves generated by wind fields far offshore propagate shoreward without much distortion, until they reach the coastal zone, where the wavelength and water depth are no longer very large with respect to wave height.

Onshore wave propagation

The asymmetry of swell waves in shallow water is clearly visible from the beach. The crest of the incoming waves becomes increasingly steep as the wave front approaches the shore. At a certain moment the wave crest catches up the wave trough; then the wave overtops and breaks. Wave steepening in shallow water implies an increasing asymmetry between onshore and offshore orbital motion of water particles. The onshore acceleration is stronger than the offshore acceleration and the onshore water motion is of shorter duration than the offshore motion, but also stronger (Fig. 2.3). The

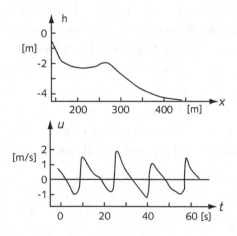

Fig. 2.3. The coastal profile near Duck (North Carolina, USA Atlantic coast), from the low-water line to 400 m offshore, representing the surf zone (zone where waves break). Below: Observed near-bottom wave-orbital velocity at 1.5 m water depth in the above profile, just before wave breaking. Notice the dominance of onshore wave-orbital velocities and also the strong acceleration from offshore to onshore velocity. Redrawn after [275].

wave orbital motion moves bottom sediments back and forth, but onshore sediment transport is higher than offshore transport. Sand is piled up against the coast by this mechanism; this goes on until a balance is reached between net onshore transport due to wave asymmetry and net offshore transport due to wave breaking and to the effect of gravity on the offshore sloping coastal profile. In reality an equilibrium profile is never attained; wave-asymmetry mainly depends on the relative wave height (ratio of wave height and depth) and this quantity changes continuously with tide and wind conditions.

Concave coastal profile

Simple models are insufficient for predicting the instantaneous coastal profile in detail. Coastal profiles have a complex shape and each profile can be quite different from other profiles at neighbouring locations along the coast. However, many coastal profiles share the common characteristic of concavity, i.e., the tendency of increased shoaling in the onshore direction [212]. A typical coastal profile is shown in Fig. 2.3; the shape is concave at the seaward edge of the surf zone and close to the shore, but not in between, where a bar is present. Observed near-bottom wave-orbital velocities, shown in the same figure, illustrate wave asymmetry; onshore orbital velocities dominate over offshore orbital velocities. The very fast reversion of offshore to onshore orbital velocity indicates that the wave crest is almost overtaking the wave trough. Flow acceleration is very strong at this moment; it has been shown that this acceleration asymmetry (also called 'acceleration skewness') contributes to net onshore sediment transport, in addition to the orbital-velocity asymmetry (also called 'velocity skewness'). The concavity of the coastal profile is related to the feedback between seabed slope and wave-asymmetry. The seabed slope increases shoreward in order to produce sufficient gravity-induced offshore transport for compensating the onshore sand transport produced by shoreward increasing wave asymmetry. This shoreward increasing wave asymmetry is related in turn to the concavity of the coastal profile. This illustrates that the shape of the coast is a direct result of mutual feedback with wave propagation. However, the equilibrium between gravity-induced off-shore transport and wave-induced onshore transport is not stable, due to wave breaking. This instability generates a variety of seabed structures that will be discussed more in detail in Chapter 7 on wave-topography interaction.

2.2.2. Tidal Asymmetry

Ocean tides are sinusoidal in time

Tidal waves in the ocean are not local disturbances; their wavelength spans entire ocean basins. They are the resonant response of the ocean to the attractive forces of the moon and the sun, which are modulated by earth rotation. The tide-generating force is the sum of many sinusoidal components, each with a particular frequency and phase. These components correspond to the different motions of the moon and the sun with respect to a fixed point in the ocean. The tidal amplitude is much smaller than the ocean depth and the ocean depth is much smaller than the tidal wavelength. The tidal propagation speed then depends only on the average water depth; it is independent of the tide and independent of time. As tidal propagation does not modify the temporal variation, ocean tides are also a linear superposition of sinusoidal motions with the same frequencies as the components of the tide-generating force, but with different phases.

Symmetry of the ocean tides

In most ocean regions (the Atlantic Ocean, in particular) the principal lunar semidiurnal tidal component is much larger than all the other tidal components. This tidal component is called M2 and has a period of approximately 12.4 h. In this case the ocean tide is quasi symmetric: The flood and ebb motions have equal strength and equal duration. Suppose now that sand grains on the seabed are displaced by flood currents over a certain average distance. The ebb-flood symmetry then implies that they will be transported back by ebb currents on average over the same distance. If the seabed is a flat horizontal plane, ocean tides will not produce any net sand displacement, because a sand grain can make no distinction between ebb and flood.

In some ocean regions other tidal components can be important too, in particular the lunar (K1) and lunisolar (O1) diurnal tidal components. The sum of the frequencies of these two components equals the M2 tidal frequency. In this case the ocean tide becomes asymmetric: The flood and ebb motions have different strength and different duration. This case will be discussed in more detail in Chapter 6 on tide-topography interaction.

In the ocean, tidal currents are often superimposed on strong non-tidal currents. Because sediment transport is a highly nonlinear function of velocity (the sediment flux may increase by a factor 10 or even more if the flow velocity doubles), the superposition of tidal and non-tidal currents affects the uptake of sediment during flood and ebb in different ways and produces an asymmetry between flood and ebb sediment fluxes.

Asymmetry of flood and ebb in shallow basins

Let us consider the most usual case of a dominant M2 ocean tide. At the ocean margin, the tidal wave enters shallow basins (coastal sea, estuary, tidal river) where the tidal amplitude is no longer very small compared to the water depth. The propagation speed now depends on the tidal wave and is no longer constant. The water depth at high water is significantly different from the water depth at low water; the high water crest of the tidal wave will thus propagate at a different speed than the low water trough. As a result, the tidal wave becomes asymmetric (Fig. 2.4). The strength and duration of the flood current are not identical to the strength and duration of the ebb current. Due to the non-linear variation of the sediment flux with flow velocity, the uptake of sediment will be different during flood and during ebb.

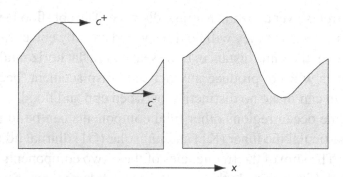

Fig. 2.4. A symmetric tidal wave generated in the ocean becomes asymmetric in shallow water, where the tidal range to water depth ratio is larger. The case sketched in the figure supposes that the HW propagation speed (c^+) is higher than the LW propagation speed (c^-), due to a greater water depth at HW compared to LW.

The average distance sediment particles travel during flood will also be different from the average distance they travel during ebb, see Fig. 2.5. The asymmetry between ebb and flood currents therefore causes a net transport of sediment, even though no net water transport may take place during the tidal period.

Feedback between tidal asymmetry and basin morphology

If the total sediment fluxes during flood and ebb in a tidal basin are not equal, sediment will be redistributed within the basin. The depth and width of the tidal basin will change. Such a morphologic change affects the propagation of the tide within the basin and therefore affects the asymmetry of the tidal wave; a change of tidal asymmetry produces in turn a change in net tidal sediment transport. Hence, the morphology of the basin and the tidal wave both evolve over time as a result of their mutual interaction. Tidal basins may nevertheless reach an equilibrium morphology, but only certain topographic configurations are possible, as will be shown in Chapter 6.

Breaking tidal waves

The breaking of wind waves is a familiar (though complex) phenomenon. One might wonder if tidal asymmetry may also result in

Fig. 2.5. Left: In the case of symmetric ebb and flood flow the same average amount of bed material will be displaced over the same distance back and forth. Right: If the flood duration is shorter than the ebb duration with a stronger flood flow than ebb flow, more bed material will be transported over a greater distance in the flood direction than in the ebb direction.

tidal wave breaking. Tidal wave breaking is far less common, but it is observed in some tidal inlets at spring tide, when the tidal range is at maximum. Tidal wave breaking produces a tidal bore; the flood advances upriver almost as a wall of water. Well developed bores occur only with spring tidal ranges exceeding 6 m at the inlet [550]. The most famous example of tidal wave breaking in Europe was the tidal bore (mascaret) in the Seine, which could reach several metres at spring tide, see Fig. 2.6. The bore was generated some 40 km from the inlet, where the tidal wave crest (high water) overtook the tidal wave trough (low water). Dredging works at the Seine inlet in the 1960s reduced the bore by increasing the propagation speed of the low tide compared to the high tide; as a result, high water could not catch up low water any more and the mascaret belonged to the past. The largest tidal bore in the world occurs in the Hangzhou Bay/Qiantang River in China at spring tide; this bore was already reported by Chinese authors over a thousand years ago [139]. Breaking tidal waves are spectacular phenomena, but in fact they are nothing else but the ultimate stage of tidal wave asymmetry.

The Qiantang River bore

Breaking tidal waves develop at tidal inlets under particular morphologic conditions. These conditions are very persistent, as shown, for instance, by the long history of the bore in the Qiantang River (Qiantangjiang). This points to the existence of a stabilising

Fig. 2.6. Tidal bore at spring-tide in the Seine at Caudebec in 1963. The bore is also visible in tidal records, see Fig. 6.32. Photograph J. Tricker, reproduced with permission of Scientific American.

morphodynamic feedback between the bore and tidal inlet morphology. Characteristic morphologic features are the funnel-shape of the inlet and the presence of a large bar (or shoal) near the inlet. In the Qiantang estuary the centre of the bar is located inshore of the inlet, see Fig. 2.7. The bar is 10 m high, has a length of over 100 km and is mainly composed of silt (grain diameter mainly between 20 and 40 μm). Most other bore-developing inlets, for instance the Amazon and Ganges-Brahmaputra, have a similar morphology; in the last two cases the centre of the bar is located seaward of the mouth.

   The large mouth bar slows down the propagation of the low-water tidal wave trough compared to the propagation of the high-water wave crest; the water depth over the bar at low water spring tide is so small that the tidal wave hardly advances, while the water depth at the subsequent high water is sufficient for fast tidal propagation. The bar also causes frictional energy dissipation, which decreases the tidal amplitude. This amplitude decrease is, at least partly, compensated by the convergence of the inlet geometry. Concentration of the landward

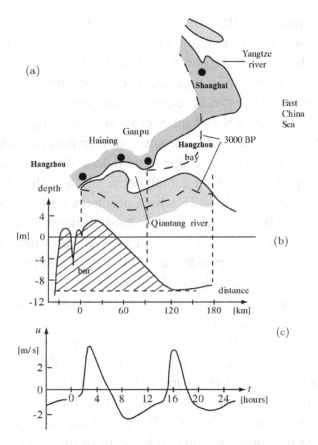

Fig. 2.7. The Qiantang tidal bore, from [464]. (a) Map of Hangzhou Bay and Qiantang River, with the present shoreline and the shoreline in 3000 BP. (b) Longitudinal depth profile along the estuarine axis. (c) Tidal variation in the current velocity at Haining, near the centre of the bar, displaying a strong flood velocity dominance. The bar and the funnel-shaped geometry of the bay are not only the cause but also the consequence of tidal asymmetry.

propagating tidal energy flux in a converging cross-section produces an increase in the tidal amplitude; at the location where high water overtakes low water the tidal range is still large enough for producing a considerable bore.

Morphodynamic feedback to tidal asymmetry

Tidal asymmetry resulting from the inlet morphology plays a major role in maintaining this morphology [198]. The development of the mouth bar mainly results from a strong asymmetry already present in the offshore tide; the duration of tidal rise at the inlet is more than two hours shorter than the duration of tidal fall. Tidal flood currents at the inlet are therefore stronger than tidal ebb currents. The mouth bar enhances the downstream sediment transport by river runoff near the inlet, by reducing the channel cross-section. High river runoff can therefore neutralise the upstream sediment transport that would occur otherwise near the inlet due to tidal flood dominance.

Tidal propagation over the mouth bar causes a further increase of tidal asymmetry (see Fig. 2.7), which promotes upstream sediment transport. Equilibrium of the inner estuarine morphology is possible only if river-induced ebb transport also increases upstream. The required increase of river-induced ebb transport results from channel convergence. A more detailed discussion of the development of mouth bars and channel convergence by morphodynamic feedback is given in Secs. 6.4.2 and 6.4.3.

The inlet bar and the funnel shape developed in the Qiantang estuary during the past millennia, possibly triggered by the sediment supply from the Changjiang (Yangtze River), just north of Hangzhou Bay, see Fig. 2.7. However, an analysis of sediment characteristics suggests that erosion of the outer Hangzhou bay is the major sediment source for bar formation and for infill of the funnel-geometry [464]. Nor is sediment supply by the Qiantang River essential for the formation of the funnel shape and the large inlet bar, as shown by numerical model simulations [979].

The example of the Qiantang estuary bore illustrates that morphologic evolution results primarily from tide-topography interaction. In Chapter 6 it will be shown that similar physical principles govern the morphologic evolution of river tidal inlet systems throughout the world.

Similarities in coastal basin morphology

The feedback between tidal propagation and basin morphology has important consequences. In the absence of such feedback, tidal basins evolve passively under the influence of tectonic motion, sea level rise and sediment supply. Tidal basins would differ greatly because of differences in original bathymetry and differences in sediment transport; one might thus expect little similarity between tidal basins. Tidal feedback processes regulate sediment infill as a function of morphology and lead the basin towards equilibrium. For instance, tidal feedback stimulates net tidally-induced sediment import in response to sea level rise, see Sec. 6.5.4. Tidal basin evolution is therefore much less dependent on original bathymetry and on sediment supply rates; there are likely to be strong similarities between tidal basins throughout the world. These similarities will be further investigated in Chapter 6. Typical planforms of different types of coastal systems occurring in nature are sketched in Fig. 2.8. Fluvial sediment supply (both rate and size) clearly plays a role, but the morphology of coastal basins also strongly depends on tides and waves. Tide-dominated coastal systems exhibit typical common morphologic characteristics and the same goes for wave-dominated coastal systems, even though the detailed morphologic structure of these coastal systems is very complex. There is no strong randomness in coastal typology; this indicates that most coastal basins are largely shaped by morphodynamic feedback processes between tides, waves and basin morphology.

2.3. Spatial Symmetry Breaking

2.3.1. *Morphodynamic Feedback to Seabed Perturbation*

Complexity of the submarine landscape

Looking at a geographical map we see that most landscapes possess an intricate morphology. This is also true for coastal sedimentary landscapes. Sometimes this morphology reflects the structure of erosion-resistant layers in the underground and sometimes it is structured by human intervention. But also in the absence of such constraints the coastal landscape morphology does not exhibit much

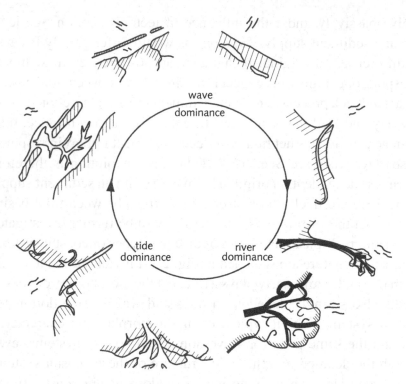

Fig. 2.8. The characteristic morphology of sedimentary coastal systems differs greatly depending on the dominant external driving force; the figure shows characteristic coastal morphologies developing in response to the dominance of fluvial sediment supply, tide action or wave action. A few intermediate morphologies for combined river-tide dominance, river-wave dominance and tide-wave dominance are also shown. Other factors (tectonic, sedimentary, ...) may play a role as well and many coastal classifications have been proposed accordingly [668]. Coastal classification makes sense, because morphodynamic feedback limits the diversity of coastal evolution. Yet every coastal system is unique, due to the great complexity at sub-system scales. In this book we will only discuss tidal inlets represented in the left half of the figure. We will call the tidal inlets in the upper left half 'tidal lagoons' and the inlets in the lower left half 'estuaries'. Adapted from [62].

uniformity. This conflicts with intuition, because (a) the characteristic spatial scales of external conditions, like wind and wave fields and tidal waves, are fairly large, (b) seabed structures tend to be smoothed by erosion and and (c) suspended sediments are dispersed by waves and currents. The almost complete absence of uniformity in coastal morphology and the high complexity of spatial patterns is a puzzling feature!

A *perfect morphodynamic balance does not exist*

Complex patterns exist both in the underwater landscape and in the sub-aerial landscape; here we focus on the former. The submarine landscape is subject to strong erosional forces due to currents; it evolves over time as a result of sediment being taken away or being deposited. The landscape will change unless a true balance exists at every location between incoming and outgoing sediment fluxes. In practice a perfect balance does not exist, at least not at every location and at every time scale; the submarine landscape never reaches a perfect equilibrium. In this section we will show that the feedback between morphology and water motion provides a mechanism for departure from a highly symmetric (uniform) morphology towards less symmetric complex morphologies. This mechanism will be illustrated through a few experiments in an highly idealised environment. However, the principle of symmetry breaking can also be experienced in many real life situations. Let us look at some examples.

Can you flatten a rippled bottom?

We go back to the seashore, a shore where the sea moves up and down the beach under influence of the tide. We arrive some time after high water. The higher parts of the beach are already drying but at several places seawater pools are left on the beach. From these pools, which consist of oblong troughs (runnels) between shore-parallel banks (ridges), the water is flowing seaward through small gullies, with a width of typically one or two metres (Fig. 2.9); the current speed is 20 to 50 cm/s. Looking at the bed of these gullies we see that the sand bed is not flat; it is covered with small ripples. The ripple wavelength is about 10–20 cm and the ripple height is of the order of 1 cm. Behind each ripple, flow vortices are visible which retain sediment in the lee of the ripple. In this way the ripple pattern moves downstream with a velocity of about 1 cm/s. Where do these ripples

Fig. 2.9.   Ridges and runnels on a meso-tidal beach at low water (Goeree, Holland). Small meandering gullies drain the runnels to the sea.

come from? Let us see what happens if we remove them! So we take off our shoes and flatten the stream bed with our feet. At first the water gets very turbid, but after a short time the bottom is visible again and locally the ripples have almost disappeared. Almost, because we see that new ripples start growing again. Within a few minutes the original situation is restored. The ripples are back! Apparently a very powerful mechanism exists through which ripples are generated starting from a flat bottom. What is this mechanism? We will see this in Chapter 4 on current-seabed interaction. For the moment we just note that apparently the channel bed morphology is subject to strong symmetry-breaking processes.

Competition between stream channels

Fortunately we have taken a shovel with us. The gully that discharges the pool into the sea makes a big meander. What happens if we shortcut this meander? We dig a shortcut channel of a size similar to the meandering gully. As soon as we have finished, the water starts flowing through the new channel. At the same time the flow through the meandering gully decreases. At the bifurcation a sill starts growing, decreasing still further the flow through the meandering channel. After some time the meandering channel is almost completely abandoned by the flow. We watch the evolution of the shortcut channel for a while. Some gradual changes in its morphology become apparent. A slight curvature develops, first at the entrance of the shortcut channel and later it also develops further downstream. This experiment (see Fig. 2.10) shows at least two things. First, a system of two competing channels is not necessarily stable, even if the two channels have a similar size. The channel which captures most of the flow will continue to grow at the expense of the other channel. In other words, if we start with a system of two symmetrical channels and deepen one of the channels at the expense of the other, then the system moves away from its original symmetrical state towards a one-channel system. Second, we see that an initially straight channel tends to develop into a meandering channel. In our experiment the incoming flow was not perfectly aligned along the axis of the shortcut channel. We see that meandering starts developing here. It shows that

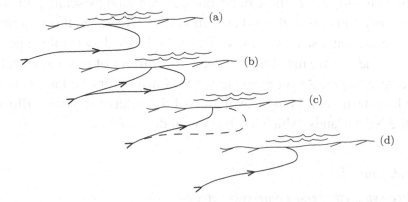

Fig. 2.10. A meandering draining gully on the beach some time after HW, before (a) and after (b) digging a shortcut. The initial meander is abandoned (c), but after a while the shortcut gully starts developing a similar meander (d).

the system increasingly deviates from along-channel symmetry after a slight perturbation somewhere along the channel.

Can we steer morphodynamic feedback?

Our beach experiments show that processes of symmetry breaking and pattern formation are inherent to the natural interaction between flow dynamics and bed sediment dynamics. In our beach experiments we changed the initial morphological state, but we did not steer the evolution of the system. The system responded to change by enhancing initial deviations from a symmetrical state. The experiments showed that the natural dynamics of the system may generate a positive feedback to deviations from symmetry. Of course, we would like to know what exactly the nature of this positive feedback is. Does such feedback always exist, or only under certain conditions? Can we predict the evolution of a system subject to morphodynamic feedback? For coastal management purposes we would like to know whether it is possible to adjust coastal development to morphodynamic feedback, or whether, how and in which cases it is possible to steer this feedback.

Learning from past morphologic evolution

A first clue to answer these questions can be found by scaling up our beach experiments to the full coastal system. However, this requires intervention at a scale which is usually far beyond available experimental means. We may look instead to the history of past large-scale coastal changes. As an example we will discuss in the last section the long-term morphologic evolution of the delta of the River Rhine in the Netherlands, which has been well documented.

2.3.2. Stability

Space-symmetry versus time-symmetry breaking

The process we call space-symmetry breaking is, in a physical sense, of a different nature than the process of time-symmetry breaking.

The two notions are similar in the mathematical sense, because both time- and space-symmetry breaking refer to nonlinearity; nonlinearity of morphodynamics with respect to temporal and spatial evolution. However, breaking of time symmetry and breaking of space symmetry have different causes. Spatial symmetry breaking refers to loss of symmetry in the morphology of the system. It can develop from a perfectly uniform spatial state, if it is in an unstable condition. Temporal symmetry breaking refers to amplification of pre-existing asymmetry in the hydrodynamic forcing of the system. Both symmetry breaking processes evolve over time through inter-action between morphology and hydrodynamics.

Response to spatial perturbation

Space-symmetry breaking relates to the growth of morphologic patterns which differ from the spatial structures existing originally in the morphology, or in the external conditions imposed on the system [631]. As an example we consider the morphodynamic system consisting of water flowing uniformly over a flat sloping seabed consisting of loose sediment grains (down-slope sheet run-off). The grains are entrained by the flow down the seabed slope and, if the slope is sufficiently long, this downslope transport is a steady, spatially uniform process. This situation does not conflict with any physical law. However, we know from experience that downslope transport of water and sediment can also take place in other ways. The flow may be concentrated in a channel or in a number of channels. The seabed may also exhibit cross-flow rip-ples, which move downstream with the flow. These other downslope transport modes obey the same physical laws as the uniform mode. It seems obvious to relate these different transport modes to dif-ferent initial morphologic conditions. But is that the whole story? The question as to how these different initial conditions come about remains unanswered. Moreover, the beach experiments discussed ear-lier show that changing the initial morphology does not necessarily change the transport mode; bottom ripples in a gully reappear very soon after they have been eliminated! But is it conceivable that a morphodynamic system switches autonomously between different transport modes?

The marble-on-a-hemisphere analogy

The answer to this question can be illustrated by an analogy. This analogy consists of a simple experiment with a perfectly round marble which we place on top of a larger perfectly round hemisphere, in such a way that it stays in equilibrium, see Fig. 2.11. Is that possible? Theoretically it is, because at the top of the hemisphere there is a location where the tangent plane is exactly horizontal. At this location the marble will stay in equilibrium. There is a practical problem, though. The location of exact horizontality is infinitely small! So, in practice, there is no chance that we can realise this equilibrium. Equilibrium at the top of the hemisphere does not conflict with any physical law and yet the system will spontaneously depart from this situation. The marble on a hemisphere is an example of an unstable equilibrium. Any perturbation of this theoretical equilibrium, even if infinitely small, will cause the marble to roll down. So we can predict with certainty that the marble will not stay on the top of the hemisphere. But can we also predict the path followed by the marble when it rolls down? And can we predict the position where

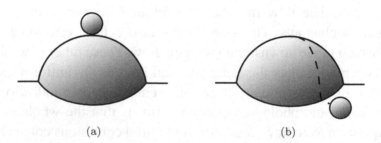

(a)   (b)

Fig. 2.11.   A perfectly round marble on top of a perfect hemisphere. Any perturbation, even infinitely small, will cause the marble to roll down. If the direction of the perturbation is unknown the path and the final location of the marble cannot be predicted.

the marble will finally arrive? The issue of predictability will be discussed later.

Instability cannot persist

Under certain conditions, uniform flow over a sloping, loose-grained sediment seabed is unstable and comparable with the situation of a marble on a hemisphere. It is an unstable mode if, for instance, the flow strength is slightly larger than the threshold for sediment motion [411,446]. In that case it is a transport mode possible in theory, but not in practice, because a range of infinitesimal perturbations of the flat seabed will tend to grow. Growth of such perturbations will finally lead to a morphology in which the flow is concentrated in one or more meandering channels with cross-flow ripples migrating over the seabed. The mechanism by which such perturbations grow out to a ripple morphology and by which channel meanders are generated is discussed in Chapters 4 and 5.

Static versus dynamic equilibrium

The marble-on-a-hemisphere analogy mimics the instability of uniform flow over a flat sediment bed only to a certain degree. An essential difference exists between the two systems, which is related to the energy dissipated in the system and to the feedback produced by the instability. If the hemisphere is fixed, no energy will be transferred from the marble to the hemisphere; the marble-on-globe state represents a static unstable equilibrium. This contrasts with the fluid flow over a flat seabed, which dissipates energy by exerting traction and lift forces on the sediment grains. A small flow perturbation caused by an incidental small disturbance of the sediment grain distribution at the seabed may grow by extracting energy from the basic flow, while enhancing the initial disturbance of the flat seabed. In this way, the initial disturbance may generate a large-scale topographic pattern and a related flow pattern. The initially flat seabed may reach after some time an equilibrium state with a topographic pattern of full-grown bedforms (ripples, dunes, bars, ridges, channels, ...).

We call this equilibrium state a dynamic (or morphodynamic) equilibrium.

Steady and non-steady equilibrium

A dynamic equilibrium is not necessarily a steady state. A non-steady equilibrium can be, for instance, a pattern of bottom ripples with a given wavelength and amplitude, that migrates with the flow. Even a seabed topography that consists of bedforms with fluctuating wavelength and amplitude, can correspond to a non-steady equilibrium. We will say that a system is in non-steady equilibrium, if under stable external conditions previous states reappear after some time. As a matter of fact, return of a system to exactly a previous state is physically impossible. The requirement for a non-steady equilibrium is the absence of a trend in the fluctuations of a system around a given steady state.

Steady coastal equilibrium states exist only in theory — not in the real world. Bedforms and channels migrate; channel meanders are cut-off and reform; connections in a channel network close and reform; marsh cliffs collapse and are rebuilt ... even under perfectly constant hydrodynamic forcing. In fact, neither steady equilibrium states nor constant hydrodynamic forcing exist in natural coastal systems. Turbulence, waves, tides, currents, river discharges fluctuate over periods ranging from seconds to decades.

As discussed earlier (see Sec. 1.1) we will use the term 'morphodynamic equilibrium' in a broad sense: It includes non-steady equilibrium states, subject to fluctuating hydrodynamic conditions. The context will generally clarify at which timescale changes in external conditions are considered fluctuations.

Forced and free instabilities

If a system can depart spontaneously (after only a slight disturbance) from an initial equilibrium state, we call the system 'instable'. The spontaneous departure of the system from its initial state is called 'instability'. One may distinguish between free and forced instabilities. We call the departure from the initial state a free instability, if the characteristics of the emerging seabed pattern depend solely on the intrinsic dynamics of flow-morphology interaction. If the characteristics of the emerging seabed pattern depend on patterns present in the hydrodynamic forcing or on the characteristics of the disturbance, we call it a forced instability.

In practice, free instabilities are not completely free; the emerging seabed pattern is often influenced by antecedent states. The memory of a previous state can fade away after some time, but it also happens that it influences the seabed pattern selected by the intrinsic dynamics of flow-morphology interaction.

Forced instabilities

Forced instabilities result from structural interventions in the system boundaries that modify the flow pattern (see Fig. 2.12). If the seabed topography was initially in a stable morphodynamic equilibrium it will be unstable after the intervention and evolve to a new equilibrium compatible with the new system boundaries. A shell on the subtidal beach causes a local perturbation of the

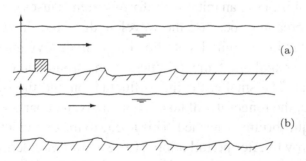

Fig. 2.12. Forced (a) and unforced or free (b) instability of the seabed. In the first case the seabed structures develop in response to flow perturbation produced by the obstacle; the length scale of the seabed pattern is related to the scale of this forced flow perturbation. In the second case the seabed pattern responds to self-produced flow perturbation.

beach morphology and generates a characteristic scour pattern in the wake of the shell. On a larger scale, a jetty extending from the beach into the sea causes a pattern of erosion and sedimentation extending in longshore direction over a distance of 5 to 10 times the jetty length. Incoming long waves which are reflected by a steep beach produce a standing wave pattern that causes an evolution towards a correspondingly undulating bar-trough pattern on the seabed. These are examples of forced instabilities of an initially unperturbed seabed morphology. If we capture the physical laws of morphodynamic evolution in a mathematical model with sufficient accuracy, then the system's response to a forced instability can be well predicted.

Free instabilities

In the case of a free instability the system evolves spontaneously from an unstable equilibrium morphology to another (stable or unstable) equilibrium morphology without being forced by any intervention in the external conditions. Instability of the initial equilibrium implies that any infinitesimal fluctuation from the equilibrium within a certain range of wavelengths, will tend to grow exponentially, with a growth rate depending on the wavelength of the fluctuation. The development of free instabilities in an initially uniform system causes spontaneous symmetry breaking, because the wavelength of the fluctuation did neither occur in the initial state nor in the hydrodynamic forcing. In practice, spatial and temporal fluctuations around an equilibrium always exist. The smaller the initial fluctuation and the smaller the growth rate, the longer it will take before a given departure from the unstable equilibrium is reached. This is due to the exponential nature of initial growth after perturbation. However, a finite departure will always be reached if sufficient time is available. Predicting morphologic evolution in the case of free instabilities will be discussed later; we will see that there are intrinsic limits to predictability, contrary to the case of forced instabilities.

2.4. Linear Stability Analysis

2.4.1. *Initial Response*

Morphodynamic feedback can only be modelled analytically for a highly idealised topography and for a limited range of spatial and temporal scales. The general morphodynamic problem cannot be solved by analytical methods, due to the existence of strong nonlinear interactions and to the broad range of space and time scales involved. Figure 1.2 gives an impression of the spatial and temporal scales at which coastal morphodynamic processes take place.

In the following we illustrate the principles of morphodynamic feedback by analysing the initial response of the system to an infinitesimal perturbation of the basic equilibrium state [479]. In the initial phase of development the perturbation is so small that the flow pattern is hardly changed and the amplitude of the flow response is almost linearly proportional to the amplitude of the bed perturbation. Higher-order nonlinear interactions are small and can be neglected. The initial development of the perturbation can therefore be described by linear equations, which can be solved, in some situations, by analytical methods.

Mathematical description

Morphodynamic evolution can be described in terms of the following quantities: (1) the spatial coordinates x, y, z, where x, y are horizontal coordinates and z is the vertical coordinate (upward positive), (2) the morphology, $z = -h(x, y) + z_b(x, y, t)$, where $z = -h(x, y)$ describes an equilibrium seabed morphology (spatial distribution of equilibrium water depth) and $z_b(x, y, t)$ the departure from equilibrium; (3) the sediment flux $\vec{q} = \vec{q}_0 + \vec{q}\,'(x, y, t)$, averaged over a period which is long compared to the periods of fluctuations in the unperturbed water motion (wave period, tidal period); $\vec{q}_0$ relates to the equilibrium state and $\vec{q}\,'(x, y, t)$ is the response to the perturbation. The x- and y-components $q^{(x)}, q^{(y)}$ are expressed as sediment volume per unit time and unit width [m^2/s].

Sediment balance equation

The equilibrium sediment flux $\vec{q}_0$ is assumed spatially uniform (independent of x, y); the departure from the equilibrium sediment flux, $\vec{q}\,'(x, y, t)$, is zero if z_b is zero. The sediment flux gradient, $\vec{\nabla}.\vec{q}' = \partial q'^{(x)}/\partial x + \partial q'^{(y)}/\partial y$, equals the amount of erosion ($\vec{\nabla}.\vec{q}' > 0$) or sedimentation ($\vec{\nabla}.\vec{q}' < 0$) per unit time and unit seabed area, if we assume that the averaging time is sufficiently long to allow us to neglect changes in suspended sediment load with respect to erosion or sedimentation. We refer to the porosity of the deposited material as p. The change in morphology per unit time follows from the sedimentation and erosion and is thus related to the sediment flux $\vec{q}'$ by:

$$(1 - p)\ \partial z_b(x, y, t)/\partial t + \vec{\nabla}.\vec{q}' = 0. \tag{2.1}$$

This sediment balance equation describes the morphologic evolution $z_b(x, y, t)$ of the seabed. It is a morphodynamic equation if we can express $\vec{\nabla}.\vec{q}\,'$ as an explicit function of $z_b(x, y, t)$. Such an analytical expression can be established only in a few, highly idealised, cases, and numerical techniques normally have to be used.

A one-dimensional idealisation

Here we will analyse the solution of Eq. (2.2) for such a highly idealised situation. We first assume that the morphology and the sediment flux are uniform in one horizontal direction, y, and we call $q = q^{(x)}/(1 - p)$. The sediment balance equation then reads:

$$\partial z_b(x, t)/\partial t + \partial q'(x, t)/\partial x = 0. \tag{2.2}$$

The porosity p is assumed to be constant. We restrict the analysis to the initial departure from equilibrium, during which z_b is very small compared to the equilibrium depth h. The depth h is approximated by a constant; this approximation is equivalent to considering only perturbations with a characteristic scale (wavelength) which is much smaller than the spatial scale of variation of the equilibrium morphology. We further assume that the sediment flux depends on the local, instantaneous flow velocity $u(x, t)$; this excludes fine sediments for which lag-effects in the suspended sediment distribution are important. Finally, we simplify the problem by considering steady boundary conditions.

Initial morphodynamic response

Now we investigate the morphodynamic response to an initial perturbation which consists of an infinitesimal undulation of the seabed, $z_b(x, 0) = \epsilon h \cos kx$, where k is the wavenumber of the undulation and ϵ a number much smaller than 1. This seabed perturbation produces a perturbation q' in the equilibrium sediment flux q_0, which can be written as a power series of the small parameter ϵ:

$$q'(x, t) = \epsilon q_1(x, t) + \epsilon^2 q_2(x, t) + \cdots, \qquad (2.3)$$

where $q_1, q_2, \ldots$ have the same order of magnitude as the undisturbed flux q_0. The first term q_1 is linearly related to the perturbation z_b; the spatial pattern of q_1 therefore contains only the wavelength of the perturbation. This implies that the *initial* response to the seabed perturbation $z_b(x, 0)$ generates no other wavenumbers than the initial wavenumber k. Therefore we may write

$$z_b(x, t) = \epsilon h f(t) \cos(kx + \Omega(t)) \qquad (2.4)$$

and

$$q_1(x, t) = f(t)\alpha q_0 \cos(kx + \Omega(t) + \phi), \qquad (2.5)$$

where $f(t)$ is a function describing the time evolution of the perturbation amplitude ($f(0) = 1$) and $\Omega(t)$ a function describing the migration of the perturbation ($\Omega(0) = 0$). The magnitude of the sediment flux perturbation is proportional to the magnitude of the seabed perturbation; however, the phase may be different. The coefficient $\alpha(k)$ and the phase difference $\phi(k)$ depend on the sediment transport mechanism and on the perturbation $u'(x, z, t)$ of the flow field. The flow field perturbation can be obtained by solving the flow equations to first order in ϵ; in Chapter 4 the parameters α and ϕ will be determined in this way for different types of seabed perturbation z_b. Here we will stay with a qualitative interpretation. The phase difference ϕ is crucial for the stability of the seabed, as will soon become clear.

Conditions for instability

Substitution of (2.3), (2.4) and (2.5) in the sediment balance Eq. (2.2) yields at order ϵ:

$$f_t = \sigma_i f, \quad \Omega_t = -\sigma_r, \quad \text{with} \quad \sigma_i = -\sigma_r \tan \phi = (k\alpha q_0/h) \sin \phi,$$
(2.6)

where the subscript $_t$ indicates differentiation with respect to time t. The first equation yields exponential growth (and therefore instability) of the perturbation if $\sin \phi > 0$ or $0 < \phi < \pi$ (assuming $\alpha > 0$). In this case, according to (2.5), the x-derivative of the sediment flux is negative at the crest of the perturbation ($kx + \Omega(t) = 2n\pi$, n is a positive or negative integer number). Sediment transport therefore converges at the crests of the seabed perturbation, which means that the perturbation amplitude will grow (Fig. 2.13). If $-\pi < \phi < 0$, the sediment flux diverges at the crests and the seabed perturbation decays. The second equation of (2.6) implies that the phase of the seabed perturbation increases or decreases linearly with time. According to (2.4), this linear phase increase/decrease corresponds to a migration of the bed perturbation with constant velocity.

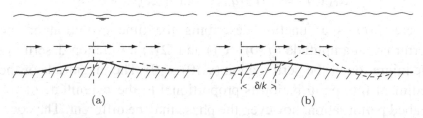

(a)   (b)

Fig. 2.13.   Seabed perturbation at time $t = 0$ (solid line) and some time later (dashed line). The arrows indicate direction and magnitude of the sediment flux. (a) If the location of the highest sediment flux coincides with the crest, then the seabed perturbation will migrate but not grow. (b) If the sediment-flux maximum is located upstream of the crest, then the perturbation will grow and migrate.

The role of inertia

The perturbation increases if the sediment flux has its maximum not too far upstream of the perturbation crest ($0 < \phi < \pi/2$) or its minimum not too far downstream ($\pi/2 < \phi < \pi$). But why should maximum or minimum of the sediment flux be shifted relative to the crests and troughs of the seabed perturbation? There are several reasons for such a phase shift, which will be discussed more in detail in Chapter 4. The influence of inertia (momentum conservation) in relation to frictional momentum dissipation is essential. Inertia delays the flow response to a seabed perturbation in remote regions (far from the seabed) relative to nearby regions (close to the seabed).

Flow over a bottom ripple

The effect of inertia can be illustrated by the flow over a bottom ripple. The water depth is smaller at the ripple crest than upstream or downstream; the flow therefore has to accelerate at the upstream ripple slope. The surface flow has not yet reached its maximum strength at the ripple crest, but it is still accelerating. However, averaged over the water column the flow cannot be accelerating at the ripple crest, because the total water flux would then be greater downstream than upstream of the crest. This implies that the flow near the bottom decelerates at the ripple crest. The near-bottom flow maximum is shifted upstream, and so is sediment bedload transport. We have $\phi > 0$, which is the condition required for initial growth of a perturbation. This instability mechanism is further analysed in Sec. 4.2.

Wavelength with the strongest initial growth

The growth rate σ_i and the migration rate σ_r of the perturbation depend on the wavenumber k because the coefficient α and the phase ϕ in (2.6) both depend on k. Seabed perturbations with a large wavenumber k (small wavelength) develop a steeper slope than perturbations with a small wavenumber; the growth of steep sloping perturbations is counteracted by gravity-induced downslope sediment transport. Seabed perturbations with small wavenumber k (large wavelength) will not induce a large spatial phase lag ϕ with the corresponding flow perturbation, compared to perturbations with a large wavenumber. It follows that strong growth is not to be expected at both extremes of the wavenumber spectrum. The strongest growth occurs at intermediate wavenumbers, which are large enough to produce a significant phase lag ϕ and small enough to avoid a strong gravity-induced down-slope transport. This dynamic balance determines the range of perturbation wavelengths which will most likely be observed in natural situations. The resulting seabed pattern thus develops according to feedback effects inherent to the physical dynamics of the system; it is not imposed by the external conditions. The wavelengths present in the seabed pattern may be entirely absent in the spatial pattern of the external forcing or in the spatial pattern of the initial morphology.

Self-organisation and pattern formation

In the previous section a qualitative description is given of the process leading to spontaneous growth of ripples or sand dunes on a uniform seabed topography. This process is triggered by a spatial shift between the initial seabed perturbation and the ensuing flow response. Such a shift may come about in several ways, depending on the length scale of the perturbation, the water depth, the presence of lateral boundaries, the average seabed slope and the character of the basic flow (steady or oscillating). Different types of flow response may generate different types of seabed instability and may lead to different types of morphologic patterns (different length scale, different orientation). Most morphologic patterns observed in the coastal environment can be related to instability produced by the flow response to seabed perturbation: Sand dunes and ripples, beach cusps, rip cells, meandering and braiding channels, tidal flats, sand banks, the equilibrium topography of tidal rivers and tidal basins, etc. These morphologic patterns have a characteristic wavelength which is not present in the external forcing; they also tend to recover spontaneously after being altered by natural events or artificial interventions.

The ability of a system to generate patterns which are not present in the external forcing is often referred to as self-organisation capacity. Pattern formation related to spontaneous symmetry breaking plays an essential role in shaping the coastal environment.

2.4.2. *Perturbation Growth Towards a Finite Amplitude*

What happens after initial growth?

Exponential growth cannot go on for ever, so after some time the growth rate has to decrease. The linear approximations (5.5, 2.5) do not hold for large perturbations and the exponential growth law (2.6) therefore ceases to be valid. This is a mathematical way to express the growing importance of physical processes which can be ignored only when the perturbation is very small, because they are related to the perturbation in a nonlinear way (quadratic or higher order). An example of such a process is the turbulence generated by flow separation in the wake behind the crest of a seabed perturbation, which increases the steepness of the lee-side slope [197]. In oscillating flow, both slopes of the perturbation will become steeper, especially near the crest. At a certain stage of evolution, sedimentation at the crest due to convergence of near-bottom flow will be exceeded by erosion due to gravity-induced downslope transport [316] (Fig. 2.14). Wave-induced seabed stirring in the zone around the crest may further stimulate downslope transport. Observations indicate that sand dunes in the North Sea, for instance, are less developed in regions of high wave intensity compared to regions of moderate wave intensity [412,586].

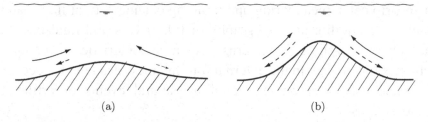

(a)   (b)

Fig. 2.14.   (a) In the first stage of development, the flow response to the seabed perturbation results in a contribution to sediment-transport convergence (solid arrows) which
is greater than the contribution of gravity-induced downslope transport to sediment-
transport divergence (dashed arrows). (b) In the final stage both contributions become
comparable.

Growth-limiting processes

Perturbation growth is also affected by a shift in sediment transport mode. An increase of the shear stress at the crest will favour
suspended-load transport over bedload transport. In Chapter 4 it will
be shown that some seabed instabilities, such as ripples and sand
dunes, depend on bedload transport for their growth. But when the
dune height increases the flow velocity at the crest may become
high enough to bring sand in suspension. Suspended load transport may then overtake bedload transport and interrupt the growth
mechanism.

There are often several processes that may limit the growth rate of
seabed patterns at an advanced stage of development. How the different growth-limiting mechanisms compare to each other is not well
known in general, because of the great complexity of the underlying
processes. It should also be mentioned that growth limitation does
not apply to all types of seabed instability. Bars in estuaries, for
example, may continue growing till they reach the highest occurring
water level.

Nonlinear perturbation growth

In the initial phase of emergence the growth rate of a perturbation does
not depend on time. This is illustrated in the simple one-dimensional

example (2.6). The linear stability analysis does not include processes that decrease the growth rate in a later stage of development. For inclusion of these processes higher order terms of the ϵ-power expansion (2.3) must be considered. Theoretically this is possible, as long as the expansion series converges sufficiently.

Inclusion of higher order terms provides information about the evolution of the perturbation at finite values of the amplitude. Nonlinear feedback in response to the initial perturbation in the equation of motion produces secondary perturbations with new wavelengths; the perturbation then loses its initial sinusoidal shape. We cannot write the evolution of the perturbation in the simple form (2.4). The most important nonlinear feedback terms will often have a quadratic character, producing secondary perturbations of one-half wavelength. If these secondary perturbations are also unstable and start growing, then the initial sinusoidal perturbation turns into a complex, asymmetric seabed pattern with finite amplitude.

Field observations show that in nature the morphology of most bedforms is indeed much more complex than a single sinus shape [197]. It appears, in particular, that the wavelength of the initially fastest growing perturbation is often not the dominant wavelength at later stages of development. Most usually an evolution towards greater wavelengths is observed, a phenomenon called 'pattern coarsening' [623]. Examples of pattern coarsening are discussed in Chapter 4. The analysis of pattern evolution can be carried out by solving numerically the nonlinear hydrodynamic equations (see for example [176, 240, 393]), or by simulating pattern formation in cellular automata models (see for example [33, 333, 940]).

Weakly nonlinear analysis

Sometimes a semi-analytic approach is possible, using a mathematical technique known as weakly nonlinear approximation. This technique consists of solving the nonlinear equations for marginally unstable conditions, which are close to stable conditions. The equations can be solved by analytical methods, because in this case only perturbations in a narrow range of wavelengths can grow. In a next step the solution is extrapolated by perturbation expansion to conditions which are further remote from the marginally unstable state. This method has been applied successfully for obtaining a semi-analytical description of the finite amplitude behaviour of perturbations in several idealised physical systems, for instance, the growth of bars in a straight river with flat bottom [752].

A simple nonlinear evolution equation

The growth rate resulting from a weakly nonlinear analysis is not constant, but depends on the perturbation amplitude. The simplest form it can take is an expression of the form $\sigma_i - f^2$, where f is a factor representing the time dependence of the perturbation amplitude. The amplitude evolution equation then reads:

$$f_t = f \left(\sigma_i - f^2 \right). \tag{2.7}$$

Initially the perturbation behaves exponentially, since we may approximate $f_t \approx \sigma_i f$ for $f \ll \sigma_i$. For negative values of σ_i, $f = 0$ is a stable solution. For positive values of σ_i the original state $f = 0$ is not stable; stable solutions are given by $f = \pm\sqrt{\sigma_i}$. In the differential equation $\sigma_i = 0$ is a bifurcation point. Moving σ_i beyond the bifurcation point induces spontaneous pattern formation. We already know that the wavelength of the pattern is a characteristic inherent to the physical dynamics of the system. Now it appears that the amplitude of the pattern is also related to inherent system characteristics, which are not imposed solely by the external conditions and which are independent of the initial state of the system.

Cyclic morphologic behaviour

In the simple one-dimensional example the system moves to a morphology characterised by a rhythmic seabed pattern with a constant amplitude; this pattern will normally migrate. Migration implies that the system does not evolve to a static equilibrium; in the final state the evolution of the seabed has a cyclic character. In reality, pattern migration is just one aspect of seabed evolution. Seabed patterns also change by nonlinear interactions and may never reach a well defined final state. Periodic or quasi-periodic behaviour of morphologic patterns is a general characteristic of many coastal systems. This will be illustrated in the next section, which deals with the instability of a two-channel system. This example also shows that the direction in which a system initially moves, starting from an unstable equilibrium, generally depends on the initial perturbation and the external conditions, as can be expected from the marble on a hemisphere analogy.

Predictability

If we accept that many features of coastal morphology originate from spontaneous symmetry breaking, then one may wonder if coastal morphology is predictable. In the literature much consideration has been given to this question, see for instance [404]. Whether a given morphology is stable or unstable depends on the external conditions, for instance, the strength of current velocities, the water depth, the wave height, the wave period and the angle of wave incidence. These conditions usually fluctuate and, as a result, stability and instability may alternate. The period of alternation can be short, on the order of hours to days or weeks, for fluctuations caused by tides or weather conditions. The period of alternation can also be very long, on the order of decades to centuries, corresponding to fluctuations caused by large-scale morphologic evolution or by sea level change.

If unstable conditions last for a sufficient time, self-organised patterns will develop. The development of small-scale patterns may take place over short periods of hours to days, but the development of large scale patterns may take periods of decades to centuries. In the foregoing it has been shown that the initial growth of morphologic patterns is strongest for wavelengths in a certain range of scales, which can be determined theoretically.

However, the initial morphology also matters. If the wavelength of the initial morphology is not very far from the wavelength of maximum instability, then the former wavelength will dominate the latter for a substantial period. Bottom cores of coastal sandbanks show that they often consist of recent deposits with a much older kernel [412]. This suggests that positive morphodynamic feedback to remnant bedforms (the remains of a coastal barrier formed at a lower sea level, for example), may dominate over the development of a new bedform pattern corresponding to the wavelength of fastest growth. Laboratory experiments and numerical models further show that initial states with highly regular bedform patterns can survive for a long time, even under morphodynamically unstable conditions [623].

Predicting coastal morphology is therefore a complex problem, requiring very accurate morphologic modelling. The periods of alternating stability and instability can be predicted only in statistical sense, but not in a deterministic sense. It is hence to be expected that accurate predictive modelling is not possible at all. In that case one may conclude that predicting morphology is subject to limitations which cannot be overcome.

2.5. Instability of a Two-Channel System

To illustrate the inherent instability of coastal morphology we will consider the instability mechanism of a two-channel system as an example. This mechanism is based on competition between mutually interacting flow paths in tidal basins. Tidal flow is seldom confined to a single well-defined channel. In natural tidal basins the flood flow usually follows different pathways. Part of the flood flow may follow the main ebb channel and bend around shoals and tidal flats. But there are often also shunt channels that cut through shoals and tidal flats. These channels compete for conveying the flood wave and this competition determines the respective cross-sectional areas of these channels. The relative importance of the channels is not always the same, and shifts may occur between the channels over the course of time. In Fig. 2.15 this is illustrated for the Western Scheldt estuary. The same phenomenon is observed in river deltas; the instability mechanism is in both situations similar, though not identical. In the following we concentrate on the latter, slightly simpler situation. We will take the Rhine delta as an example and start by describing the history of the two major branches of the delta, the Waal and the Nederrijn. This history sheds some light on the large-scale evolution of the Rhine delta and gives some clues for answering the question why it did develop in this particular way.

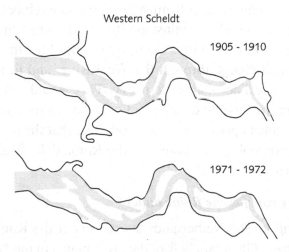

Fig. 2.15. Evolution of the Western Scheldt channel system. The main channel has several secondary channels throughout the basin. The relative importance of these channels may change over the course of time; such a change has occurred in the central part of the basin during the past century.

2.5.1. *History of the Rhine Delta*

It takes more than a few decades to change the Rhine delta

Upon entering The Netherlands, the River Rhine flows through a large coastal plain built up with fluvial deposits (sand, clay) in the upstream reaches and with marine deposits and peat in the down-stream reaches. Near the German border, about 100 km from the sea, the Rhine divides into two branches, the rivers Nederrijn and Waal (see Fig. 2.16). The land between the two branches is called Betuwe, land of the Batavians. The Batavians settled there shortly before the Roman era, more than 2000 years ago. The river branch Waal was the southern border of the Batavian territory, forming a natural barrier to the Roman conquerors. After some unsuccessful attempts to cross the river, the Roman general Drusus decided to build a dam on the Waal, near the bifurcation point with the Nederrijn. This intervention made it possible to wade across the Waal by decreasing the discharge through it, in favour of the Nederrijn (AD 0–50). After their defeat the Batavians were educated in Roman culture and engineering. A few decades later they rebelled against the Roman domination under their Batavian leader Julius Civilis. Using their new skills they managed to restore the discharge through the Waal River and to recover the southern defence line of their territory. This would not have been possible if an irreversible change had occurred in river morphology during the Roman episode; we may conclude that the time scale for substantial morphological change of the Rhine delta is greater than a few decades.

A storm surge modifies the Rhine outlet

Dike building in The Netherlands started under the Romans, but it took more than 1000 years before the river plains in the Netherlands were protected against flooding by dikes. Water boards, elected by landowners and financed through taxes, were in charge of maintaining this infrastructure, with the support of monasteries. In the Middle Ages, land reclamation extended to the lower reaches of the Delta,

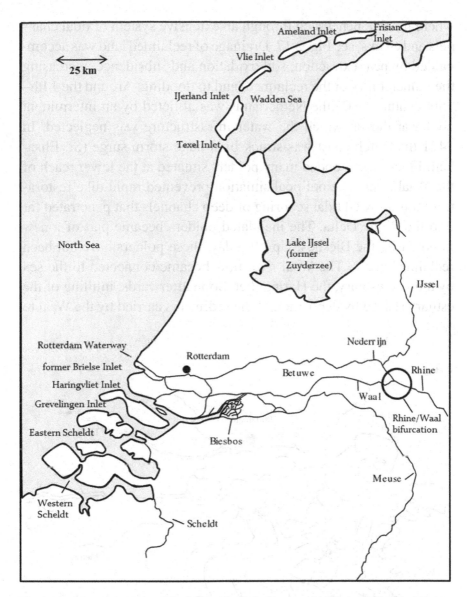

Fig. 2.16. Map of the Rhine delta in The Netherlands around 1950, indicating the bifurcation into the Waal and Nederrijn branches near the German border. The Haringvliet-Biesbos is the tidal delta created by the storm surge in 1421 and has since served as the main connection of the Waal to the North Sea. The Nederrijn got a new connection to the North Sea when the Rotterdam Waterway was constructed in 1870.

where the tide penetrated through an extensive system of tidal channels and creeks, see Fig. 2.17. Drainage of reclaimed land was accom-
panied by peat extraction, soil oxidation and subsidence, increasing
the vulnerability of the reclaimed land to flooding. Around the 14th–
15th century (AD) the Netherlands was afflicted by an intermittent
civil war during which the water infrastructure was neglected. In
1421 the Dutch coast was struck by a huge storm surge (St. Elisabeth Flood) that flooded many polders situated at the lower reach of
the Waal river. The political situation prevented rapid dike restora-
tion; this allowed tidal scouring of deep channels that penetrated far
into the Rhine Delta. The inundated polders became part of a new
flood basin, the Biesbos. Up till today, these polders have not been
reclaimed again. The Waal river now became connected to the sea
by a broad estuary, the Haringvliet. Soon afterwards, infilling of the
estuary started by deposition of the sediments carried by the Waal to

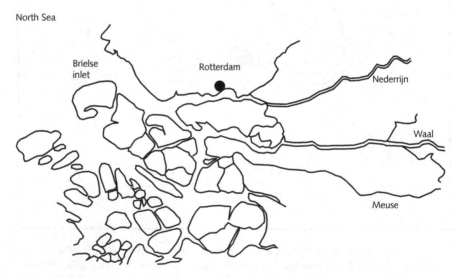

Fig. 2.17. The lower reach of the Rhine delta in the Middle Ages, before the storm surge of
1421. The Brielse inlet was the main river outlet to the North Sea. South of the Brielse inlet
the delta was formed by a mosaic of islands, polders, tidal flats and marshes, all embedded
in a complex network of tidal channels and creeks.

the Biesbos [489]. When these deposits began to obstruct the Waal flow, an artificial channel (Merwede) was dug through the Biesbos in the 19th century, in order to cope with high river floods. In 1970 the tidal intrusion in the Haringvliet became prohibited by the construction of discharge sluices at the mouth. These discharge sluices now enable the regulation of the Rhine outflow through the different Rhine branches.

Upstream and downstream impact of the Haringvliet outlet

In the centuries following the breakthrough of the Haringvliet estuary, the Waal river started carrying an increasing share of the Rhine discharge, at the expense of the other branch, the Nederrijn. During high river floods the Waal bifurcation became enlarged, while the Nederrijn bifurcation continued to shoal; in the 17th century the Nederrijn carried hardly any river discharge, see Fig. 2.18. Did the 1421 storm surge trigger the avulsion of the Nederrijn branche 100 km upstream? This question is disputed; we lack data on historical bathymetries. Avulsion processes at channel bifurcations may occur also without such a trigger — as we will see later in this section. At the end of the

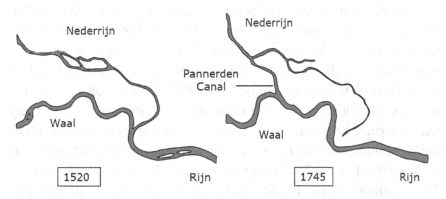

Fig. 2.18. The bifurcation of the Rhine into the Waal and Nederrijn rivers in AD 1520 and in AD 1745. In the 16th and 17th century the upstream part of the Nederrijn had become so shallow that it only carried a very small share of the Rhine discharge. A new bifurcation point was created by the construction of the Pannerden canal in 1707, shunting the upstream part of the Nederrijn.

17th century a new connection between the Rhine and the Nederrijn was created by the construction of the Pannerdens Canal.

In the meantime there were many changes in the lower delta. In the Middle Ages the Rhine Delta had several outlets, which were interconnected by an intricate system of tidal channels. In the Roman era, the largest outlet, Helenium, was located south of The Hague. In the Middle Ages it had shifted more than ten kilometres southward and was called Brielse Inlet. After the St. Elisabeth flood, the Brielse Inlet had to compete with the Haringvliet outlet. The Brielse Inlet then started shoaling and the navigation channel to Rotterdam became obstructed. Early in the 19th century a canal was constructed connecting Rotterdam to the Haringvliet. However, due to increasing draught of the ships, the canal soon became inadequate.

Creation of a new artificial Rhine outlet

In the middle of the 19th century a young engineer, named Caland, proposed the construction of a canal through the dunes, north of the Brielse inlet, connecting Rotterdam directly to the sea. A small initial short-cut channel would suffice; Caland counted on the scouring power of tidal currents and river discharge to further deepen and enlarge the canal. Today we would call this approach 'building with nature'. In fact, what Caland proposed was similar to the experiment we did on the beach at small scale. The canal, called Rotterdam Waterway, was finished in 1870. Unfortunately, Caland's concept did not work. In 1877 the entrance of Rotterdam Waterway was obstructed by large shoals; it could not compete with the other outlet branches of the Rhine. In the same period similar problems had occurred at dredged navigation channels in the Clyde and the Tyne estuaries (UK). Based on experience gained with dredging in the UK, the decision was taken to widen the Rotterdam Waterway (from 50 to 100 m) and to deepen the channel to 10 m. In addition, the harbour moles were extended further seaward and training walls were built to concentrate the current. Later the Brielse Inlet was also closed. After implementation of these improvements, the result anticipated by Caland was achieved; around 1950 the new Rhine delta was close to morphologic equilibrium. At that time the river beds in the lower delta were constrained almost everywhere within groynes and embankments and most river meanders were cut off; dredging was necessary only to maintain the depth in the harbours.

Conclusions from the Rhine delta history

What can we learn from this brief history of the Rhine Delta? In Sec. 2.3 we described an experiment with a shortcut of a meandering beach gully. This micro-scale experiment bears some resemblance to the competition between the Rhine delta branches after the creation of a new outlet. However, the spatial and temporal scales are completely different: In the order of half an hour for the beach gullies and in the order of a few hundred years for the Rhine delta. The large morphologic time scale for the Rhine delta is related to the low sediment discharge of the Rhine river; this discharge amounts to some $1-2$ million m^3 a year. The surface area of the Rhine branches is approximately $200 \, km^2$. Hence, deposition of the total sediment load would yield an average $1-2 \, m$ rise of the river bed in 200 years; this amount of sedimentation is needed to create substantial morphologic feedback.

The morphodynamic history of the Rhine Delta shows that:

- In its broad alluvial plain, the Rhine River could build a delta of several bifurcating and meandering channels;
- Several shifts occurred between different channel configurations during the past 2000 years;
- Shifts were primarily due to morphodynamic feedbacks in the fluvial system and to marine processes around the inlet; sedimentation of one channel branch and erosion of another channel branch were mutually reinforcing processes;

- Large-scale evolution of the river delta during the last millennium was influenced by human interventions (and lack of interventions);
- Modification of the natural configuration and dynamics of the delta system required large interventions of sufficient duration.

A (tidal) river delta with multiple channels is unstable, due to competetition between the channel branches. The Rhine delta is not a unique example of competing outlet branches; similar delta processes occur in other low coastal plains, for instance, the deltas of the Danube, Po, Nile, Ganges, MacKenzie, Tigris-Euphrates, Irrawaddy, Mississippi, Mekong, Brahmaputra, Yellow River, etc. [81,826,827,967]

2.5.2. *Principles of Channel Competition*

How can we understand the behaviour of delta river systems? What kind of mechanisms determine competition between channel branches? Do the same principles hold for river deltas and for tidal deltas? We will try to answer these questions by considering an idealised situation, which can be investigated with a simple analytical model. The model describes steady flow through a split channel system; the system consists of a main channel bifurcating into two parallel channels that discharge into a large basin. We perform the following physical-mathematical experiment. Initially, the flow velocity is the same in the main channel and in both bifurcating channels. There is no net sedimentation or erosion. Then we change the depth of one of the channels slightly and compute the response of the system to this perturbation using a simple sediment transport model and assuming spatially uniform depth in each of the channels. This experiment yields a stability criterion for the two-channel system. The model is by no means a realistic representation of real river deltas that are far more complex. It is meant only for demonstrating the basic principle of instability related to channel competition.

First we will address some underlying assumptions in a qualitative discussion.

The physical instability mechanism

What happens when the flow velocity is decreased in one of the parallel channels and increased in the other? When the flow enters the channel with decreased velocity, the sediment transport capacity decreases. Sediment is deposited at the channel entrance and a sill will be formed. The location of velocity decrease and related deposition will then shift to the downstream side of the sill; this morphodynamic feedback implies that the sill will grow in the downstream direction. Meanwhile a scour hole will develop at the entrance of the other channel; this scour hole also starts growing in downstream direction. Now we may ask if this sedimentation/erosion process will lead to a new equilibrium, when the sill in the first channel and the erosion in the second channel finally extend along the full length of the parallel channels. The answer depends on the flow response to the depth decrease in the first channel and the depth increase in the second channel. If the flow velocity increases in the shoaling channel and decreases in the eroding channel then the flow velocity in both channels will eventually become equal and the sedimentation/erosion process will stop. But in reality we expect the opposite response: the flow velocity will further decrease in the shoaling channel and increase in the eroding channel. The reason is that frictional momentum dissipation increases with decreasing depth. The shoaling of the first channel decreases the flow velocity in this channel compared to the other channel; the initial flow perturbation is thus enhanced and finally one of the channels will close (Fig. 2.19). This occurs when the current velocity in the shoaling channel falls below the critical threshold for sediment transport; the sill at the entrance will then grow until the channel is completely shut. This is what almost happened to the Nederrijn river branch. We can conclude that a two-channel system can easily be destabilized and may even develop into a single-channel system.

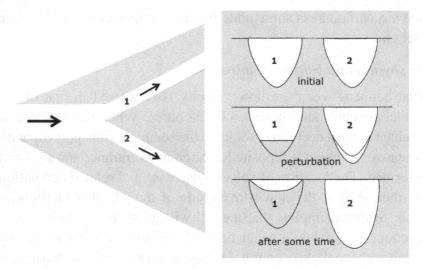

Fig. 2.19.   Instability of a channel bifurcation. Left panel: Schematic representation of a delta with two seaward flowing river branches. Right panel: The two symmetric branches are perturbed by a slight accretion of channel 1 and a slight deepening of channel 2. Instability means that the asymmetry is amplified and that finally channel 1 is completely filled.

Lateral channel accretion

In the previous section we have made the implicit assumption that channel infill in response to a decrease of flow velocity takes place through shoaling, i.e., through vertical channel accretion. In theory, channel infill may take place also through sedimentation along the channel banks, i.e., through lateral channel accretion. In the hypothetical case that channel infill is entirely stored as lateral bank accretion no substantial increase of frictional dissipation will occur (as long as the channel width remains much larger than the channel depth). This mode of channel infill therefore leaves the flow velocity unaffected; in the absence of frictional feedback the sedimentation process will come to an end. However, this case is only of theoretical interest, because in natural streams both depth and width adapt to river discharge, roughly according to a square-root law [471, 500, 749]. An initial velocity decrease therefore leads to a decrease of both channel width and channel depth. So, even if we allow for changes in channel width, a feedback process does exist as described in the previous section, leading to instability of the two-channel system.

Channel bed instability and bar formation

In wide tidal basins, multi-channel systems are an ubiquitous feature; the same is true for rivers with a very large width-to-depth ratio. In Chapter 5 it will be shown that multi-channel systems may come about as an inherent instability of the channel bed (Sec. 5.3). Their initial formation can be triggered by a lateral channel bed perturbation and by positive feedback from the ensuing perturbation of the lateral flow distribution. Further growth of the instability is stimulated by flow meandering through positive feedback between the sediment transport pattern in meandering flow and the formation of channel bed meanders. In tidal basins the formation of a multi-channel system is also enhanced by the alternation of flow direction, because the ebb flow and flood flow tend to follow different pathways through a meandering channel. It is clear that these features are not captured by the earlier one-dimensional description of channel flow. There are nevertheless many indications that channel competition takes place in wide braiding rivers and tidal basins. This competition is illustrated in Fig. 2.15, showing the shifts in relative channel importance which have taken place during the past century in the Western Scheldt basin. In some estuaries, the main and secondary tidal channels have been separated by the construction of training walls; subsequent rapid shoaling of the secondary channels has been observed, for instance, in the Seine, Mersey, Ribble and Lune estuaries [43, 587]. These observations suggest that a multi-channel system does not survive if the flow loses its two-dimensional character.

A simple model

In this section the previous considerations will be illustrated by a mathematical model referring to an idealised system of two channel branches with the initial lengths l_1, l_2, widths b_1, b_2, depths h_1, h_2 and cross-sectionally averaged flow velocities u_1, u_2 (Fig. 2.19). Upstream of the split channel system we have a single channel with width $2b$ and depth h. The total discharge through both channels, $Q = 2bhu$, is kept constant. The width of the parallel channels is allowed to vary with depth, $b_1 = b_1(h_1)$, $b_2 = b_2(h_2)$. The continuity equation reads

$$Q = 2hbu = Q_1 + Q_2 = h_1 b_1 u_1 + h_2 b_2 u_2.$$

The simplifying assumptions — uniformity of depth, neglect of inertial acceleration terms, neglect of channel curvature and cross-sectional flow distribution, sediment transport formulation — do influence the stability criterion, but cannot be assessed by the simple model.

We consider the impact of a small depth perturbation in the two split channels. We want to know if the original equilibrium situation will be restored or if the perturbation will grow and shift the system away from its initial state. Then we have in channel 1: $h_1 = h_1^0 + \epsilon h_1'$, $b_1 = b_1^0 + \epsilon b_1'$, $u_1 = u_1^0 + \epsilon u_1'$, and similar definitions for channel 2. Here, ϵ is an infinitesimal small quantity. We assume that the length of the channels is much larger than the depth ($l_{1,2}/h_{1,2} \gg c_D^{-1}$, where $c_D \approx 0.003$ is the friction coefficient, see (3.2.2)). We also assume that the flow velocity is almost uniform in this cross-section. In that case the momentum balance implies that the surface slope over the length of the channel is proportional to the frictional momentum dissipation; this dissipation is then proportional to the square of the velocity and inversely proportional to depth (see Appendix B.2). As the channels end into a large basin, we assume that the water levels at both channel ends are the same. The momentum balance then yields the relation

$$l_1 u_1^2 / h_1 = l_2 u_2^2 / h_2. \tag{2.8}$$

This gives in combination with the continuity equation

$$u_1 = 2bhu \left[b_1 h_1 + b_2 h_2 \left(\frac{h_2 l_1}{h_1 l_2} \right)^{1/2} \right]^{-1} \tag{2.9}$$

and a similar formula for u_2 with all indices interchanged.

Sediment transport formulation

We will examine now the sediment balance equations for both channels with the following assumptions:

- The sediment transport per unit width, q, is related in each channel to the cross-sectionally averaged flow velocity u by

$$q = \alpha u^{n+1} \tag{2.10}$$

  where α is a (dimensional) proportionality constant and n a number that depends on the sediment transport mode. For bedload transport the exponent $n \approx 2$; for suspended load transport experimental evidence indicates $n \geq 3$, see Sec. 3.8. More accurate formulations of the sediment flux include a threshold velocity for incipient sediment motion; here we use the simpler formula.
- Sedimentation or erosion due to convergence or divergence of the sediment flux in each channel are assumed to extend over the entire length of the split channel; the channel depth will be assumed to be uniform, except for local fluctuations corresponding to dunes or bars.
- At the upstream channel junction, each channel takes a share of the upstream sediment transport proportional to its width. This is equivalent to assuming that in the equilibrium state before perturbation, the velocity is the same in the three channels: $u_1 = u_2 = u$, $h_1 b_1 + h_2 b_2 = 2hb$, $l_1/h_1 = l_2/h_2$ (here and in the following we drop the superscript 0 for the initial unperturbed state).

From (2.9) it follows that the depth perturbations h_1', h_2' produce velocity perturbations u_1', u_2', which to first order in ϵ are given by

$$u_1' = -\frac{bhu}{(b_1h_1 + b_2h_2)^2}\left[(3b_1 + 2h_1b_{1h} - 2bh/h_1)h_1'\right.$$

$$\left. + (3b_2 + 2h_2b_{2h})h_2'\right], \tag{2.11}$$

where $b_{1h} \equiv db_1/dh_1$. The expression for u_2' is similar with all indices interchanged. For symmetric channels, the factor multiplying h_1' in the expression (2.11) is smaller than the factor multiplying h_2'. If the perturbation consists of an increase of the depth in channel 2 and an equivalent decrease in channel 1, then the velocity will decrease in channel 1 and increase in channel 2.

Stability analysis

With the above assumptions the sediment balance equations read:

$$\frac{d}{dt}(l_1b_1h_1) = -q_1^{IN} + q_1^{OUT} = -\alpha b_1\left(u^{n+1} - u_1^{n+1}\right) \tag{2.12}$$

and a similar equation for channel 2. Continuity at the channel junction requires $b_1 + b_2 = 2b$. To first order in ϵ we find, after substitution of (2.11):

$$l_1(h_1b_{1h} + b_1)dh_1/dt = -\frac{bhu}{(b_1h_1 + b_2h_2)^2}$$

$$\times\left[(3b_1 + 2h_1b_{1h} - 2bh/h_1)h_1'\right.$$

$$\left. + (3b_2 + 2h_2b_{2h})h_2'\right], \tag{2.13}$$

and a similar equation for the second channel. The solution of these coupled equations is a sum of two exponential functions, $\exp\sigma_1 t$, $\exp\sigma_2 t$. The arguments σ_1, σ_2 of these exponential functions are given by the eigenvalues of the matrix formed by the right-hand side of the above set of equations. The product of the eigenvalues is proportional to the determinant of this matrix, i.e., proportional to

$$(3b_1 + 2h_1b_{1h} - 2bh/h_1)(3b_2 + 2h_2b_{2h} - 2bh/h_2)$$
$$-(3b_1 + 2h_1b_{1h})(3b_2 + 2h_2b_{2h}).$$

Inspection of this expression shows that the determinant is negative for any value of width and depth of channel 1 and 2, because in the unperturbed state $h_1b_1 + h_2b_2 = 2hb$ and because in natural channels width and depth are positively correlated. This implies that the exponents σ_1, σ_2 cannot have the same sign; one of them must be positive. Any perturbation thus starts growing exponentially. In other words, any 2-channel configuration is unstable and will always deviate from its initial state upon perturbation. We will discuss in the next sections whether this will always lead to complete avulsion of one of the branches.

Why do river deltas and estuaries with branching channels exist?

A multi-channel delta system is usually formed during extreme river floods, after a period of aggradation of the main channel under moderate river discharges. The flood wave breaches the river bank at a certain location and seeks a new pathway through the floodplain, by following the highest topographic gradient. A new distributary channel develops, which takes part of the discharge of the original main channel. The above analysis suggests that over time one of the two channels will outcompete the other, leading to complete avulsion of one of the channels. Many river deltas had indeed more distributaries in the past than today [826]. However, partial avulsion occurs more frequently than complete avulsion [784]. Branching deltas and estuaries in a natural state are highly dynamic systems [554, 827] and do not systematically decay into a single branch system. This holds for river deltas as well as for branched channel systems in tidal environments. The point is, that the above simple model does not contain all the relevant physical processes. In the following we shall briefly describe two processes that contribute to the preservation of multi-channel systems.

Channel junction asymmetry

Channel branches generally split off at the outer bend of a channel meander, where the flow velocity is strongest and where the
channel bank is most eroded. The flow velocity in channel bends
is not distributed evenly over the channel width. A more detailed
discussion of channel meandering and the associated flow pattern is
presented in Sec. 5.4. Most of the sediment is carried where the flow
is strongest. In the simple model, we assumed that the sediment flux
is distributed over the channel branches in proportion to their widths.
This is clearly an oversimplification. In reality, the influx of sediment
is greater in the channel branch where the incoming flow velocity is
greatest. The one-dimensional model cannot directly account for this
process.

However, by slightly modifying the present simple one-
dimensional model, we can analyse the effect of an eventual inequality in channel inflow velocity. We use a procedure similar to the one
proposed by Wang *et al.*[928]. Here we assume that the highest inflow
velocity occurs at the entrance of the channel that receives the largest
part of the discharge and that this channel will consequently take the
largest part of the sediment flux. According to this assumption, we
modify the sediment influx q_1^{IN} (2.10) in channel 1 by the following
heuristic formula:

$$b_1 q_1^{IN} = \alpha u^{n+1} \frac{2b Q_1^k}{Q_1^k + Q_2^k}, \quad b_2 q_2^{IN} = \alpha u^{n+1} \frac{2b Q_2^k}{Q_1^k + Q_2^k}, \quad (2.14)$$

where $Q_1 = b_1 h_1 u_1$, $Q_2 = b_2 h_2 u_2$ are the discharges in both channels and where k is an undetermined parameter that sets the extent to
which the discharge distribution influences the sediment flux distribution over the channels.

The sediment balance for channel 1 now reads

$$l_1 b_1 \frac{dh_1}{dt} = -\alpha \left[\frac{2b Q_1^k}{Q_1^k + Q_2^k} u^{n+1} - b_1 u_1^{n+1} \right], \quad (2.15)$$

and a similar equation for channel 2 with interchanged indices. The flow velocities u_1, u_2 in channels 1 and 2 are given by (2.9). We assume that the width, depth and length of the channels 1 and 2 are initially adjusted such that there is no net sedimentation or erosion; $dh_1/dt = 0$, $dh_2/dt = 0$ in the unperturbed state. We ignore the time variation of the channel widths; we have seen in the previous example that they are not essential. This yields two additional relationships (one for each channel) that must be satisfied by the depth, width and length of the two channels. It can be shown these relationships do not influence the discussion below.

Upon a small depth perturbation $\epsilon h'_1$, the perturbed discharge Q'_1 and velocity are (to first order in ϵ) given by:

$$Q'_1 = \frac{3\epsilon Q_1}{2(1+p)} \left(\frac{h'_1}{h_1} - \frac{h'_2}{h_2} \right),$$

$$u'_1 = \frac{\epsilon u_1}{1+p} \left(\left(\frac{p}{2} - 1 \right) \frac{h'_1}{h_1} - \frac{3p}{2} \frac{h'_2}{h_2} \right). \tag{2.16}$$

The parameter p is a measure of the channel asymmetry,

$$p = \frac{b_2 h_2}{b_1 h_1} \sqrt{\frac{h_2 l_1}{h_1 l_2}}. \tag{2.17}$$

The expressions for channel 2 are similar, with interchanged indices and with p replaced by $1/p$. By substitution of these expressions in the sediment balance equations (2.15) for channel 1 and channel 2, we obtain a matrix relating the perturbations h'_1, h'_2 to their rate of change. The solution involves exponential functions with time exponents $\sigma_1 t, \sigma_2 t$. The determinant of the matrix gives the product $\sigma_1 \sigma_2$ of the time exponents. It appears that this product is proportional to

$$\sigma_1 \sigma_2 \propto Ak - n - 1, \quad A = 3 \frac{Q_2^k \left(\frac{5p}{2} - \frac{1}{2} \right) + Q_1^k \left(\frac{5}{2p} - \frac{1}{2} \right)}{\left(Q_1^k + Q_2^k \right) \left(1 + \frac{p}{2} + \frac{1}{2p} \right)}, \tag{2.18}$$

with a positive proportionality constant.

For symmetric channels ($p = 1$) we find $A = 3$; in this case the perturbations have negative time exponents for $k > (n + 1)/3$ and will therefore decay. For asymmetric channels we find larger values for the coefficient A. For instance, if $b_2 h_2 \gg b_1 h_1$, then $p \gg 1$ and $Q_2 \gg 1$, thus $A \gg 1$. In this case, the channels 1 and 2 do not strongly compete; perturbations already decay for small values of k.

This result can be understood as follows. An initial decrease of the cross-section in the smallest channel induces a decrease of current velocity in this channel and a decrease of the sediment inflow. Because the sediment inflow decreases with decreasing discharge, the relative decrease of the sediment inflow can be stronger than the relative decrease of the velocity. In this case the downstream sediment flux is higher than the sediment inflow; the channel will therefore erode and the initial channel cross-section will be restored.

Simulations with numerical models also show that strongly asymmetric river channel bifurcations can be stable [265,488]. As the sediment flux exponent n is generally smaller for bedload transport than for suspended load transport, channel bifurcations are more stable for coarse sediment beds than for sediment beds with medium and fine sediments. In the downstream (estuarine) region of rivers and in tidal basins, sediments are generally finer than in the middle and upstream parts of the river. We may therefore expect that channel bifurcations are less stable in these environments. Field observations of Van Veen [916] confirm this expectation.

A more sophisticated treatment of the partition of sediment fluxes at channel junctions was developed by Bolla Pittaluga *et al.* [91]. They considered bedload transport with lateral sediment fluxes near the channel junction produced by a lateral bottom slope. From their analysis it appears that the symmetric channel junction is stable only in case of high bedload transport and that asymmetric channel junctions are stable for low bedload transport.

Downstream deposition

A second reason for the preservation of multiple channel branches is related to downstream deposition. In the example of Fig. 2.20, the channel branch that grows at the expense of the other channel, ends in the wider confluent downstream region. There the flow velocity drops and part of the sediment load is deposited at the channel outlet. A shoal develops, that will retard the flow entering the channel upstream. The other channel branch will then take over and restore the original situation. In this scenario, the relative importance of different river branches fluctuates over time. This is a common phenomenon in

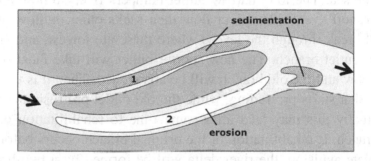

Fig. 2.20. Morphodynamic feedbacks in a channel split over a substantial length in two parallel channels. Is is assumed that the split channel system is initially in an equilibrium state. Then it is perturbed by slightly decreasing the depth of channel 1 and increasing the depth of channel 2: $h_2 > h_1$. The friction becomes stronger in channel 1 and less in channel 2. The flow velocity in channel 2 will then exceed the velocity in the upstream channel, while the flow velocity in channel 1 will be lower. A scour hole will develop at the entrance of channel 2, and a bar at the entrance of channel 1. In the course of time, these features will extend along the length of the split channel section. The flow velocity in channel 2 will increase further, while the flow velocity in channel 1 will further decrease. Accretion of channel 1 will not restore equilibrium, because channel 2 then responds with further scouring. At the downstream junction of both channels, the flow velocity is equal to its value upstream of the bifurcation. At the junction, the flow leaving channel 2 will lose its excess sediment load. Therefore a shoal develops at the end of channel 2. This generates a negative feedback to the flow increase in this channel. The flow in channel 1 may then start to increase. An increase of the flow in channel 1 results in a reversal of the erosion-sedimentation process and in a return to the initial state. This may explain the quasi-cyclic behavior in the morphodynamics often observed in branching channel systems.

river deltas [81,967]. A well documented example is the Mississippi delta, where the Mississippi outlet competes with the Atchafalaya outlet [266]. Another example is the partial obstruction of the Waal river by the fluvio-tidal delta (the Biesbos) that developed at the head of the Haringvliet estuary, when the Waal and the Haringvliet became connected in the 15th century [489].

The case of very high sediment load

River deltas with a very high sediment load (compared to their water discharge), such as the Yellow river, typically have a single outlet branch. This single branch builds a narrow delta that protrudes far into the sea. The long narrow outlet hampers river outflow during high runoff events. The river flow then seeks other outflow ways; it will break through the levees where these are lowest, and create a new outlet branch. The new shorter outlet will take most of the river flow and the old branch will be abandoned. Then it is the new outlet that starts protruding, while the old outlet protrusion erodes. Eventually this may lead after some time to rehabilitation of the old branch. If rehabilitation of the original branch occurs before its complete avulsion, the river delta will be formed by a two-branch or by a multibranch system with long-term periodic shifts in the discharge distribution over the different branches. This is a common situation for river deltas with moderate sediment load, as mentioned above.

Channel competition in tidal basins

Cyclic morphologic evolution related to channel competition is a frequent phenomenon in tidal basins (estuaries and tidal lagoons). This is mainly due to sediment deposition at places where a tidal channel widens and the current strength decreases. In the course of time these deposits tend to obstruct the channel and force the flow to change it course [460,916]. This occurs at flood deltas inside the tidal basin and at ebb-tidal deltas situated seaward of the inlet.

Wave-driven sand-spit building contributes to the obstruction of tidal channels at the inlet. Periodic shifts in the channel configuration of ebb-tidal deltas is discussed in Sec. 6.2.2.

2.5.3. *The Bornrif Cycle*

In this example, the competition between tidal channels is illustrated for a small tidal basin under natural conditions. In the mid-1990s a large ebb-delta shoal ($2 \, km^2$) joined the Wadden Sea island Ameland (for its location, see Fig. 2.16, right top). It was called 'Bornrif', after the tidal channel (Borndiep) from which it emerged. Soon after assimilation, a high beach berm developed around the shoal, preventing direct tidal submersion, except during the highest tides. The term 'rif' (Dutch for reef) stems from the presence of this berm. Tidal intrusion into the low-lying central part of the reef occurred through an inlet at the downdrift side of the shoal; the reef formed a small tidal basin with a flood delta at its centre. The length of the inlet channel grew due to downdrift inlet displacement by longshore wave-induced sand transport (see Chapter 7 on wave-topography interaction, for further explanation). A large meander also developed; it cut into the dunefoot and a restaurant at that location had to be removed, see Fig. 2.21. After ten months a second meander had developed; the old meander still existed, but was abandoned in favour of a secondary channel that had cut through the tidal flat at the inner meander bend, see Fig. 2.22 (top). One and a half year later the beach berm was broken during a storm event. A new inlet was created, much closer to the flood delta. The longer old inlet channel lost the competition with the new channel for the conveyance of flood and ebb flows and closed a year later, see Fig. 2.22 (bottom). The new inlet channel then started shifting in downdrift direction, repeating the history of the first channel. The whole morphologic cycle described above took place without human interference; it can be considered as a large-scale natural field experiment. It illustrates several feedback processes responsible for morphodynamic instability, to which we will refer later in this book.

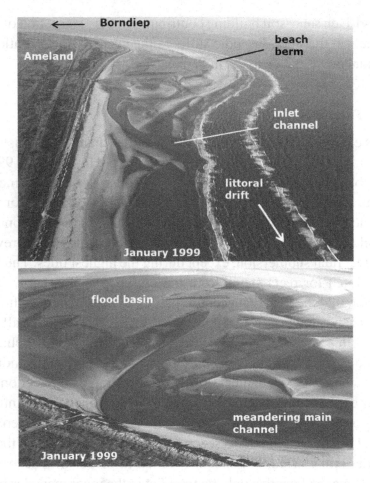

Fig. 2.21. Top: The Bornrif in January 1999, at low water. The shoal is surrounded by a beach berm, with a single tidal inlet situated at the downdrift side. The dominant wave incidence is from the north-west, at the upper right side of the picture. The centre of the shoal, which is below low tidal level, is filled and emptied by a meandering inlet channel. The outer channel bend is eroding the dune foot. A former inlet situated more closely to the centre of the shoal was abandoned. Below: Closer view of the channel meander, which in January 1999 was cutting into the dune foot. Photographs this page and next page are from an aerial monitoring campaign conducted by the Survey Department of Rijkswaterstaat Noord-Nederland.

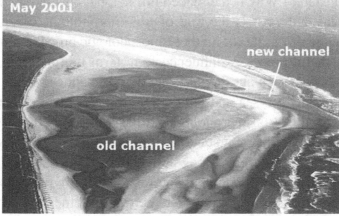

Fig. 2.22.   Top: The Bornrif at the end of October 1999. The inlet has shifted further downdrift and the length of the inlet channel has increased. Closer to the inlet a second meander has developed, which interferes with the first meander. A channel through the shoal at the inner bend is shunting the first meander; this meander is being abandoned. Below: The Bornrif in May 2001. The beach berm has been breached and a new inlet is created much closer to the centre of the shoal. The former inlet channel has been abandoned and the former inlet is closed. The new inlet has started shifting in downdrift direction. The tidal prism has decreased due to sand infill of the central shoal basin.

2.6. Summary and Conclusions

In the previous sections we have illustrated by a few examples that sea and land play a game in the coastal zone with a sometimes surprising outcome. We have also introduced the most important principles of this game: Space-symmetry breaking and time-symmetry breaking. For one particular phenomenon, the competition between parallel channels, we illustrated how these principles work. But in this chapter most questions related to 'how does the sea shape the land?' have been set aside. Taking the Bornrif development as an example, such questions include: What is the origin of migrating shoals in the coastal zone, are they a common phenomenon and how do they survive in the highly energetic coastal environment? What processes are responsible for the formation of tidal inlets, can tidal inlets become stable and if so, what is the equilibrium morphology? Why do tidal channels develop, why are they not straight but meandering and are meanders always unstable? Why is the seabed not flat, but covered with dunes and ripples and what is the role of this morphologic complexity? What are the timescales of all these morphologic processes and how do short therm processes influence long term morphologic evolution? To answer such questions we need a basic understanding of the interaction between sea and land, or more precisely, between flow dynamics and coastal morphology. In the following chapters some contributions to this understanding will be introduced. These contributions refer to current-topography interaction, tide-topography interaction and wave-topography interaction. They will be discussed separately, because the interplay between these elements is very complex and therefore beyond the scope of this book. We will mainly concentrate on idealised situations, limiting the range of morphodynamic interactions. This may sometimes give a false impression of simplicity. The Bornrif cycle can serve as an example to remind us of the complexity of the real-life physics of the coastline.

Chapter 3

Boundary Layers
and Sediment Transport

3.1. Introduction

This chapter deals with flow-topography interaction at the smallest scale: The scale of individual sedimentary grains. This interaction proceeds in two directions: the fine structure of the sediment bed (smooth, grainy, rippled) influences the flow dynamics through friction and the flow dynamics influences the bed structure through sediment transport. Sedimentary particles will be entrained by the flow when the shear stress over the bed is strong enough to dislodge the particles from their position in the bed matrix. The shear stress is mainly due to turbulent fluid motions generated at small scale bed irregularities. Bed irregularities are the result of particle entrainment and deposition processes.

In this chapter the most important factors that influence sediment transport in coastal environments will be reviewed. We first discuss the 'simple' case of steady flow over a non-cohesive sedimentary bed. Then we we pay attention to issues particular for coastal environments: the occurrence of waves (in combination with a steady current), the presence of cohesive sediments and related chemical and biotic processes. At the end of this chapter some formulations of sediment transport are discussed that are commonly used in practice.

Sediment transport processes have been widely investigated, especially since the pioneering work of Bagnold and Einstein in the

40ties of the last century [44,267]. Many formulations for describing sediment transport in practical situations have been derived since from laboratory experiments and field observations. However, there are still no theories based on first principles capable of reliable transport predictions in field situations. In the chapters on the interaction of currents, tides and waves we will make use of simple empirical formulas. These formulas have a limited applicability. For their application in practical field situations it is essential to be aware of the processes and assumptions underlying these formulas.

Since sediment transport is not the primary focus of this book, this chapter only presents a general overview of concepts and models which are most relevant for the following chapters. For a more complete overview the reader is referred to the existing literature on the topic (for example, [318,365,599,634,796,907,908]). The reader who is familiar with sediment transport processes may skip this chapter.

3.2. Near-Bed Steady Flow

3.2.1. *Flow Layers*

Turbulence

Water motion is governed by the principles of the conservation of mass and momentum. The mass and momentum balances are non-linear, nonlinearity being inherent to the principle of momentum conservation and to the free motion of the water surface. Due to this nonlinearity, the flow response to a small perturbation — even a perturbation at molecular scale — can receive a positive feedback from the flow pattern it generates and grow by extracting energy from the unperturbed flow. This is the case for flow layers with thickness δ such that the Reynolds number

$$Re = u\delta/v \qquad (3.1)$$

is in the order of 1000 or more. In this expression v is the kinematic viscosity ($v \approx 10^{-6}$ m^2/s) and u the average velocity in the flow layer.

The interaction of fluid motion with the seabed generates a cascade of unstable rotating flow structures (turbulent eddies), with dimensions ranging typically from the flow layer thickness δ down to the viscous Kolmogorov microscale

$$L_d \sim v^{3/4} \epsilon^{-1/4}, \tag{3.2}$$

where ϵ is the energy dissipation rate per unit mass [609]. This unstable water motion is known as turbulence and turbulent motion makes flow phenomena extremely complex.

The theory of turbulence is a vast topic; a comprehensive treatment is far beyond the scope of this book. We limit the discussion here to aspects of turbulence which are crucial for the interaction between fluid flow and seabed dynamics.

In practice, it is almost impossible to resolve the full turbulent flow structure from the governing equations. The question is whether this is necessary, if we are interested only in flow properties at larger scales, i.e., flow properties at the characteristic macroscopic scales of topography and external forces. In this case, instead of the detailed flow structure we only need to know the average influence of turbulence on mass and momentum conservation. This requires making assumptions about the statistical properties of turbulence; the validity of these assumptions has to be checked always against observations.

The seabed is generally rough

The statistical properties of turbulent flow over an irregular seabed cannot be derived mathematically from first principles. Our knowledge is based primarily on results from laboratory experiments and (to a lesser degree) from field studies. Many models have been developed to interpret experimental flow data and to establish statistical relationships that are valid for a broad range of field situations. There is not a unique set of relationships for describing fluid flow over an irregular seabed. It appears, for instance, that the characteristics of turbulence are different in different zones of the water column; these different zones are known as flow layers.

The existence of different flow layers is significant for the interaction of water motion with seabed sediments. This interaction can produce a positive feedback between emerging seabed ripples and dunes on the one hand and the flow pattern generated by these bedforms on the other hand. A more detailed description is given in the next chapters on current-seabed and current-channel interaction.

For turbulent flow over rough surfaces three layers can be distinguished: the roughness layer, the logarithmic layer and the outer layer, see Fig. 3.1. In the case of density stratification, due to salinity or suspended sediment, additional layers may be present.

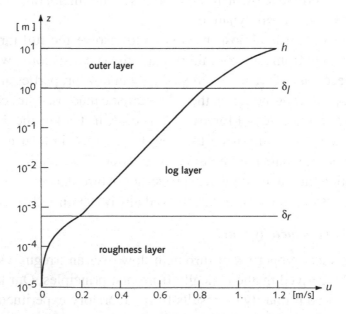

Fig. 3.1. For steady flow three boundary layers with different turbulence characteristics can be distinguished: Roughness layer $\delta_r = 2.5d$, logarithmic layer $\delta_l = 0.1\,h$ and outer layer ($h = 10\,\mathrm{m}$ in this example). A characteristic velocity profile ($\bar{u} = 1\mathrm{m/s}$) is shown for flow over a flat rough bed of medium sand (grain diameter $d = 250\,\mu\mathrm{m}$); z is the distance from the bottom.

Water flow over a seabed is turbulent if the grain Reynolds number is in the order of 70 or more [634],

$$Re_g = u_* d/\nu \geq 70. \tag{3.3}$$

In this expression d is the median grain diameter of the bed sediment and u_* the friction velocity. The friction velocity is related to the bed shear stress τ_b by

$$u_* = \sqrt{\tau_b/\rho}, \tag{3.4}$$

where ρ is the fluid density. The inequality (3.3) means that flow separation occurs at sedimentary particles with a diameter of around $70\nu/u_*$ or more. For smaller grains, a thin viscous sublayer may be present at the grain surface.

In the following sections we discuss the three flow layers, first for steady flow and then for wave flow. We ignore turbulence generated by lateral flow shear. This is not an important restriction, because in the marine coastal environment, lateral flow gradients are in general much smaller than vertical flow gradients.

Roughness layer

The bottom flow layer, also known as 'roughness layer', extends right from the sediment bed up to a thickness δ_r. For a flat bed the thickness of this layer is a few times the median grain diameter d [781],

$$\delta_r \approx (2 - 3)d. \tag{3.5}$$

We assume that under conditions of significant morphodynamic activity, the thickness of the roughness layer in coastal waters is such that the Reynolds number $u_* \delta_r/\nu$ is much larger than 1. In that case, the roughness layer is characterised by turbulent flow structures generated by small-scale bed irregularities such as sediment grains and bed ripples. These turbulent structures correspond to a very irregular process of alternate growth and separation of eddies in the wake of the bed irregularities. The turbulent fluctuations u', w' of the horizontal and vertical flow velocities are therefore correlated. Their net result for generating pressure fluctuations and vertical momentum transfer in the roughness layer, represented by $\langle u'w' \rangle$, does not average to zero.

According to the momentum balance equation, the shear stress τ in the roughness layer is equal to the sum of the surface pressure gradient and flow acceleration terms, integrated over the upper water column. The shear stress in the roughness layer is thus approximately independent of z and equal to the bed shear stress,

$$\tau \approx \tau_b = \rho u_*^2. \tag{3.6}$$

We assume that the turbulent vertical exchange of momentum over the roughness layer can be parameterised as a vertical diffusion process, according to a hypothesis of Boussinesque (1887),

$$\tau \equiv -\rho \langle u'w \rangle = \rho N u_z. \tag{3.7}$$

The turbulent eddies in the roughness layer have a characteristic vertical scale corresponding to the height of the bed irregularities, independent of z. The generalised viscosity coefficient N in the roughness layer can be represented by $N = \delta_r u_* / C$, where C is a parameter depending on the geometry of the roughness elements and on bed sediment motion.

Combining (3.6) and (3.7) we find

$$u(z) = C u_* z / \delta_r, \tag{3.8}$$

In the roughness layer, the flow velocity decreases to zero at the bottom (assuming that the seabed is at rest); the average velocity is approximately a linear function of the distance z to the bottom. For uniform steady flow, the velocity gradient depends on the layer thickness δ_r and the friction velocity u_*. For a flat bed with fixed spherical roughness elements (sand grains with diameter d), $\delta_r = d$ and $C = 8.5$ [609].

It is worth noting that the velocity profile (3.8) is similar to the profile for a flat smooth bed, where vertical momentum exchange occurs through viscosity. In that case we have a viscous sublayer of thickness $\delta_v = C v / u_*$ and $C = 11$ [609].

The thickness of the roughness layer is not constant, even under steady flow. Strong perturbations are caused by coherent turbulent motion (vortices) generated in the higher turbulent flow layers. These vortices produce an inrush of high-velocity fluid into the roughness layer ('sweeps'; $u' > 0$, $w' < 0$) or ejections of low-velocity fluid out of this layer ('bursts'; $u' < 0$, $w' > 0$). Sweeping and bursting events strongly enhance sediment motion and suspension of bed material [365,448,820] and are probably the major mechanism for suspending sand in tidal basins [794].

Influence of bed ripples

In the case of a rippled sediment bed, a fair empirical estimate of the thickness of the turbulent bed roughness layer is given by [354,523, 634,825]

$$\delta_r \approx (10 - 30)\, h_r^2/\lambda, \tag{3.9}$$

where h_r is the average ripple height and λ the average ripple wavelength. Ripple height is typically a factor 6–12 smaller than the wavelength. For ripples with an average height of 2 cm and wavelength of 20 cm, Eq. (3.9) yields a bottom layer thickness δ_r in the order of a few centimeters. For a rippled seabed with moving sand and roughness height δ_r of 1 to 2 centimeters, field observations indicate values of C between 5 and 8 [637,789].

When the shear stress at the bed increases above a certain threshold, seabed ripples are wiped out. The threshold shear stress is approximately $\vartheta \sim 0.8 - 1$, where ϑ is the Shields parameter. The Shields parameter is the dimensionless bed shear stress,

$$\vartheta = \frac{\tau_b}{g d \Delta \rho}, \tag{3.10}$$

where $\Delta \rho$ is the density difference between fluid and seabed sediments and d the medium grain size. In this so-called 'sheet-flow' regime, the seabed roughness height δ_r increases as a function of the bed shear stress [817]. The thickness of the sheet flow layer is in the order of $10d$, see Sec. 3.8.2.

Ripple characteristics and ripple formation are discussed in more detail in Secs. 4.3 and 4.3.2. The influence of ripples on sediment transport is discussed in Sec. 3.8.2.

Logarithmic layer

Above the roughness layer we have a flow layer where turbulent eddies are constrained by the distance to the bottom; as a first approximation the average eddy size increases linearly as a function of this distance. We will assume that the shear stress in the logarithmic layer can also be parameterised as a diffusion type process, see Eq. (3.7). This is a reasonable approximation under conditions of smooth spatial and temporal gradients, see also Sec. 3.4. It should be noticed that turbulent fluctuations in fluid density are ignored in Eq. (3.7); this so-called Boussinesque hypothesis is valid only for non-stratified conditions. The effect of stratification due to dense fluid suspensions is discussed in Sec. 3.5.2.

The diffusion coefficient N in the logarithmic layer is known as eddy-viscosity; observations show that it can be approximated by a linear function of the distance to the bottom (the so-called 'law of the wall'),

$$N = \kappa u_* z, \tag{3.11}$$

where $\kappa \approx 0.4$ is the Von Karman constant. As the shear stress τ is approximately constant in a zone close to the bed, it is related to the friction velocity u_* by (3.6). It follows from (3.7, 3.11) that the velocity profile takes a logarithmic form,

$$u(z) = \frac{u_*}{\kappa} \ln\left(\frac{z}{z_0}\right). \tag{3.12}$$

The parameter z_0 is the bed roughness length. It is not the height above the bottom where the actual velocity vanishes, but the height above the bottom where the logarithmic profile extrapolates to zero.

Instead of z_0, many authors use the Nikuradse equivalent roughness, defined for flat smooth beds as $k_s = 30z_0$. It is approximately equal to the thickness δ_r of the roughness layer.

Observations in the sea show that the logarithmic shape applies to approximately the lower 10–20 percent of the water column [796]. Some studies report a greater thickness of the logarithmic layer, up to about half the water depth [546]. At the transition between the roughness layer and the logarithmic layer, $z = \delta_r$, the velocities (3.8) and (3.12) should match. This results in the following equation for the bed roughness length z_0,

$$\delta_r = z_0 \exp(\kappa C). \tag{3.13}$$

For a flat bed ($C = 8.5$) we find $\delta_r = k_s = 30z_0$. If we take the thickness of the roughness layer δ_r as 2.5 times the median grain diameter d [285], we find $d/z_0 \approx 10$, which is consistent with results from field studies [552]. For a rippled bed with moving grains ($C = 5\text{-}8$), Eq. (3.13) yields for δ_r/z_0 values between 7 and 25, with $\delta_r \approx 15\,\mathrm{mm}$.

The logarithmic law (3.12) yields a good representation of the velocity profile for steady flow. In the coastal zone, currents are often modulated by an oscillating wave component. Laboratory experiments by Sleath [783] show that the logarithmic law remains a fair representation of the wave-averaged velocity profile in the turbulent boundary layer.

Outer layer

Higher up, some 10–30 percent of the total water depth above the bottom, the eddy viscosity is constrained not only by the distance to the bottom, but also by the distance to the water surface. This upper part of the water column is known as the outer boundary layer. In shallow coastal waters it often extends up to the water surface. In deeper waters the thickness is limited by the effects of earth's rotation and by the tidal timescale (in the case of tidal flow). The thickness δ_t

of the tidal boundary layer is then approximately given by [796]

$$\delta_t = 0.0038 \frac{\omega U_{max} - f U_{min}}{\omega^2 - f^2}, \tag{3.14}$$

where U_{max}, U_{min} are the maximum and minimum values of the depth-averaged tidal velocities through a tidal cycle, f is the Coriolis parameter and ω is the tidal frequency in radians.

Observations show that for steady uniform flow the velocity profile in the outer layer can be represented by a power law distribution [796],

$$u(z) = u(0)(z/h)^{1/7} \approx 1.14\bar{u}(z/h)^{1/7}, \tag{3.15}$$

where $\bar{u}$ is the depth-averaged velocity and h the water depth. Matching the velocities at the transition $z = \delta_l$ of the outer layer and the logarithmic layer yields

$$\bar{u}/u_* = (0.88/\kappa)(\delta_l/h)^{-1/7} \ln(\delta_l/z_0) \approx 5.5(h/z_0)^{1/7}, \tag{3.16}$$

where the last approximation holds for large values of δ_l/z_0. The friction factor c_D ('drag coefficient') relates the bed shear stress τ_b to the depth-averaged velocity $\bar{u}$ by:

$$\tau_b = \rho c_D \bar{u}^2, \quad c_D = (u_*/\bar{u})^2. \tag{3.17}$$

Using the relations (3.16) and (3.17) the velocity profile (3.15) in the outer layer can be written

$$u(z) \approx (9 - 10)\, u_*(z/\delta_r)^{1/7}, \tag{3.18}$$

for values of $\delta_r \sim (10 - 30)z_0$.

3.2.2. *Momentum Dissipation*

Friction factor c_D

From (3.16) we find a relationship between the friction factor c_D and the roughness length z_0:

$$c_D \approx 0.033(z_0/h)^{2/7}. \tag{3.19}$$

From observations of steady flow over flat mobile and immobile beds a best fit yields [796] $c_D \approx 0.02(d/h)^{2/7}$, where d is the median grain diameter. This is equivalent to (3.19) if we assume $z_0/d = 0.2$, and it corresponds to (3.13), if we take $\delta_r = 2.5d$ and $C = 6.3$. The friction coefficient c_D usually ranges from 0.001 to 0.01 [839], depending on several variables: Sediment type, grain diameter, the height and shape of bed irregularities (ripples), water depth, sediment transport rate and flow strength. In shelf seas and estuaries the friction coefficient is found typically on the order of 0.002–0.003 [434, 777, 839].

For instance, at Georges Bank (USA Atlantic coast, some 100 km offshore Cape Cod) the sandy sea floor (medium grain diameter 250–1000 μm) at 80 m depth is usually covered by ripples formed after high wave events, with 1–2 cm height and 15–20 cm wavelength. The maximum strength of the tidal currents $u_{1.2}$ at 1.2 m above the bed is on the order of 1 m/s. From velocity profile measurements of the semidiurnal tidal currents the friction coefficient defined by $(u_*/u_{1.2})^2$, was found to be $(3 \pm 0.1) \times 10^{-3}$, without important seasonal changes [942]. The corresponding roughness length z_0 ranges between 0.05–0.09 cm, which is consistent with a roughness layer thickness of 1–2 cm and $C \approx 8$ (see Eq. (3.8)).

The relationship (3.19) indicates that the friction factor is also a function of the water depth. Since the roughness length z_0 hardly depends on the depth [708], Eq. (3.19) can be used to estimate the depth dependency. Often the empirical formula of Manning is used instead, which reads

$$c_D = gn^2 h^{-1/3}, \qquad (3.20)$$

where g is the gravitational acceleration and n is Manning's bed roughness coefficient. The friction factor increases with decreasing depth.

Friction coefficients of 0.02 and even higher have been measured in the uprush and downrush flow of broken waves on the beach, [706]. Not only is the small water depth (typically less than 0.5 m) responsible for these high values, but the turbulence produced by wave-breaking as well. However, under highly energetic conditions the friction coefficients may also be reduced instead of increased, as indicated by field observations in the North Sea [434]. This phenomenon is attributed to near-bed stratification, due to sediment resuspension under high waves [344]. Drag reduction will also occur when a high-concentration layer of fine cohesive sediment is formed near the bed, see Sec. 3.5.

Eddy viscosity N

An estimate of the eddy viscosity N in the outer boundary layer can be derived from the momentum balance, which states that the shear stress in steady uniform (non-stratified) flow is a linear function of depth,

$$\tau(z) = \rho N u_z = \rho u_*^2 (1 - z/h). \tag{3.21}$$

By introducing the velocity profile (3.15) in this equation, it appears that N in the outer layer is approximately a parabolic function of z with a maximum at mid-depth. Its depth-averaged value can be estimated from

$$\overline{N} \approx 1.16 h u_*^2 / \overline{u} = 1.16 c_D h \overline{u}. \tag{3.22}$$

In uniform steady flow the eddy viscosity is closely related to the friction factor.

Density differences strongly reduce vertical mixing by inhibiting turbulent diffusion. Density differences due to fresh-water inflow are a common feature in estuaries and in near-coastal waters, while further offshore density stratification may occur due to temperature gradients over the water column. The eddy viscosity is strongly reduced in the presence of (even weak) density stratification, compared to homogeneous conditions [619].

Turbulent diffusion of momentum in the fluid is also influenced by the presence of suspended particles. This influence can be important close to the seabed, where suspended sediment concentrations are often high. Suspended particles have a damping effect on turbulence, as the flow dissipates energy for keeping them in suspension.

This results in an apparent decrease of the Von Karman coefficient κ [148, 178, 394], which can become as low as 0.25 [846] (see also Sec. 3.4.1).

Skin friction and form drag

Sediment transport is related to the bed shear stress. One has to distinguish the shear stress contribution related to the grain structure of the bed and the contribution related to the ripple structure [318]. The former is known as 'skin friction' or 'effective shear' and corresponds to the tractive stress exerted by the flow in the viscous sublayer or in the roughness layer on bed grains. The latter is known as 'form drag' and corresponds to energy losses of the flow in the wake of bedforms (ripples, dunes).

The skin friction includes viscous drag and pressure drag that arise due to flow around individual particles on the bed. The skin friction is particularly important for incipient sediment motion through bed load transport, which requires dislodging sediment grains from their equilibrium position in the bed matrix.

The form drag is unimportant for transporting sediment particles as bedload, because the length scale that characterises variations in the associated pressure field is much larger than the scale of individual particles [588]. Form drag makes a major contribution to frictional dissipation of momentum and is mainly responsible for sediment transport through suspended-load sediment transport. Due to its influence on the total turbulent intensity, the form drag also contributes indirectly to skin friction. For a rippled bed, form drag is typically several times larger than skin friction [796, 942].

3.3. Near-Bed Wave Flow

Inertial lag

In the previous section the vertical flow profile was discussed for steady uniform flow. The terms 'steady' and 'uniform' have to be understood in a statistical sense, i.e., averaged over the temporal and spatial scales of turbulence. Water flow in the marine environment is continuously accelerating or decelerating, either because of time varying forcing (waves, tides, etc.) or because of topography. The theoretical shapes (3.8, 3.12, 3.15) of the velocity profile seldom occur in practice. This is because the flow profile does not adapt instantaneously to variation in forcing; it takes some time before a new equilibrium is established between flow profile and turbulent stresses. This time delay is called 'inertial lag'. It is longer for large turbulent eddies than for small eddies and is therefore strongly depth dependent. For instance, observations in the Irish Sea of the tidal variation of turbulent energy dissipation as a function of depth, show an adaptation phase delay of around 2 hours between bottom ($-70\,\text{m}$) and surface for mixed winter conditions and a phase delay of more than 4 hours for stratified summer conditions [777].

The flow profile near the seabed will thus adapt faster to fluctuations in the surface pressure gradient than the flow profile higher in the water column; far away from the bed the fluid keeps it original momentum longer than close to the bed. When the flow is accelerated (or decelerated), the fluid near the bed is accelerated (or decelerated) first, followed later by the fluid higher in the water column. Compared to steady uniform flow, velocity gradients and associated turbulent stresses near the bottom are relatively larger in the acceleration phase and relatively smaller in the deceleration phase. The inertial time delay of flow adaptation plays a crucial role for the emergence of seabed structures, as will be shown later.

3.3.1. Wave-Boundary Layer

In the case of high frequency wave motion (wind-generated water waves) the layer structure is quite different from steady flow. An introduction to wave dynamics is given in Appendix D.

Due to the shortness of the wave oscillation period, a turbulent boundary layer will not develop throughout the water column. This

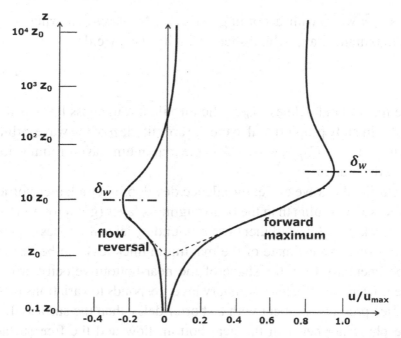

Fig. 3.2. Example of vertical velocity profiles in wave-orbital motion in the case of a rough seabed. Right curve: The instantaneous velocity profile in the phase of maximum near-bottom orbital motion. In the roughness layer (z smaller than a few times z_0), the profile is approximately linear. In the turbulent constant-stress layer above, the profile becomes logarithmic. δ_w is a measure of the upper bound of the logarithmic layer. Above δ_w starts the frictionless outer layer with the wave-orbital velocity profile $u \propto \cosh(kz)$. Left curve: The instantaneous velocity profile shortly after flow reversal near the bottom.

implies that wave-induced flow is almost frictionless (potential flow in most of the water column), except in a thin layer near the bottom, see Fig. 3.2. For a smooth bed the flow is viscous and the thickness of the layer is approximately given by

$$\delta_w \approx \sqrt{2v/\omega}, \tag{3.23}$$

where v is the viscosity and ω is the wave radial frequency; this corresponds to a layer thickness of at most a few millimetres. In the coastal environment, however, the magnitudes of bed roughness and wave-orbital velocity are such that the wave-boundary layer is turbulent during a part of the wave cycle. In this case a rough estimate of the maximum thickness of the wave-boundary layer is given by

$\delta_w \approx \sqrt{2N_w/\omega}$ with, according to (3.11), $N_w \approx \kappa u_* \delta_w$, where u_* is the maximum wave-orbital shear velocity. This yields

$$\delta_w \approx u_*/\omega. \qquad (3.24)$$

The maximum thickness δ_w of the turbulent roughness layer is thus approximately proportional to the the amplitude a of the wave-orbital excursion, $a \equiv U_0/\omega$, where U_0 is the maximum wave-orbital velocity at the seabed.

During the wave cycle, turbulence develops over a larger portion of the water column than the bed-roughness layer (grain roughness). The thickness of the turbulent wave-boundary layer δ_w varies over the wave cycle. An estimate of the maximum thickness can be derived experimentally from the shape of the near-bottom velocity profile. The flow in the turbulent boundary layer responds to variations in the surface pressure gradient with a shorter delay than the surface flow. The phase lag between the near-bottom flow and the free surface flow is in the order of $\pi/8$. At some stage of the wave cycle, the near-bottom current reaches a forward maximum, while the surface current is still lagging behind. The height δ_w where the near-bottom current velocity is maximum is indicative of the thickness of the wave-boundary layer, see Fig. 3.2. Laboratory experiments [469,869] and a theoretical wave boundary model [318] indicate

$$\delta_w/\delta_r \approx 0.09\,(a/\delta_r)^{0.82}, \qquad (3.25)$$

where δ_r is the roughness layer thickness (3.5).

Laboratory experiments also show that the velocity profile in the turbulent wave-boundary layer (for $0.2\delta_r < z < (0.2 - 0.3)\delta_w$) has a logarithmic shape during part of the wave cycle, similar as for the steady velocity profile [459,782,869]:

$$u(z,t) = \frac{u_*(t)}{\kappa} ln\left(\frac{z}{z_0}\right), \qquad (3.26)$$

where $z_0 \approx \delta_r/30$. However, Sleath [783] presented evidence for an approximately constant eddy viscosity throughout the turbulent wave layer, implying a linear velocity profile in this layer.

Momentum dissipation

In the boundary layer wave-orbital momentum is dissipated by bed shear stresses. The wave-induced bed shear stress can be related to the wave-orbital velocity by a quadratic relationship, similar to that for steady flow,

$$\tau_w = \frac{1}{2}\rho f_w |u_w(t)| u_w(t), \tag{3.27}$$

where f_w is the wave friction factor and $u_w(t)$ is the wave-orbital velocity just above the wave-boundary layer (in shallow water practically equal to the frictionless wave-orbital velocity at $z = 0$). Several empirical relations have been established for estimating the friction factor [796]; f_w depends mainly on the ratio of wave-orbital amplitude $a = U/\omega$ and bed roughness length δ_r. One of the most frequently used formulas is the empirical relationship of Swart [825]:

$$f_w = 0.00251 \exp\left(4.57\left(\frac{a}{\delta_r}\right)^{-0.19}\right), \quad \frac{a}{\delta_r} > 1.6. \tag{3.28}$$

The roughness length δ_r for a rippled seabed is more than ten times larger than for a smooth seabed; ripples increase the friction coefficient by a factor two or more. Typical values of f_w range between 0.01 and 0.03. By solving the flow equations in the wave-boundary layer, Fredsøe and Deigaard found the following theoretical expression, valid for large orbital excursions ($a > 50\delta_r$) [318],

$$f_w \approx 0.04(\delta_r/a)^{1/4}. \tag{3.29}$$

Streaming

Longuet-Higgins [537] first showed that progressive waves produce a forward drift velocity u_s (called 'streaming') in the laminar wave-boundary layer above a smooth seabed. The near-bed boundary layer is assumed to be viscous (non turbulent) in this case. Just above this layer the forward drift is maximum and given by

$$u_s \approx 3U_b/4c, \tag{3.30}$$

where U_b is the maximum wave-orbital velocity at the top of the viscous layer and c the wave propagation speed, see Fig. 3.3.

It has been suggested that the streaming phenomenon contributes substantially to sediment transport when the top layer of the seabed is fluidized by intensive wave action [307]. The streaming current extends through this highly loaded sediment layer and moves it in the onshore direction.

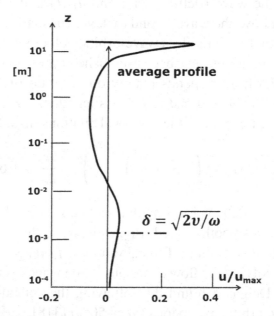

Fig. 3.3. Wave-orbital velocity profile averaged over the wave cycle for a wave propagating over a smooth seabed. For a laminar wave-boundary layer, the net velocity is in the wave propagation direction ('streaming'). The streaming velocity is maximum around the top of the viscous wave-boundary layer. The mean velocity has an opposite sign above the boundary layer, to compensate for the streaming velocity and for the net mass transport near the surface (due to positive correlation of wave-orbital velocity and water level between wave trough and wave crest).

Streaming is caused by friction in the wave-boundary layer, which advances the phase of the wave-orbital velocity u relative to the free stream velocity above the boundary layer. The phase difference between the horizontal and vertical wave-orbital velocities u and w, which is 90° far above the wave-boundary layer, therefore decreases down the water column. This results in a net wave-induced stress $\rho \langle uw \rangle$ and a corresponding gain of forward horizontal momentum in the boundary layer, see Fig. 3.4. The resulting flow acceleration is balanced by upward diffusion of mean forward momentum. Viscosity causes a decrease of the streaming velocity to zero at the fixed bed. The streaming velocity has a maximum at the top of the wave-boundary layer, where the gradient $-(\langle uw \rangle)_z$ is strongest.

The assumption of a laminar wave-boundary layer does not hold for rough seabeds. Laboratory measurements show that for rough beds the forward near-bottom drift is strongly reduced [78, 891]. Trowbridge and Madsen [854] explained this reduction of the streaming velocity by considering the time variation of turbulent viscosity in the wave-boundary layer. The time variation of the turbulent viscosity is related to the decay and build-up of the turbulent boundary layer at each wave cycle. The principle of momentum transfer to the boundary layer acts in the same way as for the constant boundary layer. However, due to the time varying thickness of the boundary layer, the forward streaming is concentrated in a thin layer above the bed, see Fig. 3.4.

For a rippled seabed, the situation is rather different. At reversal of the near orbital flow, vortices are shed off behind the ripple crests, see Fig. 4.20. The vortex shedding goes along with strong turbulent intensity, especially in the case of steep ripples. The eddy viscosity is higher near the time of orbital flow reversal and than at the times of maximum shear stress [209]. This has important consequences for the residual transport of suspended sediment, as will be discussed in Sec. 3.8.2.

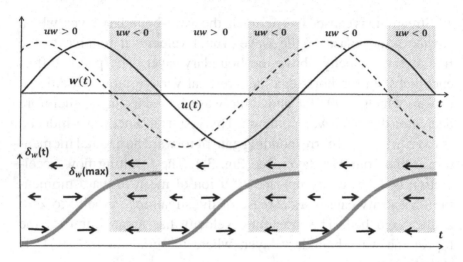

Fig. 3.4. Top graphics: The wave-orbital velocity $u(t)$ (solid line) and the vertical velocity $w(t)$ (dashed line, not at scale) in the wave-boundary layer as a function of time. A positive (negative) momentum flux uw induces an net gain (loss) of momentum in the boundary layer. The phase advance of the wave-orbital velocity u induces a gain of forward momentum in the boundary layer averaged over the wave cycle ($\langle uw \rangle > 0$), producing forward streaming, see Fig. 3.3. This net forward momentum flux is balanced by upward diffusion of net forward momentum in the boundary layer. Above the boundary layer the mean velocity is negative to compensate for the forward streaming and the forward mass transport near the wave surface. The graphics below depicts streaming in the turbulent time-dependent wave-boundary layer at different phases of the wave cycle, represented by the arrows. The thickness δ_w of the wave-boundary layer is very small at the reversal of the wave-orbital velocity and increases to $\delta_w(max)$ during the following half wave cycle. The forward streaming is therefore concentrated in a very thin layer above the bed; in the zone around $\delta_w(max)$ the mean velocity is negative.

Wave asymmetry in the boundary layer

When waves propagate from deep water into the shallow nearshore zone, the shape of the waves changes. First they become skewed (shorter crest and longer trough) and then asymmetric: rapid rise and slow fall; the wave shape resembles a saw-tooth, see for instance Fig. 2.3. This is due to the nonlinearity of wave propagation in shallow water, as explained in Secs. 2.2.1 and 6.3.1.

The skewed asymmetry in the wave-orbital motion is related to stronger velocities in the forward than in the backward direction. This will favour a net displacement of seabed material in the wave propagation direction (shoreward direction), because sediment transport is very sensitive to the strength of the flow velocity. However, net shoreward sand transport appears to be associated also with sawtooth asymmetry, often designated 'acceleration skewness' or simply 'wave asymmetry'. Acceleration skewness is associated with a faster reversal from backward to forward wave-orbital motion and a slower reversal from forward to backward orbital motion [275, 372, 451]. When this was first observed, it was not immediately clear how such an asymmetry in wave-orbital acceleration could produce a net sediment transport.

Nielsen [634] provided an explanation referring to the dynamics of the wave-boundary layer, which was confirmed later by several laboratory experiments [73, 742, 869]. These experiments show that during the fast transition from backward to forward wave-orbital motion, the turbulent wave-boundary layer is not built up as high as during the slower transition from forward to backward wave-orbital motion. The highest forward wave-orbital velocities thus occur at a thinner wave-boundary layer than the highest backward orbital velocities, see Fig. 3.5.

For frictionless wave-orbital motion, the vertical velocity profile near the bottom, in the zone just outside the turbulent layer, depends hardly on the distance from the bottom (contrarily to the steady current profile). At the overlap region $z = \delta_w$ between the turbulent logarithmic wave-boundary layer and the frictionless outer layer, the velocity in the turbulent layer, $(u_*/\kappa) \ln(z/z_0)$, has to match the orbital velocity u_b of the outer layer. A thinner wave-boundary layer thickness δ_w thus corresponds to a higher friction velocity u_* and a higher shear stress τ_b. In this way, the acceleration skewness in the wave-orbital motion results in an asymmetry between maximum shear stresses for forward and backward orbital motion in the wave-boundary layer. Hence, velocity skewness (stronger forward than backward orbital velocity) and acceleration skewness (stronger forward than backward orbital acceleration) produce both a net forward sediment flux in the wave-boundary layer. Expressions

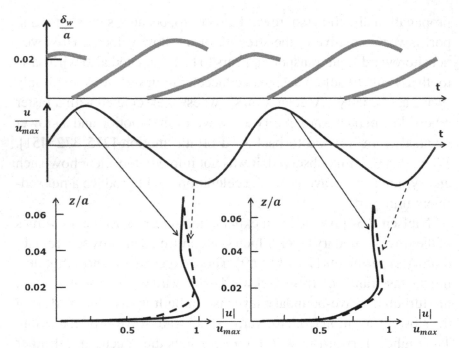

Fig. 3.5. Turbulent boundary layer for an acceleration-skewed wave, redrawn from Van der A *et al.* [869]. Middle figure: Near-bottom orbital wave velocity for two successive wave cycles. Top figure: Development of the turbulent wave-boundary layer during the wave cycles. The thickness δ_w is scaled with the amplitude a of the wave-orbital excursion. The wave boundary is thinner after the fast backward-forward velocity reversal than after the slower forward-backward reversal. Bottom figures: Vertical velocity profiles in the wave-boundary layer for different tidal phases. The near-bottom velocity shear is higher for the situation where the turbulent wave boundary is thinner.

for the influence of velocity skewness and acceleration skewness on the net wave-induced bed shear stress and bedload sediment transport are given in Secs. 3.8.1 and D.1.2.

Modelling of the wave-boundary layer [495, 855] and laboratory experiments [718] further show that velocity skewness and acceleration skewness reduce the near-bed streaming in the wave propagation direction. The near-bed streaming may even be reversed.

3.3.2. Wave-Current Interaction

In field situations, waves and currents occur often simultaneously. Because of the nonlinear nature of turbulence generation in the near-bottom boundary layer, current and waves interact. Waves and currents thus cannot be considered independently in shallow water for situations where the wave-orbital motion near the bottom (amplitude U_b) is significant compared to the steady current (depth-averaged value $\bar{u}_c$). The near-bottom turbulent boundary layer for the situation of simultaneous waves and currents is very different from the boundary layer for waves or currents alone, see Fig. 3.6. The wave motion close to the seabed strongly enhances momentum dissipation of the steady current and inversely, the steady current enhances the dissipation of wave-orbital momentum. Our knowledge of current-wave interaction is primarily based on laboratory experiments. These experiments reveal several remarkable features.

The wave-averaged component u_c of the combined wave-current velocity u_{cw}, has a logarithmic profile, which is shifted upward with respect to the logarithmic current profile without waves [353]:

$$u_c(z) = \langle u_{cw}(z) \rangle = \frac{u_{*c}}{\kappa} ln\left(\frac{z}{z_a}\right). \tag{3.31}$$

The parameter z_a is an apparent roughness length, which can be much larger than the physical roughness length z_0 [317, 634, 783]. This apparent roughness corresponds to an increase in roughness experienced by the steady current due to the additional turbulence generated by the wave motion near the seabed. The value of z_a/z_0 increases (more than linearly) with the ratio $U_b/\bar{u}_c$; z_a/z_0 also increases with a/z_0 (a is the near-bottom wave-orbital amplitude), except for small values of $U_b/\bar{u}_c$ [669]. The apparent roughness is largest for co-linear waves and currents.

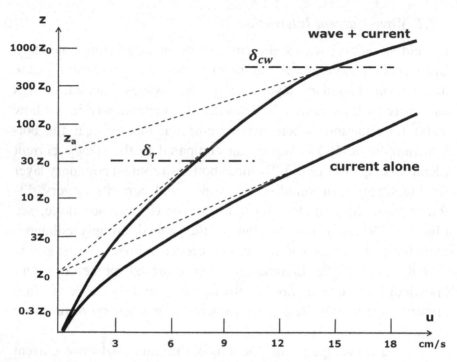

Fig. 3.6. The boundary layers for the combined wave-current case, redrawn from [317]. Left curve: Vertical profile of the flow velocity averaged over the wave period $u_c(z)$. Right curve: Vertical profile of the flow velocity $u(z)$ in the absence of waves, shown for comparison. The profile of the averaged wave-current velocity is approximately linear in the bed-roughness layer ($z < \delta_r$). In the lower part of the turbulent wave-current boundary layer the profile is logarithmic with the same intercept z_0 as for the current alone. Higher up, another logarithmic (constant stress) layer is present. The intercept z_a of the logarithmic velocity profile corresponds to an apparent roughness δ_{cw} much larger than the physical roughness length δ_r.

The expression (3.31) holds in the logarithmic (constant stress) layer above the turbulent wave-current boundary layer; the thickness δ_{cw} of this layer varies over the wave cycle. Numerical simulations show that the effective shear velocity u_{*c} is an (almost linearly) increasing function of $(U_b \bar{u}_c)^{1/2}$ [317].

Below the uplifted logarithmic profile ($z < \delta_{cw}$) another logarithmic sublayer is present, just above the turbulent roughness layer of thickness δ_r. This sublayer is similar to the logarithmic layer found in the situation without waves,

$$u_c(z) = \frac{u_{*c}^2}{\kappa u_{*cw}} ln\left(\frac{z}{z_0}\right),$$ (3.32)

where $u_{*cw} = u_{*c} + u_{*w}$ for co-linear waves and current [318].

The total shear stress τ_{cw} for the case of combined waves and current can be represented by

$$\vec{\tau}_{cw}(t) = \frac{1}{2}\rho f_{cw}|\vec{u}_{cw}(t)|\vec{u}_{cw}(t),$$ (3.33)

where $\vec{u}_{cw}(t) = \vec{u}_c + \vec{u}_w(t)$; $\vec{u}_w(t)$ is the wave-orbital velocity just above the wave-boundary layer.

Empirical expressions for the friction factor f_{cw} in the case of co-linear waves and current are given by [911]

$$f_{cw} \approx 0.08(z_a/h)^{1/3}$$ (3.34)

or by [719]

$$f_{cw} \approx \alpha f_c + (1 - \alpha)f_w, \quad \alpha = \left(\frac{\bar{u}_c}{\bar{u}_c + U_b}\right)^p,$$ (3.35)

where $f_c = 2c_D$ is the friction factor for current alone, f_w the friction factor for waves alone (3.28) and U_b the amplitude of $\vec{u}_w$. The exponent p is usually taken equal to 1.

An empirical expression for the maximum bed shear stress for rough beds is given by [796]:

$$\tau_{b,max} = \tau_{b,c}\left[1 + 1.2(\tau_{b,w}/\tau_{b,cw})^{3.2}\right],$$ (3.36)

where $\tau_{b,c} = \rho u_{*c}^2$, $\tau_{b,w}$ is the maximum shear stress for waves alone and $\tau_{b,cw} = \tau_{b,c} + \tau_{b,w}$.

3.4. Non-Cohesive Sediments

Sedimentary particles

Seabed material in coastal areas consists of very different kinds of particles. Many of these particles are produced by chemical or physical weathering of rocks and by soil erosion in inland catchment basins and are called clastic sediments. They are made of minerals, for instance, kaolinite (clay mineral), feldspar and mica (silt minerals) or quartz (sand mineral). Other particles have a biotic origin, for instance shell-debris, peat, detritus, fecal pellets and plankton.

An important characteristic by which sediment particles can be distinguished is their diameter. This diameter differs widely, from clay and silt (grain diameter approximately $10^{-6}-10^{-5}$ m) to sand (diameter approximately $10^{-4}-10^{-3}$ m) and gravel, pebbles and cobbles (diameter approximately $10^{-2}-10^{-1}$ m). The density of sediment particles ρ_{sed} is on average 2.5 to 3 times the density of water; however, heavy minerals with a much higher density may also be present.

The finest sediment particles, consisting mainly of clay minerals, behave in a particular way due their cohesive properties. The behaviour of cohesive sediments will be discussed in the next section.

3.4.1. *Suspension and Settling*

Settling velocity

Without turbulence or upwelling water motion, sediment particles will move downward, because their density is higher than the density of water. Several empirical formulas have been established for the settling velocity w_s of sand particles in still water, for instance the formula by Van Rijn [903]

$$w_s = \frac{10v}{d}\left[\sqrt{1 + 0.01d_*^3} - 1\right], \quad d_* = d\left(\frac{g\Delta\rho}{\rho v^2}\right)^{1/3}, \quad (3.37)$$

or the formula by Jimenez and Madsen [463]

$$w_s = \sqrt{gd\Delta\rho/\rho}\left[0.954 + 20.5d_*^{-3/2}\right]^{-1}. \quad (3.38)$$

Both formulas are shown in Fig. 3.7. In this expression d and d_* are the grain diameter and the dimensionless grain diameter, v is

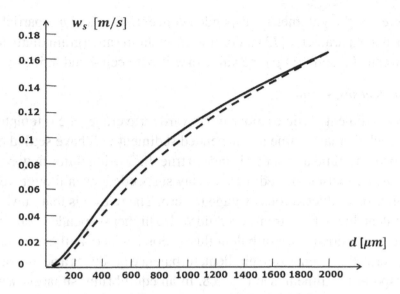

Fig. 3.7. Settling velocity at 10°C as a function of grain diameter for sand particles. Solid line: (3.37), dashed line: (3.38).

the kinematic viscosity of water ($\approx 1.35\ 10^{-6}$ m^{2}s^{-1} at 10° and $\approx$ 10^{-6} m^{2}s^{-1} at 20°), $\Delta\rho/\rho = \rho_{sed}/\rho - 1 \approx 1.6$ is the average relative density difference of sediment particles and seawater and $g = 9.8$ ms^{-2} is the gravitational acceleration.

For particles with a small dimensionless grain diameter (sand grains with a diameter smaller than 100 μm or larger particles with a lower density) the expression (3.37) can be simplified to Stokes' formula

$$w_s = g(\Delta\rho/\rho)d^2/18\nu. \tag{3.39}$$

For these particles the fall velocity is proportional to the ratio of the submerged mass and the grain diameter.

The expression (3.37) is valid under the assumption that sand grains fall freely, without interacting with each other. Close to the bed, the sand concentration can be such that this assumption does not hold. In this case, a concentration dependent correction should be applied to (3.37),

$$w_s(c)/w_s(0) = (1 - c)^n, \tag{3.40}$$

where c is the volumetric suspended concentration and n a particle-dependent parameter [721]. For fine to medium sand (grain diameter between 100 and 400 μ) the value of n is between 4 and 4.5 [50].

Turbulent suspension

Since sediment particles move downward on average, one is tempted to conclude that in time all suspended sediment will have settled on the bed. In a fluid at rest this is indeed true. In flowing fluids however, turbulence causes the sediment to stay suspended, even though vertical velocity fluctuations average to zero. The reason is that upward turbulent fluctuations transport fluid with a higher suspended concentration than downward turbulent fluctuations; hence, in the presence of a vertical concentration gradient, turbulence produces a net upward transport of sediment, see Fig. 3.8. In an equilibrium situation with steady current, the net flux of sediment through each horizontal plane in the water column will be zero. From this condition the mean concentration profile can be derived, as a function of the settling velocity w_s and the turbulent intensity (represented by the diffusivity K).

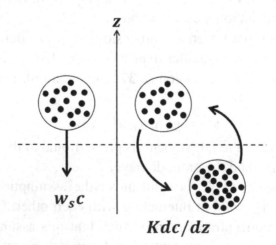

Fig. 3.8. Schematic representation of the balance between vertical sediment flux components. Upward transport by turbulent diffusion compensates downward transport due to particle settling.

Suspended sediment profile

The vertical velocity of a sediment particle w_s consists of a mean downward motion $\langle w \rangle = -w_s$ and a fluctuating component w' due to turbulence, with zero mean, $\langle w' \rangle = 0$. The sediment concentration $c(z)$ can also be represented by a mean concentration $\langle c \rangle$ and a fluctuating component c'. The mean vertical flux through any given horizontal plane equals $\langle cw \rangle = -\langle c \rangle w_s + \langle c'w' \rangle$. The last term is the flux caused by turbulent diffusion; it is generally assumed that in the absence of stratification, the turbulent motion of water particles can be characterised by a random walk. In this case the flux is approximately proportional to the vertical gradient of the mean concentration,

$$\langle c'w' \rangle \approx -K d\langle c \rangle / dz, \tag{3.41}$$

where z is the distance from the bottom and K is the turbulent diffusivity. The condition that the mean vertical sediment flux equals zero implies

$$w_s c = -K dc/dz. \tag{3.42}$$

For simplicity we have left out the brackets denoting turbulence time averaging. In the logarithmic boundary layer the turbulent diffusivity K can be represented by an expression similar to that for the turbulent viscosity N [177],

$$K = \beta_S \kappa u_* z, \tag{3.43}$$

where κ is the Von Karman constant, u_* the friction velocity and β_S the inverse turbulent Schmidt number. The Schmidt number takes into account the influence of suspended particles on turbulent diffusion, caused by density stratification and mutual interaction between particles. It depends on the ratio $w_* = w_s/u_*$, approximately as $\beta_S \approx 1 + 1.5w_*^2$ [480]. Usual values of β_S range between 1 and 1.5, in the absence of density stratification [177, 857]. Above a rippled bed much higher values of β_S may occur, up to $\beta_S \approx 3$ [209, 556]. This is related to a stronger correlation between u', c' than between u', w' in the flow separating downstream of the ripple crests [985], see Figs. 3.9 and 4.20.

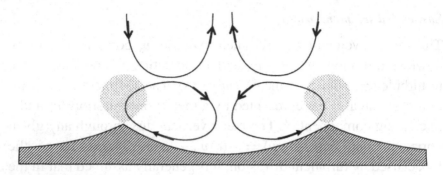

Fig. 3.9. Schematic representation of the residual circulation produced by flow separation downstream of ripple crests in an oscillating current. Sediment suspension by this mechanism is basically different from suspension by turbulent diffusion.

As mentioned earlier, suspended sand particles also influence the eddy viscosity N (the turbulent diffusion of momentum in the fluid) [954]. At high concentrations, this results in an apparent reduction of the Von Karman coefficient κ. The product $\beta_S \kappa$ is therefore less sensitive to the suspended sand concentration than κ and β_S separately [846].

If the settling velocity w_s is assumed to be constant, the solution of (3.42) yields a power-law distribution for the suspended sediment equilibrium profile in a stationary current, see Fig. 3.10,

$$c(z) = c_{ref}(z/z_{ref})^{-w_s/\beta_S \kappa u_*}. \tag{3.44}$$

Here, z_{ref} is a bed reference level and c_{ref} the corresponding suspended concentration. The exponent

$$R_s = w_s/\beta_S \kappa u_* \tag{3.45}$$

is the so-called Rouse number. Sediment tends to be almost uniformly distributed over the water column if the settling velocity is low and the turbulent diffusivity is high; in the opposite case the sediment concentration has a strong vertical gradient. For medium-to-coarse sand most of the suspended sediment is confined to a near-bed layer with a thickness of 10 to 20 times the ripple height [165].

Near the surface, the diffusion coefficient K approaches zero. Therefore a factor $(1 - z/h)$ should be included in the expression

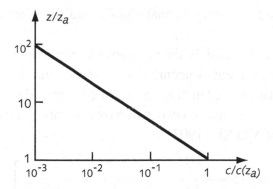

Fig. 3.10. Suspended sediment equilibrium profile in a double-logarithmic plot, according to (3.44), for typical values $w_s = 0.02$ m/s and $u_* = 0.03$ m/s.

(3.43). In most cases this is not relevant, however, because of the smallness of near surface suspended sand concentrations.

Near-bed concentration

Many models, most with a strong empirical character, have been proposed for describing sediment suspension and sediment transport by waves and currents [206, 796, 903]. Numerical simulations with a simplified mathematical model show that particles are picked up from the bed and carried into suspension when low-speed turbulent streaks start to break up and generate small incoherent vortices that increase mixing. Low-speed streaks are generated at the end of the acceleration phase when the flow is maximum, and break down during the deceleration phase of the wave cycle [922].

Realistic mathematical descriptions of the sediment uptake process need to include explicitly the turbulent fluid and particle motions in the near-bed zone. Models based on large-eddy simulation techniques (LES) have been developed which are capable of reproducing many features which are also observed in the laboratory. LES models confirm the important role of coherent turbulent structures, especially in the presence of bed ripples [985]. These models also show that significant differences exist between the sediment uptake processes for straight-crested two-dimensional ripples and for three-dimensional ripples.

However, LES models do not provide explicit expressions for the near-bed sediment concentration. Expressions for the near-bed concentration to be used in practice have been derived from laboratory experiments. In the case of suspension by currents alone one may use the formula of Van Rijn [903]

$$c_{ref} = 0.015 \frac{d\,\tau_*^{3/2}}{z_{ref}d_*^{0.3}}, \quad z_{ref} = min\left[\frac{h}{100}, \delta_r\right]. \tag{3.46}$$

In these expressions the following notations are used: $\tau_* = (\tau_s - \tau_{cr})/\tau_{cr}$, d = particle diameter, d_* = dimensionless particle diameter (3.37), δ_r = thickness roughness layer and h = water depth.

This expression is valid for current velocities above a certain critical threshold, producing a skin friction shear stress, τ_s, which is larger than a critical value τ_{cr}. This critical value corresponds to the minimum stress required to lift a sediment particle out of its position in the seabed. For sandy beds, Van Rijn [907] recommends

$$\vartheta_{cr} = 0.115 d_*^{-0.5}, d_* < 4,$$
$$\vartheta_{cr} = 0.14 d_*^{-0.64}, 4 < d_* < 10, \tag{3.47}$$

where d_* is the dimensionless grain diameter (3.37) and $\vartheta_{cr} = \tau_{cr}/gd\Delta\rho$ the critical Shields parameter. The formula (3.46) takes into account the bed shear stress caused by bed irregularities, such as bed ripples and dunes; δ_r is then approximately half the average height of bed ripples or dunes. Bottom-ripple steepness may play a more important role than ripple height [396], but this feature is not explicitly included.

For strong currents the near-bottom reference concentration varies approximately with the third power of velocity, according to (3.46). Observations indicate that the near-bed concentration saturates at a volumetric concentration of approximately 0.3 [210]; this may occur already when the Shields stress is larger than 0.75. Van Rijn [908] provides prescriptions how to use (3.46) under conditions of combined current and waves.

Wave induced suspension

Wave action strongly contributes to suspension of sediment; it is often the dominating factor. As mentioned before, wave-generated turbulence and diffusivity are basically different from current-induced turbulence and diffusivity; this is due to the shortness of the wave period relative to the time scales of turbulent motion [468, 634, 783].

Under moderate wave action the seabed is often rippled. The formation of ripples strongly enhances the suspension of sediment, by the process of vortex shedding, especially for steep ripples (steepness $h_r/\lambda > 0.12$ [210]). Sediment is lifted up into the flow at the upstream side of a ripple and advected downstream by the current and upward by the vertical velocity component of streamwise vortices [317]. This process is schematically depicted in Fig. 4.20. Although it can be simulated in numerical models (for instance, [555]), the influence of this process on sediment suspension is represented in most sediment transport models by empirical relationships.

In the frictional bottom boundary layer, waves and currents interact non-linearly and produce together a higher skin friction than the sum of the individual contributions. The strong accelerations associated with wave-orbital motion also influence seabed erodibility [417].

In the near-coastal zone, wave-orbital velocities are generally stronger than steady current velocities. In this case it is often assumed that wave action is mainly responsible for sediment suspension (expressed by a wave-stirring function) and transport is mainly caused by the steady current. However, such a simplification should be used cautiously. For instance, residual sediment transport may also be produced by waves alone as a result of wave asymmetry and wave-induced currents [595].

3.4.2. Graded Sediment

Bed armouring

The bed concentration formula (3.46) is valid for sand with a well-defined median grain diameter d. In practice the sediment bed often contains a mixture of different grain sizes, so-called 'graded sediment'. The sediment fraction with the finest grains is suspended more easily than the coarsest fraction. If the current shear stress is just higher than the critical stress for suspension of the fine sediment fraction, but much lower than the critical shear stress for the coarse fraction, the fine sediments are progressively removed from the bed top layer. After some time, the bed top layer consists of coarse particles only. The sediment uptake by the flow then stops, because the coarse top layer shelters the fine particles below. This phenomenon is called 'armouring' of the sediment bed [235,643,779]. For larger shear stresses, the fine and the coarse fractions can both be set in motion. It appears that the critical shear stress for moving both the fine and coarse fractions is approximately equal to the critical shear stress for a non-graded bed with the average grain size of the mixture. The critical shear stress for mobilising all the grains of a non-cohesive graded bed is thus smaller than the critical shear stress for a non-graded bed with the coarse fraction only.

Sand-mud mixture

If the fine sediments in the top layer are cohesive, the critical shear stress for erosion may drastically increase. In a loosely packed water-saturated seabed, pore pressure fluctuations disrupt contacts among grains, facilitating entrainment of particles by bed shear stress [599]. Bed erosion is reduced if the pores between larger sediment particles are filled with fine cohesive particles, as the underlying layers become isolated from the water column and wave-induced pressure fluctuations hardly penetrate the seabed. The addition of 30 percent mud to a sand bed can increase the critical shear stress by as much as a factor of 10. The erosion rate of a sandy seabed can be reduced by a factor of 5 or more by addition of more than 3 percent mud [413, 607]. Incorporation of sand grains in a cohesive bed also increases the critical erosion shear stress while the erosion rate is reduced.

The erosion of a layered mixed bed may be modelled as a sequence of sandy and muddy erosion events. Through the processes of bed consolidation and bed armouring the transport dynamics of non-cohesive and cohesive sediments become interrelated. This is a serious complicating factor for modelling sediment transport in coastal and estuarine waters [516]. In the next section a short overview will be given of major processes influencing the transport of fine cohesive sediments.

3.5. Cohesive Sediments

3.5.1. *Suspension and Settling*

Flocculation

For small non-cohesive sediment particles the mean settling velocity follows approximately Stokes' law (3.39). This expression was derived theoretically for Reynolds numbers much smaller than 1, i.e., for very fine unaggregated sediment particles (grain diameter in the order of 10^{-6} m $\equiv 1 \, \mu$m). The settling velocity according to Stokes' law is so small that in theory fine sediments would always remain in suspension. The fraction of fines that always remain in suspension is called washload; it consists of colloids — organic molecules, bacteria and very fine clay minerals with a diameter in the order of 0.1 μm or less. Through their negative electrical charge they bind with water molecules and other dilute substances.

In water containing salt ions, the electro-chemical properties of clay particles favour mutual adhesion and adhesion with other small particles, like silt and fine sand grains. This adhesion process is called flocculation. The settling velocity of flocs is much larger than the settling velocity of the composing particles. Increase of settling velocity due to flocculation is the major cause of mud deposition in estuaries.

Flocculation is affected by a number of factors [260,599]:

• *Organic polymers*

Organic molecules like polysaccharides, proteins, nucleic acids and lipids are ubiquitous in the marine environment; they are exuded by all living organisms (from bacteria to fishes) and are termed 'extracellular polymeric substances' (EPS). They adhere to particles through ion-exchange reactions and cation bridging [351] and form large fibers that bind particles together. Bacterial colonisation may also stimulate floc growth [9,528].

• *Concentration*

Flocculation is enhanced when particles collide more often [496]. Observations show that the size of flocs increases with increasing concentration of suspended fine particles, see Fig. 3.12. The largest flocs only grow if the suspended sediment concentration is sufficiently high. Large flocs, called macroflocs, are formed as aggregates of smaller flocs, called microflocs, see Fig. 3.11. Macroflocs have the greatest settling velocity, notwithstanding their low density. However, macroflocs are fragile and can easily break up when the shear stresses in the fluid become high. Deposition at the seabed is also counteracted by hindered settling. In the lower part of the vertical the sediment concentration can be so high that flocs impede each other's downward motion. This causes a decrease of the effective settling velocity [656], see Fig. 3.13.

• *Turbulence*

Turbulence enhances the collision frequency of sediment particles and thus favours floc growth. However, when turbulence is too strong, flocs are broken up by turbulent shear stresses — especially the large flocs. This occurs frequently in the shear layer close to the seabed. The settling velocity is then decreased and sediment particles will easily go back into suspension [260,590]. For a given strength of

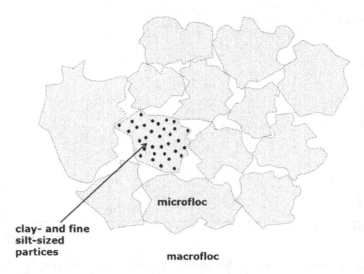

microfloc

**clay- and fine
silt-sized
partices**

macrofloc

Fig. 3.11. In tidal environments, flocs appear in two kinds: Microflocs and macroflocs. The macroflocs are aggregates of microflocs, loosely bound by extracellular polymeric substances (EPS). Microflocs are aggregates of fine sedimentary particles, mainly clay minerals, small amounts of fine silt (feltspar, mica) and organic detritus. They are bound by electrochemical forces and by the ubiquitous EPS. Microflocs can survive the shear stresses produced by tidal currents; their size is typically smaller than the length scale at which turbulent eddies dissipate (the Kolmogorov length scale (3.2)). This length scale decreases with increasing current shear stress to values of typically $100-150\,\mu$m. Macroflocs can attain much larger diameters, up to order 1 mm and even larger. They are formed by collisions among microflocs, due to turbulence, Brownian motion and differential settling. However, they are broken up when tidal currents become strong and the length scale of turbulence dissipation decreases. Macroflocs settle much faster than microflocs. Most of the downward movement of fine sediment in the water column is related to settling macroflocs.

turbulence a balance is established between floc formation and floc breakup.

Laboratory experiments show that the length scale L_d of decaying turbulent eddies (the Kolmogorov microscale (3.2)) sets a limit to the floc size [919]. The Kolmogorov microscale depends on the energy dissipation rate per unit mass ϵ, which can be measured or derived from $k - \epsilon$ turbulence models. A rough estimate is provided by the approximate expression

$$\epsilon = u_z \tau / \rho \approx u_*^3 (1 - z/h)/\kappa z. \tag{3.48}$$

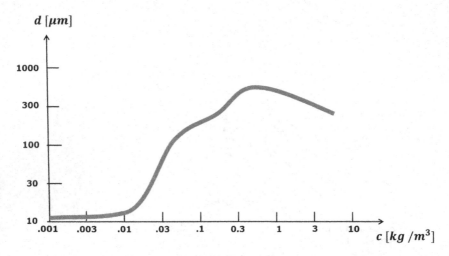

Fig. 3.12. Example of maximum floc size as a function of concentration, in the French estuaries Seine and Gironde, redrawn after [602]. In other estuaries, for example the Ems–Dollard, floc sizes up to 1–2 mm have been observed around slack water [899].

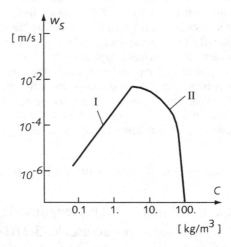

Fig. 3.13. Flocculation as a function of the concentration, after [592]. High sediment concentrations produce larger flocs and higher settling velocities (upgoing branch I, fit to experimental data: $w_s = 5.10^{-4}c^{1.3}$ m/s). Floc settling is hindered at high suspension density (downgoing branch II, fit to experimental data $w_s = 2.6\ 10^{-3}(1 - 0.008\ c)^{4.65}$ m/s).

Flocculation is often related to the velocity shear rate $G\left[s^{-1}\right]$ in the fluid, defined as

$$G = \nu/L_d^2 = \sqrt{\epsilon/\nu}, \qquad (3.49)$$

where ν is the kinematic viscosity. Although the expression (3.48) cannot be used at $z = 0$, it shows that the velocity shear rate G is largest near the seabed, where the Kolmogorov microscale is smallest. Maximum floc growth, or highest settling velocity, occurs for intermediate values of the velocity shear rate $G = G^*$. Different field studies report values for G^* in the order of $10\ s^{-1}$ [666,919].

• *Salinity and pH*

Cl^- ions facilitate the coagulation of clay particles by decreasing the the energy barrier produced by their negative electrical charges [269]. In field situations it is often difficult to isolate the influence of salinity on flocculation from other influences, for instance metallic or organic coatings [432]. It has been shown that acidity also influences flocculation processes. With increasing acidity (decreasing pH) the floc size increases [598].

Settling velocity

Flocs are easily disturbed; their size and composition are strongly variable. Therefore it is quite difficult to establish a generally valid expression from which the settling velocity of fine cohesive sediment can be derived. The settling velocities measured in laboratory experiments and during field campaigns in different estuaries reveal a very wide spread of almost a factor 100, as illustrated in Fig. 3.14.

Several recent studies propose different methods for estimating the settling velocity [513,957]. Here we only mention the simplest formulas, established by Soulsby *et al.* [799]. They result from a parameter fit to a large set of data from the turbid Tamar and Gironde estuaries. The data are consistent with the following assumptions: (1) the settling velocity of microflocs (diameter $<160\ \mu m$) depends only on the

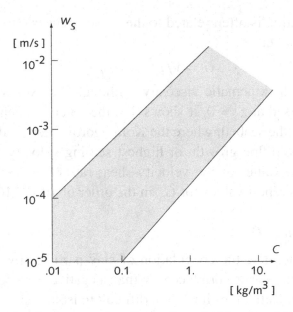

Fig. 3.14. The domain of measured median settling velocities of mud flocs in different estuaries as a function of concentration, after [899]. At each concentration the measured values span a range exceeding one order of magnitude.

velocity shear rate G; (2) the settling velocity of macroflocs depends both on G and on the suspended concentration. The formulas for the settling velocities of macroflocs w_{sM} and for microflocs $w_{s\mu}$ read:

$$w_{sM} = \frac{gB_M}{G}\left(\frac{c}{\rho}\right)^k \left(\frac{Gd_\mu^2}{\nu}\right)^{0.33} \exp\left[-\left(\frac{u_{*sM}}{\sqrt{\tau/\rho}}\right)^{0.463}\right],$$

$$w_{s\mu} = \frac{gB_\mu}{G}\left(\frac{Gd^2}{\nu}\right)^{0.78} \exp\left[-\left(\frac{u_{*s\mu}}{\sqrt{\tau/\rho}}\right)^{0.66}\right], \qquad (3.50)$$

where the index M designates the macroflocs and the index μ the microflocs and c is the suspension concentration in kg/l. Parameter values for the Tamar and Gironde estuaries are: $B_M = 0.13$, $B_\mu = 0.6$, $k = 0.22$, $u_{*sM} = 0.067\,m/s$, $u_{*s\mu} = 0.025\,m/s$, $d_\mu = 10^{-4}\,m$, $d = 10^{-5}\,m$. In order to apply this formula, the distribution of suspended particles between macroflocs and microflocs should be known. Field data indicate that the concentration of macroflocs relative to microflocs mainly depends on the suspended sediment concentration; macroflocs dominate for large concentrations (>1 g/l), microflocs for small concentrations (<10 mg/l).

Laboratory experiments show that the floc settling velocity is also influenced by the presence of suspended fine sand [562]. A small fraction of fine suspended sand reduces the growth potential of macroflocs; the settling velocity is therefore smaller than for a pure mud suspension. On the contrary, the settling velocity of microflocs is increased.

The formation of macroflocs in an estuary takes some time; a stable distribution of macroflocs and microflocs is reached after typically one or two hours [919]. The formulas (3.50) should not be applied for situations where the flow conditions are strongly variable over periods of less than one hour.

The settling velocity of microflocs in the Tamar and Gironde estuaries is typically in the order of 0.5–1 mm/s; the settling velocity of macroflocs is typically a factor 5 larger. It follows that settling of fine cohesive sediment is mainly related to the settling of macroflocs; this has also been observed in other estuaries [561,899].

Hindered settling

Hindered settling occurs in the dense suspension close to the seabed. There are several theories to describe hindered settling; here we follow Winterwerp [200,955]. In the dense suspension the flocs fill almost all the space. When the space is entirely filled, settling becomes impossible. The sediment concentration at which the space is completely filled is called the gelling concentration c_{gel}. The gelling concentration depends on the nature of the flocs. In some situations a value of 40 g/l is found, but generally the value is higher; $c_{gel} \approx 100$ g/l.

For taking into account the influence of other flocs in the suspension, the settling velocity should be corrected for several factors. Contacts with other flocs depend on c/c_{gel}; floc interaction decreases the settling velocity by a factor $(1 - c/c_{gel})^m$. The weight of the floc in the dense suspension is lower than the weight in clear water by a factor $(1 - c/\rho_s)$, where ρ_s is the density of the suspension. The viscosity of the suspension is increased relative to the clear water viscosity ν [262]. This increase amounts to a factor $(1 + 2.5c/c_{gel})$, according to experimental evidence. This leads altogether to the following expression for the hindered settling velocity of fine cohesive sediment:

$$w_s(c) = w_s(0)\frac{(1 - c/\rho_s)(1 - c/c_{gel})^m}{1 + 2.5c/c_{gel}}, \qquad (3.51)$$

where for the exponent the value $m = 2$ is used. A similar but more elaborated formula was proposed by Camenen [134]; this formula includes the possibility that the flocs do not entirely fill all the space when the gelling point is reached.

3.5.2. Fluid Mud

The concentration of fine cohesive sediments in suspension can become very high close to the seabed. Potential mechanisms are [582]: (1) Convergence of residual fine sediment transport; (2) Fluidisation of consolidated mud deposits by waves and (3) Settling of mudflocs into the near-bottom suspended layer, at a rate higher than the dewatering rate. Under these conditions fine sediments become trapped in a colloidal suspension, called fluid mud. This occurs, in particular, in periods around neap tide and near the seawater intrusion limit in turbid estuaries. Fluid mud layers are frequently observed at harbour entrances and in artificially deepened navigation channels [224, 662].

Due to hindered settling, the floc settling velocity decreases for concentrations around $10\,\mathrm{kg/m^3}$. The maximum in the settling velocity curve (Fig. 3.13) implies that mass settling to the fluid mud layer converges to a concentration exceeding $10\,\mathrm{kg/m^3}$ [263]. The suspended sediment concentration in fluid mud may even reach $250\,\mathrm{g/l}$.

Buoyancy suppresses turbulent exchange between layers of different density, stimulating the formation of a lutocline — a sharp interface between layers with different suspended sediment concentrations, see Fig. 3.15. The density difference between the layers is thereby further increased as an auto-enhancing process. Winterwerp [954] argues that the density increase may finally exceed the sediment carrying capacity of the lower layer, leading to the collapse of this layer and the formation of mud banks.

Fluid mud layers may be entrained by near-bottom currents over considerable distances without being dispersed over the whole water column [483]. If the seabed is sloping, the fluid mud slides down as a turbidity current. In shallow water, fluid mud absorbs wave energy. The strong wave damping observed at muddy coasts (for example the coasts of Guiana and Surinam) is caused by the presence of extensive

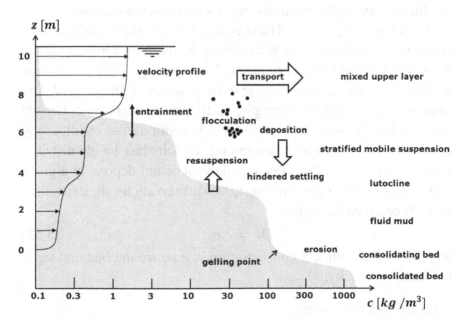

Fig. 3.15. Schematic representation of suspended sediment layers in a turbid estuary. Turbulence is strongly suppressed at the interface between different layers, due to density stratification. This happens already at relatively low concentration differences of a few hundred mg/l [362, 525].

fluid mud layers [851]. If near-bottom currents are absent or very weak, the fluid mud layer consolidates and becomes part of the bed substrate.

A seabed covered with fluid mud offers almost no resistance to the flow in the overlying layer. Prandle [688] related the friction coefficient to the mud content of the sediment bed in many UK estuaries; he found evidence for a strong decrease of the friction coefficient with increasing mud fraction. According to Winterwerp [957], the reduction in flow resistance is mainly due to the suppression of turbulence at the lutoclines. The friction factor c_D (3.17) can be reduced by a factor as large as 10.

3.5.3. *Deposition and Erosion*

Deposition

Settling of flocculated particles in the water column does not always lead to seabed deposition. The reason is that the shear stress experienced by flocs close to the bottom may be too high for keeping the flocs intact; if flocs break up the settling velocity of the constituent particles becomes too low and the settling process is interrupted. The properties (size, density, strength) of all flocs in suspension change due to collisions as well as erosion and deposition. As a result, at any time and location in water, floc settling velocity has a wide distribution, and there exist flocs that can deposit or not deposit at a given bed shear stress [956]. According to field observations, the deposition rate De can then be written

$$De = cw_s, \qquad (3.52)$$

where c is the sediment concentration just above the bed and w_s is the settling velocity.

Consolidation and erosion

As discussed already, bed erosion is primarily caused by turbulent shear stresses over the bed, with an important role of large turbulent eddies ('sweeps' and 'bursts'). Erosion takes place if the flow shear stress at the seabed τ_b exceeds a threshold value τ_{cr}, that depends on the degree of consolidation of the seabed. The rate of erosion can be expressed as [592,663]

$$Er = \mu(\tau_b/\tau_{cr} - 1)^m, \tag{3.53}$$

where μ, τ_{cr}, m are parameters that need to be determined from experiment or from modelling.

The seabed top layer is eroded first; it often consists of fresh unconsolidated deposits with a low critical erosion shear stress τ_{cr}. Then the more consolidated and erosion resistant underlying layers become exposed. Knowledge of the critical erosion shear stress τ_{cr} of these layers and knowledge of the parameters μ, m requires modelling of the consolidation history of the seabed.

The critical shear stress for erosion depends to a large degree on the water content of the sediment bed. A high water content implies a low bulk density of the sediment bed. An empirical formula gives a linear relationship between the critical shear stress and the bulk density ρ_b of the sediment bed [21],

$$\tau_{cr} = 5.4 \, 10^{-4}\rho_b - 0.28, \tag{3.54}$$

where τ_{cr} is expressed in Pa and ρ_b in kg m^{-3}. This formula should be used with caution, however. The bulk density depends not only on the water content, but also on the characteristics of the sedimentary particles — for instance the sand/mud ratio. The shear stress further depends on salinity [658] and on the presence of organic substances that bind particles together.

There are several theories for seabed consolidation [151,956]; pioneering work was done by Gibson *et al.* [343]. These theories describe the compression of soft soils by expulsion of pore water under the weight of overlying sediment. Semi-empirical relationships are introduced that relate the pore pressure and the effective shear stress (by particle contacts) to the concentration and density of the sediment particles. Till present, there is only little practical experience with these theories for field situations involving tides and waves [516].

Figure 3.16 shows an empirical curve of the minimum (critical) current strength needed to initiate bed erosion, as a function of grain diameter for different types of sediment particles [685]. This figure does not take into account the influence of waves on bed erosion, which may be particularly important, for instance on tidal flats. When waves are present, erosion takes place at much weaker currents than in the absence of waves [180,219,746]. This is attributed, in particular, to the wave-induced cyclic loading and pore-pressure amplification. Cyclic loading may lead to liquefaction (shear-induced structural breakup) or fluidisation (pore pressure induced break-up) of the upper bed layer [20,593]. Under extreme conditions (storms) the erosion strength of the upper sediment bed may be entirely lost.

Channel bank erosion

In the previous sections we have described erosion as a process of channel bed scouring by the removal of individual sediment particles

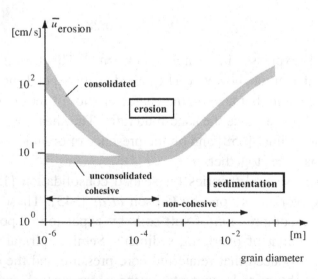

Fig. 3.16. Erosion and deposition regimes as a function of average current strength and average grain size. Consolidated cohesive sediment beds can resist erosion until high current velocities. Redrawn after [685].

from the surface of the channel bed. This picture is correct for sandy channel beds, but not for muddy beds.

The first reason is that in muddy beds, particles are tied together by cohesive forces. Shear stresses exerted by the flow may pull out clumps of sediment from the channel bed, instead of individual grains. These large aggregates are carried away by the flow and it can take a long time before they are broken up [599]. This happens, in particular, for consolidated muddy beds under high flow shear stresses. The large shear stresses required for the erosion of consolidated muddy beds are reflected in Fig. 3.16.

The second reason is related to gradients in pore pressure. Pore pressure gradients can easily build up in muddy beds, due the limited soil permeability. This contrasts with sandy soils, which are well drained. Gradients in pore pressure are built up under conditions of alternating drying and wetting. This happens at the banks bordering tidal channels. These banks often contain sufficient cohesive material for reducing their overall permeability. Under drying conditions, negative pore water pressures (suction forces) build up, that produce an increase in apparent strength of the bank material. For vegetated banks, evapotranspiration may contribute to enhancing these negative pressures. When the tide level rises, the banks are submerged, the soil becomes saturated and the water pore pressure increases. As long as the water level in the channel remains high, it provides a counter pressure to the pore water pressure in the bank. However, when the water level drops, the soil matrix is weakened by the excess pore pressure. If the bank slope is steep, the bank material experiences a strong down slope gravitational force. The resulting tension produces cracks that weaken the slope stability further. Finally the weight of the bank material cannot be supported any longer by internal cohesion forces and part of the bank collapses. If the currents along the bank are strong enough, the collapsed bank material is subsequently removed. A typical example is shown in Fig. 3.17.

The above scenario occurs frequently at the outer banks of channel meanders. In a meandering channel, the greatest current strength

Fig. 3.17. Bank failure in the Western Scheldt estuary. Photo by Marcel Taal.

occurs along the outer bend. The underwater bank toe is scoured and the bank slope is increased, until it collapses by the previously described mechanism. The mechanisms responsible for the onset of channel meandering are discussed in Sec. 5.4.

Up till now there are no theoretical models capable to provide reliable quantitative predictions of bank erosion. A method often used to assess bank stability is based on the Mohr–Coulomb theory [655], which relates the critical tangential gravitational stress to the normal stress, the internal friction angle and the internal cohesion of the bank material. In morphodynamic models, the rate of bank retreat is often simply related by an empirical coefficient to the excess velocity along the bank compared to the average velocity in the channel [440] or to a power of the flow shear stress along the bank [255].

Bank erosion processes play an important role in the development of channel meanders and the migration of tidal channels. Channel bank erosion is also termed lateral erosion, because it contrasts with vertical erosion — the removal of sediment from the bed surface. Surface erosion hardly occurs when the bed consists of consolidated cohesive sediment or when the bed is protected by vegetation. In such situations, lateral erosion by currents and waves is a more effective process. In tidal flat and marsh areas, bank erosion and related channel migration are a major mechanism for reactivating old deposits and for enhancing tidal basin morphodynamics [328], see also Sec. 5.6.2.

3.6. Biotic Activity

The previous considerations hold for muddy beds in artificially deepened navigation channels and harbour entrances and for muddy deposits in highly energetic environments. In the shallow sheltered zones of estuaries and lagoons the seabed is populated with organisms that greatly influence the seabed properties. Biogenic modification of the seabed also occurs at greater depth in shelf seas, prohibiting ripple formation during spring and summer [942].

A fluid mud layer with a density of $1100 \, \text{kg} \, \text{m}^{-3}$ is almost entirely composed of water, so the erosion threshold should be very low. However, a critical shear stress as high as $0.5 \, \text{Pa}$ was measured in the field [21], pointing to biogenic stabilization of the mud through the adhesion of extracellular polymeric substance (EPS). Other observations show that the critical erosion shear stress of soft fine-grained sediment in the intertidal zone increases from typically $0.2 - 0.5 \, \text{Pa}$ to more than $3 \, \text{Pa}$ when biofilms are present [504]; these biofilms consist mainly of EPS exuded by diatoms.

Measurements in the Dutch and German Wadden Sea show that biotic activity can either stabilise or destabilise the seabed. In the case of stabilisation, the critical shear stress for erosion can become 3 times higher and the erosion rate 10 times lower. In the case of destabilisation, the critical erosion shear stress can be halved and the erosion rate increased by a factor 3 [93,191]. Seasonal variation in the density of stabilising microphytobenthos can increase net sediment deposition on a mudflat by a factor 2 and interannual variation in the density of bioturbating clams can decrease sediment deposition by a factor 5 [962].

The colonisation of mudflats by halophytic vegetation enhances their capacity to trap fine sediments and stimulates their lateral expansion [382,626]. Plant roots are known to stabilise the soil, by creating a structural network and by increasing the organic content of the soil. Stems and leaves also contribute to reducing the turbulent shear stresses exerted on the sediment bed by currents and waves, but this does not hold for plants with stiff stems and leaves [918]. Bacteria, when breaking down the organic material in the soil, can release gases that destabilise the soil.

Biota influence the stability of the soil in many other ways. The net effect of biota on seabed erosion may be positive or negative, but no existing model provides a reliable simulation of these processes.

Biostabilisation

The main bio-stabilisers are microphytobenthos (microalgae, bacteria) by forming biofilms on tidal flats [19, 867]. Stabilization occurs through secretion of EPS that bind sediment particles together at the mud surface and smooth the surface. Stabilisation is also enhanced by infill of the inter-particle voids and hydrogelation of the underlying sediments [542]. Benthic macrofauna may increase sediment stability by binding particles with secretions intended to construct their tubes, but most often they destabilize sediment. The resistance of mudflats against surface stress is also increased by exposure to sunlight [20].

Marine organisms often bind their fecal material into fecal pellets [25]. Most benthic macrofauna produce mucous-coated feces, but the durability of the feces or fecal pellets varies greatly [332]. If fecal pellets are not immediately resuspended, the surrounding sediment will become muddier, organically enriched and more cohesive. Fecal pellets or pseudofaeces produced by cockles may indirectly increase surface sediment stability by stimulating the growth of microphyto-benthos. In most cases, however, the presence of pellets decreases sediment stability.

Bioturbation

Bioturbation (tracking, digging and burrowing) activity of macro-fauna loosens the bed material, maintains the porosity, prevents com-paction and thus increases the water content. An increase in water content of fine-grained sediments from 50 percent to 60 percent can result in a decrease of the sediment stability of about 25 percent [685]. The clam *Macoma baltica*, the mudsnail *Hydrobia ulvae* and the amphipod *Corophium volutator* destabilise the surface sediment by grazing on microphytobenthos and so counteract the formation of algae mats. The cockle *Cerastoderma edule* is responsible for bioresuspension; the continuous expulsion of mud from the sediment increases the sand/mud ratio of the sediment [610]. Bioturbation also increases oxygen penetration in the sediment and enhances micro-bial processes like nitrification, denitrification, sulfate reduction and pyrite ($FeS2$) oxidation because of the higher oxygen availability in sediment that would be anoxic otherwise [536]. The clam *Tellina fabula* is a bioturbator occurring predominantly in sandy substrates; it makes the seabed more prone to erosion [94]. The sea urchin *Echinocardium cordatum* feeds from organic material at the seabed surface; it brings fine sediment particles down and consequently coarsens the seabed top-layer [94].

The bioturbation depth or mixing zone is roughly 10 centimetres in both shallow water and the deep sea [472]. Animals that feed at depth will have much greater effects on sediment mixing than organisms that feed and defecate at the surface [124]. The biomass and abun-dance of the infauna are usually poor predictors of particle-mixing rates. A few large subsurface deposit feeders can affect sediment movement far in excess of their biomass or numerical contribution to the community [332].

While biofilms smooth the bed and decrease flow friction, bio-turbating animals roughen the bed and increase friction. Many ben-thic macrofauna produce tubes (e.g. worm tubes) that protrude a few millimetres to centimetres above the bed. The lugworm *Areni-cola marina* creates a pit-and-mound landscape, consisting of feeding funnels and fecal mounds, which remains intact under relatively mild hydrodynamic conditions. Another common tube-building worm in the coastal seas of the northern hemisphere is *Lanice conchilega*. There are different views regarding their influence on the sediment dynamics of the seabed. *Arenicola marina* is thought to produce addi-tional roughness and turbulence, which enhances the winnowing of fine material from the sediment matrix [419]. Flume experiments with *Lanice conchilega* show that it reduces the near-bottom flow and facilitates the deposition of fine sediment between the pits and mounds; the ripples on top of the sediment surface are lower in regions where it occurs in high densities [98].

Bio-geomorphological modelling

From the previous paragraphs it clearly appears that biota can have an important influence on sediment dynamics in coastal environ-ments. The inverse is also true: the hydrodynamic and sedimentary characteristics of coastal environments determine to a great extent the development of biotic activity. This two-way coupling between physical and biological dynamics can have a profound influence on the natural development of the coastal environment as a self-organising system. During the past decade a lot of research has been carried out on the influence of biogeomorphological feedbacks on the natural evolution of different coastal systems. Examples are the mutual interaction between vegetation dynamics and the morpho-logical evolution of salt marshes [290, 834, 836] and the coupled dynamics of tube-building worms and tidal dune development in the North Sea [96].

A detailed treatment of these new developments transcends the scope of this book. Research on bio-geomorphological modelling is still in an exploratory phase. The establishment of quantitative relationships of the mutual interaction of biota and hydrosedimentary dynamics is a difficult task, which has been undertaken so far only for a limited number of habitat types. In practice, the influence of biotic activity is implicitly included when physical parameters in morphodynamic models are tuned to simulate observed morphological processes. It has to be realised that this often implies a limitation to the predictive capabilities of these models.

3.7. Bed Level Evolution

3.7.1. *Morphodynamics*

The evolution of the seabed in interaction with the flow dynamics is called morphodynamics. The ancient Greek word 'morphos' means 'shape', in our case the shape is the topography of the seabed. Coastal morphodynamics is the dynamics of the seabed topography in the coastal zone. The topography (bed elevation) of the seabed relative to a horizontal reference level is represented by the function $z_b(x, y, t)$.

The seabed topography evolves because of erosion and sedimentation. This can be written in the form

$$(1 - p_b)\partial z_b(x, y, t)/\partial t = De - Er, \qquad (3.55)$$

The seabed porosity p_b is assumed to be constant; in the following we will incorporate the factor $(1 - p_b)$ in the definition of z_b. The deposition and erosion rates, De and Er, are expressed as sediment volume per unit time. Formulas for De, Er are given by (3.52) and (3.53), respectively.

Erosion is related to the uptake and transport of sediment by currents and waves and deposition is related to the settling of sediment when and where the transport capacity of the flow is insufficient to keep the sediment in suspension, see Fig. 3.18. The sediment balance

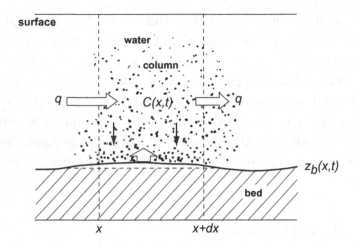

Fig. 3.18. Schematic representation of bed evolution as a sink/source term in the sediment balance of the water column.

for a water column of depth D is given by

$$\frac{\partial}{\partial t}(D\bar{c}) + \nabla\cdot\vec{q} = \frac{\partial}{\partial x}\left(DK^{(x)}\frac{\partial\bar{c}}{\partial x}\right)$$

$$+ \frac{\partial}{\partial y}\left(DK^{(y)}\frac{\partial\bar{c}}{\partial y}\right) + Er - De, \quad (3.56)$$

where D is the instantaneous water depth, $K^{(x)}$ and $K^{(y)}$ are diffusion coefficients, $\bar{c}$ is the depth averaged sediment volume concentration and $\vec{q}$ is the volumetric sediment flux vector. The diffusive fluxes $K^{(x)}\partial c/\partial x$ and $K^{(y)}\partial c/\partial y$ are generally smaller than the advection flux $\vec{q}$. In simple diagnostic models the diffusive fluxes are usually ignored. The lateral diffusive flux (perpendicular to the advection flux) can be substantial, however, especially for fine sediments that are well mixed over the water column.

The suspended load, $C \equiv D\bar{c}$, can vary strongly over short periods. However, the contribution of the term $\partial C/\partial t$ is very small compared to the erosion and sedimentation fluxes at the large time scale of morphological evolution.

The bed evolution equation then takes the simple form $((z_b)_t \equiv \partial z_b / \partial t)$

$$(z_b)_t + \vec{\nabla} \cdot \vec{q} = 0. \tag{3.57}$$

This equation defines the relationship between long-term bed changes and flow dynamics. When applying this formula, the porosity factor $1 - p_b = \rho_{bed} / \rho_{sed}$ has to be incorporated in the bed-level variation $(z_b)_t$.

3.7.2. *Suspension and Settling Lag*

The amount of sediment in suspension does not adjust immediately to a change in the flow strength. When the velocity is suddenly reduced, it takes some time before sediment particles have settled and a new equilibrium suspension has established; this time is known as settling lag T_{sed}. When the velocity is suddenly increased, more sediment is eroded from the seabed. Again it takes some time before an equilibrium is reached; this is the resuspension time lag T_{er}.

In a uniform channel the suspended load will tend to an equilibrium concentration, also called 'saturation concentration', which depends on the shear stress and the flow transport capacity, which both depend on the flow velocity u. We therefore write the depth averaged suspended saturation load as a function of u, $C_{sat} = C_{sat}(u)$. If the flow velocity increases, the seabed erodes, and if the flow velocity decreases, we have deposition.

In a reference frame that follows the flow (the depth-averaged flow if the sediment is well mixed over the vertical, or the near-bottom flow if the suspended sediment is concentrated in the lower part of the water column), the sediment advection flux is small. If we focus on short-term deposition-erosion, the term $\nabla \cdot \vec{q}$ can now be ignored in the sediment balance equation (3.56). We thus may write

$$dC/dt = Er - De,$$

where the time derivative d/dt is taken in the moving frame. If we assume that the suspended load C adapts to the saturation load C_{sat}

following an exponential relaxation law, we have

$$\frac{dC}{dt} = \frac{1}{T_{sed}}(C_{sat}(u) - C), \quad \frac{dC}{dt} = \frac{1}{T_{er}}(C_{sat}(u) - C). \quad (3.58)$$

The first equation refers to the situation where the current velocity decreases ($C_{sat}(u) < C$) and the second equation to the situation where the current velocity increases ($C_{sat}(u) > C$). The settling time lag T_{sed} can be derived from the settling velocity formulas (3.37) or (3.38). The resuspension time lag T_{er} depends on the consolidation history of the sediment bed and has to be determined experimentally.

The saturation load C_{sat} can be derived from the suspended sediment profile (3.44). If we choose the formula (3.46) for the concentration near the bed we find

$$C_{sat}(u) = \frac{0.015\, d\, \tau_*^{3/2}}{d_*^{0.3}} \frac{1 - (z_{ref}/D)^{(R_s - 1)}}{R_s - 1}, \quad (3.59)$$

where D is the water depth, R_s the Rouse number (3.45), d the medium grain size and d_* the dimensionless grain size (3.37); z_{ref} is the reference depth and τ_* the relative excess shear stress, both defined in (3.46).

For cohesive sediment the concept of saturation concentration is often not applicable. This is because the seabed usually consists of sedimentary particles of different sizes (graded seabed). When the fine sediments are winnowed from the top layer, the underlying fine sediments are shielded by the remaining larger particles in the top layer (seabed armouring). In that case the change in sediment concentration has to be determined from the sediment balance equation,

$$C_t + \vec{\nabla}\vec{q}_s = Er - De. \quad (3.60)$$

For non-graded fine sediment beds and sufficiently strong currents, the saturation concentration may become very large. Fine cohesive sediments then form a fluid mud layer.

3.8. Sediment Transport

Sediment transport by currents and waves has been a major research topic in the scientific coastal community during the past decades. The previous sections illustrate that sediment transport involves a great diversity of phenomena. In spite of numerous field observations, laboratory experiments and model studies, all these phenomena cannot yet be captured by a single fundamental theory. However, we can use empirical formulas, which have been tested and refined in many field and laboratory studies. Some formulas are more elaborate than others, involving parameters that have to be fine-tuned for practical application. An overview of recent advances in sediment transport research can be found in Van Rijn *et al.* [913], with many literature references. Below we discuss some basic concepts of sediment transport and formulations that can be used in different coastal situations.

Transport modes

When the flow-induced skin shear stress is strong enough to expel sediment particles from the seabed, these particles start moving with the current. Part of the transport takes place over the seabed (by rolling or leaping) and another part is transported while being kept in suspension by turbulence. The former situation corresponds to bedload transport and the latter situation to suspended-load transport. Bedload becomes suspended load when the jumps become large, but the division between bedload and suspended load is not clearly defined. Important criteria for distinguishing bedload and suspended-load transport are:

- bedload transport becomes rapidly saturated when the flow is kept constant, while suspended load generally does not saturate;
- settling and resuspension lags are generally very small for bedload transport, while settling and resuspension lags can be very large for suspended-load transport.
- The bedload sediment flux is designated by q_b and the suspended sediment flux by q_s. Both types of transport may occur

simultaneously; in that case the total sediment flux is taken as the sum of both contributions,

$$q = q_b + q_s. \tag{3.61}$$

The ratio between bedload and suspended-load transport depends on the strength of the shear stress at the seabed and on the sediment diameter, see Fig. 3.19. Under conditions of strong currents and high waves, most of the transport will take place in suspension. Under moderate conditions, the coarse sediment fraction will be transported as bedload and the fine fraction as suspended load.

We will show that in the case of a rippled seabed, asymmetric waves may produce transport of fine sediment in a direction opposite to the transport direction of the coarser material contained in the ripples. This illustrates that the mechanisms for bedload and suspended load are not identical and that it makes sense to make a distinction between these two transport modes.

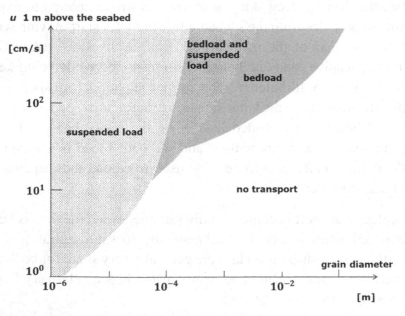

Fig. 3.19. Relationship between flow velocity (1 m above the bed), grain size and transport mode for uniform non-cohesive sediment.

In the previous sections we have described the major characteristics of the interaction of waves and currents with the seabed. This interaction is very complex; there is no theory that provides a complete description applicable for any field situation. Models for erosion and sedimentation rely on empirical relationships derived from laboratory experiments and field observations. The same holds for sediment transport models [264]. Many formulas for sediment transport have been proposed, but no formula is better than the others for all situations. In the next paragraphs we will review some formulas currently used in analytical and numerical models.

The sediment concentration c is expressed as sediment volume per unit fluid volume. This is a dimensionless quantity, that should be multiplied by the sediment density ρ_s to obtain the concentration in [kg/m^3]. The sediment flux q is expressed as a transported sediment volume per unit time through a section of unit width; the dimensions are [m^2/s]. It is often more convenient, however, to express the sediment flux in dimensionless units. In this case we will use Φ as the dimensionless sediment flux; it is related to q by

$$\Phi = \frac{q}{\sqrt{gd^3\Delta\rho/\rho}}, \qquad (3.62)$$

where d is the average sediment grain diameter and $\Delta\rho$ the density difference between sediment and water, $\Delta\rho = \rho_s - \rho$. For sand particles $\Delta\rho \approx 1.65$. The factor $gd^3\Delta\rho$ represents the submerged weight of a sediment particle with volume d^3.

3.8.1. *Bedload Transport*

Bedload depends on the shear stress exerted by the flow on individual sediment particles at the seabed. This shear stress relates to the skin friction and not to seabed features (form drag). Below a certain critical shear stress τ_{cr}, no sediment particles are removed from the seabed. Just above the critical shear stress, bedload transport is mainly due to the downdrift motion of small-scale seabed structures [871], which are formed by interaction of the bed and the near-bed flow. A more detailed discussion of this interaction is presented in Sec. 4.2.

Meyer-Peter and Mueller [596] found that the bedload transport per unit width in rivers, Φ_b in dimensionless units, can be expressed as

$$\Phi_b = m|\vartheta - \vartheta_{cr}|^n, \tag{3.63}$$

where ϑ is the Shields parameter (non-dimensional shear stress (3.10)) and ϑ_{cr} the critical Shields parameter corresponding to τ_{cr}. For the parameters m and n they proposed the values 8 and 1.5. In other studies values were found in the range 4–12 for m and 1.4–1.65 for n, where the lowest values of m go with the lowest values of n [830]. An estimate for the critical shear stress for the onset of sediment motion at a non-cohesive bed can be obtained from (3.47). Typical values for the near-bottom critical velocity for medium to coarse sand range between 0.2 m/s and 0.4 m/s, corresponding to $\vartheta_{cr} \approx 0.055$ [258].

In the case of a rippled seabed, bedload transport is often associated with ripple migration. According to this assumption bedload transport can be estimated from the displaced ripple volume per time unit.

Bottom slope effect on bedload transport

The seabed is in general not flat; the seabed topography is designated by $z_b(x, y, t)$ relative to a horizontal reference frame. Bedload transport is sensitive to the seabed slope; the bedload flux is deviated toward the downhill direction because of gravity. This effect can be incorporated in the formula for sediment transport (3.63) by a modification of the critical shear stress for particle motion. Assuming that the bed topography varies only in the flow direction, $z_b = z_b(x)$, the bedslope effect is often modelled as [797]

$$\vartheta_{cr} = \vartheta_{cr}^{noslope} \left(1 + \frac{(z_b)_x}{\tan \varphi_r} \right), \tag{3.64}$$

where $(z_b)_x$ is the local bed slope and φ_r the angle of repose. This is the angle at which spontaneous avalanching may occur in the absence of bed shear stress. Experiments of avalanching in still water with sediment from Dutch estuaries in the grain-size range $d = (1 - 5).10^{-4}$ m indicate the relationship [192]

$$\tan \varphi_r = 5.4 \ 10^5 d^{1.88}. \tag{3.65}$$

This yields $\tan \varphi_r = 0.1$ for $d = 0.2$ mm and $\tan \varphi_r = 0.6$ for $d = 0.5$ mm.

For the slope effect in an arbitrary direction, we define the Shields vector

$$\vec{\vartheta} = \frac{\vec{\tau}_b}{gd\Delta\rho}, \tag{3.66}$$

where $\vec{\tau}_b$ is given by (D.47) in the case of combined current and waves. The gravity term, which is assumed proportional to the seabed slope $\vec{\nabla} z_b$, can be introduced according to

$$\vec{\Phi}_b = \Phi_b^{noslope} \left(\frac{\vec{\vartheta}}{|\vec{\vartheta}|} - \gamma \vec{\nabla} z_b \right), \tag{3.67}$$

where $\Phi_b^{noslope}$ is the bedload flux without slope. Instead of (3.67), we will also use the equivalent formula for bedload transport

$$\vec{q}_b = \alpha_b |\vec{u}|^3 \left[\frac{\vec{u}}{|\vec{u}|} \left(1 - \frac{u_{cr}}{|\vec{u}|} \right) - \gamma \vec{\nabla} z_b \right], \tag{3.68}$$

for $u > u_{cr}$, where u_{cr} is the critical flow velocity for bed erosion and α_b a coefficient of order 10^{-4} $[m^{-1}s^2]$. The critical flow velocity for bed erosion can be neglected if $|u| \gg u_{cr}$.

Experimental data for the slope parameter γ are scarce. Taking $\gamma = 1/\tan \varphi_r$ gives $\gamma \approx 1.6 - 10$. A field study of the infilling of excavated holes in the surf zone indicates a value $\alpha_b \gamma \sim (2-6) \ 10^{-4}$ $[m^{-1}s^2]$ for sediment with grain size 0.3 mm [616].

Bedload transport by waves

It is generally assumed that bedload transport by waves can be represented by an expression of the type (3.68), averaged over the wave period,

$$\langle q_b \rangle = \alpha_b \left[\langle u_w^3 \rangle \left(1 - \frac{u_{cr}}{u_w^{max}} \right) - \gamma \langle |u_w|^3 \rangle (z_b)_x \right], \qquad (3.69)$$

where $u_w(x, t)$ represents the wave-orbital velocity. The factor α_b is of the same order as for a steady current. In the case of a sinusoidal wave, the first term in brackets of (3.69) vanishes; only the downslope transport term remains. Wave-driven bedload transport thus requires non-sinusoidal waves.

We distinguish two types of non-sinusoidal waves: (1) velocity-skewed waves with stronger wave-orbital velocities in the propagation direction than against, and (2) acceleration-skewed waves with stronger wave-orbital acceleration in the propagation direction than against. Wave skewness and acceleration skewness are mathematically defined here as

$$S_u = \frac{\langle u_w^3 \rangle}{\langle |u_w|^3 \rangle}, \quad A_u = -\frac{\langle (\mathcal{H} u_w)^3 \rangle}{\langle |(\mathcal{H} u_w)|^3 \rangle}, \qquad (3.70)$$

the Hilbert transform $\mathcal{H}$ being defined as

$$\mathcal{H}[u] = \frac{p.v.}{\pi} \int_{-\infty}^{\infty} \frac{u(t')}{t - t'} dt',$$

$$\mathcal{H}[\sin(\alpha t)] = -\cos(\alpha t), \quad \mathcal{H}[\cos(\alpha t)] = \sin(\alpha t).$$

If we assume that u_w is represented in (3.69) by the free-stream near-bottom orbital velocity, then bedload transport is non-zero for skewed waves and positive in the direction of strongest wave-orbital velocity. However, for waves with only acceleration skewness, bedload transport according to (3.69) is zero.

Bedload transport including acceleration skewness

In Sec. 3.3.1 we have shown that free-stream acceleration skewness produces a net wave-averaged shear stress at the seabed. Acceleration skewness therefore contributes to net bedload transport. Hence, we should use in the bedload transport expression (3.69) the wave-orbital velocity u_w^{bed} in the wave-boundary layer, instead of the free-stream velocity u_w. This can be done by substituting in (3.69) the empirical formula (D.23) given in Appendix D.1.2 for relating u_w^{bed} to the free-stream wave-orbital velocity u_w.

Nielsen [635] proposed an empirical formula for wave-induced bedload transport, similar to (3.69), that takes acceleration skewness in the free-stream wave-orbital velocity explicitly into account. Leaving out, for simplicity, the down-slope transport contribution and the factor $(1 - u_{cr}/u_w^{max})$, this formula reads

$$\langle q_b \rangle = \frac{\alpha_b}{\rho f_w / 2} \langle \tau_b u_w \rangle, \quad \tau_b = \frac{1}{2} \rho f_w \left[\cos \varphi_\tau u_w + \frac{\sin \varphi_\tau}{\omega} (u_w)_t \right]^2. \tag{3.71}$$

The angle φ_τ was empirically determined at $\varphi_\tau \approx 50°$. In terms of the free-stream skewness and acceleration skewness parameters the transport formula of Nielsen can be written as

$$\langle q_b \rangle = \alpha_b \frac{2 + \cos 2\varphi_\tau}{3} \langle |u_w|^3 \rangle \left[S_u + \frac{\sin 2\varphi_\tau}{2 + \cos 2\varphi_\tau} A_u \right]. \tag{3.72}$$

A similar formula is obtained if we replace in (3.69) the free-stream velocity u_w by the boundary-layer velocity u_w^{bed} (D.24), given by Henderson [388]. This expression was derived by assuming a frequency-independent phase shift φ_{bed} between u_w^{bed} and the free-stream wave-orbital velocity u_w. Substitution yields

$$\langle q_b \rangle = \alpha_b \cos(\varphi_{bed}) \langle |u_w|^3 \rangle [S_u + \tan(\varphi_{bed}) A_u],$$

The phase angle is of order $\varphi_{bed} \sim 25°$, which gives a result is similar to (3.72),

$$\langle q_b \rangle \approx 0.6\alpha_b \langle u_w^3 \rangle \left[1 + 0.5 \frac{A_u}{S_u} \right]. \tag{3.73}$$

Bedload transport by a combination of current and waves

Bedload transport by a steady current in coastal environments is strongly enhanced by wave action. We call θ_{cw} the angle between the steady current (taken along the x-axis) and the wave propagation direction. For this situation the formula (3.63) can be extended by substituting in the Shields vector (3.66) the expression (D.47) for the total shear stress of combined current and waves, $\vec{\tau}_{cw}(t)$. The net influence of wave action is given by the average $\langle \vec{q}_b \rangle$ of the instantaneous bedload flux over the wave cycle.

The general expressions are rather complex, in particular for the case of asymmetric waves. We consider the special case of waves with velocity skewness (no acceleration skewness):

$$u_w(t) = u_1 + u_2 = U_1 \left[\cos(\omega t) + \Delta \cos(2\omega t) \right],$$

$$\Delta = U_2 / U_1 \ll 1.$$

With this expression for the wave-orbital velocity, the following approximate formula can be derived for the x- and y-components of the dimensionless bedload flux $\vec{\Phi}$ [795]:

$$\Phi_b^{(x)} \approx m Max \left[\vartheta_c^{1/2} (\vartheta_c - \vartheta_{cr}), (0.95 + 0.19 \cos 2\theta_{cw}) \vartheta_w^{1/2} \vartheta_c \right.$$

$$\left. + 0.23 \Delta \vartheta_w^{3/2} \cos\theta_{cw} \right],$$

$$\Phi_b^{(y)} \approx \frac{0.2m\vartheta_w^2}{\vartheta_w^{3/2} + 1.5\vartheta_c^{3/2}} \left[\vartheta_c \sin 2\theta_{cw} + 1.2\Delta\vartheta_w \sin\theta_{cw} \right]. \tag{3.74}$$

In these formulas $m \approx 10$ and $\vartheta_c \equiv |\vec{\vartheta}_c|$, $\vartheta_w \equiv \langle |\vec{\vartheta}_w| \rangle$; $\vec{\vartheta}_c$ and $\vec{\vartheta}_w$ are defined by the relations

$$\vec{\vartheta}_{cw} \equiv \frac{\vec{\tau}_{cw}}{gd\Delta\rho} = \vec{\vartheta}_c + \vec{\vartheta}_w \left[\cos(\omega t) + \Delta \cos(2\omega t) \right], \quad \vec{\vartheta}_c \equiv \left\langle \vec{\vartheta}_{cw} \right\rangle.$$

If the maximum value of $|\vec{\vartheta}_{cw}|$ does not exceed ϑ_{cr}, then $\Phi_b^{(x)} = \Phi_b^{(y)} = 0$.

3.8.2. Suspended-Load Transport

Instantaneous transport

The advection transport of sediment in suspension is given by the integral

$$\vec{q}_s(t) = \int_0^D \vec{u}(z, t)c(z, t)dz, \qquad (3.75)$$

where the seabed is taken at $z = 0$ and where D is the total instantaneous water depth. For the computation of this integral we need to know the instantaneous distribution of the suspended sediment concentration. Therefore we have to solve the sediment advection–diffusion equation (3.56), with a given suspended sediment concentration or suspended sediment flux at the seabed boundary. The solution of this equation gives the temporal and spatial distribution of erosion and sedimentation. Approximate empirical expressions for suspended-load transport are not needed, as in the case of bedload transport.

Solving the sediment advection–diffusion equation is often quite complex. An example is the model developed by Jacobsen, Fredsoe and Jensen, for the simulation of bar development on the shoreface by breaking waves [450]. In order to deal with the interaction of waves and currents, the model solves the wave-boundary layer dynamics. The simulation of wave breaking requires modelling of the water-air interface. Shear stresses are related to the current velocity profile by equations containing (semi)empirical constants, which are tuned trough comparison of simulation results with field data. Settling and resuspension are determined from empirical relationships, which also involve adjustable parameters. With present computation possibilities, this approach is feasible only for simplified representations of real field situations and for short time scales.

In the most simple case of only a steady current, the integral (3.75) can be integrated directly. Using the expressions (3.18), (3.44) and (3.46), we find

$$q_s = m d u_* \left(\frac{\vartheta_{cr}}{\tau_{cr}} \right)^{3/2} (\tau_s - \tau_{cr})^{3/2}, \quad m = \frac{0.15}{(R_s - \frac{8}{7}) d_*^{0.3} \vartheta_{cr}^{3/2}},$$

$$(3.76)$$

where d is the medium grain size, d_* the dimensionless grain size (3.37), δ_r is the roughness height, τ_s the skin shear stress and τ_{cr} the critical shear stress for erosion. If the Rouse number $R_s = w_s / \kappa u_*$ (3.45) is close to or smaller than 8/7, a factor $\left[1 - (\delta_r / D)^{(R_s - 8/7)} \right]$ has to be included in the definition of m. For a flat, non-cohesive medium-sand bed, $\tau_s \approx \tau_b$. The suspended sediment flux q_s (in non-dimensional form) can then be written as

$$\Phi_s \equiv \frac{q_s}{\sqrt{g d^3 \Delta \rho / \rho}} = m \vartheta^{1/2} (\vartheta - \vartheta_{cr})^{3/2}. \qquad (3.77)$$

If we take grain size $d = 300 \mu m$, $R_s = 3.1$ ($w_s = 0.04$ m/s, $u_* = 0.032$ m/s corresponding to $\bar{u} = 1$ m/s and $\delta_r = 1$ mm) and $\vartheta_{cr} = 0.04$ (3.47) we find $m \approx 5$. For stronger roughness $\delta_r = 1$ cm the result is $m \approx 10$.

The main difference with bedload transport formulas, such as (3.68), is the velocity dependence, which for high current velocities involves a 5th power of $|u|$ instead of a 3rd power ($\vartheta \propto u^2$, $R_s \propto u^{-1}$).

Wave-averaged transport

Suspended load transport does not respond instantaneously to changes in the near-bed velocity, in particular changes due to wave-orbital motion. In the case of skewed waves, the stronger forward orbital motion under the wave crest brings more sediment into suspension than the weaker return motion under the wave trough. This may induce a net forward (onshore) suspended sediment flux. However, if the time scale of sediment resuspension and settling is longer than the wave period, a large part of the resuspended particles will remain in suspension after reversal of the orbital motion. This may result in a backward (offshore) transport [165,356,473,717,742,920]. Similarly, when a steady weak current (current velocity smaller than the wave-orbital velocities) is superposed on the wave-orbital motion, suspended sediment may experience a net transport in the direction opposite to the current.

Transport in the opposite direction of the strongest orbital wave motion or in the opposite direction of a superposed weak steady current, occurs in particular for fine sand, and in situations where the seabed is rippled. Sediment laden vortices, released from the ripple crests during the phase of strongest near-bed velocities, are swept over the ripple crest when the wave-orbital motion is reversed. The net sediment motion is in the direction of weakest near-bed velocities, as shown in Fig. 3.20.

In order to account for this effect, empirical formulations for the wave-induced sediment flux have been developed. In such formulas

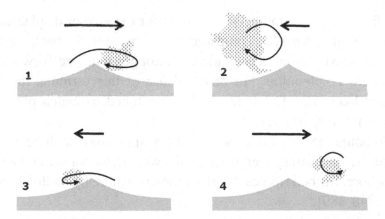

Fig. 3.20. Suspension transport by an asymmetric wave over steep ripples. 1: The strong onshore orbital flow produces a large vortex and a large cloud of suspended sediment behind the ripple crest that is shed off at flow reversal. 2: The weaker orbital return flow transports the suspended cloud offshore. 3: The weaker offshore orbital flow produces a small vortex behind the ripple crest with a smaller suspended cloud. 4: The smaller cloud is transported onshore by the strong onshore orbital return flow. The net effect is an offshore transport of suspended sediment, against the direction of the strongest wave-orbital motion.

the sediment flux is averaged separately over the positive (forward orbital motion) and negative parts of the wave cycle. A general formula, which also includes the effect of acceleration skewness in the wave-orbital motion is given below [235, 870],

$$q_s \propto \frac{\sqrt{gd\Delta\rho/\rho}}{T}\left[T_c(\Omega_{cc} + \alpha_c\Omega_{tc}) - T_t(\Omega_{tt} + \alpha_t\Omega_{ct})\right], \quad (3.78)$$

where $T = T_c + T_t$, with T_c and T_t the durations of the wave crest and wave trough periods respectively, and where

Ω_{cc} = volumetric sand load suspended during the wave crest period and transported during the crest period [m];

Ω_{ct} = volumetric sand load suspended during the wave crest period and transported during the trough period [m];

Ω_{tt} = volumetric sand load suspended during the wave trough period and transported during the trough period [m];

Ω_{tc} = volumetric sand load suspended during the wave trough period and transported during the crest period [m].

The factors $\alpha_c > 1$ and $\alpha_t < 1$ account for the greater displacement of the sediment load Ω_{tc} compared to Ω_{ct}, because the backward $\rightarrow$ forward wave-orbital acceleration is stronger than the forward $\rightarrow$ backward acceleration. For practical use of this formula, the suspended loads $\Omega_{cc}, \Omega_{tt}, \alpha_c\Omega_{tc}, \alpha_t\Omega_{ct}$ are tuned to match observed sediment transport rates.

Although net suspended sediment transport may be directed offshore, the shear stress averaged over the wave cycle for skewed waves is onshore. Seabed ripples therefore move onshore by bedload transport [748, 889].

Sheet flow

Under strong wave action, seabed ripples are washed out and the top layer of the seabed is fluidised. This fluidised top layer is called 'sheet flow layer', see Fig. 3.21. Sheet flow occurs for values of the Shields parameter ϑ larger than 0.5 (medium sand) or larger than 1

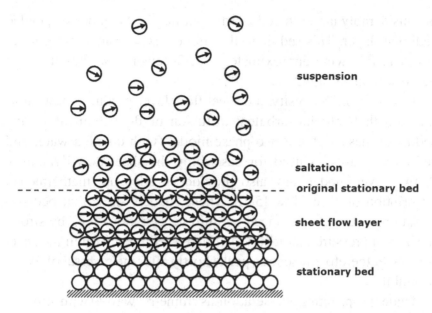

suspension

saltation

original stationary bed

sheet flow layer

stationary bed

Fig. 3.21. Schematic representation of sheet flow. Strong wave-induced stresses fluidise the top layer of the seabed by disrupting the contacts between sand grains. The concentration in the sheet flow layer is very high, with permanent collisions among sand grains. The sand flux in the sheet flow layer is much higher than the sand flux in the overlying saltation and suspension layers.

(coarse sand) [336]. The sediment transport in the sheet flow layer can be substantial, even though this layer is very thin. The layer thickness is typically in the order of 10 times the average sediment grain diameter d. The thickness h_{sheet} increases almost linearly with increasing shear stress as $h_{sheet} \approx (5-10)\,\vartheta\,d$, according to laboratory experiments [559,817,951]. The concentration in the sheet flow layer is in the order of 1000 kg/m^3 just above the stationary bed. The concentration decreases strongly with height, to less than 100 kg/m^3 at the top of the sheet flow layer.

In the sheet flow layer the sediment particles are continuously in contact with each other, creating intergranular stress. The friction caused by particles collision decreases the velocity in the sheet flow layer to typically half of the velocity in the upper layer. The velocity in the sheet flow layer is high enough to produce a sediment flux which is considerably larger than the sediment flux in the upper suspended sediment layer. This sediment flux increases with increasing shear stress τ, following approximately a 2/3-power law, similar as for bedload transport [242].

Due to its high density, the sheet flow layer produces flow stratification that inhibits turbulent diffusion of flow momentum and sediment mass. Often a two-phase model (with coupled water and sediment phases) is used for describing the sheet flow dynamics. A two layer model may also provide an adequate mathematical description of sheet flow [560]. These models show that bed fluidisation is strongly enhanced by fluid accelerations (i.e., by strong horizontal pressure gradients) [417]. The sediment load in the sheet flow layer therefore reaches a peak shortly after the reversal of wave-orbital flow.

Under propagating acceleration-asymmetric waves (with stronger forward than backward acceleration) the thickness of the sheet flow layer is greater during onshore orbital flow than during offshore flow [308]. Incipient sediment motion is probably stimulated by strong pressure gradients exerted on the seabed during the phase of forward acceleration. This results in an onshore directed net sediment flux.

Under velocity-skewed propagating waves with stronger forward (onshore) orbital motion than backward (offshore) orbital motion, the highest suspension occurs just before the forward $\rightarrow$ backward flow reversal. In the case of a fine grained sandy seabed the suspended sediment remains in suspension after flow reversal. The transport of this high sediment load by the backward orbital flow then produces a net offshore sediment flux, against the direction of strongest wave-orbital motion [242, 560].

The wave-averaged sheet-flow transport can be represented by the formula (3.78). This formula can only be used if field data are available for tuning the parameters Ω_{cc}, Ω_{tt}, $\alpha_c \Omega_{tc}$, $\alpha_t \Omega_{ct}$. Otherwise the intra-wave sediment dynamics must be modeled, in order to take settling-resuspension lags into account [356]. In the case of minor settling-resuspension lags, the simpler formula (3.77) can be used, by taking the average over the wave period. The contribution of acceleration skewness can be incorporated by including in (3.77) a factor $(1 + m_a A_u / S_u)$, where A_u is the acceleration skewness parameter (3.70) and m_a an adjustable parameter.

3.8.3. *Total-Load Transport*

It is sometimes difficult to distinguish in practice between bedload and suspended-load transport. If we are only interested in crude estimates of the sediment flux the distinction may not be necessary. A simple expression for the total load flux that can be used in this case is the empirical formula of Engelund and Hansen [285]:

$$\Phi = m\vartheta^{5/2}, \quad m = \frac{0.04}{c_D}, \tag{3.79}$$

where c_D is the friction factor. The different expressions for bedload (3.63), suspended-load (3.77) and total-load transport (3.79) are compared in Fig. 3.22 for 300 μm sand, as a function of the depth-averaged velocity u. The sediment flux increases greatly at velocities higher than 1 m/s, mainly due to a strong increase of the suspended load. According to these formulas, suspended-load transport dominates bedload transport at velocities higher than 1 m/s.

All sediment transport formulas published in the literature, which are generally based on laboratory and field experiments, exhibit a strong increase of the sediment flux with increasing flow velocity. For bedload transport, most formulas indicate an increase following a third power $n = 3$ of the velocity. For suspended load transport, most studies indicate a higher power, $n = 4 - 5$.

The magnitude of the sand flux depends not only on the flow velocity u, but also on particular local conditions regarding grain size distribution, fraction of fine cohesive material, presence of waves (wave asymmetry, wave propagation direction), bedforms and biotic

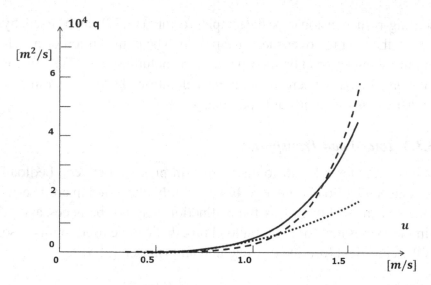

Fig. 3.22. Comparison between formulas for bedload transport (dotted curve (3.63) with
$m = 8$), suspended-load transport (dashed curve (3.77)) and total-load transport (solid curve
(3.79)) as function of the depth-averaged velocity u. The curves are drawn for medium
sand (300 μm), no waves and no bed slope. The following values were used: Roughness
height $\delta_r = 1$ cm, friction coefficient $c_D = 0.002$, critical Shields parameter for incipient
sediment motion $\vartheta_{cr} = 0.04$. For $u > 1.2$ m/s suspended-load transport strongly increases
with velocity and dominates bedload transport.

activity. In order to estimate the range of variation of the sand flux
for these different conditions, we consider the factor α_3, which is the
ratio of the total sand flux and the third power of the flow velocity,
$\alpha_3 = q/\langle u^3 \rangle$. The brackets $\langle \, \rangle$ indicate averaging over the wave cycle.
The value of α_3 was estimated by Camenen and Larroude [132] for
all published sand transport formulas, considering a large range of
velocities. A comparison of these sediment transport formulas shows
that the estimate

$$q = \alpha_3 \langle u^3 \rangle / \rho_{sed}, \quad 0.5 < \alpha_3 < 1.5 \; \left[kg m^{-4} s^2 \right] \qquad (3.80)$$

gives a reasonable average of the sand fluxes measured in laboratory
and field studies [132, 908].

For fine to medium sediment ($d < 300 \, \mu m$) and $|u|$ on the order
of $1 \, m/s$ (much larger than the critical erosion velocity), one may use

the formula

$$q = \alpha \langle u|u|^{n-1} \rangle, \qquad (3.81)$$

with the median value $\alpha = (3 - 5) \times 10^{-4} \left[m^{2-n} s^{n-1} \right]$ and values of n between 3 and 5.

In the case of strong bed relief the down-slope transport should be taken into account:

$$\vec{q} = \alpha \left\langle |\vec{u}|^{n-1} \left[\vec{u} - \gamma |\vec{u}| \vec{\nabla} z_b \right] \right\rangle. \qquad (3.82)$$

Total load transport by a combination of current and waves

The shoreface is a region where sediment transport is strongly influenced by wave action. When offshore generated waves enter shallow water (water depth much smaller than the wavelength) they interact with the seabed. The waves become asymmetric with stronger onshore than offshore orbital motion and stronger onshore than offshore acceleration. This results in a wave-induced onshore sediment transport outside the breaker zone, which is counteracted by downslope sediment transport (see Chapter 7). Sediment is transported both as bedload and suspended load. For this situation the sediment flux formulas (3.74) and (3.78) can be used.

Another frequently used formula for sediment transport across the shoreface has been derived by Bailard [46], based on a theory proposed by Bagnold [45]. The formula for bedload transport is related to the work required to lift a sand grain out of a sloping seabed matrix:

$$\vec{q}_b = \alpha_b \left[\langle |\vec{u}|^2 \vec{u} \rangle - \gamma_b \langle |\vec{u}|^3 \rangle \vec{\nabla} z_b \right]. \qquad (3.83)$$

This formula is identical to (3.68) after averaging over the wave period. Acceleration skewness can be taken into account by introducing a factor $(1 + 0.5 A_u / S_u)$, see (3.73).

The formula for suspended-load transport is related to the phenomenon of auto-suspension (particles settle less easily if the streamlines diverge from the bed):

$$\vec{q}_s = \alpha_s \left[\langle |\vec{u}|^3 \vec{u} \rangle - \gamma_s \langle |\vec{u}|^5 \rangle \vec{\nabla} z_b \right]. \tag{3.84}$$

This formula ignores the effects of settling-resuspension lags and acceleration skewness. The coefficients in (3.83) and (3.84) are given by

$$\alpha_b = \frac{\epsilon_b c_D}{g \tan \varphi_r \Delta \rho / \rho} \, , \quad \gamma_b = \frac{1}{\tan \varphi_r} \, , \quad \alpha_s = \frac{\epsilon_s c_D}{g w_s \Delta \rho / \rho} \, , \quad \gamma_s = \frac{\epsilon_s}{w_s}.$$

In these expressions u is the cross-shore near-bottom velocity, which contains a fluctuating component (the wave-orbital velocity) and a steady component, which is generally smaller than the fluctuating component. The brackets stand for time averaging over the wave period. The coefficients ϵ_b, ϵ_s are efficiency parameters with values in the order of 0.1 and 0.02 respectively. In the bedload formula φ_r is the angle of repose (3.65). Application of the formula requires that the bed slope $\beta = |\vec{\nabla} z_b|$ is much smaller than $\tan \varphi_r$. In the suspension transport formula the seabed slope particularly affects the fine sediment fraction; this fraction settles less rapidly and is kept more easily in suspension in periods of seaward wave motion than landward motion. The formula for suspended load only makes sense if the bed slope β is much smaller than the ratio for autosuspension $w_s/|\vec{u}|$.

Conditions for significant sediment transport

The strength of currents in natural systems fluctuates over time as a result of storms, tides, waves and river floods. The great sensitivity of sediment transport to the flow velocity implies that conditions with strong currents and high waves contribute more to morphologic evolution than periods with weak currents and low waves. Spring tides are more significant than neap tides and storm weather is more significant than calm weather. Periods of storms and high river floods have a great impact, but their frequency is low. Short-term morphologic change under extreme conditions in general deviates from the long-term morphologic trend. This trend will often be restored (at least partly) under more average significant conditions.

On the inner shelf and the shoreface, currents are generally not strong enough to move much sediment in the absence of waves. According to Soulsby [796], high-energy conditions which occur on average over several weeks each year provide the greatest contribution to sand movement on the inner coastal shelf and shoreface. Normative wave and current conditions correspond to the middle-sized waves combined with the middle-sized currents occurring during these periods.

3.9. Summary and Conclusions

Sediment transport is an essential link in coastal morphodynamic processes. Feedback processes that drive the formation of patterns in coastal morphology involve primarily flow structures at the scale of these patterns; seabed patterns and flow structures at smaller scales have a minor influence. But this is not true for sediment transport. Sediment transport depends strongly on the small-scale interaction between flow, sediment particles and small-scale bed patterns, such as ripples.

Whether or not a sediment grain is extracted from the seabed depends on the properties of the bed matrix and on turbulent flow properties in a thin layer above the bed. The size, form and density of the particle matters, as well as the porosity of the seabed, the porewater pressure and the presence of chemical or biological particle bindings.

The process of particle settling on the seabed depends on small-scale interactions as well, with a major role of turbulence in the near-bed layer. This holds in particular for the settling of flocculated fine sediment. The strength of floc binding determines whether the floc resists near-bed shear stresses and settles or whether it disintegrates in smaller constituents which are resuspended.

Because these small-scale processes of particle entrainment and particle settling are highly variable in space and time, it is practically not feasible to include them in morphodynamic models. These models therefore use parametric formulations for sediment transport based on empirical relations established from field observations and laboratory experiments.

Many empirical sediment transport formulations can be found in the literature. None of these formulations has a general validity. Most formulations have been derived for particular situations. The practical use of sediment transport formulas requires knowledge of flow and sediment characteristics and characteristics of the seabed substrate.

Flow characteristics refer especially to shear stresses in the near-bed layer. Bed roughness is an important parameter, for which several estimates exist, applicable for flat seabeds or for rippled seabeds. The turbulence characteristics and shear stresses in the near-bed layer are very different for steady flow and wave-orbital flow. Research of the past decade has greatly improved our understanding of the dynamics of the wave-boundary layer. It was shown that the wave-boundary layer can behave very differently for different wave types, with major consequences for the magnitude and direction of sediment transport. The absence or presence of bed ripples plays an important role as well.

A major fraction of the sediment in many estuaries and tidal lagoons consists of fine cohesive particles. Most of these fine sediments are clustered in flocs bound by organic molecules. Their size and settling velocity depends on flow turbulence. Flow turbulence is decreased when vertical gradients in floc density are high. The mutual dependency of floc density and turbulence leads to the formation of high-concentration turbidity layers, which have a strong impact on sediment transport. Major advances have been made during the past decade to include these processes in estuarine sedimentation studies.

Sediment transport formulas for suspended load are based either on an empirical formula for the near-bed equilibrium suspended concentration or on an empirical relation between bed shear stress and erosion. These empirical formulas have to be adapted to the characteristics of underlying bed layers when the top layer is eroded. No generally applicable formulations for particle suspension are available for situations where the top layer and underlying bed layers have different characteristics.

Sedimentation of tidal flats is often strongly influenced by biotic activity. Here again, we have to deal with mutually dependent processes, because biotic activity is also influenced by sedimentation. Only qualitative descriptions of this mutual interaction are available, for the time being. Their inclusion in morphodynamic models is still a fairly pristine research topic.

Morphodynamic processes can be very different under conditions of bedload and suspended load. It is generally assumed that the total sediment transport is the sum of bedload and suspended load. However, in the case of a graded seabed (mixture of sediments with different grain sizes) bedload and suspended-load transport interact. Winnowing of fine sediment from a graded seabed leaves a coarse-grained top layer which prevents suspension of underlying fine sediments. This so-called bed armouring invalidates suspended-load transport formulas. This illustrates that generally applicable sediment transport formulations do not exist. Choosing the right sediment formulation for simulating morphodynamic processes in the field is often not obvious; transport formulas generally have to be tuned to observations.

Chapter 4

Current-Seabed Interaction

From Ripples to Sandbanks

4.1. Introduction

Why is the seabed not flat?

Should we not expect the energy permanently transmitted by currents to the seabed to erode and flatten sedimentary bottom structures? Seabed maps and *in situ* observations of sedimentation-erosion processes contradict this expectation. In some cases, the morphology of a sedimentary seabed is related to pre-existing structures, for instance, when scour holes are formed near obstacles. For most seabed structures there are no such straightforward relationships; their existence cannot be explained by topographic or hydrographic constraints. Laboratory experiments show that sedimentary structures, such as ripples or dunes, develop on the sediment bed even under conditions of almost perfect initial uniformity. Scientists have always been challenged by this phenomenon, but were long unable to give a plausible explanation.

Seabed instability

It is only in the second half of the past century that the major underlying principles of seabed pattern formation were first explained; these principles relate to the nonlinearity of flow-topography interaction. The same principles apply to a broad range of phenomena of pattern formation in nature, including, for instance, the development

160

of sea waves. At present many topographic features of the seabed are thought to originate from nonlinear flow-topography interaction and related seabed instability. However, the dynamics of this interaction are quite complicated. Mathematical descriptions presented in this chapter are greatly simplified. It is difficult to ensure that all essential physical processes are adequately included and hence the results should be interpreted with care. These models should therefore not be considered as substitutes for physical reality, but as tools for improving our ability to interpret phenomena observed in the field.

The importance of correctly interpreting information about seabed topography is considerable. Although the accuracy of sounding techniques has greatly improved, the cost of frequent seabed surveys can be generally prohibitive. Interpreting seabed maps helps to identify areas of rapid change with related temporal and spatial scales, which is highly relevant to navigation, sand mining, pipeline installation and other offshore activities.

Dune and sandbank families

The process of sediment transport is related to the fine-scale structure of the seabed; ripples on the seabed are both a cause and a result of sediment transport. Seabed ripples are the smallest of a large family of rhythmic bedform patterns, known as the 'dune family', and which includes structures called sandwaves and megaripples. We will use the term 'dune' instead of 'sandwave', because the genesis of dunes (sandwaves) in river flow and tidal flow results basically from the same physical mechanisms. Moreover, the term sandwave is also used for large-scale shoreline undulations, see Sec. 7.5.1.

The members of the dune family share the characteristic that the formation process is strongly related to the vertical structure of the flow field (Fig. 4.1(a)). Rhythmic bedform patterns also exist at much larger scales. These structures are known as the 'sandbank family', with members such as 'bars' and 'ridges'; they are more specifically related to the horizontal structure of the flow field

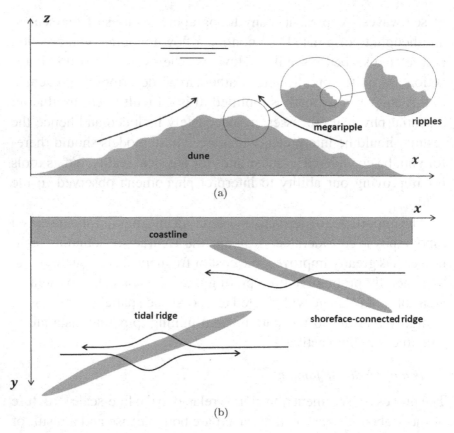

Fig. 4.1. Schematic representation of the sand dune and sandbank families. Upper panel (a): Vertical-longitudinal section of a large dune on the seabed with superimposed megaripples, on which again smaller bed ripples are running. Arrows: Residual circulation induced by the dune. Lower panel (b): Plan view of a linear ridge on the sloping inner shelf, bending towards the coastline ('shoreface-connected ridge') and a tidal ridge on the outer shelf, inclined to the left (Northern hemisphere) relative to the main tidal flow axis. Arrows = flow deviation induced by the ridges.

(Fig. 4.1(b)). Differences among the sandbank family members relate to large-scale topographic features, such as average seabed slope and lateral flow constraints (channel boundaries). Several (but not all) of these structures occur both in steady and oscillating currents. Often different families and family members coexist, in which case they are superposed [503], as in Fig. 4.2. They induce different kinds of

Fig. 4.2. A field of wave-induced bed ripples superimposed on a field of megaripples.

modifications to the unperturbed flow field that interact with each other; compound structures therefore exhibit additional complexity.

In the first sections, the principles of current-topography interaction are described first for the smaller scale feedbacks, responsible for the formation of ripples and dunes. In the last section, the larger scale feedbacks will be discussed, responsible for the formation of different types of sandbanks, such as tidal ridges and shoreface-connected ridges. In between, a section is dedicated to seabed structures related to grainsize sorting. The discussion of the morphodynamic feedbacks in laterally bounded flow, responsible for the formation of alternating bars, tidal flats and channel meandering, is postponed to the next chapter.

4.2. Dunes

4.2.1. *Qualitative Description*

Dependence on flow strength

The seabed under flowing water is seldom flat. The 'regular' seabed is characterised by bottom structures with a great variety of shapes and

Table 4.1. Characteristic dimensions of dune-type bedforms at different flow regimes [92, 197, 708].

	Current ripples	Wave ripples	Megaripples	Dunes
Spacing λ [m]	0.1–0.5	0.3–1	1–10	5–300
Height H [cm]	1–2	3–15	10–50	50–500
Geometry	2D–3D, highly variable	2D straight in troughs	sinuous to 3D, scour pits	straight to sinuous
flow velocity [m/s]	0.15–0.5	0.2–0.8	0.7–1.5	0.5–1.5

horizontal and vertical scales, depending on sediment type, sediment diameter and flow velocity (see Table 4.1). Ripples and dunes are the most common structures; intermediate bedforms called 'megaripples' also occur. They are observed in rivers, estuaries, tidal inlets and in coastal seas. The seabed becomes rippled when the flow is relatively weak (15–50 cm/s); the lower value holds for fine sand. Ripples have a typical wavelength of 10–50 cm and a typical height of a few centimeters (generally less than 3 cm); ripples generated in flows dominated by wave action are a factor 2 larger. Dunes are formed when the flow is somewhat stronger — typically in the order of 50 cm/s to 1 m/s. They appear in water depths greater than about 1 m. Their size increases with increasing water depth; the size hardly depends on flow strength [305]. Ripples and dunes may exist simultaneously; in that case the ripple pattern is superimposed on the much larger dune pattern. Megaripples are often observed in shallow waters; they are typically several decimetres high and their length ranges between 1 and 10 metres. The appellation megaripple is a bit misleading, because megaripple fields have often a patchy structure.

The flow velocity required for the formation of ripples and dunes depends primarily on sediment characteristics. Higher flow velocities are required for coarse grained seabeds and for seabeds containing fine cohesive materials. In these cases there will be little ripple formation. Ripple and dune formation is also influenced by waves; strong wave action (in zones where waves break or collapse, for instance) inhibits ripple and dune formation [585]. At the incipient stage of formation, ripple and dune fields exhibit generally a two-dimensional pattern, with long straight crest lines perpendicular to the dominant flow direction. When flow conditions remain sufficiently stable, ripple and dune fields develop toward a mature state with increased height and wavelength. Under oscillating flow the two-dimensional structure is maintained or even reinforced. Under steady flow however, the crest lines become sinuous with numerous interruptions and the troughs becomes highly irregular, with scour pits and spurs.

Dunes on the North Sea bottom

Dunes occur in regions with moderate tidal flow, where maximum flow velocities exceed 0.5 m/s [425]. Dunes (sandwaves) are observed, for instance, in almost the entire Southern Bight of the North Sea (see Fig. 4.3), in particular in the region where the average maximum tidal bottom stress is between 0.5 and 2 N/m^2 [585] (see Fig. 4.4). The seabed of the Southern Bight is sandy with medium grain diameter of 250–500 μm in the southern part and medium grain

Fig. 4.3. Occurrence of dunes in the Southern Bight of the North Sea (shaded area), from [434]. The region of occurrence coincides roughly with the region where maximum tidal bottom stresses are between 0.5 and 2 N/m^2, see Fig. 4.4. The dash indicates the location of the seabed map of Fig. 4.5.

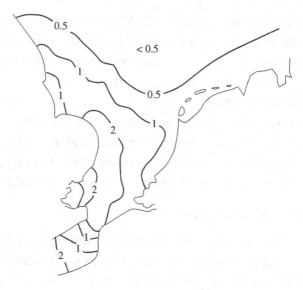

Fig. 4.4.   Distribution of maximum bottom stress (N/m^2) due to semi-diurnal tidal currents in the Southern Bight of the North Sea, from [674].

diameter of 125–250 μm in the northern part. Seabed sediments north of the Southern Bight are very fine, with a medium grain diameter below 125 μm. Figure 4.5 shows a detailed map of a 1 by 5 km strip of the North Sea bottom, obtained with a multibeam sonar. The seabed is covered with dunes, which are part of a large dune field. A field of megaripples is superimposed on the dunes. The orientation and wavelength of both dunes and megaripples are well defined in a statistical sense, but the detailed structure of dunes and megaripples has a three-dimensional character.

Scale relationships for dunes

How are bedforms related to flow and sediment characteristics? Many investigations have been devoted to this question. However, measuring bedforms is not easy, because they rarely appear as isolated two dimensional structures. We often encounter compound structures, with ripples, megaripples and smaller dunes superimposed on larger dunes. To make it more difficult, flow conditions are typically

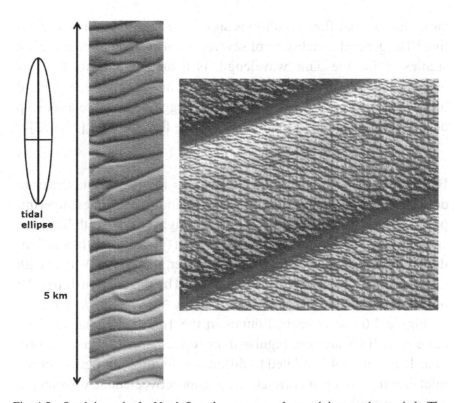

Fig. 4.5. Sand dunes in the North Sea; the crests are clear and the troughs are dark. The left image shows a bottom strip of 1×5 km situated about 50 km off-shore the North-Holland coast at 25–30 m depth. The location is indicated in Fig. 4.3. Average dune spacing is about 250 m; the dune height is about 5 m. The tidal current ellipse is very elongated; the main axis is parallel to the surveyed bottom strip. The maximum tidal current strength is between 0.5 m/s at neap tide and 0.9 m/s at spring tide. Maximum flood flow (to the north) is stronger than maximum ebb flow — about 30 percent at spring tide. The tidally averaged mean flow is also directed northward. Sanddunes are clearly not rectilinear and bifurcate frequently. The normal to the dune crests makes a small cyclonic angle with the main tidal current axis. A more detailed picture of the dunes is shown in the right image; it reveals magaripples running over the sand dune. The normal to the megaripple crests make a rather large anticyclonic angle with the normal to the dune crest, and a small anticyclonic angle with the tidal current axis. It also appears that the sand dunes are asymmetric: The dune through (dark) is lying close to the dune crest (clear) at the northern side. The lee side of the sand dunes thus faces the north, which is indicative of northward dune migration.

variable, so what flow conditions should be considered representative? The general conclusion of several extensive phenomenological studies is that the dune wavelength is related mainly to the flow depth, while the wavelength of ripples and megaripples is not. Field observations and laboratory experiments suggest a linear relationship between the wavelength λ of dunes and the flow depth h [11,972]

$$\lambda \approx 6h. \qquad (4.1)$$

However, the scatter in the data is high, $\lambda \approx (1 - 30)h$, and there are doubts about the general validity of this relationship. Dunes superimposed on tidal sandbanks in the Southern Bight of the North Sea have similar wavelengths near the sandbank crest (at small depth) and near the swales (at large depth) [641]. Moreover, the typical wavelength of these dunes is much larger than indicated by the relationship (4.1); observed wavelengths are on the order of $\lambda \approx (10 - 50)h$ [898].

Figure 4.6 shows seabed dunes in the Texel inlet channel. The dune crest lines are less regular than further offshore in the North Sea. This is probably related to differences in flow strength over the inlet channel. No clear correlation appears between dune wavelength and water depth.

The commonly observed height H of dunes is of the order of

$$H \approx h/6. \qquad (4.2)$$

The scatter in the relationship for the dune height is rather less than for the dune wavelength [305].

The migration direction and migration rate of dunes depend on tidal flow asymmetry and on the strength of the dominating flow direction; observed migration rates range from a few metres per year to several hundred metres per year or sometimes even more. In unidirectional or in asymmetric tidal flow the dune profile is also strongly asymmetric, with gentle stoss-side slopes (on the order of a few degrees) and steep lee-side slopes (on the order of ten degrees up to the angle of repose of about 30 degrees, indicative of flow separation) [197].

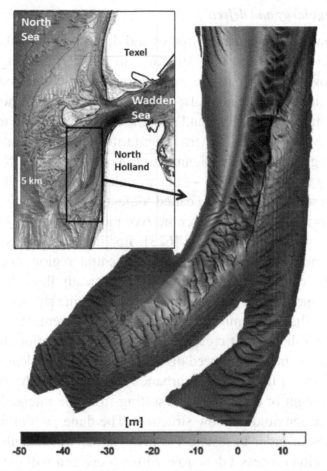

Fig. 4.6. Seabed topography of the southern Schulpengat channel (right) in the ebb-tidal delta of Texel Inlet (top left) based on detailed bottom soundings (see also Fig. 5.4). The main channel (depth ca. 30 m) turns to the south-west and terminates in a shoal (depth ca. 10 m), where flow divergence is visible in the dune pattern. The channel bed is covered with dunes; the crests are sinuous and bifurcating. In this deeper northern part the average dune wavelength is 150 m and in the southern shoaling area less than 100 m. A channel bifurcates to the south along the coast; the dune wavelength here is also on the order of 150 m although the depth (ca. 10 m) is less than in the main channel. Almost everywhere in both channels the southern dune slope is steeper than the northern slope, indicating ebb-dominant tidal flow. The lingoid shape of the dune crests indicates that the ebb flow is strongest at the centre of both channels.

Pattern regularity and defects

The formation mechanism of ripples and dunes is essentially related to interaction with the flow structure in the vertical-longitudinal plane. However, the long-crested pattern of ripple and dune fields in oscillating flow suggests also some coherence in the horizontal flow structure. In fact, long and straight crests evolve due to differences in migration speed where irregularities in the crest alignment occur. Irregularities in the alignment of dune crests in the North Sea are clearly visible in Fig. 4.5 — in particular loose crest endings and forks. These irregularities are called 'defects'.

Defects are thought to trigger the evolution of the dune (or ripple) field to a more regular pattern [623]. For instance, crests migrate faster at their loose ends than at their central region, because of the lower crest height at the extremities. In steady flow, this higher migration speed is the main cause of crescent dune shapes. In oscillating flow, the higher migration speed of crest extremities promotes merger with the nearest crest in the dominant migration direction. Differences in migration speed disturb locally the wavelength of the dune (or ripple) field. This disturbance conflicts with the preferential wavelength of dune growth resulting from the interaction with the vertical-longitudinal flow structure. The dune pattern therefore reorganises in order to restore the preferential wavelength; in this reorganisation process a dune row with a loose end will merge with a neighbouring dune row or it will be annihilated [623]. Werner and Kocurek [941] provided evidence that the reorganisation of bedform patterns in response to defects may induce a continuous increase of pattern wavelength and to the evolution of a ripple field into a megaripple field and finally into a dune field. However, the evolution stops when the defects have disappeared and the bedform pattern has reached a high degree of regularity.

Orientation of crestlines

One should expect that crestlines are, on average, always perpendicular to the dominant current orientation. However, smaller bedforms often have a different orientation than the larger bedforms on which they are superimposed [299, 502], as illustrated in Fig. 4.5. Several hypotheses have been formulated for the obliqueness of the average crest angle. Large bedforms may deflect the near-bed flow and this deflection may produce a different orientation of smaller superimposed bedforms [389, 557]. It has been observed, for instance, that in the swales between large linear bedforms, the flow tends to be aligned with the swale direction. Another suggestion is that obliqueness of large bedforms relative to the main flow direction may result from non-alignment of flood and ebb flow [733]. In the case of the compound bedforms of Fig. 4.5 it is also possible that the megaripple field has evolved under storm conditions with flow direction and strength which differ from the average. It is not clear, however, if these hypotheses apply to all situations where oblique crest lines are observed.

4.2.2. Feedback Mechanism

Seabed instability

In a tidal environment the ripple and dune structure of the seabed is visible at low water. The obvious and widespread appearance of these structures has long been a puzzling phenomenon which has challenged many researchers. It is only a few decades ago that an explanation based on first physical principles was given. Although the full complexity of ripple and dune mechanics has not yet been explained, at present general agreement exists that bedform generation results from the instability of bed-flow interaction. Initial formation of ripples and dunes may start from a plane seabed when the flow strength just exceeds the minimum required for the formation of bedforms.

Vertical flow structure

It is generally assumed that the development of ripples and dunes is related primarily to the vertical flow structure rather than the horizontal flow structure; the latter is primarily associated with the development of sandbanks. Therefore we will first investigate the influence of a local seabed structure on the vertical flow distribution. Because of the similarity of dune genesis mechanisms for tidal flow and steady unidirectional flow, we will concentrate on the latter, simpler case. Before introducing equations, we will discuss a qualitative conceptual model. We consider a dune of very small height as a two-dimensional perturbation (uniform in the direction perpendicular to the flow) of an otherwise perfectly plane seabed. We will show that this perturbation induces a modification of the flow field such that sediment transport converges at the crest of the perturbation. In other words, from the interaction between the seabed and the flow field results a positive feedback promoting the growth of the bottom perturbation.

Inertial delay in the vertical plane

We consider the initial phase of dune growth, where the near-bottom flow follows the seabed and where no flow separation occurs at the lee of the dune. We also assume that the flow over the dune is subcritical; the flow velocity is much smaller than the propagation speed of surface disturbances (small Froude number). This is the normal situation in tidal environments; supercritical flow will be discussed later.

We define two flow regions: A flow region delimited by a streamline just above the bottom at height $h_1(x)$ and an upper flow region delimited by a streamline at height $h_2(x)$, see Fig. 4.7. In the initial situation, when no dune is present, h_1 and h_2 are independent of x. Then a shallow dune with profile $z_b(x)$ is deposited on the seabed. We assume that the height of the upper flow region is such that the streamline at h_2 is not directly affected: $|(h_2)_x| << |(z_b)_x|$. The flow

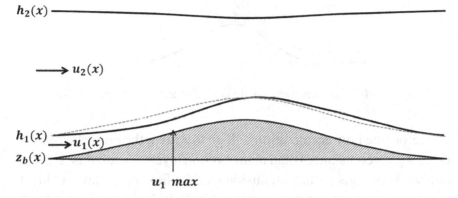

Fig. 4.7. Flow over a shallow dune. The figure depicts schematically a near-bottom stream-line $h_1(x)$ and a higher streamline $h_2(x)$ (eventually the free water surface) which is not strongly affected by the dune disturbance. The dashed near-bottom streamline corresponds to a hypothetical situation in which inertial lag does not exist; in that case the momentum balance is symmetrical upstream and downstream of the dune crest, and so is the stream-line. The solid streamline $h_1(x)$ displays the effect of inertial lag. The flow does not respond instantaneously to the influence of increased friction; horizontal and vertical acceleration are both delayed when passing the dune. Inertia concentrates the flow in the near-bottom layer at the upstream side of the dune and the opposite happens at the downstream side of the dune. The near-bottom flow will thus be stronger at the upstream slope of the dune than at the downstream slope. The near-bottom flow velocity therefore reaches a maximum upstream of the dune crest, and at the dune crest the near-bottom flow is already decelerating. The sediment transport capacity is decreasing at the crest and sediment will be deposited. The dune crest therefore will tend to grow.

accelerates when passing over the dune; we call $u_1(x)$ the average flow velocity in the bottom layer and $u_2(x)$ the average flow velocity in the upper layer. Because h_1 and h_2 are streamlines, the fluxes $[h_1(x) - z_b(x)] u_1(x)$ and $[h_2(x) - h_1(x)] u_2(x)$ are independent of x. At the dune crest $(z_b)_x = 0$; we thus can eliminate $(h_1)_x$ by differentiation $((h_2)_x \approx 0)$:

$$(u_1)_x = -\frac{u_1(h_2 - h_1)}{u_2(h_1 - z_b)}(u_2)_x. \tag{4.3}$$

The gradients $(u_1)_x$ and $(u_2)_x$ have either opposite signs at the dune crest or they are both zero. The additional friction caused by the dune is felt more strongly in the bottom layer than in the top layer; it slows down the flow in the bottom layer more than in the upper layer.

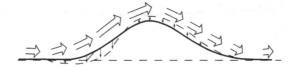

Fig. 4.8.   Transport convergence at the dune crest and resulting dune growth.

Therefore the flow acceleration will go to zero in the upper layer later (i.e., at greater distance) than in the bottom layer; we call this 'inertial delay'. This leads to the conclusion that the flow in the bottom layer must decelerate at the dune crest. This flow deceleration produces a convergence of the sediment flux at the dune crest, if we assume that the sediment flux reacts instantaneously to the near-bottom flow velocity. Sediment is deposited at the dune crest and the crest height will grow, see Fig. 4.8. Hence, an initial seabed perturbation in the form of a linear dune perpendicular to the flow induces a modification of the flow structure which contributes to growth of the dune. The crucial assumption is that the flow adaptation needed to reach an equilibrium flow profile takes more time (in a frame moving with the fluid) at the surface than near the bottom. This will be the case if the near-surface flow and near-bottom flow interact by turbulent diffusion of momentum, such as in a non-stratified outer flow layer.

4.2.3. *Dune Genesis in Steady Flow*

In this section we will look in more detail at the mechanism of dune genesis starting from an initially flat sand bed. Therefore we follow the analysis of Engelund and Fredsøe [284, 314], who developed a mathematical description for dune development from which analytical expressions can be derived. We present this model because it provides a more profound understanding of the feedback process that triggers dune development. More elaborate numerical models which were developed afterwards are based on essentially the same principles. For the sake of simplicity we restrict the analysis to the case of steady flow; dune development in tidal flow involves basically the same physical mechanism. The case of tidal flow will be discussed later.

The model of Engelund and Fredsøe investigates the stability of a small perturbation of an initially flat seabed. This approach is called 'linear stability analysis', see Sec. 2.4.

We consider the following situation. The channel bed is initially flat, the water depth h is constant and the bed is composed of non-cohesive granular material. The flow discharge is constant in the x-direction with a near-bottom velocity u above the critical value u_{cr} for grain motion. Sediments are transported mainly as bed load and the number of moving grains is directly related to the flow strength u, for $|u| > u_{crit}$. We further assume that the net transport is influenced by the seabed slope, according to a linear relation. We then may approximate the sediment flux by the formula (3.68). A small perturbation $z_b(x, t)$ of the seabed, uniform in lateral direction, will then evolve according to the bed evolution Eq. (3.57):

$$(z_b)_t + q_x = 0; \quad q = \alpha_b |u|^3 \left[\frac{u}{|u|} \left(1 - \frac{u_{cr}}{|u|} \right) - \gamma (z_b)_x \right],$$

(4.4)

where α_b is a constant of order $(1-4) \cdot 10^{-4}$ [m^{-1}s^2] and where γ is the slope transport coefficient of order 1–4, see Secs. 3.8.3 and 3.8.1. The subscripts t and x stand for partial differentiation with respect to these variables. A porosity factor $1 - p_b \sim 0.7$ is incorporated in the perturbation height z_b.

The channel bed perturbation is very small relative to the water depth, $|z_b| << h$. We will further assume that the seabed perturbation has an undulating character with wavelength $\lambda = 2\pi/k$; in complex notation

$$z_b(x, t) = \hat{z}_b(t) \exp(ikx).$$

(4.5)

The velocity u can be decomposed in a component u_b corresponding to the near-bottom velocity in the unperturbed state (flat channel bed) and a small deviation $u'(x, t)$ produced by the perturbation

$z_b(x, t)$. Because the perturbation is very small, the velocity deviation u' near the bottom can be represented by a polynomial expansion in $\epsilon = \hat{z}_b(0)/h$,

$$u = u_b + u' = u_b + u_1'\epsilon + u_2'\epsilon^2 + \cdots \qquad (4.6)$$

The coefficients u_1', u_2' are complex functions of x and t.

Substitution in Eq. (4.4) and retaining only terms of first order in ϵ gives

$$\hat{z}_b(t) = \hat{z}_b(0) \exp(-i\sigma t),$$

with

$$i\sigma = \frac{q_x}{z_b} = 3iq_0'u_1\frac{k}{h} + q_0\gamma k^2. \qquad (4.7)$$

In this expression

$$u_1 = \frac{\hat{z}_b(0)u_1'}{z_b u_b}; \quad q_0 = \alpha_b u_b^3; \quad q_0' = q_0\left[1 - \frac{2u_{cr}}{3|u_b|}\right]. \qquad (4.8)$$

The exponent σ is a complex quantity. The real part of σ — more precisely $\Re\sigma/k$ — represents the speed at which the perturbation moves with the flow. The imaginary part, $\Im\sigma$, represents the rate at which the perturbation decays or growths,

$$\Im\sigma = 3q_0'\Im u_1\frac{k}{h} - \gamma q_0 k^2. \qquad (4.9)$$

The amplitude of a small perturbation will start growing exponentially if $\Im u_1$ is positive and greater than $\gamma k h\, q_0/(3q_0')$. This corresponds to an upstream shift of the maximum positive flow perturbation relative to the perturbation crest. In this case, the near-bed flow velocity slows down over the perturbation crest; the sediment flux converges and the crest will accrete.

In the following we will examine this condition more in detail. Therefore we will consider a simplified flow model from which an explicit expression of the velocity perturbation u_1 can be derived.

Flow response to a small seabed undulation

We now consider two flow layers, as in Fig. 4.7. However, the bottom layer is now very close to the bed, such that friction strongly dominates inertial delay. The streamline $h_1(x)$ corresponding to the top of the logarithmic bottom layer (the dashed line in Fig. 4.7) then runs almost parallel with the dune height $z_b(x)$. In the upper layer the dune influences the flow by lifting the height of the bottom layer, but it does not affect the eddy viscosity. The resulting flow perturbation in the upper layer can be modelled to first order without detailed knowledge of the flow in the bottom layer. We reproduce below the procedure followed by Engelund and Fredsøe and their underlying assumptions.

The laterally uniform seabed bed perturbation produces a flow pattern that can be described by the vorticity in the vertical plane, $\zeta = w_x - u_z$. The vertical axis is chosen upward from the bottom and $u = u(x, y)$, $w(x, y)$ are the velocities in respectively the longitudinal and vertical direction. The vorticity equation follows from the Boussinesq Eqs. (A.5), (A.7) and (A.8). Because of the assumption of laterally homogeneous flow, $v = 0$, the Boussinesq equations can be simplified to

$$u_x + w_z = 0,$$

$$uu_x + wu_z + \frac{1}{\rho}p_x = N(u_{xx} + u_{zz}) - E_x,$$

$$uw_x + ww_z + \frac{1}{\rho}p_z + g = N(w_{zz} + w_{xx}) - E_z, \qquad (4.10)$$

where N is the eddy viscosity and E the turbulent energy density. The vorticity equation is obtained by subtracting the z-derivative of the second equation from the x-derivative of the third equation. With the help of the first equation we find

$$u\zeta_x + w\zeta_z = N(\zeta_{zz} + \zeta_{xx}). \qquad (4.11)$$

This equation states that an excess or a deficit of vorticity due to convergence or divergence of the vorticity fluxes (the two left-hand terms) is dissipated by turbulent diffusion (the two right-hand terms).

The eddy viscosity N in the upper layer (above the logarithmic layer) is taken as a constant. The velocity u in the expression (4.4) refers to the velocity at the lower boundary $h_1(x)$ of the upper layer. As this boundary runs parallel with the dune, we replace $h_1(x)$ by $z_b(x)$, by shifting the origin of the z-axis. It will appear later that the upper boundary $h_2(x)$ corresponds to the water surface; therefore we write $h(x)$ instead of $h_2(x)$. The height of the upper layer in the presence of the channel bed perturbation z_b is thus given by $h - z_b$.

For solving the vorticity equation it is convenient to consider the streamfunction $\psi(x, z)$, which is defined by the condition that the flux between two streamlines is constant:

$$\int_{z_1}^{z_2} u(x, z)dz = \psi(x, z_1) - \psi(x, z_2), \quad \text{or} \quad \psi_z = -u. \quad (4.12)$$

The continuity equation $u_x + w_z = 0$ is satisfied by setting $\psi_x = w$ and the vorticity can therefore be expressed as

$$\zeta = w_x - u_z = \psi_{xx} + \psi_{zz}.$$

In terms of the streamfunction the vorticity equation reads

$$\psi_z(\psi_{xx} + \psi_{zz})_x - \psi_x(\psi_{xx} + \psi_{zz})_z$$
$$= N\left(\psi_{zzzz} + 2\psi_{zzxx} + \psi_{xxxx}\right). \quad (4.13)$$

Boundary conditions

The flow has to satisfy boundary conditions at the top of the upper layer ($z = h$) and at the top of the bottom layer ($z = z_b$). At the top of the upper layer we have the conditions:

- the top of the layer $z = h$ corresponds to a streamline,

$$\left[2g\eta + u^2 + w^2\right]_x = 0 \quad \text{(Bernouilli)} \quad \text{and} \quad w = u\eta_x, \quad (4.14)$$

where η_x is the streamline slope and $F = u/\sqrt{gh}$ is the Froude number at $z = h$;

- no momentum of the flow perturbation is transfered through the top of the layer,

$$Nu_z' = 0. \tag{4.15}$$

The boundary conditions at the top of the bottom layer $z = z_b$ are:
- the bottom boundary is a streamline to first order in ϵ,

$$w = (z_b)_x \, u_0(z_b); \tag{4.16}$$

- momentum transfer of horizontal momentum to the bottom layer is proportional to the square of the velocity $u^2(z_b)$ at the top of the bottom layer, with the same friction coefficient as for the unperturbed flow:

$$(u_z + w_x)/(u^2 + w^2) = u_{bz}/u_b^2, \tag{4.17}$$

where u_b and u_{bz} refer to the unperturbed velocity u_0 and its z-derivative at $z = 0$ (the top of the bottom layer in the unperturbed situation). At $z = z_b$ we have to first order $u_0 = u_b + u_{bz} z_b$ and $(u_b)_z = u_{bz} + u_{bzz} z_b$, where u_{bzz} is the second z-derivative of u_0 at $z = 0$.

Origin of flow convergence at the dune crest

It appears that, due to this latter boundary condition, the perturbed flow depends crucially on the vertical variation of the unperturbed flow $u_0(z)$ at the top of the near-bottom layer. A strong vertical variation of the basic flow implies a strong increase in flow velocity $u_b + u_{bz} z_b$ along the upstream flank of the perturbation and a related strong increase of momentum dissipation. The boundary condition (4.17) states that a balance must exist between transfer of momentum to the bottom and vertical momentum diffusion Nu_z in the flow region above the upstream flank. This is realised by a flow adjustment u_1 which increases the velocity shear $(u_0)_z$ by decreasing the near-bottom flow velocity $u_0(z_b)$ along the upstream flank. The flow perturbation u_1 therefore produces in the region around the crest a

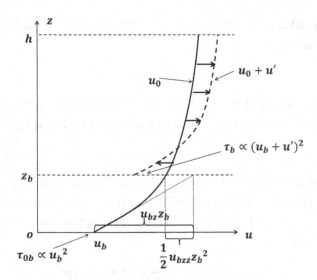

Fig. 4.9. Schematic representation of the flow adjustment mechanism u' at the upstream dune flank. The flow adjustment u' decreases the velocity at $z = z_b$ and increases the gradient u_z in order to match the increase of bottom friction τ_b relative to τ_{0b}.

downstream velocity decrease and a corresponding convergence of sediment transport, see Fig. 4.9.

The condition (4.16) generates a similar adjustment to the basic flow. However, this adjustment is only significant for perturbations with very small wavelengths.

Infinitesimal perturbation and linearisation

The streamfunction is expanded as a power series in $\epsilon = \hat{z}_b/h$, in a similar way as (4.6):

$$\psi(x, z) = \psi_0(z) + \psi_1'(z)\epsilon + \psi_2'(z)\epsilon^2 + \cdots \qquad (4.18)$$

ψ_0 represents the basic flow, in the absence of bed perturbation; ψ_1' represents the first order perturbation of this basic flow caused by the presence of the perturbation $z_b(x)$. The first order perturbation must be linear in z_b; we therefore define $\psi_1'(x, z) = \psi_1(z)u_b z_b(x)/\epsilon$.

Substitution in the vorticity Eq. (4.13) and retaining only first order terms yields

$$(\psi_1)_{zzzz} - \left[2k^2 + \frac{iku_0}{N} \right] (\psi_1)_{zz}$$

$$+ \left[k^4 + \frac{iku_0}{N} \left(k^2 + \frac{(u_0)_{zz}}{u_0} \right) \right] \psi_1 = 0. \qquad (4.19)$$

This linear differential equation is easily solved if the coefficients could be considered constant. Therefore it is assumed that u_0 and $(u_0)_{zz}$ are just weakly varying functions of z. This assumption clearly does not hold in the bottom layer, but this region is excluded from the model. In that case the solution has the form:

$$\psi_1 = ae^{\kappa z/h} + be^{-\kappa z/h} + ce^{\chi z/h} + de^{-\chi z/h}, \qquad (4.20)$$

with

$$\kappa^2 \approx (hk)^2 + \frac{h^2(u_0)_{zz}}{u_0}, \quad \chi^2 \approx (hk)^2 - \frac{h^2(u_0)_{zz}}{u_0} + \frac{ih^2ku_0}{N},$$

and a, b, c, d constant coefficients. The approximations assume $4N|(u_0)_{zz}| \ll ku_0^2$; this is justified because usually $N \ll hu_0$ and $h^2|(u_0)_{zz}| \gg u_0$. Therefore we also have $|\Re\chi| \gg |\Re\kappa|$ and $|\Re\chi| \gg 1$.

Flow perturbation at the top of the near-bottom layer

The coefficients a, b, c, d are determined by the boundary conditions (4.14)–(4.16). These boundary conditions are rewritten at first order in z_b in terms of the stream function ψ.
At $z = h$:

$$hF^2(\psi_1)_z = \psi_1; \quad (\psi_1)_{zz} = 0. \qquad (4.21)$$

At $z = 0$:

$$\psi_1 = -u_b z_b;$$

$$u_b \left(u_{bzz} z_b - (\psi_1)_{zz} + (\psi_1)_{xx} \right) = 2u_{bz} \left(u_{bz} z_b - (\psi_1)_z \right). \qquad (4.22)$$

It is important to distinguish between the value and derivatives of u_0 at the top of the bottom layer and within the upper layer. The former are written as u_b, u_{bz}, u_{bzz} and the latter as u_0, $(u_0)_z$, $(u_0)_{zz}$ respectively. The value and derivatives of u_0 within the upper layer depend on z. However, the variation of u_0 in the upper layer is weak. We therefore consider them as constants and take the average values over the upper layer. In general, the average value of $(u_0)_{zz}$ in the upper layer is much smaller (a factor of order 10) than the value at the top of the bottom layer.

Substitution of (4.20) in the above boundary conditions gives

$$F^2(a\kappa e^{\kappa} - b\kappa e^{-\kappa} + c\chi e^{\chi} - d\chi e^{-\chi})$$
$$= ae^{\kappa} + be^{-\kappa} + ce^{\chi} + de^{-\chi},$$
$$(\kappa^2 - k^2)(ae^{\kappa} + be^{-\kappa})$$
$$+ (\chi^2 - k^2)(ce^{\chi} + de^{-\chi}) = 0,$$
$$a + b + c + d = 1,$$
$$\kappa^2(a+b) + \chi^2(c+d) - 2u_{bz}(\kappa a - \kappa b + \chi c - \chi d)$$
$$= u_{bzz} - 2u_{bz}^2 - k^2.$$

The full solution yields lengthy expressions for the coefficients $a-d$. It can be simplified by using the inequalities $N/(hu_b) \approx c_D << 1$ and $h|\Re\chi| \gg 1$.

We have not yet defined the near-bottom level at which the velocity in the expression (4.4) should be evaluated. We assume that this level corresponds to the top of the bottom layer. According to the definition of ψ_1 we have $u_1 = -h\psi_{1z} = [\kappa(b-a) + \chi(d-c)]$. Substitution of the approximate solutions for a, b, c, d gives

$$\Re u_1 \approx f(\kappa),$$

$$\Im u_1 \approx \frac{u_0}{|u_0|}\sqrt{\frac{2N}{h^2k|u_0|}}\left[p_1^2 + \frac{1}{2}(p_3 - p_2) + (hk)^2 + p_1 f(\kappa)\right],$$

$$(4.23)$$

with

$$\kappa = h\sqrt{k^2 + (u_0)_{zz}/u_0} \quad \text{or} \quad \kappa = ih\sqrt{-k^2 - (u_0)_{zz}/u_0},$$

$$p_1 = hu_{bz}/u_b, \quad p_2 = -h^2(u_0)_{zz}/u_0, \quad p_3 = -h^2 u_{bzz}/u_b,$$

$$f(\kappa) = \kappa \frac{\cosh \kappa - \kappa F^2 \sinh \kappa}{\sinh \kappa - \kappa F^2 \cosh \kappa}.$$

Perturbation growth rate

The expression (4.9) together with (4.23) yields an estimate of the growth rate:

$$\Im\sigma = \frac{3|q_0'|}{h^2}\sqrt{\frac{2N}{h|u_0|}}\left[\sqrt{kh}\left[p_1^2 + \frac{1}{2}\right.\right.$$

$$\left.\left. \times (p_3 - p_2) + (hk)^2 + p_1 f(\kappa)\right] - p_4(hk)^2\right], \quad (4.24)$$

where q_0' is the sediment flux defined in (4.8) and

$$p_4 = \gamma|q_0/3q_0'|\sqrt{h|u_0|/2N}.$$

The growth rate is a function of the perturbation wavelength, as shown in Fig. 4.10. For small wavenumbers (large wavelength) the growth rate is positive. Perturbations with wavenumbers in this range are exponentially amplified in their initial stage of formation. Perturbations with larger wavenumbers decay. The existence of a maximum in the growth rate curve is due to the stabilising effect of the down-slope transport. For high values of hk the growth rate becomes positive again. Bed perturbations with a very small wavelength (large wavenumber) are strongly amplified according to the formula (4.24). Such small structures may correspond to bed ripples. They depend on the flow perturbation in the bottom layer, which is not resolved in the above analysis. The results for high hk are therefore disregarded.

Dune wavelength

For given values of p_1, p_2, p_3 and p_4, the growth rate $\Im\sigma$ depends only on hk. This implies that the wavelength of maximum growth λ

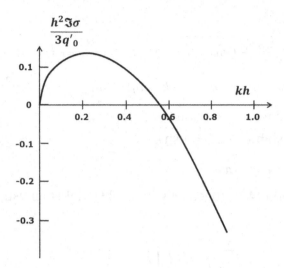

Fig. 4.10. The scaled growth rate $\Im\sigma$ according to (4.24) as a function of the dimensionless wavenumber kh. In this example, the unperturbed flow is characterised by $p_1 = hu_{bz}/u_b = 1$; $p_2 = h^2(u_0)_{zz}/u_0 = -1$; $p_3 = -h^2 u_{bzz}/u_b = 10p_2$; $p_4 = \gamma|q_0/3q_0'|\sqrt{h|u_0|/2N} = 16$. This latter value corresponds to $N/hu_0 \approx c_D \approx 0.002$ and $\gamma q_0/q_0' = 3$. For the Froude number F a small value is taken of 0.1. The wavenumber corresponding to the largest growth rate is the most likely wavenumber of dunes occurring in nature at their initial stage of formation.

is proportional to the layer height h, with a proportionality constant depending on these parameters. For small to moderate values of p_1 and p_2 the growth rate has a maximum for values of hk typically in the range of 0.1–1, see Fig. 4.11. This corresponds to dune wavelengths in the order

$$\lambda \sim (6 - 60)h, \tag{4.25}$$

where h is the flow depth or the height above the bottom at which the flow is not disturbed by the presence of the dune.

The wavelength (4.25) estimate is much larger than the depth h. This implies that for steady flow, h should correspond to the free surface. In the case of tidal flow, the thickness of the upper layer does not necessarily correspond to the water depth h. For water depths of 20 m or more it might be related instead to the thickness δ_t of the tidal boundary layer (3.14). In this case the wavelength of strongest

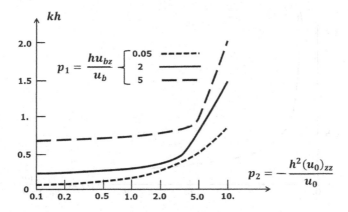

Fig. 4.11. The dimensionless wavenumber of maximum growth as a function of the first and second derivative of the unperturbed velocity at the bottom, p_1 and p_2. Other parameters are $p_3 = 10p_2$; $p_4 = 16$ and $F = 0.1$. For larger values of p_1 and p_2 the model does not produce finite wavenumbers of maximum growth.

growth does not depend on depth. This would be in agreement with dune observations in tidal environments discussed earlier, see for example [641] and Fig. 4.6.

For smaller values of the slope parameter γ or larger values of the eddy viscosity N we find smaller wavelengths (larger wavenumbers), as shown in Fig. 4.12. The same dependence is found in numerical model simulations [438]. Considering that h is the water depth, the wavelength estimate (4.25) is of the same order as the large dunes (sometimes called mega-dunes) observed in natural rivers [310].

Growth time scale and migration rate

The e-folding time scale for growth, $1/\Im\sigma$, follows from (4.24). Choosing $h = 5$ m and $q'_0 = 10^{-4}$ m^2/s, and taking other values as in Fig. 4.10, we find a time scale of about 150 hours. Growth from an almost flat bottom will take at least several times more, which is longer than the dune growth time scales usually observed in natural flows.

The migration rate $\Re\sigma/k$ follows from substitution of (4.23) in (4.7). The result is

$$\Re\sigma/k = 3q'_0 f(\kappa)/h. \qquad (4.26)$$

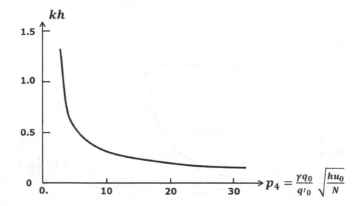

Fig. 4.12. The dimensionless wavenumber of maximum growth as a function of $p_4 = \gamma|q_0/q_0'|\sqrt{h|u_0|/N}$ for $p_1 = hu_{bz}/u_b = 1$ and $p_2 = h^2(u_0)_{zz}/u_0 = -1$, $p_3 = -h^2u_{bzz}/u_b = 10p_2$. For increasing values of the stabilising parameter γ or decreasing values of the eddy viscosity N, exponential growth only occurs for small wavenumbers hk, i.e., for perturbations with large wavelengths λ.

Using the same flow values as for the growth rate, we find a downstream migration rate in the order of 0.4 m/hour.

Limitations of the model

The mathematical model applies to the initial stage of dune genesis and not to their later development. In this initial stage the dune dimensions are small enough for justifying the neglect of nonlinear higher order terms in the model. The model only describes bedforms that are influenced by the flow perturbation in the upper layer. This is the case for dunes with a wavelength of the same order or larger than the height of the upper layer. The model does not apply for smaller bedforms, such as ripples.

Upon further growth of the dunes, higher order terms in the mathematical description have to be taken into account. The structure of fully developed dunes appears to be more complex than the simple 2D-structure assumed in the model. The leeside slope of fully developed dunes often approaches the angle of repose (avalanching angle), although dunes with more gentle leeside slopes are also frequently observed [76]. A theoretical description of the equilibrium shape of former dune type has been given by Fredsøe and Deigaard [318].

An analysis of dune growth for finite dune amplitudes indicates that the results of the linear stability analysis remain qualitatively valid. The growth rates predicted in the weakly nonlinear case are slightly smaller than for the linear case and the wavelength of maximum growth is slightly larger, while the sensitivity to gravity-induced down-slope transport is less [462].

4.2.4. *Dune Formation in Tidal Flow*

The process of dune formation under tidal (reversing current) conditions is basically the same as for steady flow. This is obvious if we assume that acceleration and deceleration of tidal currents do not strongly affect the vertical flow profile. Then ebb flow and flood flow can be considered as successive periods of stationary flow, with reversed flow direction. Dune growth occurs in both periods, similar as for steady flow without flow reversal, and the wavelength of strongest growth is given by the same expressions (4.24).

Seabed dunes are often called 'sandwaves'. As indicated earlier, we avoid this term because of the morphodynamical similarity with river dunes.

An essential difference between unidirectional flow and tidal flow is the migration speed of dunes; in the former case migration is much faster than in the latter case. For a symmetric tidal wave and in the absence of a steady current, the tidally averaged structure of the flow perturbation u' will be symmetric around the dune crest.

According to (4.6) and (4.8) we have near the bottom $u'(x, t) = \Re(z_b u_b u_1)/h$, with z_b given by (4.5) and u_1 given by (4.23). We represent tidal motion by successive intervals of positive and negative velocity ($\pm u_b$, $\pm u_0$). After taking the tidal average we find for the near-bottom flow perturbation

$$\langle u'(x)\rangle = -\frac{z_{b,max}}{h} u_b \Im u_1 \sin kx. \qquad (4.27)$$

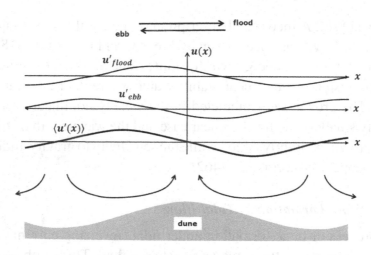

Fig. 4.13. Near-bed steady circulation $\langle u'(x)\rangle$ generated by oscillating flow over a sinusoidal bed perturbation. The fluid excursion is assumed much larger than the perturbation wavelength and fluid acceleration effects are ignored. The flow perturbation $u'(x)$ is maximum upstream of the perturbation crest, both during flood (upper curve; unperturbed current from left to right) and during ebb (middle curve; unperturbed flow from right to left). The sum of the flow perturbations for ebb and flood (lower curve) is directed to the crest and away from the trough.

This expression shows that the average flow pattern in the tidal case corresponds to a series of circulation cells, with near-bottom flow directed towards the dune crests (see figure 4.13).

The above description is too simplistic, however. Tidal acceleration and deceleration have a significant influence on the vertical velocity profile. The time scale T_u for adaptation of the flow profile depends on the eddy viscosity N and the depth h and can be approximated by $T_u \approx h^2/N$. Using for N the estimate (3.22) we find $T_u \approx h/c_D \overline{u}$. The ratio R of the tidal time scale $\omega^{-1} = T/2\pi$ and the inertial adaptation time scale T_u is thus given by $R \equiv 1/T_u\omega = c_D u/h\omega$. For $c_D = 0.0025$, $\overline{u} = 1$ m/s and $h = 20$ m the value of R is close to 1. Therefore, during most of the tidal period the vertical distribution of shear stress will be different from the equivalent steady flow. The tidal variation of the shear stress interferes with the spatial variation related to flow acceleration and flow deceleration over the bottom perturbation. On the continental shelf (great depth,

$R \leq 1$) this effect will be more significant than in tidal basins (small to moderate depth, $R > 1$).

The results of numerical and analytical investigations of seabed instability under tidal conditions show many similarities with the steady flow case [339, 424, 426]. The tidally averaged flow distribution over the bed waves exhibits the same structure of residual eddies, with near-bottom flow directed towards the dune crests (Fig. 4.13). The fastest growing wavelength depends linearly on the depth h and the proportionality constant is in the range given by (4.25). This preferred dune wavelength depends only weakly on the tidal excursion $L = U_0 T / \pi$ (U_0 is the maximum tidal flow velocity) and on the ratio R of tidal period and inertial adaptation time scale.

The linear stability analysis for dune formation under steady flow was extended by Besio *et al.* [75] to tidal conditions, by solving numerically the perturbed vertical flow profile for a time-independent eddy viscosity N varying with depth z. Roughness was assumed to be due to seabed ripples 30 cm long and 5 cm high. The stability analysis included both bed-load and suspended-load transport. The results for the fastest growing wavelength were compared with observed dune wavelengths in the North Sea and showed good agreement [75, 159].

Dune development in the sea is also influenced by waves; it appears that intense wave action tends to inhibit dune development [585]. Fredsøe and Deigaard [318] showed that this is due to suspension lag effects, which are particularly important in the case of large tidal flow velocities.

Suspended-load transport

By using (3.68) as the sand transport formula we have implicitly assumed that dune formation proceeds by bedload transport. Observations show that dunes do not grow in very fine sand ($d < 130\mu$) [12]. This may be explained by considering the much stronger delay of the suspended-load response to variations in flow strength, compared to the bedload response. Although the near-bed velocity reaches a maximum upstream of the dune crest, one may expect that the suspended load reaches a maximum downstream, in particular if the flow velocity is high [169]. In this case the sediment flux is still increasing when the flow passes the dune crest; the divergence of the suspended sediment flux at the dune crest implies that sediment is eroded from the crest. Dune development is generally inhibited if suspended-load transport dominates over bed-load transport; for fine sediment beds this is usually the case.

Simulations with a 3D process-based numerical model support this conclusion [97]. These simulations show that suspended-load transport cannot be ignored; it counteracts the development of tidal dunes more strongly for short wavelengths than for large wavelengths. The fastest growing dune wavelength in the presence of suspended-load transport is therefore substantially larger than for bed-load transport alone. It further appears that a correct estimate of the dune wavelength is obtained only if the tidal variation of the eddy viscosity is taken into account.

In regions of the North Sea where dune fields are observed, their existence is also predicted by numerical models [425]. However, these models predict dune fields as well in regions where they do not occur [917]. Possible reasons are (1) limited availability of mobile medium sand, (2) the presence of a small fraction of cohesive sediments which increases the critical shear stress for erosion (see Sec. 3.5.3), (3) the presence of a seabed with mixed fine- and coarse-grained sediments, which becomes armoured by winnowing of the fine sediments from the top-layer (see Sec. 3.4.2), or (4) the influence of biotic activity on the bed roughness and bed cohesion (see Sec. 3.6). An exploratory model simulation shows that inclusion of the influence of biota on the bed roughness and on the sediment composition of the seabed top-layer, diminishes the discrepancy between observed and predicted dune field regions [95].

4.2.5. Antidunes

Supercritical flow

When the flow velocity in a flume with sandy bed is gradually increased, dunes become very irregular and then disappear; the upper bed layer starts behaving like a fluid sand sheet which is entrained by the near bottom flow, see Sec. 3.8.2. Sheet-flow conditions appear when the Shields parameter (3.10) exceeds the value $\vartheta \approx 0.5$; this corresponds to flow velocities close to 1 m/s for medium sand. When the flow velocity is still further increased the sand sheet becomes unstable; the sheet layer begins to develop bed waves, which tend to move against the flow direction. These bed waves are called antidunes. The water surface is undulated in a similar way as the bed wave, but the water surface amplitude is larger: The water depth is greater at the dune crest than at the dune trough. This happens when the flow becomes near-critical or supercritical; the Froude number $F = u/\sqrt{gh}$ is close to or greater than 1.

The surface wave is unstable and it breaks frequently; when it breaks the bed-wave is also destroyed. But after a short delay it reappears. Conditions of near-critical or supercritical flow correspond to fast flowing water in very small water depths. Such conditions are rather exceptional in natural coastal environments, but they sometimes occur in gullies on tidal flats during ebb flow and in migrating channels with sliding banks. Antidunes can also be observed in the backwash of breaking waves or in the outflow of beach runnels during ebb, see Fig. 4.14.

Difference with subcritical flow

The flow response to an initial bed perturbation is totally different for near-critical conditions then for subcritical conditions. Water is piled up at the dune crest, in contrast to the water level dip occurring in subcritical flow; the flow is not accelerated at the upstream dune flank, but decelerated. The flow velocity is lowest at the crest of the perturbation and highest at the trough; the phase of the longitudinal velocity distribution is, to a first approximation, at 180° to the bottom perturbation.

The piling up of water above the dune crest is due to the inability of the flow to move faster over the dune crest. When the Froude number is greater than 1, a change of surface level inclination

Fig. 4.14.   Antidunes appearing in near-critical ebb flow from a beach runnel. The surface-wave amplitude is greater than the bed-wave amplitude. The antidunes are formed in fluidised sand and are highly unstable.

cannot propagate upstream (the upstream propagation of a surface perturbation is blocked for any wavenumber $k > 0$, see (4.38)). This inhibits the flow acceleration that would be required to increase the speed of the flow over the dune crest.

Suspended load transport

The impact of a bed perturbation on the flow distribution is described by the Engelund-Fredsøe model, equations (4.13–4.17) and (4.23). We now neglect the friction-related flow perturbation which is on the

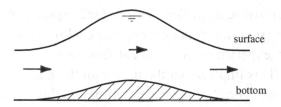

Fig. 4.15. The general pattern of near-critical flow over an antidune. Water depth is greater at the crest than at the trough and the opposite holds for the flow velocity.

order of $\sqrt{N/hu_0} \ll 1$; the result is, to first order in ϵ,

$$u(z) = u_0 \left(1 + \frac{\cosh kh - hkF^2 \sinh kh}{\sinh kh - hkF^2 \cosh kh} kz_b \right). \qquad (4.28)$$

When $F = \sqrt{\tanh(kh)/kh}$ the denominator vanishes; this corresponds to an unstable resonant situation where the upstream propagation of a surface wave with wavenumber k is blocked by the current.

For values of the Froude number in the range

$$\sqrt{\tanh(kh)/kh} < F < \sqrt{\coth(kh)/kh}$$

the phase of the flow velocity is exactly opposite to the phase of the bed perturbation z_b. If we assume bedload transport, then the sediment flux and the near-bed velocity have an identical phase; the bedload transport is thus at its minimum right at the crest of the perturbation. Therefore no accretion or erosion of the crest will take place; bed perturbations cannot grow through bedload transport. For the high flow velocities under consideration, the assumption of bedload transport is physically unrealistic, however; the flow is strong enough to bring sediment particles in suspension.

Suspension transport introduces a delay between transport rate and shear stress, because deposition and resuspension take time. Hence, the suspended sediment flux at the crest of the perturbation is not yet at its minimum; this minimum will be reached after the crest. Sediment transport decreases at the crest and the crest will therefore accrete. Bed perturbations may thus grow in supercritical flow because of the transport delay related to deposition and resuspension.

Most of the upstream flank of the antidune experiences a sediment-flux decrease (sediment flux convergence), while most of the downstream flank experiences a sediment-flux increase (sediment flux divergence). This implies that the upstream flank accretes while the downstream flank erodes; the perturbation will thus move upstream.

4.3. Ripples

4.3.1. *Bottom Ripples in Steady Flow*

Height and wavelength of bottom ripples

Ripples play a crucial role in the friction experienced by currents flowing over a sedimentary seabed. The friction coefficient for flow over a rippled bottom is typically a few times larger than for a flat bed (see Sec. 3.3.1). Ripples are produced by the flow itself, through a dynamic interaction between bottom profile and flow structure. After the general description presented in Sec. 4.2.1, we will now discuss the process of ripple formation more in detail, and in particular the initial stage of ripple formation.

The process of ripple formation on a sandy sediment bed starts as soon as the current strength is sufficient for setting sediment grains in motion; the critical velocity for initiation of grain motion is 10–20 cm/s for a cohesionless bed of fine-medium sediment, see Fig. 3.17. Flume experiments indicate that the ripple wavelength λ is not much influenced by the shear velocity u_*, but that it mainly depends on the median grain diameter d of the bed sediment according to the relation [972]

$$\lambda \approx (500 - 1000)\, d. \tag{4.29}$$

The largest λ/d ratios hold for fine sand grains of about $100\,\mu$m and the smallest for large grains of about $500\,\mu$m [798]. Flume experiments and field observations show that ripple formation does not occur in sediment with a grain size larger than about $1000\,\mu$m (very coarse sand) or a grain size finer than about $100\,\mu$m.

Shortly after emergence, the ripple pattern exhibits great regularity with almost straight crests perpendicular to the flow. Ripple heights

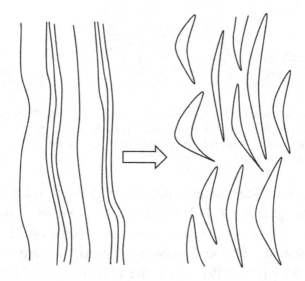

Fig. 4.16. Current ripples evolve from an initial straight-crested sinusoidal pattern to a complex three-dimensional pattern. Flow is from left to right.

are observed in the range 10–20 mm; the height slightly increases with grain diameter [708,798]. When the water depth is less than one metre there is no distinction between ripples and dunes. Ripples and dunes appear as distinct bedforms at greater depth and flow velocities higher than 40–50 cm/s. When they occur together, bottom ripples are generally superimposed on dunes. Observations show that in this case the streamlines in the outer flow layer follow the dune bed waves, but not the ripple bed waves. The flow separates at the ripple crest and the trough between the ripple crests is 'filled' with a lee vortex [76,708], see Fig. 4.17. When the velocity is increased above a value on the order of 1 m/s the ripples disappear and the bed surface layer is entrained by the near bottom flow as a sand sheet, as discussed in Sec. 3.8.

Mechanisms for ripple formation

Several processes contribute to the genesis of ripple patterns. When sediment grains start moving, small scattered sediment pileups will

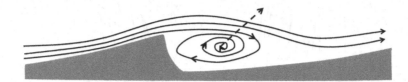

Fig. 4.17.   Flow separation and vortex generation at the lee of a ripple. The outer flow layer does not feel the ripple profile, because it experiences the trough between the ripples as being filled with a vortex. Vortices have a three-dimensional structure and are ejected after a short time in the flow [365].

first appear on the seabed; these pileups grow by trapping other moving sediment grains [179]. The random distribution of the pileups suggests that they originate from turbulent bursting processes [365]; this is consistent with the observation that ripples are not formed in laminar flow [707, 949]. When the pileups reach the roughness height their spacing becomes more regular; the term sand wavelet is sometimes used to designate this initial regular pattern [179]. The height of the sand wavelets is much smaller than the final ripple height and so is the wavelength. The observed initial wavelengths can be represented by the relationship [508]

$$\lambda \approx (100 - 200)\, d. \tag{4.30}$$

Upon further growth the spacing of the wavelets increases as a result of coalescence. Small bed features migrate with a speed inversely proportional to their height; differences in wavelet height yield different migration speeds, which lead to merging of wavelets [707]. Finally, a ripple pattern develops with an average wavelength given by (4.29). One may conclude that the final ripple pattern is produced by mutual interaction between ripples, and therefore cannot be described adequately by the initial, linear instability mechanism.

There are several hypotheses about the regularity of the ripple spacing. Observations indicate that new sediment pileups are produced downstream of existing pileups. Flow separation in the lee of these small sediment mounds may produce a vortex which causes erosion downstream of the initial mound and formation of a following mound. In this way a train of bottom irregularities is built up from which a ripple pattern may subsequently develop [800]. This process classifies ripple formation as a forced seabed instability rather than a free seabed instability.

Ripple formation as an inherent bed instability

The initial development of a wavelet pattern may also be attributed to an unforced instability, similar to the initial development of dunes [150,182,720]. Such an assumption is supported by other similarities between ripples and dunes, such as initial shape and migration. Dune formation is related to the response of the upper flow layer to perturbation of the sediment bed, while ripple formation may be related to the response of the near-bottom flow in the logarithmic layer to seabed perturbation. This would explain why dune dimensions are related to the flow depth, while ripple dimensions are not.

The mathematical description of flow perturbation in the near-bottom layer is in principle similar as for the upper layer, but more complex. We indicate here the most essential steps of the analysis of Fourrière, Claudin and Andreotti [310], without entering into details. They consider the zone where the shear stress perturbation τ' is independent of the vertical coordinate z, such that the velocity perturbation u' is well represented by a logarithmic profile, $u' \sim \ln(z/z_0)$. In this zone the shear stress depends just weakly on the the bed roughness length z_0, if z_0 is much smaller than the ripple wavelength λ ($\lambda/z_0 >> 10^4$). The expression for the shear stress perturbation reads (in complex notation) [447]

$$\tau' = \tau_0(A + iB)kz_b(x, t), \qquad (4.31)$$

where τ_0 is the undisturbed shear stress, $z_b(x) = \hat{z_b}\exp i(kx - \sigma t)$ is the ripple profile and A and B are slowly varying functions of z_0 with approximate values $A \approx 2.2$ and $B \approx 0.53$ for $10^{-5} < kz_0 < 10^{-3}$. The shear stress has a positive phase of $\approx \pi/12$ relative to the ripple profile; it is maximum upstream of the ripple crest.

The physical reason for this phase advance is the same as for dune formation and qualitatively explained in Fig. 4.7. The decrease of near-bottom shear stress over the crest causes ripple accretion by convergence of sediment transport.

Ripple wavelength of maximum growth

The wavelength of maximum growth depends on the stabilisation process of ripple accretion. For dune stabilisation, down-slope avalanching by gravity effects is thought to be the primary cause. However, for ripple stabilisation settling delay is probably more important, because even bedload transport does not adapt instantaneously to changes in bottom shear stress. There is a delay due to the inertia of grains leaping along the bed; the inertial length scale L_{sat} is estimated in the order of 10 grain diameters [310]. The adaptation of the bedload sediment flux q can be represented by the relaxation equation

$$q_x = (q'_{sat} - q)/L_{sat}, \qquad (4.32)$$

where q'_{sat} is the equilibrium bedload flux related to τ'.

If the flow velocity is sufficiently far above the threshold value for particle motion, we may use the simplified Meyer-Peter and Mueller transport formula, $q \propto |\tau|^{3/2}$. This yields for the perturbation of the saturated bedload flux q'_{sat} the expression

$$q'_{sat} = q_0 \frac{3\tau'}{2\tau_0} = \frac{3}{2} q_0 (A + iB) k z_b(x, t), \qquad (4.33)$$

where q_0 is the bedload flux in the unperturbed situation. The instantaneous perturbation of the bedload flux q' follows from substitution in (4.32):

$$q' = \frac{3}{2} q_0 \frac{(A + iB) k z_b(x, t)}{1 + ik L_{sat}}. \qquad (4.34)$$

For the time exponent σ we find, from the bottom evolution equation (3.57),

$$\sigma = -i \frac{q'_x}{z_b} = \frac{3}{2} q_0 k^2 \frac{A + iB}{1 + ik L_{sat}}. \qquad (4.35)$$

The growth rate is given by

$$\Im\sigma = \frac{3}{2}q_0 k^2 \frac{B - AkL_{sat}}{1 + k^2 L_{sat}^2}. \tag{4.36}$$

Because in general $k^2 L_{sat}^2 << 1$, the wavelength λ of maximum growth can be approximated by

$$\lambda \approx 3\pi A L_{sat}/B \approx 40 L_{sat} \approx 400d, \tag{4.37}$$

where d is the median grain diameter. This estimate is in reasonable agreement with the observed wavelength of emerging ripples.

Fully grown ripples

The previous linear stability model only describes the initial stage of pattern formation. Ripple steepness increases very soon after initial formation, causing flow separation and a recirculation bubble at the lee of the ripple, see Fig. 4.17. Flow separation yields a positive feedback to ripple growth, already in an early phase of ripple formation. The spatial coherence of the vortices is weak, however; a vortex which is initially uniform along the crest line, is easily broken up by small disturbances. The vortex structure becomes three-dimensional and perturbs the linearity of the ripple crest [708]. Fully grown current ripples generally exhibit a complex pattern, with linguoid, cuspate or honeycomb structures, see Fig. 4.18. Crest lines are at varying angles to the main flow direction and only the average trend of the crests lines is perpendicular to the flow. Understanding this behaviour requires detailed modelling of the interaction between flow over seabed topography and the structure of turbulent motion.

After initial emergence at wavelengths in the order of hundred or a few hundred grain diameters, ripples grow further in length and height, mainly by amalgamation of smaller ripples [365,508]. During this process, which is often referred to as 'pattern coarsening', the ripple influence extends higher into the flow. When the wavelength becomes comparable to the flow depth, the ripple influences the flow over the whole water depth, even if the ripple height is much smaller

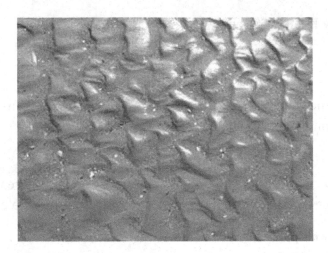

Fig. 4.18. Linguoid ripple pattern, formed in a rip channel on the beach which has fallen dry at low water. The gentle slopes at the right (stoss-side) and the steep slopes at the left (leeside) indicate that the flow was coming from the right.

[310]. At this stage the ripple may generate surface waves. According to linear wave theory (see Appendix D.1.1) they propagate at speed

$$c_\pm \approx u \pm \sqrt{g\tanh kh/k} = \sqrt{gh}\left(F \pm \sqrt{\tanh kh/kh}\right), \quad (4.38)$$

where k is the ripple wavenumber, h the water depth, u the flow velocity near the water surface and F the corresponding Froude number. When the ripple wavenumber further decreases and reaches a value such that $c_- = 0$, or

$$F_{res} \equiv \sqrt{\frac{\tanh kh}{kh}} = F, \quad (4.39)$$

the surface wave becomes resonant with the ripple wavelength. The resonance phenomenon inhibits further growth of the ripple wavelength. This is consistent with observations showing that ripple wavelengths are usually observed with $F_{res} < F$ and that they tend to stabilise when $F_{res} \sim F$ [310]. It also shows that the largest ripples have a wavelength depending on the water depth; they may thus be called dunes.

The time scale for reaching an equilibrium ripple (or dune) pattern depends on the flow strength; if the velocity is close to the critical value for incipient sediment motion, the time scale may be quite long (up to 100 hours) [708]. The equilibrium pattern has a dynamic character, with quasi-cyclic or chaotic fluctuations. However, the average wavelength and height remain approximately unchanged [378].

Megaripples and mega-dunes

The previous considerations suggest that one may observe bed undulations in natural flows with a large range of wavelengths. These different wavelengths are not related only to the sediment grain size (for ripples) or to the flow depth (for dunes). The actual wavelength of the bed undulation depends also on the stage of evolution. Because flow conditions are usually highly variable, bed undulations in their final state of evolution might be more the exception than the rule.

Megaripples are found both in current-dominated and in wave-dominated environments [331]. Observations in the surf zone of a tidal beach (depth between 0.3 and 1.8 m) show that the formation of megaripples starts at small wavelengths of typically one metre or less [168]. This wavelength is larger than the empirical wavelength (4.29) usually found for ripples in flume experiments. No strong correlation is found between the wavelength and water depth; for this reason they are not classified as dunes. One might speculate that the wavelength of megaripples is related to the increase of the turbulent boundary layer thickness due to ripples or due to wave-current interaction (see Sec. 3.3.2) [426, 720]. However, clear evidence for this hypothesis is missing. The designation 'megaripple' stems from the observation that they are often carry ripples with a much smaller wavelength.

Observations in the surf zone by Clarke and Werner [168] show that megaripples may grow in a few days into bed waves with wavelengths of several metres and that they finally reach a size typical of dunes. The pattern coarsening process proceeds through mutual interaction and merging of smaller bedforms, where the presence of defects (irregularities) in the bedform pattern plays an important role [333,941]. This generation process of megaripples and dunes is basically different from the linear instability mechanism described earlier in this section.

The same observations also show that the formation of megaripples is inhibited when flow conditions are strongly variable [168]. This suggests that ripples only grow into megaripples when flow conditions are sufficiently stable. The high temporal variablity of hydrodynamic conditions may explain why a gradual evolution of megaripples is not always observed in the surf zone [509].

The bed undulations corresponding to the instability mechanism described by the model of Fredsøe and Engelund, have a much larger wavelength than the megaripples or dunes described above. They are often covered with dunes, and may thus be called 'mega-dunes'. Although mega-dunes are in the region $F_{res} > F$, some observations indicate that they may also result from the coarsening of an initial ripple pattern [310,365]. The small growth rate (4.24) of mega-dunes supports the idea that the feedback process of the Fredsøe and Engelund model is preceded in some cases by other processes, such as ripple development. For the time being, there is no consensus regarding this question.

4.3.2. *Wave-Induced Ripples*

In the coastal environment bed ripples are generated not only by currents, but also by waves. Wave ripples are formed when wave orbital velocities are not too high — typically below 0.5 m/s for a fine to medium sandy seabed. Often two distinct regimes of wave ripples are observed [107,373,524,849,852]:

- Anorbital ripples: Small ripples with a wavelength $\lambda \approx 500d$ (d is the medium bed grain diameter) and ripple height h_r of the order $h_r \approx (0.1 - 0.12)\lambda$.
- Orbital ripples: Large ripples with average wavelength λ proportional to the wave orbital amplitude $a = U_0/\omega$ [492] ($U_0 =$ amplitude of the wave orbital velocity, $\omega =$ wave radial frequency). The proportionality constant λ/a is in the range 1.2–1.6 and the

ripple steepness h_r/λ in the range 0.14–0.17. Orbital ripples are often found in sub-tidal zones under conditions of fairly constant hydrodynamic forcing.

Other empirical relationships for the ripple wavelength and ripple height involve also the wave mobility parameter $\Psi = \rho U_0^2/gd\Delta\rho$, where $\rho/\Delta\rho$ is the immersed sediment weight ([133, 355]). However, the dependence on Ψ is substantial only for high mobility numbers [644].

According to laboratory experiments and field observations, two and three-dimensional ripple patterns both occur. Three-dimensional ripples occur mainly in fine sand and have generally a smaller wavelength than two-dimensional ripples [644].

Transient ripples

In many field situations seabed ripples are observed with intermediate height and length [798]. They correspond to transient states and are called suborbital ripples [946]. Suborbital ripples occur typically in settings where the hydrodynamic forcing is highly variable, for instance in intertidal zones [41].

Wave ripple patterns in field situations are often not in equilibrium with the prevailing wave climate. A well established ripple pattern does not respond quickly when wave conditions change, especially in the case of a decreasing wave-orbital excursion. Adaptation can take hours, but also days. At the observatory off the US New Yersey coast at 12 m depth, a storm-generated ripple field persisted several days during the calm after-storm period, before adapting to the new situation with a three times smaller equilibrium wavelength [849].

Field observations on Sable Island Bank (Nova Scotian Shelf, depth 30 m) showed the following sequence of seabed forms during a storm event [524]: (1) relict wave-dominant ripples with worm tubes and animal tracks formed during the preceding fair-weather period; (2) irregular, sinuous, asymmetrical current-dominant and intermediate wave-current ripples under bedload transport; (3) regular, nearly straight or sinuous asymmetrical to slightly asymmetrical wave-dominant ripples under saltation/suspension; (4) upper-plane bed under sheet-flow conditions; (5) small, crest-reversing, transitory ripples at the peak of the storm; and (6) large-scale lunate megaripples which developed when the storm decayed.

Difference with current-induced ripples

Wave-induced ripples differ from current-induced ripples in at least two aspects:

- Wave ripples migrate much less than current ripples; the ripples are therefore more symmetric, the crest line exhibits greater continuity and the overall ripple pattern has greater regularity (compare Figs. 4.18 and 4.19).
- In a steady current, the generation mechanism of ripples involves a single length scale: The sediment grain size; in an oscillating current, ripple generation is also influenced by a second length scale: The wave-orbital excursion. This second length scale is crucial for the generation of orbital ripples; it is of less importance for anorbital ripples.

Fig. 4.19. Typical wave-ripple pattern on the beach. The pattern is more regular than for current ripples. The ripples are symmetric with rounded crest, due to symmetric wave orbital motion.

For current-induced ripples, the wavelength of fastest growth is mainly determined by a balance between the destabilising effect of inertial lag and the stabilising effect of the bedload adjustment time lag, with a secondary role for gravitational stabilisation [150, 310]. For wave ripples, the bottom slope term in the transport formula is important, to limit the growth of perturbations with a very short wavelength [309].

Wave ripple formation

Wave ripple formation is triggered by an instability mechanism similar to the mechanism for current ripples and dunes (see Fig. 4.7). The near-bottom orbital flow profile does not adjust instantaneously to a bottom perturbation throughout the turbulent boundary layer, as a consequence of inertial lag. The maximum shear stress at the bottom is thus shifted upstream of the ripple crest, resulting in convergence of sediment transport at the ripple crest and local accretion. Averaged over the wave period, residual eddies are generated in the boundary layer, which produce a steady near-bed flow (streaming) to the ripple crests [150, 309, 780]. The picture is qualitatively the same as for tidal flow over dunes (see Fig. 4.13). The steady streaming moves sediment particles towards the ripple crests and produces growth of an initial ripple perturbation.

Wavelength of orbital ripples

The ratio of the orbital ripple wavelength λ and the wave-orbital excursion a is an approximately fixed number of about 1.5. There are several explanations.

In Sec. (3.25) it was shown that the thickness δ_w of the wave boundary layer is almost linearly proportional to the wave-orbital amplitude a. If we assume that the ripple wavelength λ scales with the thickness of the wave boundary layer, then it follows that the ripple wavelength also scales with a. According to (3.25), the wave boundary thickness is an increasing function of the bed roughness length. This is consistent with the observation that the ripple wavelengths for a coarse grained seabed are larger than the wavelengths for medium sand [373,852].

Sleath [780] presented another argument for the selection of a ripple wavelength close to the wave-orbital amplitude. When a sediment particle is picked up from the seabed by wave action, it leaps along the seabed following more or less the wave-orbital velocity. Its settling location after half a wave cycle will be affected by the steady boundary-layer streaming toward the nearest ripple crest. However, it take generally at least several wave cycles before it reaches the ripple crest. If the leaps are smaller than the ripple wavelength, there is a high probability that successive settling locations are each time closer to the nearest crest. If the leaps are much larger than the ripple wavelength, it will pass several crests. The correlation between successive settling locations is much smaller in this case. The effect of steady streaming circulation is thus strongest for ripple wavelengths much larger than the wave-orbital amplitude. However, for large ripple wavelengths, the steady streaming is weak. The ripple wavelength λ for which the steady streaming contributes most to ripple growth, thus corresponds to a value of λ/a close to one.

Vortex ripples

In steady flow, sediment is entrained as bedload over the gently sloping stoss side of each ripple; the vortex generated at the steep lee side remains attached to the ripple crest or will be swept to the next ripple. In wave orbital flow, vortices are generated alternatively at both sides of the ripple, depending on the wave phase. These vortices are swept back over the ripple crest when the flow changes direction, see Fig. 4.20.

Visual observations show that the shedding of vortices at the ripple crests plays a major role in the development of orbital ripples [317]. Vortex shedding can be simulated in LES (Large Eddy Simulation) models [985] and in lattice Boltzmann numerical flow models [158]. In this latter type of models, fluid flow is simulated by streaming

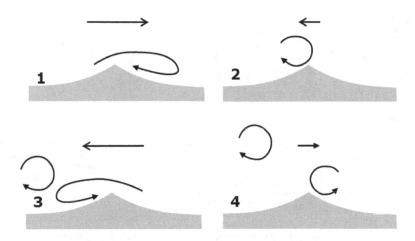

Fig. 4.20.   Vortex-shedding from a ripple crest during a wave cycle.

and mutual collisions of fictive fluid particles; this method is well suited for flow over ripples due to easy implementation of irregular geometries. It was applied by Nienhuis *et al.* [636] for simulating the generation of orbital ripples under wave forcing. The results show that the maximum length of the flow separation zone downstream of a ripple crest equals the wavelength λ when the ratio when λ/a equal 1.3. The authors conclude that this ratio maximizes the bed shear stress that transports sand from troughs to ripple crests and therefore corresponds to the wavelength of maximum growth.

Marieu *et al.* [565] simulated the generation of orbital ripples by solving the Reynolds-averaged Navier-Stokes equations with a $k - \omega$ turbulence model (k stands here for the turbulent kinetic energy and k/ω for the eddy viscosity). They considered a situation where bed-load transport dominates suspended-load transport and where ripple steepness is limited by avalanching. The small ripples generated initially in the model coalesce into a stable pattern of orbital ripples with wavelength $\lambda \approx 1.5a$. The coalescence involves ripple annihilation and ripple merging, see Fig. 4.21. This figure shows that the irregularities (defects) present in the ripple pattern during early development have disappeared in the final regular pattern

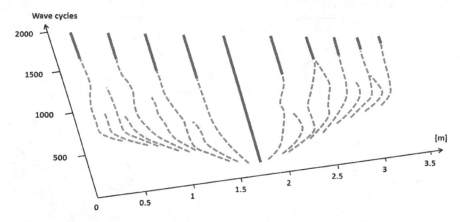

Fig. 4.21. Numerical simulation of wave-induced ripple formation. Grainsize $d = 0.3$ mm, maximum wave-orbital velocity $U_0 = 0.26$ m/s, wave period $T = 6$ s ($a = U_0/\omega = 0.25$ m). The simulation starts with a single initial Gaussian bump at $x = 1.75$ m on a flat seabed. Small ripples appear after a few hundred wave cycles with wavelength of the order of $400\,d \approx 0.12$ m; the crests are indicated by dotted lines. In the course of time, some of the small ripples grow and others merge or disappear. Finally a stable ripple pattern emerges; crests are indicated by solid lines. The wavelength is approximately $\lambda = 1.5a = 0.38$ m. Redrawn after [565].

Bio-induced ripple structures on tidal flats

As discussed in Sec. 3.6, biotic substances can substantially influence soil erosion resistance. Soil stabilising biofilms, consisting of extra-cellular polymeric substances (EPS) exuded by diatoms, often appear on muddy intertidal flats during spring. EPS dissolve when they are permanently covered by water; biofilms are absent in hollows or runnels on the tidal flat which remain filled with water during low tide. This spatially different behaviour of EPS promotes the formation of bedforms on the tidal flat. The colonisation of mounds (or ridges) by EPS-exuding diatoms protect them from wave erosion, while hollows (or runnels) are easily eroded by wave action. It has been shown that this feedback can generate a spatial pattern of mounds and hollows (or a pattern of ridges and runnels) on the tidal flat [937]. The spatial dimensions of the observed pattern is similar to the dimensions of wave-induced ripple patterns, but it is less regular: The wavelength is on the order of a few decimetres up to a metre and the height on the order of a few centimetres.

4.4. Sorted Bedforms

Graded sediment

In the previous sections we have assumed that the grain size distribution of sediment is uniform in space and unaffected by sediment transport processes. This assumption is reasonable in situations where large-scale sorting of grain size fractions has previously taken place, in response to gradients in shear stresses exerted on the seabed by currents and waves. In many field situations, however, the assumption of uniform grain size distributions does not hold. This is the case, for instance, in storm-dominated coastal zones where medium and fine sands are available in limited amounts. Such situations often exist on the inner shelf, outside the surf zone, where the seabed has a bimodal (or multimodal) sediment distribution, consisting of a mixture of fine to medium sand and coarse to gravelly sand. Seabeds composed of sediment fractions of different sizes are often called 'graded sediment beds'. In Sec. 3.4.2 we have seen that the composition of graded sediment beds is variable in time, due to winnowing of the fine sediment fraction when the shear stress at the seabed is strong enough for eroding the fine sediments but too weak for eroding the coarse fraction.

Sediment transport

The presence of a mixture of fine and coarse sediment affects sediment transport in several ways: (1) Winnowing of fine sediment from the seabed leaves a top layer of coarse sediment, which hides the underlying fines from further resuspension. This 'armouring' of the seabed limits the availability of fine sediment. (2) The seabed roughness depends on the sediment composition of the top layer; when the top layer consists predominantly of coarse sediment, the roughness will be higher than for a sediment bed with a fine top layer.

(3) Seabed roughness affects the near-bed turbulence and the near-bed current structure; the turbulence level increases with increasing roughness. (4) High near-bed turbulence levels inhibit (although do not completely rule out) settling of fine sediments on coarse sediment beds; coarsening of the seabed top layer through winnowing of fines is therefore a self-reinforcing process.

Coarsening of the seabed top layer increases roughness not only by skin friction, but also by form drag (see Sec. 3.2.2). If bed shear stresses are high enough for moving coarse sediment as bedload, seabed ripples are formed which are higher and steeper than on a fine sediment bed [634,798,907]. The roughness and near-bed turbulence levels of coarse sediment beds are thus substantially increased compared to fine sediment beds.

Segregation of fine and coarse sediment

Because coarsening of a graded seabed through winnowing of fine sediments is a self-reinforcing process, spatial inhomogeneities of the bed top layer will expand: These inhomogeneities may grow into alternating patches of coarse and finer sediment. Rhythmic patterns of alternating coarse and fine sediment patches have been observed on the inner shelf of many coastal zones [622]. They have an elongated structure, perpendicular to the residual (wave-averaged) sediment transport direction. They are typically a few hundred metre wide and up to several kilometres long. An illustration of sorted sediment patterns is presented in Fig. 4.22, based on observations in the wave- and storm-dominated coastal zone of Aquitaine (France). This study demonstrated that sorted sediment patches on the inner shelf may persist over several decades without great change [581].

Modelling of sediment segregation

A plausible generation mechanism of this type of sorted bedfoms was identified by Murray *et al.* [621,622], in a model study of sediment transport with two sediment fractions of different grain size. They used a transport formula of the type (3.84) for wave-suspended

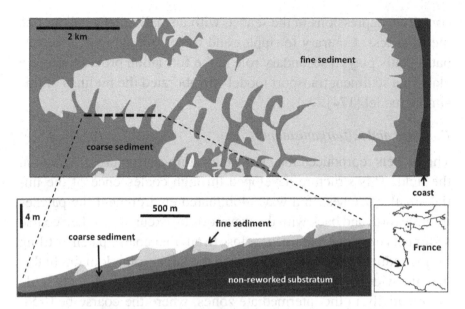

Fig. 4.22. Sorted bedforms in the nearshore coastal zone of Aquitaine (France), at water depths of 24–50 m. Light grey = fine-grained sediment (medium grain size $d =$ 0.16–0.4 mm), dark grey = coarse-grained sediment ($d = 0.5–5$ mm). Upper panel: Plan view; lower panel: Cross-section along a transect. Medium to fine sand patches (0.5–2 m thick) are cut by smaller, elongated coarse sediment depressions. The average spacing of the sand patches is of the order of 500 m. Sand patches predominantly overlay the coarse-grained blankets on the seaward extremities, while coarse-grained blanket wedges partly overlay the seaward extremities of the sand patches. The coarse depressions have a tendency to elongate (about 200 m in 10 years); the sandy patches move slowly shoreward (about 60 m in 10 years). Redrawn after [581].

sediment, with different drag coefficients c_D and different settling times $T_{settle} = h_{eff}/w_s$ for the fine and coarse sediment fractions (h_{eff} is an effective suspended concentration profile height, w_s is the settling velocity). The drag coefficient and the effective settling time were approximated by linearly increasing functions of the coarse fraction in the sediment bed. Starting from a random distribution of inhomogeneities in the initial bed sediment distribution, this simple model produced segregation and growth of coarse and fine sediment domains, similar to sorted bedforms observed in field situations. The mechanism is basically an erosional process; the coarse patches constitute depressions in the seabed with a small relief of about one metre or less. Contrary to ripples and dunes, topographical perturbation only plays a secondary role in the formation process. A more elaborate sediment transport model corroborated the findings of this simple model [174].

Feedback and self-organisation

The models reproduced the elongate rhythmic structure observed in the field. This structure developed through coalescence of the initial smaller patches. In a wave-dominated environment the patches migrate forth and back with the wave-orbital excursion. When coarse sediments from one patch settle close to another coarse patch, settling of fine sediment becomes inhibited in the larger joint domain. In this way the width of the coarse patches increases. Fine sediments settle preferentially in the intermediate zones, where the coarse bed sediment fraction is lower. The width of the patches cannot become much larger than the wave orbital excursion if there is no residual migration through wave asymmetry or a residual current. Patch growth by coalescence in lateral direction is promoted by directional variability of the wave propagation direction and variations in the direction of steady currents. Gravity-induced down-slope transport of fine sediment limits the topographical relief of the patches. Hence, the composition of the coarse domain, its residual migration speed (increasing with increasing coarseness) and sorted bed form spacing are linked to the geometry of sorted bed forms.

In the case of a strong steady flow component, the patch width can, in theory, increase indefinitely. Repeated storms can also generate widely spaced patterns [347]. However, in field situations, natural variability in wave height and current direction may destroy the regularity of the pattern. Burial of coarse-grained patches under fine sediment may occur when ripples have been wiped out by a storm, inducing negative feedback to the growth of coarse patches in the calm after-storm period [175]. The positive sorting feedback may be revived under subsequent high-energy wave conditions. This triggers a new cycle of pattern formation, influenced by the history of previous patterns [623].

Pattern defects

The development of initial bedforms into mature structures involves processes which can be quite different from the initial generation mechanisms. The initial exponential growth rate generally decreases under the increasing influence of negative feedback processes. This may lead in some cases to a stable final state, but more often to periodic or quasi-periodic (non-deterministic) asymptotic behaviour, even under constant external forcing conditions (see Sec. 2.4.2). Natural variability in external conditions is an additional factor that makes pattern evolution unpredictable at long term.

In the case of sorted bedforms, initial inhomogeneities in the spatial sediment distribution develop into a more or less rhythmic pattern by successive mergers of smaller patches. Model simulations show that this process increases gradually the size and spacing of the patches.

These simulations also indicate a relationship between the rate of patch evolution and the regularity of the pattern. An initially regular pattern of straight long-crested bands of coarse-grained sediment hardly responds to a change in forcing conditions, while an initially less regular pattern evolves much faster under the same change in forcing conditions [435]. The regularity of the pattern is manifested in the density of defects: Irregularities consisting of crest terminations, eyes and forks, see Fig. 4.23. It appears that the presence of defects promotes the evolution of an initial pattern toward a new pattern when forcing conditions are changed. The numerical experiments show that the pattern adjustment progresses through stages characterized by defect propagation and defect creation, and that the rates of pattern adjustment are determined by the initial defect density.

In our discussion of ripple formation (Sec. 4.3) we have seen that megaripples and dunes may develop from an initial ripple field by a pattern coarsening process involving successive mergers of

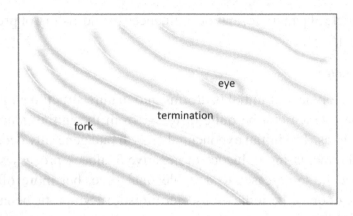

Fig. 4.23.   Bedform pattern with 3 types of defects: Termination, eye and fork.

individual ripples. As mentioned earlier (see Sec. 4.2.1), the presence
of defects probably plays a role in this pattern coarsening process as
well [623,941].

4.5. Sandbanks

4.5.1. *Qualitative Description*

Ripples and dunes belong to the small and medium scale structures of
the seabed. There are also much larger bedforms: Sandbanks. They
did not appear in the previous analysis of bed-flow interaction. Their
large dimensions suggest that sandbank dynamics is probably more
related to the horizontal flow structure than to the vertical structure.
This hypothesis will be examined in this section, using similar considerations as before.

Sandbank categories

Different types of sandbanks may be distinguished [259]:

- Tidal ridges, also called tidal sandbanks. Their length is in the
  order of several tens of kilometers (in some cases even 100 km),
  their width is typically 5 to 15 kilometres and their height some
  tens of metres. The crest line is almost straight and makes a small

cyclonic angle to the main flow direction. Tidal ridges migrate very slowly; the profile is asymmetric, with the steepest slope at the lee side (down the propagation direction). Sediment is coarsest around the crest and finest around the troughs (swales). Tidal ridges occur on the open shelf in areas where tidal currents are strong. They often sit on a erosion-resistant basement, consisting of coarse gravel or compacted cohesive sediment. In wide tidal embayments different flood and ebb dominated channels are often separated by elongate bars; these elongate bars resemble offshore tidal ridges and may result from similar morphodynamic processes [558].

- Shoreface connected ridges. Scales are comparable to tidal ridges, the height is often somewhat lower. They migrate along the coast and the crest line is oriented in downdrift shoreward direction (attached to the shoreface). The downdrift seaward flank has a steeper slope than the landward flank; sediment is coarsest around the trough and finest on the seaward flank. Shoreface connected ridges occur on the inner shelf, outside the breaker zone.
- Headland-associated banks. Elongated tidal banks of a few kilometres long, formed in the vicinity of headlands, on fairly steep sloping coasts.
- Ebb tidal deltas. Systems of sandbanks and channels at the seaward side of tidal inlets. Their size and morphology is mainly related to the tidal flow through and around the inlet, but wave-driven longshore transport also plays an important role.
- Delta lobes. Shore-parallel spits attached to river delta protrusions on wave-dominated micro-tidal coasts.

Forced and free seabed patterns

The occurrence and the scale of the last three categories of sandbanks are linked to the flow pattern generated by existing local coastal features; in that respect they differ from the first two categories, which are not directly related to local topographic features. The latter three sandbank categories may be considered as forced seabed patterns,

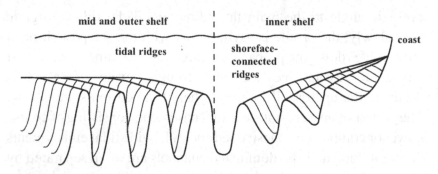

Fig. 4.24. Schematic cross-shore section of the inner and mid shelf with shoreface-connected ridges and tidal ridges for the Northern Hemisphere. The vertical scale is strongly exaggerated relative to the horizontal scale. The shoreface-connected ridges are situated on the gently sloping inner shelf and curve towards the coast in downdrift direction. The tidal ridges are located on the open shelf and oriented cyclonically with respect to the main tidal flow axis.

while the former two are related to seabed instability. A particular feature of these former two sandbank types is their great length compared to their width; therefore they are sometimes called linear sandbanks or linear ridges (see Fig. 4.24).

In this section we investigate the mechanisms through which these ridges may arise, as large-scale instabilities of the seabed of the inner and mid shelf. Some aspects of ebb-tidal deltas are discussed in the chapter on tide-topography interaction. Beside sandbanks, other large-scale topographic structures exist on the seabed such as isolated mounds or troughs; such structures are related to past geological events, for instance, ice scour or earth crust motions, and will not be considered here.

Dynamic equilibrium

Linear sandbanks as described above are found on many continental shelves, for instance, in the southern North Sea, see Fig. 4.25. These sandbanks are in dynamic equilibrium with the existing flow and sand transport regimes if they satisfy the following conditions:

• they occur in more or less regular patterns;

Fig. 4.25. Bathymetric map of the Dutch part of the North Sea (Southern Bight); dark is deep and clear is shallow. Depths range between 10 m (close to the coast) to about 40 m offshore. The North Sea basin has been subsiding for several million years at an average rate of the order of a cm per century [981]. In this period thick sediment layers have been deposited of mainly medium- to fine-grained sand. The overall topography exhibits an East-West asymmetry, with a smaller depth along the Dutch coast than along the UK coast. The sediment composition has a South-North gradient: Medium-grained sand towards the south and fine-grained sediments towards the north. In the central part of the Southern Bight large ridge-form sandbanks can been seen, with length in the range 25–100 km, width in the range 5–10 km and height in the range 5–25 m. The ridge orientation is generally South-North; the crests make a small cyclonic angle to the major tidal current axis. Close to the coast, one can observe ridges of similar dimensions but with a different orientation. These ridges, which are oriented towards the coast, are called shoreface-connected ridges. More difficult to distinguish are the dunes that cover most of the southern part of the Bight. The dunes are oriented approximately perpendicular to the main tidal current axis; they have a similar height as the tidal ridges but the wavelength is much smaller (approximately a few hundred metres), see Fig. 4.5.

- they are predominantly composed of the same type of sediments as neighbouring shelf zones;
- they occur in dynamic sedimentary areas, characterised by strong currents and cross-flow oriented ripple and dune cover.

Field observations

The first investigation of ridge fields was made by T. Off in the early 1960s.[647]. Many investigations have followed since, revealing ridge fields as a common characteristic of sandy shelf seas. Sandbank fields have been observed, in particular, on the Atlantic inner shelf of North America [346, 821] and South America [301, 661], the North Sea inner and mid shelf [26, 412, 503], the East China Sea [534, 930] and the Yellow Sea [532, 533]. Similar sandbank fields are found on the same shelves in deeper water, where current-induced sand transport is too weak for maintaining sandbank morphology [71, 346], for instance in the northern North Sea and in the Celtic Sea [568]. These relict sandbank fields are in a regressive stage, although there is still evidence of active ripple and dune formation. Seismic soundings in the East China Sea have revealed the existence of extensive sandbank fields buried under Holocene deposits [72].

Relict or modern?

Sandbanks probably developed already at the earliest stages of sea level rise, but there is still uncertainty about their origin. Their initial formation may be related to the reworking of relict shoreline deposits during the Holocene transgression, since such deposits are found in the core of many present sandbanks [412]. However, the orientation of present sandbanks often does not match ancient shorelines; the evolution of sandbanks far off the present shoreline may have become independent of the initial stage. In the following we will further examine this point of view. It will be shown that sandbank development can be explained by morphodynamic feedback, which does not require any relict structure as an initial trigger. This does not preclude that sandbanks may actually originate from relict deposits; it only means that the same sandbanks could have developed from a flat seabed and that the morphology of mature sandbanks does not necessarily bear a causal relationship to preexisting seabed structures.

4.5.2. *Tidal Sand Ridges*

Seabed instability

Tidal ridges are immense structures on the shelf sea bottom. A single tidal ridge may contain as much as a few billion cubic metres of sand. One may wonder by which forces these submarine giants have been shaped. No geological events have been identified that provide a convincing explanation for their appearance on the outer shelf. But is it conceivable that a structure so huge can generate itself out of a flat bottom, without specific external influences? J. Huthnance [437] was the first to advance such a daring hypothesis and to provide a theoretical underpinning for tidal ridge formation as the result of seabed instability. An important indication for the dynamical origin is the observation that the ridges are always inclined relative to the main flow direction [136, 301, 412, 641, 661, 664, 821, 921]. This implies that the flood and ebb flow have to cross the ridge, thereby creating different flow patterns on the upstream and downstream flanks. The importance of upstream-downstream flow asymmetry for bedform generation has been demonstrated for ripples and dunes. In the case of tidal ridges the asymmetry appears in a different way, but the same factors play a role: Momentum conservation (or its equivalent: Vorticity conservation) and bottom friction. Flow perturbation by tidal ridges is also influenced by earth's rotation, because of the large spatial scale of the perturbation. For that reason, flow perturbations corresponding to cyclonic and anticyclonic orientations of the ridge relative to the main flow direction, are not symmetric. In the following we will first give a qualitative description of the process of tidal ridge formation; then we will discuss two more mathematical approaches proposed by Zimmerman [989] and Huthnance [437].

Flow over a tidal ridge

We consider a shallow ridge, at an angle to the undisturbed tidal flow, and examine the path followed by a fluid parcel when crossing the ridge. It makes no difference whether we choose the flood or the ebb tidal phase; the process of ridge growth is symmetric with respect to the flow direction. The path followed by the fluid parcel is shown in Fig. 4.26. In fact, two paths are shown. The dashed path would be followed by the fluid parcel if we neglected inertial delay. In that case the streamline depends on the following three factors:

- Continuity of fluid discharge across the ridge. As the water depth at the ridge crest is smaller than at the trough, the cross-ridge

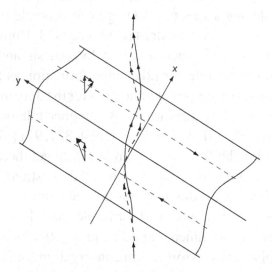

Fig. 4.26. Streamlines crossing a ridge with an anti-clockwise orientation relative to the undisturbed flow. The dashed line corresponds to the path a fluid parcel would follow in the absence of inertial delay. The solid line represents the path of a fluid parcel which is delayed by inertia. At the upstream flank the delayed flow has an along-ridge component in positive y-direction and at the downstream flank an along-ridge component in negative y-direction, relative to the non-delayed flow. The vector diagrams show the actual velocity vector that results from the superposition of the non-delayed velocity vector (pointing to the top of the page) and the along-ridge velocity deviation due to the inertial lag. The length of the actual velocity on the upstream ridge flank is increased, while the length of the velocity on the downstream flank is decreased. The sediment flux is convergent at the ridge crest; hence, the crest will accrete.

velocity component increases at the uphill slope and decreases at the downhill slope.

- Bottom-friction torque. When a fluid parcel crosses the ridge, its ridge-parallel velocity component will decrease in response to increased bottom friction. Due to the ridge inclination a cross-flow array of fluid parcels will not decelerate simultaneously; the parcels closest to the ridge will decelerate first. The fluid array will thus rotate towards the crest line.

- Earth's rotation. When moving uphill the water column is compressed and therefore the planetary vorticity (due to earth's rotation) is increased. Vorticity conservation forces the fluid to adopt an opposite rotation; in the Northern Hemisphere this opposite rotation is clockwise (anti-cyclonic). If the ridge is turned cyclonically relative to the flow direction, then the fluid rotates towards the ridge-normal direction; in the opposite case the fluid rotates away from the cross-normal direction.

In Fig. 4.26 the ridge has a cyclonic orientation relative to the undisturbed flow direction. The dashed streamline is the path a fluid parcel would follow if the flow momentum adapted instantaneously to the above three factors. This hypothetic streamline is perfectly symmetric relative to the ridge crest.

Feedback mechanism

The adaptation of the ridge-parallel flow momentum to the frictional torque and to earth's rotation is delayed due to inertia (advection of vorticity). The streamline bends towards the ridge-normal direction with a spatial lag, as represented by the solid particle path in Fig. 4.26. This streamline is therefore not symmetric relative to the ridge-crest. Before crossing the crest, the actual streamline is deviated relative to the non-inertial streamline in the ridge-parallel downstream direction; after crossing the ridge the streamline is deviated in ridge-parallel upstream direction. These deviations indicate a secondary circulation around the ridge; this circulation is caused by the flow perturbation due to frictional torques and vorticity conservation. The strength of the flow velocity at the upstream flank of the ridge is greater than at the downstream flank; this is because the circulation contributes asymmetrically to the unperturbed velocity at the upstream and downstream flanks of the ridge (see Fig. 4.26). The flow will therefore carry more sediment towards the ridge crest than away from it and we may expect the ridge amplitude to grow.

Instability of the seabed

As first postulated by Smith [788], the presence of a residual circulation around the bank is essential for its growth and maintenance. Oceanographic surveys at tidal sandbanks confirm the existence of flow circulation around tidal sandbanks [412, 414, 586, 728]. Theoretically, sandbank circulation is initiated already at ridges of very small height. This means that the seabed is unstable; a perturbation of infinitesimal amplitude may grow, provided the horizontal dimensions are such that the flow response to the perturbation is mainly determined by frictional torques and momentum conservation. In the following a more mathematical description of the flow response to a ridge perturbation will be presented.

4.5.3. Zimmerman's Qualitative Ridge Formation Model

Instant vorticity balance

The vertical structure of the flow was ignored in the previous qualitative description; it does not play a crucial role in the feedback of flow perturbation to ridge growth. This will also be the starting point for the mathematical analysis in which mass and momentum balances will be considered for depth-averaged variables. The analysis is simplified by eliminating pressure gradient terms. This is achieved by considering the vorticity balance instead of the momentum balance (see Appendix A.5). The depth-averaged balance of the potential vorticity $(\zeta + f)/D$ reads (A.18)

$$\frac{\partial}{\partial t}\left(\frac{\zeta + f}{D}\right) + \vec{u}\cdot\vec{\nabla}\left(\frac{\zeta + f}{D}\right) = -\frac{r}{D^2}\zeta + \left(\vec{u}\times\vec{\nabla}\frac{r}{D}\right)\cdot\frac{\vec{e}_z}{D}. \quad (4.40)$$

Here $\vec{u} = (\bar{u}, \bar{v})$ is the depth-averaged flow velocity vector, $\zeta = \bar{v}_x - \bar{u}_y$ the vorticity, D the instantaneous water depth, f the Coriolis parameter and $\vec{e}_z$ a unit vector in the vertical upward direction. In this equation a linearised form of the quadratic bottom friction is used, $\vec{\tau}_b = \rho r\vec{u}$, where r is the linear friction parameter with dimension [m/s]. The second term on the r.h.s. of Eq. (4.40) shows that potential vorticity is generated by bottom friction gradients. This happens when the flow crosses isobaths at an angle; the flow is then rotated perpendicular to the isobaths. The first term on the r.h.s. stands for vorticity dissipation by bottom friction. The second term on the l.h.s. of (4.40) represents local increase or decrease of potential vorticity, due to tidal advection.

Tide-averaged vorticity balance

Equation (4.40) shows that vorticity produced by bottom friction has opposite signs for ebb and flood. This implies (1) that ebb-vorticity and flood-vorticity have opposite signs and (2) that tidal vorticity advection has the same sign at ebb and flood. When averaging (4.40) over the tidal period, the tidal vorticity advection term remains as the only tidal contribution in the balance equation of residual vorticity. This tidally averaged balance thus shows that tidal vorticity advection is responsible for the production of residual vorticity at the tidal ridge. In the following it will be shown that the production of residual vorticity is equivalent to the production of a mean flow circulation around the ridge.

Sand bank circulation

Residual circulation can be evaluated by integrating the depth-averaged momentum balance equation (see A.15)

$$\vec{u}_t + (\vec{u} \cdot \vec{\nabla})\vec{u} + f\vec{e}_z \times \vec{u} + \frac{1}{\rho}\vec{\nabla}p + \frac{\vec{\tau}_b}{\rho D} = 0, \qquad (4.41)$$

along a depth contour. The circulation $\mathfrak{C}(t)$ along a depth contour is related to vorticity by Gauss' law

$$\mathfrak{C}(t) \equiv \oint \vec{u}(t) \cdot \vec{dl} = \int\int_{\Sigma} \zeta \, dxdy, \qquad (4.42)$$

where Σ is the area enclosed by the depth contour. To integrate the momentum balance equation we use the relation

$$(\vec{u}\cdot\vec{\nabla})\vec{u} = (\vec{\nabla}\times\vec{u})\times\vec{u} + \tfrac{1}{2}\vec{\nabla}\cdot\vec{u}^2 = \zeta\vec{e}_z\times\vec{u} + \tfrac{1}{2}\vec{\nabla}\cdot\vec{u}^2. \qquad (4.43)$$

We use again a linear approximation of the bottom friction term, $\vec{\tau}_b/\rho D \approx (r/h)\vec{u}$, where r is the linearised friction coefficient and h the tide-averaged water depth. The result then is

$$\frac{\partial \mathfrak{C}}{\partial t} = -\oint (\zeta + f)(\vec{u} \times \vec{dl}) \cdot \vec{e}_z - \frac{r}{h}\mathfrak{C}. \qquad (4.44)$$

We integrate this equation over the tidal period with the assumption that the tidal amplitude is much smaller than water depth; we find

$$\langle \mathfrak{C} \rangle \equiv \oint \langle \vec{u}(t) \rangle \cdot \vec{dl} = -\frac{h}{r} \oint <\zeta\bar{u}_\perp> dl, \qquad (4.45)$$

where $\bar{u}_\perp$ is the tidal component perpendicular to contour element dl (outward positive). We now consider the area Σ which corresponds to the tidal ridge above mean bed level. Then it follows from Eq. (4.45) that tidal vorticity advection over the ridge produces a residual circulation around the ridge, because the sign of the product $\zeta\bar{u}_\perp$ does not change around the contour and is the same for ebb and flood (see

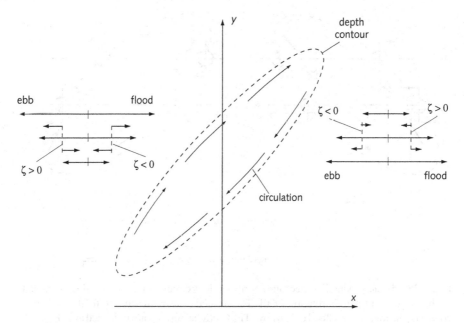

Fig. 4.27. Schematic representation of the tidally averaged vorticity balance, producing residual circulation. Lateral shear of the tidal velocity at both sides of the sandbank is due to the combined effect of sandbank friction and bank inclination. The corresponding vorticity has opposite signs at both sides of the sandbank and also changes sign from flood to ebb. The product of the cross-ridge flow component and vorticity is always positive. (The outward normal direction to the contour is taken positive.)

Fig. 4.27). The strength of the residual circulation thus equals the amount of vorticity advected through the ridge contour during a tidal period.

Ridge inclination

The sign of the vorticity produced by the bottom friction torque depends on the ridge orientation relative to the flow direction. It appears that the residual circulation is clockwise if the sand ridge crest is oriented anticlockwise relative to the tidal current; in the opposite case the residual circulation is anticlockwise, if we assume that the bottom friction torque is large enough to overcome the effect of earth's rotation. This principle is illustrated in Fig. 4.28.

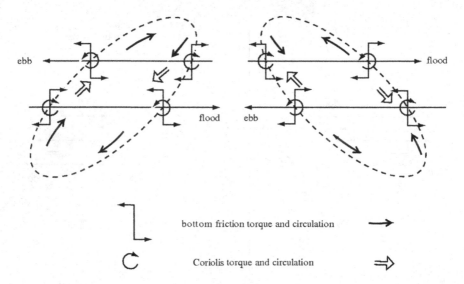

bottom friction torque and circulation

Coriolis torque and circulation

Fig. 4.28. Basic production mechanisms of residual circulation around a tidal sandbank in the presence of earth's rotation [990]. The dashed lines represent the tidal ridge depth contours, the long arrows flood or ebb flow. The frictional torques are indicated as in Fig. 4.27 and produced in the same way. The rotating arrows represent the Coriolis torques. For ridges with cyclonic orientation, frictional torques and Coriolis torques have the same direction; for ridges with anticyclonic orientation the direction is opposite. In the former situation the circulation induced by earth's rotation enhances the circulation induced by friction; in the latter case the circulation induced by friction is decreased by earth's rotation. Tidal ridges observed in the field have a cyclonic orientation relative to the main tidal flow axis; however, the inclination is sometimes quite small.

In addition to this residual circulation, a current along the depth contour with a quarter-diurnal (M_4) period is also initiated. The contribution of this higher harmonic component to ridge growth is smaller than the contribution of the residual circulation and will be ignored for simplicity.

Sediment transport towards the crest

As already mentioned, the secondary circulation around a tidal ridge produces spatial variations of net sand transport, with important consequences for ridge growth. This is illustrated by a simplified sediment transport formula, where gravity effects are ignored:

$$\vec{q}(x, y, t) = \alpha_b |\vec{u}|^2 \vec{u}. \tag{4.46}$$

We evaluate the tide-averaged sand transport $\langle q \rangle$ in a direction perpendicular to the sand ridge crest, which we will call the x-direction. The main tidal flow (u_{M2}, v_{M2}) is assumed to be semi-diurnal and ebb-flood symmetric. Interaction of this tidal flow with the sand ridge produces a residual flow component v_{M0} in y-direction, parallel to depth contour. If we consider topographic structures of small height compared to the water depth then the secondary flow components are an order of magnitude smaller than the main tidal flow. The first order contribution of the flow perturbation to the tide-averaged sand transport across the sand ridge is given by:

$$\langle q \rangle = 2\alpha_b \langle u_{M2} v_{M2} v_{M0} \rangle. \tag{4.47}$$

This expression shows that non-zero tide-averaged sediment transport across the ridge requires $u_{M2}, v_{M2} \neq 0$; the angle θ of flow incidence on the ridge must be different from 0 or $\pi/2$. For $0 < \theta < \pi/2$, the residual circulation v_{M0} at the upstream slope of the ridge is aligned with the ridge in downstream direction (v_{M2}, v_{M0} same sign); at the downstream slope the direction is reversed (v_{M2}, v_{M0} opposite sign). This occurs during both ebb and flood. From (4.47) it then follows that the residual sediment transport is always directed towards the ridge crest. Sediment will be deposited at the ridge crest and the crest height will increase. This means that there is a positive feedback between a linear topographic structure which is inclined relative to the flow axis, such as a sand ridge, and tide-induced sediment transport. Such a seabed perturbation will thus grow spontaneously as an instability of the seabed.

Symmetry of cross-ridge flow

An essential difference with the generation mechanism of dunes is the absence of asymmetry in the cross-ridge velocity. In the case of tidal ridges the growth process is not produced by net uphill flow, but by asymmetry between uphill and downhill sediment load. The reason is that the total flow strength (vectorial sum of unperturbed flow and sandbank circulation) is greater at the upstream ridge flank than at the downstream flank.

Influence of gravity

The foregoing suggests that tidal ridge type perturbations of any horizontal dimension (within the validity range of the depth-averaged flow model) will be amplified and thus may develop on the seabed. However, the growth rate is not the same for all tidal ridge dimensions. Depth gradients are larger for small-scale seabed structures than for large structures. Hence, the strongest vorticity and the strongest residual circulation are produced by seabed structures of small horizontal scale. However, small-scale seabed structures have steeper bottom slopes; due to gravity, sediment particles are carried by the flow more easily downhill than uphill. This applies to both bedload and suspended-load transport. The gravity effect implies that small-scale structures will grow less easily than larger scale structures; their height will remain small.

Influence of earth's rotation

Flow over large scale topographic structures is influenced by earth's rotation, through the principle of conservation of potential vorticity. When the flow approaches a sand ridge, water depth will decrease and therefore potential vorticity conservation requires that the flow is deflected in anticyclonic direction (to the right on the Northern hemisphere). If the sand ridge crest is oriented cyclonically relative to flow (either ebb flow or flood flow), then the flow will also deflect anticyclonically due to bottom friction: Friction and earth's rotation effects add up. However, if the sand ridge crest is oriented anticyclonically relative to the flow, then potential vorticity conservation and bottom friction torque counteract each other, so less vorticity is created. Thus strongest residual circulation is produced at a sand ridge oriented cyclonically relative to the flow direction, see Fig. 4.28. This is consistent with the observation that tidal ridges observed on the Northern Hemisphere are generally inclined to the left relative to the main flow direction [300, 664, 822], while on the Southern Hemisphere the inclination of tidal ridges is predominantly to the right relative to the main flow direction [661].

4.5.4. *Huthnance's Analytic Ridge Formation Model*

Flow response to a small bed undulation

An analytical derivation of the previous qualitative results was first made by Huthnance [437]. The essential steps of his flow-seabed interaction model are reproduced below. In this model we start with a flat sea bottom, at constant water depth h_0. Next, the sea bottom is disturbed with a small ridge-type undulation $z_b(x, t)$ with wavelength $\lambda = 2\pi/k$ in x-direction and amplitude $\hat{z}_b$ uniform in y-direction:

$$\Re z_b(x, t) = \hat{z}_b e^{\sigma_i t} \cos(kx - \sigma_r t) = \hat{z}_b \Re e^{i(kx - \sigma t)}. \tag{4.48}$$

Here $\sigma_r/k = \Re \sigma/k$ is the migration rate of the bottom perturbation and $\sigma_i = \Im \sigma$ is the growth rate (positive or negative). The depth-averaged undisturbed flow $\vec{u}_0 = (u_0, v_0) = U(\sin \theta, \cos \theta)$ makes an angle θ with the ridge-crest direction y, see Fig. 4.29. Bottom topography and fluid flow are uniform in y-direction; therefore the

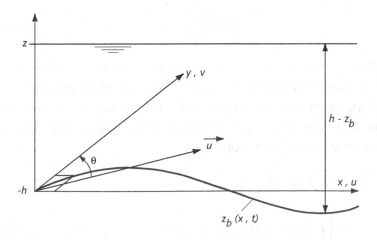

Fig. 4.29.   Definition of symbols and axes for fluid flow across a sequence of banks.

flow components in x and y-direction can be derived respectively from the continuity equation,

$$(h_0 - z_b)u = h_0 u_0 = h_0 U \sin \theta \qquad (4.49)$$

and the ridge-parallel momentum balance equation

$$v_t + u v_x + v v_y + f u + g \eta_y + c_D \frac{\sqrt{u^2 + v^2}}{h_0 - z_b} v = 0. \qquad (4.50)$$

The undisturbed flow obeys the equation

$$v_{0t} + f u_0 + g \eta_{0y} + c_D \frac{U v_0}{h_0} = 0, \qquad (4.51)$$

where $U = \sqrt{u_0^2 + v_0^2}$.

Infinitesimal steady flow perturbation

We will first consider the case of steady flow instead of tidal flow, i.e., u_0 and v_0 are constants and $\vec{u}_t = 0$. The bottom perturbation induces a small perturbation of the flow velocity $\vec{u}_1 = (u_1, v_1)$, which is on the order of ϵU, with $\epsilon = \hat{z}_b / h_0 \ll 1$. Development of the continuity equation (4.49) to order ϵ gives

$$u_1 = \frac{z_b}{h_0} U \sin \theta. \qquad (4.52)$$

The flow perturbation u_1 in cross-ridge direction is in phase with the perturbation z_b; this means that there is no upstream-downstream flow asymmetry in cross-ridge direction, as we already expected from the qualitative analysis. Development of the ridge-parallel momentum balance equation to order ϵ gives

$$u_0 v_{1x} + f u_1 + \frac{c_D}{h_0}$$

$$\times \left[U v_0 \frac{z_b}{h_0} + U v_1 + v_0 (u_1 \sin \theta + v_1 \cos \theta) \right] = 0. \quad (4.53)$$

The linearity of this equation implies that the along-ridge flow perturbation v_1 can be written in the form:

$$v_1 = -\chi \frac{z_b}{h_0} U = -|\chi| \frac{z_b}{h_0} U e^{-i\phi}, \tag{4.54}$$

where χ is a yet unknown complex function of x and t; ϕ is the spatial phase lag of the flow component v_1 relative to the bottom perturbation z_b. If $\phi = 0$ the alongshore flow perturbation is symmetric around the ridge crest; in that case there will be no sediment transport gradient at the ridge crest and no accretion or erosion will occur. If $\phi = \pi/2$, the alongshore flow perturbation is perfectly asymmetric around the ridge crest; the along-ridge flow perturbation has opposite signs at the upstream and downstream ridge flanks. This corresponds to a flow circulation around the ridge; it has been shown before that such a flow circulation causes asymmetry of sediment transport at the ridge crest leading to ridge growth or ridge decay.

Flow adaptation length scale

Substitution of (4.52 and 4.54) in (4.53) yields for $\chi(x, t)$ the equation

$$\chi \left[ikl \sin\theta + (1 + \cos^2\theta) \right] - \frac{fl}{U} \sin\theta - \cos\theta(1 + \sin^2\theta) = 0, \tag{4.55}$$

where the length scale l is given by

$$l = h/c_D. \tag{4.56}$$

The along-ridge velocity perturbation is found from substitution in (4.54),

$$v_1 = -\frac{\cos\theta(1 + \sin^2\theta + p\tan\theta)}{1 + \cos^2\theta} U \frac{z_b}{h_0} \cos\phi \, e^{-i\phi}, \tag{4.57}$$

where

$$p = \frac{fl}{U}, \quad \tan\phi = \frac{kl \sin\theta}{1 + \cos^2\theta}. \tag{4.58}$$

This velocity perturbation depends crucially on the length scale l. This length scale corresponds to the inertial adaptation lag of the

flow to the ridge-related bottom friction; this can be seen by comparing the scales of the inertial (third) and frictional (last) terms in the
momentum balance Eq. (4.50). As discussed earlier, it is precisely this
inertial adaptation lag that produces flow asymmetry relative to the
ridge crest. Maximum flow circulation around the ridge crest occurs
when the phase lag $\phi = \pi/2$. According to (4.58), this corresponds
to a bottom perturbation of infinitely small wavelength.

Convergence of the sediment flux

However, growth of ridges with very small wavelength will not occur
in reality, because the bottom slopes are too steep. This can be seen
by using a sediment transport formula which takes into account the
stabilising influence of gravity on bottom slope steepening (3.67),

$$\vec{q} = \alpha_b |\vec{u}|^n \left[\frac{\vec{u}}{|\vec{u}|} - \gamma \vec{\nabla} z_b \right]. \tag{4.59}$$

The critical velocity for sediment motion, u_{cr}, has been ignored for
simplicity, considering that significant sediment transport relates to
conditions of much stronger flow velocity, $|u| \gg u_{cr}$. The expression
(4.59) can be evaluated by substitution of $\vec{u} = \vec{u}_0 + \vec{u}_1$ with $\vec{u}_1$ given
by (4.52) and (4.57). Ridge growth is related to the sediment transport
gradient,

$$z_{bt} + \vec{\nabla} \cdot \vec{q} = 0. \tag{4.60}$$

We only need to consider the x-component of the sediment flux,
because there is no gradient in the y-direction. The growth rate is
given by

$$\Im\sigma = \Re\left[z_{bt}/z_b \right] = -\Re\left[\vec{\nabla} \cdot \vec{q}/z_b \right]. \tag{4.61}$$

To first order in ϵ we find

$$\vec{\nabla} \cdot \vec{q} = \alpha_b U^{n-1} \left[(\cos^2\theta + n\sin^2\theta) u_{1x} \right.$$
$$\left. + (n-1)\sin\theta\cos\theta\, v_{1x} - \gamma U z_{bxx} \right], \tag{4.62}$$

and after substituting (4.52), (4.57) and (4.58),

$$\Im\sigma = \alpha_b \gamma U^n \left(\frac{1+\cos^2\theta}{l\sin\theta}\right)^2 \left(\xi \sin^2\phi - \tan^2\phi\right), \qquad (4.63)$$

$$\text{with} \quad \xi = \frac{n-1}{\gamma c_D}\left(\frac{\sin\theta\cos\theta}{1+\cos^2\theta}\right)^2 (1+sin^2\theta + p\tan\theta).$$

As $c_D \ll 1$ and $\xi \gg 1$, growth rates are positive, except for ϕ close to $\pi/2$. According to (4.58), this implies that only very small wavelengths are completely suppressed by the bottom-slope term in the sediment transport formula.

Tidal ridge growth rate

The maximum growth rate $\Im\sigma$ is found for $\Im\sigma_\phi = 0$, or

$$\tan^2\phi = \xi^{1/2} - 1. \qquad (4.64)$$

This yields

$$\Im\sigma = \alpha_b U^n \frac{\gamma}{l^2}\left[\sqrt{\frac{n-1}{\gamma c_D}}\cos\theta\sqrt{1+\sin^2\theta + p\tan\theta}\right.$$

$$\left. - \frac{1+\cos^2\theta}{|\sin\theta|}\right]^2. \qquad (4.65)$$

Because $p > 0$ growth rates are larger for positive values of θ than for negative values, see Fig. 4.30. This implies that low ridges with cyclonic orientation grow faster than ridges with anticyclonic orientation, confirming what we already found.

Ridge inclination θ

The angle θ at which growth is maximum is between 35 and 40 degrees, taking the values: $n = 4$, Coriolis parameter $f \approx 10^{-4}\text{s}^{-1}$, depth $h \approx 20\,\text{m}$ and friction coefficient $c_D \approx 3.10^{-3}$, with only a weak dependance on these parameters. This ridge angles observed in tidal seas are generally somewhat smaller.

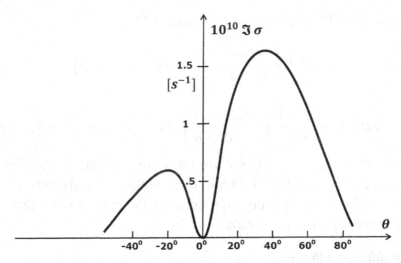

Fig. 4.30. Growth rate as function of ridge inclination θ relative to the flow axis, for $p = 1$ (mid latitude, northern hemisphere), friction coefficient $c_D = 0.003$ and sediment transport parameters $n = 4$ and $\gamma = 1$. The depth ($h = 20$ m) and velocity ($U = 0.5$ m/s) only influence the scale of the curve. The growth rate is positive because for each θ-value a particular wavelength is chosen which corresponds to maximum growth, according to (4.64). For positive angles, corresponding to cyclonic ridge orientation relative to the flow axis, the growth rate is larger than for negative angles.

For the inertial phase lag of the secondary ridge-parallel current v_1 relative to the bottom perturbation z_b, we find $\phi \approx 74°$; this implies that we are not very far from maximum flow asymmetry ($\phi = 90°$). At the ridge-crest v_1 is negative; the flow over the ridge is thus always deflected towards the cross-ridge direction.

Ridge wavelength

The wavelength λ is given by (4.58, 4.64)

$$\lambda = 2\pi/k = \frac{\sin\theta}{(1 + \cos^2\theta)\sqrt{\xi^{1/2} - 1}} \frac{2\pi h}{c_D}. \tag{4.66}$$

For $p = 1$ and $\theta = 36°$ we find $\lambda \approx 0.65h/c_D \approx 220h_0$. In water depth of 30 m the model predicts an initial tidal ridge wavelength of 6–7 kilometres. For wavelengths larger than the value of maximum growth, the growth rate (4.63) remains positive and decreases

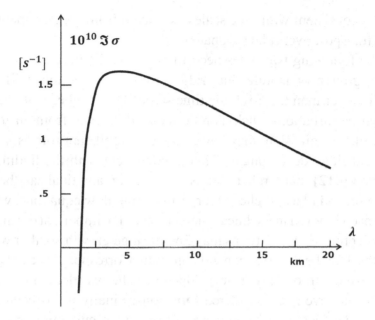

Fig. 4.31. Growth rate as function of the wavelength λ, for $p = 1$ (northern hemisphere), friction coefficient $c_D = 0.003$ and sediment transport parameters $n = 4$ and $\gamma = 1$. The depth ($h = 20$ m) and velocity ($U = 0.5$ m/s) only influence the scale of the curve. The maximum growth rate occurs at approximately 4 km, but for larger wavelengths the growth rate is not much smaller.

only slowly to zero, while for smaller wavelengths the growth rate becomes rapidly negative, see Fig. 4.31. This implies that perturbations with larger wavelength than (4.66) will also experience substantial growth. Therefore one may expect that in natural situations ridges may occur with wavelengths of the order of (4.66) or larger.

Fully developed tidal ridges

The foregoing analysis only applies to the initial phase of tidal ridge development. Field observations show that tidal ridges may grow quite high, attaining a large fraction of the water depth. The time scale $1/\Im\sigma$ for ridge growth follows from (4.65). For the previous example with $p = 1$, $h = 30$ m, $c_D = 0.003$, maximum tidal velocity of 1 m/s and sediment transport parameter $\alpha_b = 4.10^{-4}$ (3.68), we find a growth time scale (amplification by a factor e) of a few hundred years.

This is consistent with time scales estimated from measurements of sand transport over tidal sandbanks [921].

The following two notes need to be made, however. In the first place, growth of mature tidal ridges from a flat bottom will take much longer than the e-folding time scale. This implies that ancient seabed perturbations, such as relict coastal barriers from an initial stage of transgression, may have played a significant role as kernel for tidal ridge development. This is consistent with the finding by Houbolt [412] that a relict core is present in many tidal sandbanks. In the second place, higher-order terms in the flow equations, which have been ignored in the linear analysis, become important during the growth process. These terms may favour ridge growth at other wavelengths than the initial exponential growth. Correspondence between the wavelength of fully grown ridges and the wavelength of initial ridges may even be considered fortuitous. Finally it has been suggested that factors such as wave action, ridge curvature and tidal ellipticity [423] need to be taken into account.

Tidal flow

So far we have ignored the tidal nature of the flow. The time dependence of tidal flow can be included in the analysis by considering

$$\vec{u} = (u_0 I(t), v_0 I(t)),$$

with, for instance $I(t) = \cos \omega t$. The mathematical treatment is similar, but Eq. (4.55) now changes into

$$\frac{l}{U} \chi_t + \chi \left[iklI \sin \theta + |I|(1 + \cos^2 \theta) \right]$$

$$- \frac{fl}{U} I \sin \theta - I|I| \cos \theta (1 + \sin^2 \theta) = 0. \qquad (4.67)$$

An analytical solution can be found only in a few special cases, for instance, when representing the tide by a square-wave function,

$$I = 1 \quad \text{for} \quad 0 < t < T/2, \quad I = -1 \quad \text{for} \quad T/2 < t < T.$$

The results are similar to the case of stationary flow except for a correction factor multiplying ξ in Eq. (4.63). The correction factor is close to 1, especially if the ridge wavelength is much smaller than the tidal excursion, i.e., $h/c_D \ll UT/\pi$. The conclusion is, that in case of sufficient flow strength (U on the order of 1 m/s), the wavelength of ridges in tidal flow and in stationary flow depends linearly on water depth with approximately the same proportionality factor.

Comparison with observations

One of the best studied tidal ridge fields is situated in the Southern Bight of the North Sea at depths of typically 30 m, see Fig. 4.32.

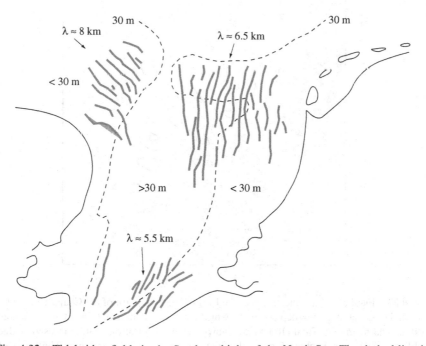

Fig. 4.32. Tidal ridge fields in the Southern bight of the North Sea. The dashed line is the 30 m depth contour. The height of the ridges is typically 10 m, but some ridges attain almost 30 m. The orientation is cyclonic relative to the main flow axis with an angle between 20° and 30°. The wavelength λ corresponds to the average spacing between the ridges in each ridge field. The ridge fields do not cover the entire Southern Bight; however, their location does not bear any obvious relationship with the distribution of seabed grain diameter, strength of tidal currents or wave action.

These ridges are rotated at an angle of 20°–30° in cyclonic direction relative to the main flow axis. The observed wavelengths of 6–8 km are close to the estimate of the linear stability analysis according to Huthnance's model.

The sand ridge field about 80 km north of the Changjiang River mouth (China) has a particular radial planform. This radial planform can be related to the local characteristics of the (mainly semi-diurnal) tidal wave [969], illustrated by the tidal ellipses in figure 4.33. Tidal ellipses are elongated in the northern sector and almost circular in the south-eastern sector. The sandbanks were formed in mid-Holocene

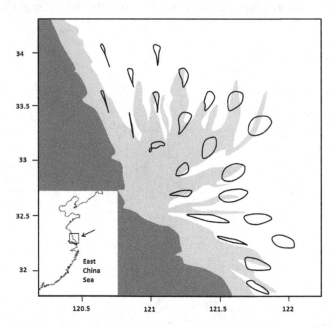

Fig. 4.33.   Field of radial sand ridges in the East China Sea, north of the Changjiang River mouth. Dark = land, grey = depth <10 m, white = depth >10 m. Coordinates are in degree latitude and longitude. Tidal ellipses are shown at several locations within the sand ridge field. The tidal range is about 5 m; tidal velocities are of the order of 1 m/s. The crests of the ridges situated around the centre of the field almost reach the water surface. The depth in the troughs is 10-20 m. The seabed sediment consists mainly of silt (grainsize less than 100 μm) in the north and fine sand in the south. The radial structure of the velocity field is produced by a locally generated tidal Poincaré wave, which is trapped to the coast. The radial sand ridge field is probably formed in interaction with this particular tidal velocity field. Figure redrawn after[969].

times by reworking of thick layers of fine sediments deposited by the Changjiang (Yangtze) and Huanghe (Yellow) rivers in Holocene and pre-Holocene eras.

In the northern sector the tidal current crosses the ridges at a small angle. Their formation can be explained by the previously discussed Zimmerman-Huthnance mechanism [971]. The formation mechanism of the ridges in the southern sector is less clear. The ridges are less elongated and crossed by shallow channels; this may be related to the rotational character of the tidal currents. It has been suggested that the orientation of the ridges in the central and southern part of the sandbank field is influenced by relict incised fluvial channels [930]. The ridge orientation could also be influenced by the slope of the inner shelf, as discussed in the next section.

Wave influence

In periods of high waves a fully developed sandbank will regress; some 20% of the bank volume may be lost [503], while the bank is restored during calm weather conditions. These observations indicate that waves play an important role in controlling the final bank size. However, wave-induced sand suspension does not only affect bank growth in a negative sense. Sediment transport by tidal currents is substantially increased by high wind waves [921]. This suggests that sandbank dynamics is underestimated both at the constructional and the regressional stages, if only tidal currents are considered.

4.5.5. *Shoreface-Connected Ridges*

Occurrence

Figure 4.25 shows that the bottom of the North Sea is undulated, with large ridges not only far offshore but also near to the shore, on the inner shelf. Similar nearshore ridges are observed elsewhere along the North Sea coast [27] and in other shelf seas, for instance, the Atlantic coasts of North America [821, 822] and South America [301, 661] and the coasts of the Yellow Sea and East China Sea [533].

The dimensions of the nearshore and offshore ridges are similar; the nearshore ridges have a wavelength of a few kilometres up to 10 kilometres and their height is in the order of ten metres. They extend from the shoreface in offshore direction over the inner shelf, in water depths of 10 to 30 metres and their slope is not much larger than the inner shelf slope. However, a marked difference with the offshore tidal ridges is the orientation of these ridges. Nearshore ridges have generally no anticyclonic orientation relative to the main flow; they always bend to the coast with an angle of 30°–50° and they extend up till the shoreface [760]; for that reason they are called shoreface-connected ridges.

Instability of the seabed slope

In spite of the apparent similarity of tidal ridges and shoreface-connected ridges they are thought to be formed by a completely different mechanism. Tidal flow, bottom friction and gravity-induced downslope transport, which are essential factors for tidal ridges, only play a minor role for the dynamics of shoreface-connected ridges. In contrast to most other seabed structures, the delayed flow response to gradients in bottom friction is not the primary instability factor. On the inner shelf, seabed instability is related to the average cross-shore slope. The presence of a seabed slope induces flow asymmetry between the upstream and downstream flanks of an oblique seabed perturbation and a cross-ridge transport gradient at the perturbation-crest. The principle is illustrated in Fig. 4.34. A qualitative description of this generation mechanism is given below.

Feedback mechanism

We assume a stationary coast-parallel flow, with the coast on the right-hand side. The flow will be deflected by any ridge-type perturbation of the seabed, provided the crest line makes an oblique angle with the flow. A deflection in cross-ridge direction is produced by bottom friction torques, but in the absence of bottom friction a similar deflection results from continuity and vorticity conservation. Continuity

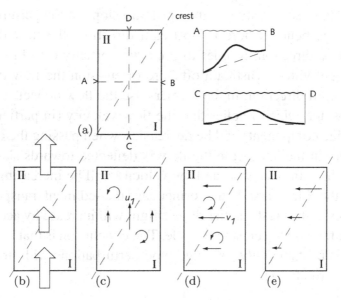

Fig. 4.34. Development of shoreface-connected ridges as an instability inherent to a sloping shoreface with longshore flow. We consider a rectangular domain situated around the ridge crest (a), which divides the domain in an upstream part I at the right lower half, and a downstream part II at the upper left half. The right boundary of the domain is parallel to the shoreline (x-direction); without the ridge, the depth increases linearly from right to left (y-direction). The unperturbed longshore flow enters at the bottom and leaves at the top (b); if the flow strength is not altered by the ridge, the water flux (open arrows) over the ridge is smaller than the incoming and outgoing water fluxes. Continuity of the water flux can be restored by an additional longshore velocity component u_1 at the ridge crest (c). However, this additional velocity component corresponds to a change of vorticity over the crest, because continuity requires $u_{1y} \neq 0$. This change of vorticity cannot be compensated by friction-induced vorticity, because it does not have the right sign. Continuity of the longshore water flux over the ridge can only be achieved by an offshore flow component v_1 at the ridge crest (d). Vorticity is conserved if the longshore variation v_{1x} is equal to u_{1y}; this requires that the longshore velocity perturbation u_1 is small compared to the cross-shore perturbation v_1. The offshore water flux over the ridge should be approximately the same in the shallow nearshore zone as in the deep offshore zone. This implies that the offshore velocity has to decrease in offshore direction. As a result, the flow will be decelerating at the ridge crest (e); this flow deceleration entails convergence of the sediment flux at the ridge crest. Hence, sediment is deposited at the crest and the ridge-perturbation will grow.

requires flow acceleration at the upstream slope of the perturbation. For a ridge bent offshore in upstream direction, the flow deflects in offshore direction in order to preserve vorticity (see Fig. 4.34); for a ridge with downstream offshore orientation the flow deflects in coastward direction. In the former case the flow deflects towards deeper water, which implies that the flow velocity (in particular its downridge component) will be decreasing when passing the crest of the ridge. In the latter case the flow is deflected towards shallower water, which implies that the flow velocity will be increasing when passing the ridge-crest. In the former case the sediment transport rate decreases at the crest and the ridge height will increase by accretion. In the latter case the crest will erode. The conclusion is that a sloping seabed is unstable only for ridge-type perturbations with upstream offshore orientation.

Unidirectional storm-driven flow

We will now consider an initial ridge-perturbation inclined offshore towards the flood direction. In this case the flood flow will be deflected seaward over the ridge; during flood the perturbation will grow. The ebb-flow, however, will be deflected landward over the same ridge-perturbation; during ebb the perturbation will decay. There-fore shoreface-connected ridges will not develop under symmetric tidal conditions. They can only be formed if one flow direction along the coast is dominant over the other. Along the Dutch coast northward flow dominates over southward flow, mainly because the wind climate is dominated by southwesterly winds. This may explain why south-west running shoreface-connected ridges have developed along the Dutch coast (see Fig. 4.35) as predicted by the previously described instability mechanism. We may expect that shoreface-connected ridges are best developed on coasts where storm winds coming from a preferred oblique direction are the major cause for longshore sediment transport. Such storm-dominated coasts are, for instance, the Atlantic coast of North and South America; along these coasts shoreface-connected ridges are frequent and well developed.

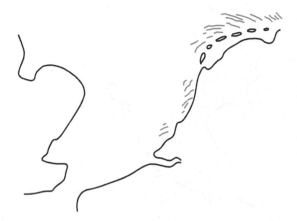

Fig. 4.35. Shoreface-connected ridges along the Dutch coast. The ridges occur in a zone where storm-driven net sediment transport is greater than tide-induced net sediment transport, see Fig. 4.36.

Along the North Sea coast, tidal influence is stronger than along the American Atlantic coast, see Fig. 4.36. This may explain why shoreface-connected ridges are less prominent features (lower height, sometimes absent) along the North Sea coast. The stratigraphy of shoreface-connected ridges along the Dutch coast reveals not only storm deposits, but also tidal deposits [589]. A tidally induced growth mechanism might contribute to the growth of these ridges, even if the ridge inclination relative to the flow axis is opposite to the ridge inclination which is favoured by earth's rotation, see Fig. 4.30.

Sediment distribution

As discussed above, shoreface-connected ridges may develop as a result of unidirectional flow along the slope of the inner shelf. This flow is responsible not only for ridge-growth, but also for ridge-migration. Observed migration rates are on the order of only a few (generally less than 10) metres per year. The downstream slope of the ridges is steeper than the upstream slope, which is a normally observed feature of migrating bedforms. Observations also show that at the downstream slope the seabed sediment is finer than at the upstream slope [823]. This is consistent with ridge migration by storm

Fig. 4.36. Southern bight of the North Sea and the zone (shaded) where net tide-induced sediment transport is greater than storm-induced net sediment transport [881]. Along most of the Dutch North Sea coast and along the coast of the Wadden Sea storm events contribute more to net sediment transport than tides.

currents. During these conditions fine sediments are suspended more easily than coarse sediments. When the flow is accelerated on the upstream ridge-flank it will suspend a relatively larger fraction of fine sediments than coarse sediments. This produces a coarsening of the upstream flank of the ridge. When the flow decelerates at the downstream flank, part of the previously suspended load will settle. As this load contains relatively more fine than coarse material, the seabed sediment on the downstream flank of the ridge (which is seaward oriented) will be finer than on the upstream flank.

Initial formation

Instability of the inner-shelf slope is not the only explanation for the occurrence of shoreface-connected ridges. Other phenomena may play a role as well. Shoreface-connected ridges are often found at locations where, during the Holocene transgression, the coast has received substantial sediment input from rivers or from erosion of Pleistocene deposits. In these regions coastal barriers have probably developed, which may not have kept pace with sea level rise. Remnants of these coastal barriers on the inner shelf might form the origin of the present shoreface-connected ridges [64,412]. There are also indications that shoreface-connected ridges are related to former inlets. Abandoned ebb-tidal deltas along a retreating coast may have provided the sediment source for these ridges [583]. These hypotheses do not necessarily conflict with the theory of formation through instability of the inner shelf-slope. Relict seabed structures might have acted as an initial seabed perturbation which has been remodelled to a shoreface-ridge pattern by the described earlier instability mechanism. Without this morphodynamic feedback relict seabed structures would probably not have persisted.

4.5.6. *Trowbridge's Model for Shoreface-Connected Ridges*

Flow response to an infinitesimal bed undulation

The first model of seabed instability leading to shoreface-connected ridges was proposed by J. H. Trowbridge [856]. Using a similar approach we will show that shoreface-connected ridges can be generated as an instability of a sloping seabed in the case of uniform frictionless stationary longshore flow. The present model is greatly simplified and far from realistic; many processes influencing the dynamics of ridge generation are left out. The model only aims to illustrate the salient features of pattern generation on a sloping seabed. A justification of several simplifying assumptions, such as ignoring bottom friction, is given by Trowbridge [856] and Calvete [126]. The symbol conventions are shown in Fig. 4.37. The x-axis is in longshore direction, the y-axis points in offshore direction. We will consider a uniform basic flow u_0 in longshore direction and a sinusoidal bottom perturbation z_b parameterized as

$$\Re z_b(x, y, t) = \hat{z}_b \Re e^{i(kx \sin\theta + ky \cos\theta - \sigma t)}. \tag{4.68}$$

Here k is the wavenumber of the perturbation and θ is the angle between the crest line and the x-axis. The time evolution depends on the complex number σ with $\Re\sigma/k$ being the migration rate and

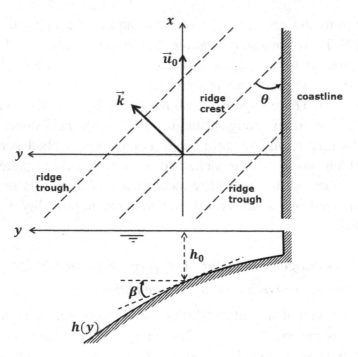

Fig. 4.37. Definition of symbols and axes for longshore flow over a ridge with oblique orientation to the shoreline. Upper panel plan view and lower panel vertical cross-shore section.

$\Im\sigma$ the exponent for growth or decay. We will consider an emerging ridge for which $\epsilon \equiv \hat{z}_b/h_0 \ll 1$.

Concave equilibrium shore profile

It will be assumed that the depth h only depends on the cross-shore coordinate y and that it can be approximated by an exponential function

$$h(y) = h_0 \exp(\beta y/h_0). \qquad (4.69)$$

This expression represents an upward convex depth profile, while in reality the shoreface profile is more often concave. If the width of the shoreface h_0/β is much greater than the wavelength $\lambda = 2\pi/k$, we may consider $h(y)$ as a fair approximation of a locally linear depth increase. It will be assumed that the cross-shore depth profile (4.69) is in equilibrium and that the processes responsible for maintaining this equilibrium are not dependent on the flow perturbations caused by the emerging ridge (4.68). For the inner shelf slope this is a plausible assumption; for the lateral slopes of a flow channel it is probably not. Bank-connected ridges in flow channels do exist and are designated as alternating bars or point bars. However, these bars are thought to arise from a completely different type of current-topography interaction, as will be discussed in the next chapter.

Frictionless flow perturbation

The depth-averaged flow velocity $\vec{u}$ is to first order in ϵ given by

$$\vec{u} = (u_0 + \epsilon u_1, \epsilon v_1), \qquad (4.70)$$

where $\vec{u}_0$ is the unperturbed longshore flow, and $\epsilon \vec{u}_1$ the flow perturbation due to the emerging ridge. We ignore bottom friction and earth-rotation effects. In this case there are no vorticity producing terms in the momentum balance equation. Together with the assumed uniformity of the basic flow (u_0 independent of x and y), the vorticity is given by

$$\zeta = v_{1x} - u_{1y} = 0. \qquad (4.71)$$

This equation is solved by defining the flow potential Φ such that

$$\Phi_x = u_1, \quad \Phi_y = v_1. \qquad (4.72)$$

The continuity equation reads

$$\vec{\nabla} \cdot ((h - z_b)\,\vec{u}) = \vec{u} \cdot \vec{\nabla}(h - z_b) + (h - z_b)\vec{\nabla} \cdot \vec{u} = 0. \qquad (4.73)$$

To first order in ϵ this gives

$$h(u_{1x} + v_{1y}) + h_y v_1 = u_0 z_{bx},$$

and after substitution of (4.72),

$$h_0(\Phi_{xx} + \Phi_{yy}) + \beta \Phi_y = iku_0 z_b \sin\theta e^{-\beta y/h_0}. \qquad (4.74)$$

The solution has the form $\Phi = \chi z_b \exp(-\beta y/h_0)$ and the complex number χ is found by substitution in (4.74), yielding

$$\Phi = -u_0 z_b \sin\theta \frac{\beta\cos\theta + ikh_0}{(kh_0)^2 + \beta^2\cos^2\theta} e^{-\beta y/h_0}. \qquad (4.75)$$

Sediment flux convergence

We evaluate the erosion-sedimentation pattern resulting from sediment flux gradients over the perturbed seabed from

$$(z_b)_t + \vec{\nabla}\cdot\vec{q} = 0\,, \qquad (4.76)$$

where the seabed porosity factor has been incorporated in the sediment flux $\vec{q}$. From (4.68) it follows that the (positive or negative) growth rate of the perturbation, $\Im\sigma$ is given by

$$\Im\sigma = \Re((z_b)_t/z_b) = -\Re(\vec{\nabla}\cdot\vec{q}/z_b). \qquad (4.77)$$

For reasons explained below, the sediment transport $\vec{q}$ is written as

$$\vec{q} = C\,\vec{u}, \qquad (4.78)$$

where C and $\vec{u}$ represent the volumetric sediment load (load divided by sediment density) and the average current velocity in a near-bottom layer in which the bulk of the sediment flux is concentrated. Gravity-induced down-slope transport is ignored.

Wave-induced sediment suspension

As shoreface-connected ridges are observed almost exclusively on storm-dominated shelves, it is assumed that they are generated on the inner shelf during storm conditions. Under such conditions sediment concentrations near the bottom depend more on wave stirring of the seabed than on the mean current velocity u_0. Because the wave height and wave orbital velocities depend on water depth, we may expect that C is also a function of depth, $C = C(h)$. The bed perturbation also influences the sediment concentration through water depth, but for emerging ridges this is a higher order effect that will be neglected.

Growth rate

The expression of the sediment flux gradient can be simplified by using (4.73):

$$\vec{\nabla} \cdot \vec{u} C(h) = (h - z_b)\vec{u} \cdot \vec{\nabla} \frac{C(h)}{h - z_b}. \qquad (4.79)$$

To first order in ϵ the result is

$$\vec{\nabla} \cdot \vec{u} C(h) = u_0 \frac{C}{h}(z_b)_x + h v_1 \left(\frac{C}{h}\right)_y$$

$$= u_0 \frac{C}{h}(z_b)_x - \Phi_y \frac{h_y}{h}(C - hC_h). \qquad (4.80)$$

Using (4.77), (4.78) and (4.75) we find for the growth rate

$$\Im \sigma = \beta \frac{u_0}{h_0^2} \cos \theta \sin \theta \frac{(kh_0)^2 + \beta^2}{(kh_0)^2 + \beta^2 \cos^2 \theta}(C - h_0 C_h). \qquad (4.81)$$

This expression is evaluated at an offshore distance y corresponding to $h(y) = h_0$. The derivative $C_h = \partial C/\partial h$ is negative, because wave stirring decreases with increasing depth; the factor $(C - h_0 C_h)$ is thus positive and could be approximated by the sediment load close to the breaker line. The growth rate is positive for positive values of the angle θ, i.e., for ridge perturbations with an upstream offshore oriented crest line. This is what we already expected from the qualitative description of ridge growth. We may conclude that a sloping seabed is unstable if it is subject to a slope-parallel current and wave-stirring.

Ridge inclination

The angle θ for which the growth rate is fastest corresponds to vanishing of the derivative of (4.75) with respect to θ. This yields

$$\cos \theta = \left(2 + (\beta/kh_0)^2\right)^{-1/2}. \qquad (4.82)$$

For ridges with a wavelength which is on the order of the width of the inner shelf ($k \approx \beta/h_0$) we find $\theta \approx 45°$. The angle can be larger if the wavelength is much larger than the inner shelf.

Growth and migration rates

From (4.81) a rough estimate can be derived of the growth rate (e-folding amplification time scale) of emerging shoreface-connected ridges. A typical value of the sediment load under storm conditions (wave height 4 m, wave period 10 s) is on the order of $C \approx 10^{-4}$ m [796]; we suppose that these conditions occur on average 20 days a year. We further assume that under these conditions the sediment load is inversely proportional to the square of depth and that $u_0 \approx 0.5$ m/s. We take the depth $h_0 = 15$ m and the slope $\beta = 10^{-3}$ and consider a ridge wavelength of the same order as the width of the inner shelf. The time scale for ridge growth $1/\Im\sigma$ is then found on the order of a few thousand years. The migration rate $c_{ridge} = \Re\sigma/k \approx \Im(u_0(z_b)_x C/z_b kh) \approx u_0 C \sin\theta/h$ is found to be on the order of 5 m per year. Shore-face connected ridges are very slowly evolving structures according to the model and this is consistent with observations.

Ridge wavelength

The expression (4.81) shows that the growth rate decreases with increasing wavenumber k. From this we might conclude that the growth rate is strongest for perturbations with infinite wavelength. Observations show that shoreface-connected ridges have a finite wavelength, which is on the order of the width of the sloping inner shelf (5–10 km). The simple model we have presented fails to reproduce this wavelength as fastest growing instability mode. This is due to the choice of the convex depth profile (4.69); this choice is physically less realistic but simplifies the model by producing linear equations. For small wavelengths only the local depth dependence counts, so the choice of a convex or concave profile does not make a great difference. For large wavelengths, comparable to the width of the shoreface, the difference is crucial. The maximum growth rate of the perturbation corresponds to the situation with greatest offshore flow deceleration. For a convex profile the depth continues increasing in offshore direction. As the offshore deflection of the current increases with wavelength, the flow deceleration will be strongest for very large wavelengths. In reality the width of the sloping inner shelf is limited; the offshore flow deceleration is restricted to this limited width. The ridges with the fastest growth rate then correspond to a wavelength which is comparable to the width of the inner shelf [126]. A finite-amplitude analysis with more refined models, including bottom friction and earth-rotation effects, gives predictions not only of the wavelength, but also of the growth and migration rates and of the final amplitude of shoreface-connected ridges, which are consistent with observations [127]. This analysis also shows that the ridge wavelength and migration rate are not strongly influenced by finite-amplitude effects. These results further support the hypothesis that surface-connected ridges can be generated as free instabilities of the sloping inner shelf bed under the action of storm-driven flow.

4.6. Summary and Conclusions

The often ephemeral appearance of sedimentary bedforms in natural flows has long been a mystery. This is probably so because understanding their origin was not a matter of high urgency. This changed by the increasing exploitation of mineral resources. Exploration of the underground required interpretation of relict bedforms in bottom cores, for information on the eventual presence of mineral deposits. Insight in the formation and behaviour of bedforms became also an issue for the installation of pipes and cables on the seabed. Finally it was realised that by better understanding the interaction of the sea with its substrate, we can define more effective measures for protecting the environment and for safeguarding coastal settlements.

Sedimentary seabeds host a great diversity of natural bedfoms, from small (ripples) to very large (tidal ridges). These bedforms arise from patterns in the seabed forcing by currents and waves; these patters in turn arise from the flow disturbance generated by the bedforms themselves. Spontaneous emergence is a consequence of this mutual interaction. During the past 50 years much research has been dedicated to this interaction. This research has revealed several mechanisms through which bedforms can arise. A logical classification of bedforms is based on the type of generation mechanism.

In this chapter we have discussed four mechanisms for the emergence of bedforms. The corresponding bedform types are: Dunes, tidal ridges, shoreface-connected ridges and sorted bedforms.

We have focused on the question: How do these different bedforms arise and under which conditions? Linear stability analysis is a powerful tool for answering this question. Other tools are simulations based on cellular automata techniques and simulations with process-based numerical models. Each of these tools has its advantages and disadvantages; a combined approach is therefore necessary. In all cases, observations in the field and in laboratory settings are crucial for improving the models and for testing the reliability of the results.

The second question is: How do bedforms evolve after inital emergence? This question is more complicated than the first one. Linear stability analysis cannot answer this question. This is because after initial formation, the strength of the interaction between bedforms and flow increases; several new interaction processes start overruling the mechanism of initial formation. Moreover, the natural conditions under which initial formation took place, usually do not remain stable in the course of time. Often there is even not a single well-defined stage of final development. Therefore we have to rely on the cellular automata tool or on process-based numerical models. In this chapter, the second question has been dealt with in less detail than the first question.

The first bedform type includes dunes and ripples, because it appears that they can emerge from the same type of interaction. The most generally accepted formation mechanism is related to the inertial delay in the vertical flow structure around a bed perturbation. Perturbations of a flat uniform sediment bed may arise by the impact of turbulent eddies. When inertial delay effects induced by the bed perturbation are limited to the logarithmic boundary layer, the resulting bedform is a ripple. Under certain conditions ripples can grow; they become dunes when they start affecting the flow structure over the whole water column. This limits further growth; the wavelength of dunes is related to the water depth. The wavelength of ripples at the initial stage of formation is related to the sediment grain size. Wave-orbital ripples might be considered a separate category, because their wavelength depends on the wave-orbital excursion.

Spontaneous emergence of ripples and dunes can be simulated in mathematical models. The model results depend on the formulation of sediment transport, which relies on empirical knowledge. Predicting the spatial distribution of these bedforms still poses problems; for instance, the occurrence of dunes on the continental shelf is less widespread than predicted by the models.

There are other complications. Dunes are often covered with ripples, affecting the interaction with the flow. Intermediate structures exist, called megaripples, which may represent intermediate compound development stages. But they could also depend on other formation mechanisms. One may conclude that the formation mechanism of ripples and dunes is not yet fully elucidated.

Greater consensus exists regarding the formation of tidal ridges on tide-dominated continental shelves. These are very large linear bedforms, which make a small angle to the dominant tidal flow direction. When crossing the ridge, the tidal flow is locally deviated. The spatial lag of the flow deviation relative to the ridge crest generates a residual circulating flow around the ridge. The modified flow pattern induces convergence of the sediment flux at the ridge crest at the initial stage of formation. This flow pattern is influenced by earth rotation; for this reason the ridge inclination relative to the tidal flow direction is opposite in the northern and southern hemispheres. So far there is no field evidence of tidal ridges conflicting with the theory.

The third category are shoreface-connected ridges, which occur on the inner shelf of storm-dominated coasts. They resemble tidal ridges by their great length and their inclination relative to the dominant current direction. However, their existence is due to a different feedback mechanism. This feedback is related to the onshore upward slope of the inner shelf, close to the coast. The deviation of a storm-driven flow crossing the ridge modifies the vorticity, due to the cross-shore bed slope. Vorticity conservation induces an additional flow component, which promotes sediment accumulation at the ridge crest, if the ridge is orientated onshore in the downstream direction. The appellation 'shoreface-connected ridge' refers to this onshore orientation. The core of shoreface-conected ridges often contains relics of ancient shorelines. Again, there is no field evidence contradicting this formation mechanism.

The last category of bedforms dealt with in this chapter are the so-called 'sorted bedforms'. These bedforms are not primarily topographic structures, but patterns in sediment distribution. They are observed in nearshore zones where sediments of different grainsize coexist. In this case the feedback mechanism is related to the influence of grainsize on near-bottom turbulence levels. High near-bottom turbulence in zones where the seabed is predominantly composed of coarse-grained sediments inhibits settling of fine-grained sediments. The influence of grainsize on near-bottom turbulence levels thus promotes segregation of fine- end coarse-grained sediments. This generates spatial patterns of alternating coarse-grained and fine-grained sediment patches on the seafloor.

The spontaneous emergence of bedforms is termed 'self-organisation'. A characteristic of self-organised bedforms is their rhythmic pattern with a particular wavelength. This wavelength is not imposed by pre-existing geographic constraints or by pre-existing spatial scales in current and wave fields. They result from the feedback dynamics involved in their generation processes. The observed wavelengths of rhythmic bedform patterns in their initial stage corresponds to the wavelength for which the amplification by feedback processes is strongest.

Confidence in the theoretical explanation of the generation mechanism of bedforms is mainly based on the correct prediction of the order of magnitude of these wavelengths. It appears that the wavelength of maximum growth is often conditioned by gravity-induced down-slope transport. There is still considerable uncertainty about the formulation and magnitude of this counteracting transport mechanism, which limits the predictive capability of bedform-generation theories.

Rhythmic bedform patterns often display long crest lines. Recent research has greatly improved understanding of the observed regularity in crest lines. It appears that irregularities in crest lines (called 'defects') tend to disappear during the evolution of bedform fields after their initial formation. Simulation models based on the cellular automata approach show that defects migrate at different speeds within the bedform field, due to size differences and shadow effects. These differences in migration speed promote coalescence of bedforms where defects occur. This also provides an explanation for the observed increase in wavelength of bedform fields during their evolution, a phenomenon termed 'pattern coarsening'.

Increase of the computational power of numerical simulation models has greatly advanced our understanding of bedform dynamics during the past decade. However, our understanding is nor yet complete and in certain situations existing models still fail. Further progress is hampered in particular by shortcomings in the formulation of sediment transport. Improving our understanding of sediment transport processes is therefore an important topic of future research.

Chapter 5

Current-Channel Interaction:

Alternating Bars, Channel Meanders
and Tidal Flats

5.1. Introduction

The morphodynamics of tidal lagoons, estuaries and rivers differs from seabed morphodynamics by the presence of lateral flow boundaries. The generation of small bedforms, like ripples or dunes, is not substantially influenced by lateral flow constraints. These bedforms are essentially related to the vertical flow structure and are present both in bounded and unbounded flow systems. In this chapter, we focus on large bedforms related to lateral flow boundaries, which do not occur in the open sea.

Large-scale bedforms in laterally bounded flow will be called 'bars'. The morphology of these bars is different from the marine sand ridges and their generation mechanism is also different. Bars in rivers and tidal basins can emerge as free instabilities, but their further development is strongly influenced by topographic constraints imposed on the flow. These topographic constraints are different from place to place; hence, the bar patterns observed in bounded flows are less regular than the ridge patterns in the sea.

Similar as in the sea, small and large scale bedforms coexist; ripples, megaripples and dunes are often superimposed on each other and on the much larger bars. The interaction of flow and sand transport patterns related to these bedforms results in complex morphodynamics.

The basic processes responsible for bar formation in rivers and tidal channels are similar. In tidal lagoons and estuaries several additional phenomena influence the current-topography interaction, such as tidal variation of water levels and currents, wave action, density gradients, cohesive sediments and intensive biotic activity. We therefore discuss first the river case, which is conceptually simpler, and afterwards the tidal case.

The development of bars in rivers and tidal channels induces flow meandering. In this scenario, channel meandering is initiated by bar development. However, flow meandering can also arise directly from instability of the channel banks, by a feedback process leading to amplification of a small initial channel bank displacement. In this other scenario, bar formation is forced by the development of channel meanders. Both scenarios will be discussed.

The last sections of this chapter deal with the development of tidal flats and marshes. Their initial development is often related to the formation of channel bars. It will be shown that later stages of their development are strongly influenced by waves and biotic activity.

5.2. Morphology of Rivers and Tidal Channels

5.2.1. *Bedforms in Rivers*

Alternating bars

The most prominent manifestation of current-topography interaction in rivers is the presence of alternating bars and channel meanders. Alternating bars are associated with flow meandering; they are situated at the inner meander band. The development of these bars and the development of flow meanders are related processes. The bars are alternately located at opposite channel banks, forming a sequence of flow meanders, see Fig. 5.1. Alternating bars are accretional features; opposite to the bar, the channel is scoured and cuts into the outer channel bank.

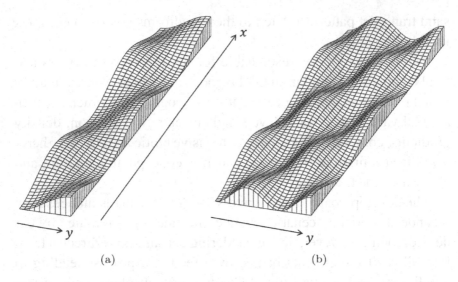

Fig. 5.1. Left panel (a): Schematic representation $z_b(x, y) \propto \cos(\pi y/b) \cos(kx)$ of alternating bars in a straight channel (width $= b$). The flow meanders between the bars; the channel depth is maximum opposite to the bars; the depth is minimum halfway the bars. Right panel (b): Schematic representation $z_b(x, y) \propto \cos(3\pi y/b) \cos(kx)$ of multiple-row bars in a straight channel.

Alternating bars at the inner bend of river meanders are called 'point bars'. Because the flow is always in the same direction (contrary to the tidal case), they have a crescentic elongate shape. In topographically constrained rivers, alternating bars are not necessarily associated with river meanders.

Observations suggest that the wavelength λ (spacing between successive bars at the same channel side) of alternating bars depends on the channel width b. The empirical relationship

$$\lambda = 6b \tag{5.1}$$

provides a reasonable estimate of observed bar wavelengths, but the scatter in the observed lengths is large ($\lambda/b = 2-10$) [197, 563]. Other empirical expressions also involve the channel depth h, the longitudinal bed slope and the bankfull discharge. Alternating bars are observed only if the aspect ratio of the river channel b/h is sufficiently large, i.e., greater than 10 [153].

Braided rivers with multiple bars

Wide rivers display often a more complex shoal pattern than a single row of alternating bars. Several meandering channel branches may coexist, separated by multiple rows of shoals. Rivers divided over several streams are called 'braided rivers'. Braided rivers are characterised by high Froude numbers during periods of morphodynamic activity. Bars separating different streams in braiding rivers are called 'multiple-row bars' or 'en-echelon braid bars', see Fig. 5.1.

The multiple row bars are typically more elongate than alternating bars, and the slopes are steeper. Levee deposits — rapid deposition of the coarsest sediments when the flow expands over the river bank — may contribute to their development. Observations show that rivers with a high aspect ratio b/h (typically greater than 100 [660, 973]) divide over different streams. These different streams form an intricate network of channels and shoals. The development of such a network can be initiated as an instability of the interaction between flow and bed topography. The mechanism is thought to be similar to that for a single row of alternating bars.

5.2.2. Bedforms in Tidal Channels

Flood and ebb channels

Alternating bars also occur in tidal channels, where they have often developed into tidal flats. The characteristic morphology of tidal basins consists of a meandering channel system with tidal flats situated at the inner channel bends. This is illustrated in Fig. 5.2 for the Western Scheldt and Eastern Scheldt tidal basins in the Netherlands. The first comprehensive explanation of tidal basin morphology was given by Van Veen in 1950 [916]; the headlines are reproduced below.

The bending of flood and ebb flows around tidal flats is counteracted by inertia; the flow tends to follow its original direction. The curvature of the flood streamline is shifted in flood direction and the opposite happens for the ebb streamline. Flood flow and ebb flow thus tend to scour different channels, as shown in Fig. 5.3. At some

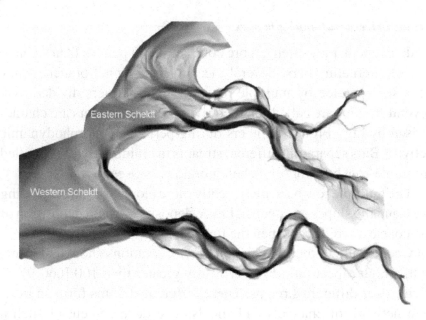

Fig. 5.2.   Tidal basins of the Rhine–Meuse–Scheldt Delta: The Western Scheldt (tidal range 4.5 m; length of the estuarine part 90 km) and Eastern Scheldt (tidal range 3.5 m; basin length 40 km). The tidal flow in these wide basins has shaped a pattern of interlaced flood-dominated and ebb-dominated channels. The Eastern Scheldt was greatly affected by the construction of a storm surge barrier at the inlet and compartment dams at the landward end in 1986. Darkness indicates depth.

places the flood and ebb channels are separated by a shoal, whereas at other places they are concentrated at opposite sides of the same channel. The sediment transport is on average directed landward in the flood channels and seaward in the ebb channels. The flood and ebb streamlines intersect at inflexion points between tidal flats. At these locations the net sediment transport is minimum. At places where flood and ebb streamlines intersect, sediment is deposited and a sill is formed. These sills constitute a hindrance to navigation. In estuaries and tidal basins with an important navigation function, dredging of sills is usual practice. Dredging has to be repeated frequently, because the natural accretion of sills is an ongoing process.

The inertial influence on the bending of the streamlines is stronger for the flood flow than for the ebb flow. The reason is that the tidal

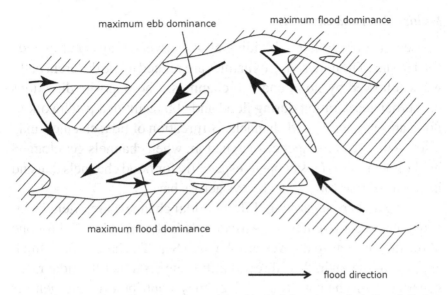

Fig. 5.3. Meandering tidal channel network with tidal flats (grey shaded) and distinct flood and ebb-dominant channels. The streamlines of the flood flow and the ebb flow do not coincide. The location of maximum flood flow at the outer channel bends is shifted landward by inertia and the location of maximum ebb flow is shifted seaward. Part of the flood flow spills over the channel bend and rushes into the adjacent tidal flat; a small flood channel is scoured by which the tidal flat is fed with sediment.

elevation is lower at the phase of maximum ebb flow than at the phase of maximum flood flow. The ebb flow is therefore more constrained by the channel bathymetry than the flood flow. The flood flow can more easily spill over the channel bend and run into the the tidal flat. This effect is enhanced by the fact that the water level on the tidal flat has not yet reached the level of the incoming flood wave. Small flood channels ('flood barbs') are scoured by the inrush of (part of) the flood flow into the tidal flat. Such flood channels running from the outer channel bend into the upstream tidal flat are commonly observed in estuaries and tidal basins. These flood channels provide important sediment feeding of the tidal flats and are therefore responsible for the accretion and maintenance of the tidal flats. These tidal flats in turn curb the flood and ebb flows, which means that the morphodynamics of the system is self-sustained.

Multiple bars and channels

A phenomenon similar to braiding of wide rivers, may occur in wide tidal basins. The underlying dynamics are related to the different pathways followed by flood and ebb channels. The opposite directions of inertial acceleration during flood and ebb result in a patchwork of flood and ebb dominated channels. Bifurcation of tidal channels may also occur as a consequence of avulsion, when channels get clogged by sedimentation. This happens especially for flood channels that end in shoals or that form swatchways across bars.

Elongate bars separating different tidal channels are often designated 'channel margin bars'. An example is shown in Fig. 5.4 for one of the major inlets to the western Wadden Sea. The bars make a small angle with the main flow direction; this suggests that the tidal ridge generation mechanism (Sec. 4.5.2) may contribute to the genesis and maintenance of these bars [558]. The combination of cross-bar tidal flow with along-bar residual circulation produces convergence of sediment transport at the bar crest both during ebb and flood, as demonstrated in Fig. 4.28. On the open shelf the maximum growth rate of tidal sandbanks corresponds to wavelengths (bank spacing) of several kilometres. The elongate estuarine bars have a much smaller width; this may be due to topographic constraints imposed on the flow. Another interpretation for the location and the levee-type form of these bars is the convergence of sediment transport to zones where flood and ebb dominated channels meet.

Meander evolution

Channel meandering is an ever ongoing process in natural tidal basins. Meanders tend to grow if space for growth is available; during the growth process the channel length increases and frictional energy losses become more important. The water level difference over the meander increases, favouring tidal flow over the inner-bend shoal at the expense of tidal flow through the main channel. A bypass channel through the shoal ('swatchway') may develop, shunting the main

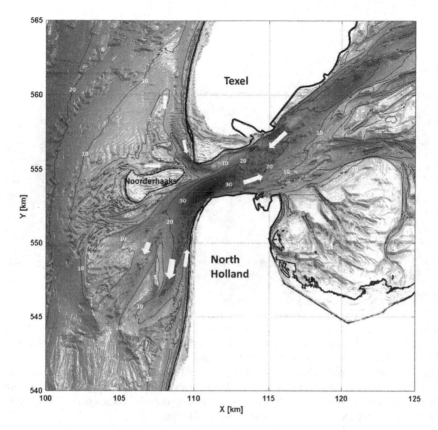

Fig. 5.4. Seabed topography of the inflow and outflow zones of Texel inlet (see Fig. 2.16 for its location), from detailed bottom soundings [278]. Depth contours are indicated in [m]. Increasing darkness corresponds to increasing water depth, except along the shoreline. The depth at the inlet attains 40 m. The large ebb-delta shoal (Noorderhaaks) remains partly dry at mean high water. The maximum depth-averaged ebb-tidal velocity at the inlet is almost 1.5 m/s, for an average tide. Tidal flow in the ebb-delta is generally ebb-dominated, except in channels situated close against the North-Holland coast and Texel coast. Residual flow is indicated by white arrows. At several places channels are separated by elongated shoals (light boundaries along several channels), which make a small angle with the tidal flow; a close-up is shown in Fig. 4.6.

channel; the main channel will silt up and may even get abandoned. The instability of a two-channel system was discussed in Sec. 2.5 and illustrated in Figs. 2.21 and 2.22. These figures show that the main channel, which initially cuts straight through the tidal flat, starts developing a meander. At a later stage the meander is abandoned and

Fig. 5.5.   Tidal flat system of the Baie du Mont Saint Michel (France, the Mont Saint Michel in the foreground). The maximum tidal range exceeds 14 m; the picture is taken at low water. The tidal channel system is highly dynamic; channels migrate across the whole bay on a multi-decadal time scale; channel meanders are cut off and redevelop at a time scale of years to decades. Marshes develop only near the boundaries of the bay, where the channels are less active. The substrate consists of 'tangue': Fine, non-cohesive sediment containing shell debris, fine sand and clay minerals. Photo Patrick Dontot.

the whole process may repeat. In this scenario tidal basin morphology is continuously evolving. Figure 5.5 shows the highly dynamic tidal flat system of the megatidal Baie du Mont Saint Michel, where tidal channels migrate permanently across the bay while forming new meanders and abandoning old ones.

Delta sand bodies

Flood channels generally end on shoals where the channel widens; here the flow expands, the flow velocity drops and sediment is deposited. These deposits are called 'deltas' or 'delta sand bodies'; they have a characteristic lobate form, the lobes representing deposition zones. Flood deltas are situated at the interior of tidal basins while ebb deltas are situated seaward of the inlet. In contrast to the rhythmic pattern of alternating and multiple-row bars, the delta bodies are isolated structures; their morphology is strongly related to the overall topography of the tidal basin. The development of these sand bodies influences the tidal flow pattern, not just locally but even at basin scale. Their development therefore interacts with overall deposition and erosion patterns in the basin and produces shifting of channels and bars. This mutual feed-back generates permanent morphologic evolution with a long term quasi-cyclic character. Morphologic cycles with periods ranging from several tenths up to hundred years have been documented for many tidal delta complexes [271, 808]. The morphodynamics of flood deltas will be discussed more extensively in Chapter 6 on tide-topography interaction.

5.3. Alternating Bars in Rivers

Stability of a straight flat channel bed

It is generally accepted that bar formation provides a major trigger for initiating large-scale topographic structures in alluvial channels, including channel meandering. This hypothesis will be examined in this section. We will address the following questions:

- which mechanism is responsible for breaking the symmetry of a perfect straight uniform channel?
- what type of rhythmic perturbation is amplified by channel bed instability, under which condition?
- which dynamic balance determines the wavelength of the strongest growing perturbation?

The method used is a linear stability analysis, similar as in the previous sections. Several investigations of this type have been reported in literature, examining different assumptions regarding bed-flow interaction, for instance, [86, 125, 315, 486, 660, 754, 766]. We will restrict the analysis here to a very simple model, which highlights the most essential features of alternating bar dynamics. The validity of this simple model is restricted to a limited range of natural channel flows; it is not applicable for flows which are strongly meandering or influenced by geographic constraints.

Alternating bar perturbation

We consider a straight channel with an initially flat bottom and a uniform, constant and unidirectional flow. We slightly disturb this uniform channel in a particular way: We impose a long wave bed undulation of infinitesimal amplitude corresponding to an alternating pattern of a shoal (initial stage of a bar) at one channel side, which opposes a pool at the other channel side (see Fig. 5.1). The water depth is greater over the pool than over the shoal and therefore more water will flow over the pool than over the shoal. At the upstream slope of the shoal the flow is therefore deviated to the deeper opposite side of the channel and the inverse occurs at the downstream slope. Due to this flow deviation, sediment is taken away from the upstream slope of the shoal and brought to the downstream slope. The result is a downstream migration of the shoal-pool pattern, see Fig. 5.6.

Feed-back mechanism

But this is only part of the story. Due to friction the flow velocity is decreased over the shoal and increased over the pool. Friction-induced deceleration of the flow when moving onto the shoal is counteracted by inertia; the same holds for friction-induced acceleration at the downstream shoal slope. Flow deceleration lags behind the decrease of depth. Hence, the location of minimum flow velocity is not at the crest of the perturbation, but downstream of the crest. The flow at the crest is still decelerating, implying sediment transport convergence and sediment deposition. Sediment deposition at the crest of the perturbation will cause growth of the amplitude of the perturbation. Growth of the perturbation amplitude enhances the flow deceleration at the emerging bar crest and therefore stimulates further bar growth. The opposite process of self-enhanced scouring takes place at the other side of the channel. The highest velocity

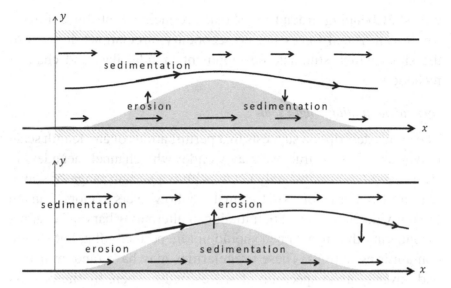

Fig. 5.6.   Development of an alternating bar. The figures show a plan view of a channel with a local perturbation of small amplitude corresponding to a shoal situated at the right bank. The incoming flow at the left (arrows designate the flow velocity) is steady and unidirectional and in both figures the solid line schematically represents a streamline. In the top panel it is assumed that the friction is weak and that the flow adapts to the shoal topography without inertial delay. The flow accelerates at the upstream slope of the perturbation and at the downstream slope it decelerates. This causes erosion at the upstream slope and accretion at the downstream slope: The perturbation will thus move downstream. The streamlines are symmetric with respect to the shoal structure; no net erosion or deposition takes place at the crest of the perturbation. In the bottom panel inertial and frictional effects are taken into account. Maximum flow velocity does not occur at the shoal crest, but upstream. This is caused by inertial lag of the flow response to the shoal topography. The streamlines are displaced downstream and they are asymmetric with respect to the shoal structure. The flow is decelerating not only at the downstream slope, but also around the shoal crest. In the zone around the shoal crest deceleration of the longitudinal flow component causes deposition of sediment and growth of the shoal crest.

does not occur opposite to the crest of the perturbation, but farther downstream. Therefore erosion will take place in the pool region. We conclude that the combined effect of friction and inertia stimulates the initial growth of a channel bed perturbation consisting of a shoal at one side of the channel and a pool at the other side, see Fig. 5.6. When the amplitude of the alternating bar pattern increases, the flow will start bending around the shoal, producing centrifugal forces. These centrifugal forces induce secondary circulations directed to the shoal which stimulate development of tidal flats and channel meanders.

Inertial adaptation length scale

The above description suggests that perturbations of any length scale may grow. If this is true, we may wonder why channel meanders in rivers (and tidal basins) only occur within a certain range of wavelengths, from order one km for small/shallow rivers to more than ten km for wide/deep rivers. For initiation of alternating bars and channel meandering the channel bed should not be perfectly flat, but should contain irregularities. These irregularities may have random nature and cover a broad spectrum of wavelengths. Then, by the process described above, feedback of channel-bed irregularities to the flow structure may produce amplification. The point is, however, that not all irregularities will grow as fast. Bed undulations with a wavelength such that the flow deceleration at the crests of the undulations is greatest, will overrule other irregularities and become dominant morphologic features.

Let us examine in more detail the flow over a shoal, represented by a positive bed undulation at one side of the channel, opposite to a trough at the other side. Flow deceleration at the shoal is a lag effect due to inertia. The lag is greatest for undulations with a small wavelength. At first sight, this would imply that undulations with small wavelength grow faster than undulations with large wavelength. There is a limit, however. The wavelength λ of the bed undulation should be large enough to allow adaptation of the flow velocity perturbation u' to the increase/decrease of friction at the slopes of the shoal-pool perturbation. The adaptation length scale follows from the requirement that the perturbations of inertia (represented by the term $(uu_x)' \approx 2\pi uu'/\lambda$ in the momentum balance equation) and friction (represented by the term $(c_D u^2/h)'$, where c_D is the friction coefficient and h the average channel depth) are of the same order of magnitude. This yields the estimate

$$\lambda \approx \pi h / c_D \approx 1000h. \qquad (5.2)$$

According to this requirement, undulations with a wavelength (spacing) on the order of λ will experience fastest growth. A similar estimate was found in other studies [375] and it matches the order of magnitude of observed meander lengths of North American rivers [758]. According to the analysis of Savenije [749] the aspect ratio (width-to-depth ratio b/h) of natural rivers is well represented by

$$b/h = \pi^2 / 2 \tan \varphi_r, \qquad (5.3)$$

where φ_r is the angle of repose. Using for φ_r the relation (3.65), the length scale (5.2) can be expressed as $\lambda \approx 20b$ for grain size $d = 0.2$ mm and $\lambda \approx 100b$ for grain size $d = 0.5$ mm. This is substantially larger than the relation (5.1).

The length-scale estimate 5.2 ignores the influence of lateral flow diversion from the shoal to the pool, which is essential for flow deceleration over the shoal. This implies that the wavelength λ of fastest growth is not just a function of depth h but also a function of width b. We will show in the next paragraph that for wide channels the wavelength λ of fastest perturbation growth depends on channel depth h and channel width b approximately as $\sqrt{hb}$. This result was derived by Andersen in 1967 [22]. The analysis by G.P. Williams of a large database of river geometries yielded the empirical relationship [950]

$$\lambda \approx 30(hb)^{0.65}. \qquad (5.4)$$

5.3.1. *A Simple Model for Alternating-Bar Genesis*

Flow response to an alternating bed undulation

We will now investigate the processes which initiate alternating bars in more detail, to clarify the assumptions behind the previous arguments. Therefore we perform a linear stability analysis, similar to the methods of previous studies by Callander [125], Parker [660], Fredsoe [315], Seminara [768] and others. We consider a channel of constant width b and depth h_0 and uniform depth-averaged flow u_0 in longitudinal direction. We assume $b \gg h_0$ and we look at the response (u, v) of the depth-averaged flow to a small channel bed perturbation. The channel bed is perturbed with a small longitudinal undulation which has opposite sign at opposing channel banks, see Fig. 5.1; in complex notation this reads:

$$z_b = \epsilon h_0 \cos(\pi y/b) \exp\left[i(kx - \sigma t)\right], \qquad (5.5)$$

with $\epsilon \ll 1$. The physical bottom perturbation is given by the real part $\Re z_b$. The wavenumber k is related to the bar wavelength λ by $k = 2\pi/\lambda$. The parameter σ in the exponent is a complex quantity; $\Im \sigma$ is the growth rate and $\Re \sigma/k$ the migration rate of the bar. The perturbed depth $h(x, y, t) = h_0 + \eta(x, y, t) - z_b(x, y, t)$, where η is the surface elevation with respect to a horizontal reference level.

The depth-averaged momentum balance equations read:

$$uu_x + vu_y + g\eta_x + c_D \frac{u\sqrt{u^2 + v^2}}{h} = 0,$$

$$uv_x + vv_y + g\eta_y + c_D \frac{v\sqrt{u^2 + v^2}}{h} = 0,$$

and the mass balance equation:

$$(uh)_x + (vh)_y = 0,$$

where u and v are the current velocity components in longitudinal and transverse directions respectively, and c_D is the friction coefficient.

All variables are scaled with respect to the undisturbed situation as follows:

$$u \to u_0 u, \quad v \to u_0 v, \quad \eta \to h_0 \eta, \quad h \to h_0 h, \quad z_b \to \epsilon h_0 z_b,$$

$$x \to h_0 x/c_D, \quad y \to by/\pi.$$

The x-scaling is derived from (5.2); the scaling of x and y is chosen such that the order of magnitude of flow properties and their gradients are the same. The scaled equations read:

$$uu_x + B^{-1}vu_y + F^{-2}\eta_x + \frac{u\sqrt{u^2+v^2}}{h} = 0,$$

$$uv_x + B^{-1}vv_y + F^{-2}B^{-1}\eta_y + \frac{v\sqrt{u^2+v^2}}{h} = 0,$$

$$(uh)_x + B^{-1}(vh)_y = 0, \qquad (5.6)$$

with

$$F^2 = u_0{}^2/gh_0, \quad B = c_D b/\pi h_0.$$

Infinitesimal perturbation and linearization

The velocities u, v and the surface elevation η include small perturbation components u', v' and η' that can be expressed as a power series in ϵ:

$$u' = \epsilon u_1 + \epsilon^2 u_2 + \cdots, \quad v' = \epsilon v_1 + \epsilon^2 v_2 + \cdots,$$

$$\eta' = \epsilon \eta_1 + \epsilon \eta_2 + \cdots.$$

The eventual dependence of the friction coefficient c_D on depth and velocity is ignored for simplicity. Now we only retain terms which are of first order in ϵ:

$$u_{1x} + F^{-2}\eta_{1x} + 2u_1 - \eta_1 + z_b = 0, \qquad (5.7)$$

$$v_{1x} + F^{-2}B^{-1}\eta_{1y} + v_1 = 0, \qquad (5.8)$$

$$-z_{bx} + \eta_{1x} + u_{1x} + B^{-1}v_{1y} = 0. \qquad (5.9)$$

We focus on rivers with hydraulic characteristics comparable to tidal basins and estuaries. We thus consider streams with small Froude numbers (channel depth not much less than 5 m, current velocity not exceeding 2 m/s). Small Froude numbers, $F^2 \ll 1$, imply

$|\eta_{1x}|, |\eta_{1y}|, |\eta_{1}| \ll 1$. In this case we get the simpler equations

$$u_{1x} + F^{-2}\eta_{1x} + 2u_{1} + z_{b} = 0, \qquad (5.10)$$

$$v_{1x} + F^{-2}B^{-1}\eta_{1y} + v_{1} = 0, \qquad (5.11)$$

$$-z_{bx} + u_{1x} + B^{-1}v_{1y} = 0. \qquad (5.12)$$

The solution has the following form

$$u_{1} = \tilde{u}z_{b}, \quad v_{1} = \tilde{v}z_{b}\sin y, \quad \eta_{1} = \tilde{\eta}z_{b}. \qquad (5.13)$$

This ensures that the transverse velocity $v = 0$ at the channel boundaries $y = 0$ and $y = \pi$. Substitution gives

$$(ik + 2)\tilde{u} + iF^{-2}k\tilde{\eta} = -1, \qquad (5.14)$$

$$(ik + 1)\tilde{v} - B^{-1}F^{-2}\tilde{\eta} = 0, \qquad (5.15)$$

$$ik\tilde{u} + B^{-1}\tilde{v} = ik, \qquad (5.16)$$

where k is the dimensionless wavenumber $k = 2\pi h_{0}/(c_{D}\lambda)$. By eliminating $\tilde{v}$ and $\tilde{\eta}$ we obtain

$$\tilde{u} = -\frac{1 - B^{2}k^{2}(1 + ik)}{2 + ik + B^{2}k^{2}(1 + ik)}. \qquad (5.17)$$

For small values of Bk, i.e., large wavelength-to-depth ratios ($\lambda \gg 2b$), the phase ϕ of u with respect to z_{b} is given by

$$\tan\phi = -\frac{\Im(1/(ik + 2))}{\Re(1/(ik + 2))} = k/2.$$

The strongest sedimentation at the bar crest occurs when the deceleration of the current is maximum: $\phi = \pi/2$. This corresponds to an infinitely small wavelength λ, which contradicts our previous assumption. Hence, the terms with Bk in Eq. (5.17) cannot be ignored.

Small Froude numbers

The simplification that consists of neglecting terms of order F^{2} requires some comments. It is justified for most estuaries and tidal basins, because the current velocities seldom exceed a few metres per second, even during stages of high river discharge or during storms. The channel depth is in general larger than a few metres. The Froude number $F = u^2/gh$ is therefore much smaller than 1 under all conditions. In natural rivers, the Froude number is generally also much smaller than 1 under average conditions. However, river morphology is to a great extent ruled by conditions of very high river discharges [518], where the Froude number can be close to 1. The simple model discussed here is meant primarily to clarify the morphology of tidal channels. It is also valid for sandy-bed rivers in coastal delta plains. Application of the model to rivers in general requires some caution.

Sediment transport including gravity effects

For the analysis of sedimentation and erosion around a sandy channel bed perturbation, we use the simplified sediment transport formula (3.82), assuming that the transported sediment load responds instantaneously to changes in the current velocity,

$$\vec{q} = \alpha |\vec{u}|^n \left(\frac{\vec{u}}{|\vec{u}|} - \gamma \vec{\nabla} z_b \right). \qquad (5.18)$$

We have also assumed that $|\vec{u}|$ is much larger than the critical velocity for bed erosion. The last term in this formula represents gravity-induced downslope sediment transport; this term is included to avoid the growth of bedforms with unrealistic high steepness. We evaluate the gradient of the bed-load sediment flux to first order in ϵ, yielding (in dimensional form)

$$\vec{\nabla}.\vec{q} = \alpha u_0^n [nu'_x/u_0 + v'_y/u_0 - \gamma(z_{bxx} + z_{byy})]. \qquad (5.19)$$

Now we substitute v'_y from (5.12); $x-$ and $y-$derivatives follow from (5.5). The result is

$$\vec{\nabla}.\vec{q} = \alpha u_0^n \left[ik \left((n-1)\frac{u'}{u_0} + \frac{z_b}{h_0} \right) + \gamma z_b (k^2 + (\pi/b)^2) \right]. \qquad (5.20)$$

Evolution of the bed perturbation

The growth of the bed perturbation follows from the sediment balance equation

$$z_{bt} + \vec{\nabla}.\vec{q} = 0. \tag{5.21}$$

This balance equation expresses the fact that for each channel segment a difference between incoming and outgoing sediment fluxes produces a change of bed elevation. The bed porosity factor is left out as it is not essential for our discussion. From the sediment balance equation we can find an expression for the migration and growth rates $\sigma = \Re\sigma + i\Im\sigma = i z_{bt}/z_b$ of the bed perturbation, by substituting (5.5), 5.20 and (5.17) in (5.21). The result is

$$\sigma = \frac{\alpha c_D u_0^n}{h_0^2} \left[\frac{-k(n-1)(1 - B^2 k^2(ik+1))}{ik + 2 + B^2 k^2(ik+1)} \right.$$

$$\left. + k - i\gamma c_D(k^2 + B^{-2}) \right], \tag{5.22}$$

where $k = 2\pi h_0/(c_D\lambda)$, and $B = c_D b/(\pi h_0)$.

Wavelength of maximum bar growth

The real part of σ is the bar migration rate and the imaginary part the growth rate. Both depend on the wave number k. If it is assumed that the perturbation with the fastest growth rate will overrule all other perturbations, then the corresponding wavenumber can be found by maximizing the growth rate $\Im\sigma$ with respect to $\xi = B^2 k^2 = (2b/\lambda)^2$:

$$\Im\sigma = \frac{\alpha \pi^2 u_0^n}{c_D b^2} \left[(n-1)\frac{\xi + 2\xi^2}{(2+\xi)^2 + p\xi(1+\xi)^2} - \gamma c_D(\xi + 1) \right], \tag{5.23}$$

with $p = B^{-2} = (\pi h_0/c_D b)^2$. This is identical to the expression derived for $\Im\sigma = 0$ in [752]. For very large aspect ratios, $b/h_0 \gg (\gamma c_D)^{-1}$, the solution is independent of n and γ:

$$\lambda \approx 2.6\sqrt{\frac{h_0 b}{c_D}} \approx 46\sqrt{h_0 b}, \tag{5.24}$$

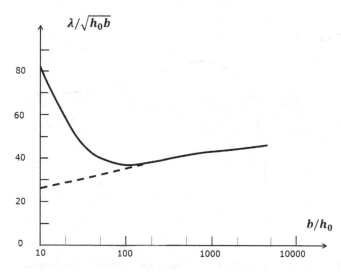

Fig. 5.7. Wavelength corresponding to the maximum growth of alternating bars, using the sediment transport formula (5.18) with $n = 4$. Solid line: Solution for $\gamma = 1, c_D = 0.003$. Dotted line: Solution for $\gamma = 0$ (no gravity-induced down-slope sediment transport).

where we have taken for the friction coefficient $c_D = 0.003$. The analytical expression of the wavelength λ of maximum bar growth for arbitrary aspect ratio is complicated. It is graphically displayed in Fig. 5.7 as a function of the aspect ratio b/h_0. The wavelength of maximum bar growth depends on downslope sediment transport only for small values of the aspect ratio ($b/h_0 < 100$). For values of b/h_0 of 100 or more, the wavelength is also independent of the exponent n in the formulation of the sediment flux. It appears that in this case the detailed formulation of the (bedload) sediment flux does not influence critically the initial bar development.

Growth and migration rates

The bar growth rate is obtained by substituting the value of the wavelength of maximum growth (Fig. 5.7) in the expression (5.23). The result is shown in Fig. 5.8. The e-folding growth time is large in comparison to the tidal period: A few days for small channels ($h \sim 5$ m, $b \sim 500$ m) and order one week for large channels ($h \sim 10$ m, $b \sim 2000$ m). For narrow channels with aspect ratios in the order of

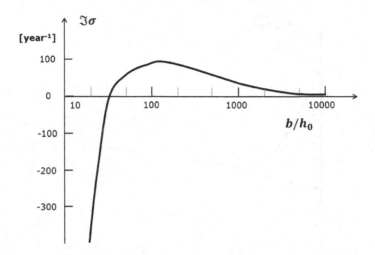

Fig. 5.8.   Growth rate in units $[year^{-1}]$ of the fastest growing alternating bar, as a function of the width-to-depth ratio b/h_0, for a channel of 10 m depth. The following parameter values are used: $\alpha = 4 \times 10^{-4}m^{-2}s^3$, $\gamma = 1$, $c_D = 0.003$, $n = 4$. No alternating bars are predicted for narrow channels, $b/h_0 < 33$, because the growth rates are negative. In the case of a 10 m depth channel, alternating bars grow fastest (e-folding in 4 days) if the width-to-depth ratio is slightly larger than 100.

30 or less, the maximum growth rate is negative. This is due to the stabilizing effect of transverse sediment transport down the slope of the bar perturbation. Alternating bars thus cannot emerge spontaneously in narrow channels; this is consistent with observations [153].

The migration rate $\Re\sigma/k$ follows from the expression (5.22),

$$\Re\sigma/k = \frac{\alpha u_0^n}{h_0}\left[1 - \frac{(n-1)((1-\xi)(2+\xi) - p\xi^2(1+\xi))}{(2+\xi)^2 + p\xi(1+\xi)^2}\right],$$

$$(5.25)$$

where $\xi = (2b/\lambda)^2$ and $p = (\pi h_0/c_D b)^2$. In this expression we substitute the value of the wavelength of maximum growth given in Fig. 5.7. It appears that alternating bars migrate in downstream direction with a speed in the order of 5 metres a day. This holds for straight rivers; in meandering rivers, alternating bars (point bars) are constrained by the river meanders, which migrate generally at a lower speed. In tidal channels with ebb and flood currents of approximately equal strength, alternating bars do not migrate.

Comparison with river data

We compare here the relationship found for the alternating-bar wavelength from the linear stability analysis with observed wavelengths in the field. It is not obvious beforehand that the wavelength of initial maximum growth corresponds to the wavelength selected by the nonlinear processes that come into play after the stage of initial growth. However, experimental and theoretical studies suggest that subsequent nonlinear processes induce only a moderate change of the initially selected wavelength [506, 756].

Data on the wavelength of alternating bars and associated meanders are available for many rivers. As alternating bars are typically situated at the inner river bends, bar spacing is closely related to meander spacing. We thus may compare observed meander length scales [440, 758] with the result of the linear stability analysis.

In general, the observed wavelengths match fairly well a linear relationship of λ with $\sqrt{h_0 b}$, even if the depth and width are not large [440]. The ratio $\lambda/\sqrt{h_0 b}$ is found to be in the order of 50 for large rivers ($b/h_0 \sim 100$) and in the order of 65 for narrow rivers ($b/h_0 \sim 20$) [315, 518]. In flume experiments a ratio of 72 is found [22]. These observations are in reasonable agreement with the result of the linear stability analysis (Fig. 5.7). However, the scatter in the data is large. This suggests that other processes that are not taken into account in the simple linear stability model, may play a role.

Settling lag effect

In the analysis of the alternating-bar instability we assumed that the transported sediment load was permanently saturated; it adapted instantaneously to changes in the bed shear stress (or, equivalently, to changes in the fluid velocity). In reality, as settling and resuspension processes take some time, the transported sediment load is not saturated, but lags behind the saturated state. For fine sediment, the time lag can be so long that saturation is never realised. Here we will assume that most of the sediment consists of medium sand, which is mainly transported in a thin layer above the seabed. In this case the transported sediment load will not be far from saturation and the settling and resuspension time lags T_{sed} will be very small compared to the tidal period T. With this assumption the following analysis is relevant both for river conditions and tidal conditions.

We write the sediment flux as

$$\vec{q}_s = \vec{u}C, \tag{5.26}$$

where $C(x)$ (expressed as a volume of unit width and length) is the locally transported sediment load in the water column in a thin layer above the seabed and $\vec{u}$ the local near-bed velocity. We ignore the effect of transverse and gravity-induced downslope transport components by focusing on alternating bar perturbations with large wavelengths.

The suspended load C equals the saturation load C_{sat} in the absence of spatial and temporal gradients. We assume that the sediment composition of the seabed is uniform and that sufficient erodible sediment is available at any location. The saturated load then depends uniquely on the local near-bed velocity u:

$$C_{sat}(x) = \alpha |\vec{u}|^{n-1}. \tag{5.27}$$

For a field of sinusoidal bottom perturbations with wavelength $\lambda = 2\pi/k$, we write, in complex notation,

$$C_{sat}(x) = C_0 + C'_{sat}, \quad C'_{sat} = C'_{sat,max}e^{ikx},$$

were C'_{sat} is the perturbation of the suspended load related to a small perturbation u' of the current velocity. We neglect here the time dependence of C'_{sat} related to the evolution of the bottom perturbation, because of the long evolution timescale.

Due to settling lag, the perturbation of the sediment load C' adapts gradually to the saturation value C'_{sat}, while moving with the current u. In a moving frame, C_{sat} follows the field of bottom perturbations of wavelength $\lambda = 2\pi/k$. Considering that for small bed perturbations, u is approximately constant and equal to u_0, we write for the moving frame

$$C'_{sat}(t) = C'_{sat,max}e^{iku_0t}.$$

Now we assume that in a frame moving with velocity u_0 the transported sediment load adapts to the saturation load according to a relaxation law with a response time T_{sed}:

$$\frac{dC'}{dt} = \frac{1}{T_{sed}}(C'_{sat} - C'). \tag{5.28}$$

Following the assumption $T_{sed} \ll T$, we ignore the tidal variation of u_0. The solution of this equation then reads

$$C' = C'_{sat}(t) \cos \beta e^{-i\beta} \approx C'_{sat}(x,t)(1 - ikl)\cos^2\beta, \tag{5.29}$$

with $\tan \beta = ku_0T_{sed} = kl$ and l is the spatial settling lag, $l = u_0T_{sed}$. We only consider small lag distances l compared to the perturbation wavelength, $kl \ll 1$. We also assume that the bed perturbation is small and that it can be written in the form (5.5) of an alternating bar field. The saturation concentration $C'_{sat}(x)$ (5.27) is expanded in the velocity perturbation $u' = \tilde{u}z_b = \epsilon h_0 \tilde{u} \cos(\pi y) \exp(i(kx - \sigma t))$, yielding

$$C'_{sat} = \alpha(n - 1)u_0{}^{n-1}\tilde{u}z_b. \tag{5.30}$$

This result is substituted in the sediment flux (5.26). After taking the gradient of the sediment flux we can extract the growth rate $\Im\sigma = \Re z_{bt}/z_b$ from the sediment balance Eq. (5.21). We then find

$$\Im\sigma \approx \frac{\alpha u_0{}^n}{h_0}k(n - 1)\cos^2\beta(\Im\tilde{u} - kl\Re\tilde{u}). \tag{5.31}$$

Because we restrict the analysis to settling lags that are small compared to the wavelength, we may approximate $\cos^2\beta \approx 1$. The

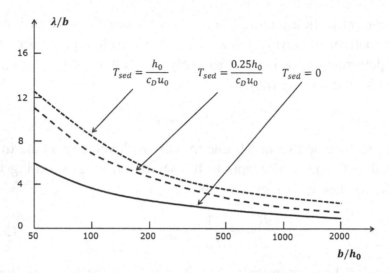

Fig. 5.9. Wavelength corresponding to the maximum growth of alternating bars, as a function of the aspect ratio b/h_0, for different values of the sedimentation lag time T_{sed}. Solid line: No sedimentation time lag; dashed line: Sedimentation time lag $T_{sed} = h_0/(4c_D u_0)$; dotted line: Sedimentation time lag $T_{sed} = h_0/(c_D u_0)$. For $u_0 = 1\,m/s$, $h_0 = 5\,m$, $c_D = 0.003$, the dashed line corresponds to a sedimentation lag of about 7 minutes and the dotted line to a sedimentation lag time of almost half an hour. Even a small sedimentation lag time produces a substantial increase of the wavelength of maximum growth.

velocity perturbations $\tilde{u}$ is given by (5.17). Substitution yields

$$\Im\sigma \approx \frac{\alpha\pi^2 u_0{}^n}{c_D b^2}(n-1)\frac{\xi + 2\xi^2 - l\xi[(\xi-1)(2+\xi) + p\xi^2(1+\xi)]}{(2+\xi)^2 + p\xi(1+\xi)^2},$$

(5.32)

with $p = (\pi h_0/c_D b)^2$ and $l = c_D u_0 T_{sed}/h_0$.

The wavelength of maximum growth rate is found by maximising this expression with respect to the variable $\xi = (2b/\lambda)^2$. The wavelength of the initially fastest growing alternating bar perturbations is shown in Fig. 5.9 as a function of the width-to-depth ratio, for different values of the settling lag T_{sed}. This figure shows that settling lag favours the growth of alternating bars with a larger wavelength than the preferred wavelength in the absence of settling/resuspension lag. Even for a small time lag of 7 minutes the wavelength increases by a factor 2. In fact, the time lag T_{sed} is a function of the bar height. In the initial stage, when the bar is low, the time lag is small (no time lag in the limiting case of zero bar height). The time lag increases with increasing bar height. More detailed numerical analyses [315, 934] also show that settling and resuspension lag effects in rivers produce larger alternating bar wavelengths than mere bedload transport, if the shear stress is not too large.

Multiple bars

The simple model for alternating bars can be extended to the case of multiple bar-channel systems. Therefore we describe the schematic topography of Fig. 5.1 by the function (5.5) by

$$z_b = \epsilon h_0 \cos(m\pi y/b) \exp\left[i(kx - \sigma t)\right], \qquad (5.33)$$

with $m = 3$ for a two-channel system, $m = 5$ for a three-channel system, etc. This function is substituted in the previous linear stability analysis for a single row of alternating bars, replacing the function (5.5). This yields expressions for the velocity perturbation $\tilde{u}$, the sediment flux gradient $\vec{\nabla}.\vec{q}_b$ and the growth rate $\Im \sigma$ that are identical to respectively (5.17), (5.20) and (5.22), with b replaced by b/m, B replaced by $B_m = c_D b/m\pi h_0$ and p replaced by $p_m = B_m^{-2}$. The result for the initially fastest growing wavelength is thus the same as in Fig. 5.7, with b replaced by b/m on both axes.

The growth rate is positive only for sufficiently wide channels, $b > 30 m h_0$. This result can be derived directly from (5.22), by neglecting terms with $B_m^2 x^2$ (i.e., excluding very wide channels). When putting $k = 2 \tan \phi$, we find

$$\Im \sigma = \frac{\alpha c_D u_0{}^n}{h_0{}^2}\left[(n-1)\sin^2\phi - \gamma c_D(4\tan^2\phi + B_m^{-2})\right]. \qquad (5.34)$$

The greatest growth rate corresponds to $\cos^2\phi = 2\sqrt{\gamma c_D(n-1)}$. For $\phi \ll 1$ this yields a bar wavelength given by

$$\lambda = 2\pi/k \approx (\gamma c_D/4(n-1))^{1/4} 2\pi h_0/c_D \approx 270 h_0, \qquad (5.35)$$

for $\gamma = 1, n = 4, c_D = 0.003$. The corresponding growth rate is found to be

$$\Im\sigma = \frac{\alpha c_D u_0^n}{h_0^2}\left[(n-1) + 2\sqrt{(n-1)\gamma c_D} - \gamma c_D B_m^{-2}\right].$$

$$(5.36)$$

This growth rate is positive only if

$$b/h_0 > m\pi\sqrt{\gamma/(n-1)c_D} \approx 33m. \qquad (5.37)$$

The different alternating bar modes have each a different growth rate. These growth rates can be read from Fig. 5.8. The m-mode for which mb is close to $100h_0$ will grow fastest. It will therefore dominate the channel topography. For instance, for $30 < b/h_0 < 200$, we expect a single channel bending through a row of alternating bars. For $200 < b/h_0 < 500$, we expect two channels branches bending through a double row of alternating bars. This is in reasonable agreement with field observations and laboratory experiments of river systems compiled in [315,660].

5.3.2. Development of River Meanders

What happens after initial bar formation?

The initial tidal flow pattern gets more and more disturbed during the growth of alternating bars and multiple-row bars. Centrifugal forces come into play when the flow is substantially deviated by the bars. The lateral flow boundaries will shift by scouring at the outer channel bends. The convergence and divergence zones of the sediment transport will change accordingly and may produce new shoal and pool structures. The flow response to the bars thus becomes different from the initial stage and the initial bar growth mechanism will be overtaken by other processes. Extending the linear stability model to include higher order terms (using techniques such as weakly nonlinear finite amplitude analysis) will not capture the important morphological changes that occur in reality.

The initial development of alternating bars promotes the development of channel meanders. The flow in channel meanders is more complicated than the almost rectilinear flow at the onset of bar formation. The flow has a three-dimensional character, with fluid parcels following a spiral pattern. In wide tidal channels, earth rotation plays a role in addition to the centrifugal forces. Tidal variation of water level and water surface slope are further complicating factors. Only the gross features of the resulting flow patterns and sediment transport patterns can be captured by simple analytical models.

Flow in a channel bend

Here we present a simplified mathematical description of the flow in a bending channel that details some of the qualitative explanations given above.

The influence of the bend on the fluid flow is most transparent in a coordinate system that follows the meandering channel, see Fig. 5.13. The longitudinal coordinate s measures the distance along the channel axis; the transverse coordinate n measures at each value of s the

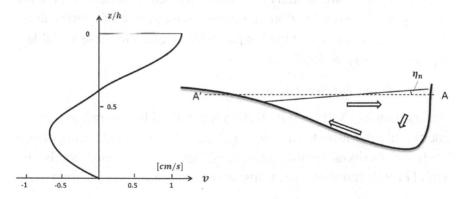

Fig. 5.10. Left panel: Transverse circulation in a channel bend, according to (5.40), for depth $h = 10\,\mathrm{m}$, longitudinal surface velocity $u_0 = 1\,\mathrm{m/s}$, bend radius $R = 5000\,\mathrm{m}$, Coriolis parameter $f = 10^{-4}\,\mathrm{s}^{-1}$ and eddy viscosity $N = c_D h \bar{u} = 0.0025\,h u_0$. Right panel: Typical bathymetry of a transverse section at a channel bend, with residual transverse circulation. The planform of the channel bend is shown in Fig. 5.13.

distance perpendicular to the channel axis. The vertical coordinate z equals $-h$ at the bottom and η at the water surface. Figure 5.10 displays schematically a typical cross-section at a channel bend. We call $\mathbb{C}$ the curvature of the channel at each location s and we assume that the radius of curvature $R \equiv 1/\mathbb{C}$ is much larger than the channel width b. If $x(s)$, $y(s)$ are the coordinates of the channel axis, then R is given by the formula $R = (x_s^2 + y_s^2)^{3/2}/|x_s y_{ss} - y_s x_{ss}|$. The subscripts s stands for partial derivation to this variable: $x_s \equiv \partial x/\partial s$, $y_s \equiv \partial y/\partial s$. We also assume that the channel width b is constant and that the flow is constant in time. The equations of motion in the new coordinate system then become

$$uu_s + vu_n + \mathbb{C}uv - fv + p_s/\rho - (Nu_z)_z = 0,$$

$$uv_s + vv_n - \mathbb{C}u^2 + fu + p_n/\rho - (Nv_z)_z = 0,$$

$$(vh)_n + (uh)_s + \mathbb{C}vh = 0. \qquad (5.38)$$

The symbols have the following meaning: $u(s, n, z)$ is the velocity along the channel axis, $v(s, n, z)$ the transverse velocity, f is the Coriolis acceleration and N is the turbulent viscosity. The pressure gradients p_s, p_n are related to the water surface inclination and to the density gradient (inclination of isopycns) in the case of stratification. The subscripts s and n stand for partial derivation to these variables: $u_s \equiv \partial u/\partial s$, $u_n \equiv \partial u/\partial n$, etc.

Transverse vertical circulation

The equations (5.38) can be further simplified by assuming that the inertial acceleration terms uv_s, vv_n can be neglected in comparison with the frictional terms. This is justified only for wide bends and small Froude numbers. The transverse momentum balance then reads

$$-\mathbb{C}u^2 + fu + p_n/\rho - (Nv_z)_z = 0. \qquad (5.39)$$

This equation shows that a non-uniform vertical profile of the along-channel flow velocity $u(z)$ implies vertical structure both in the centrifugal acceleration and in the Coriolis acceleration. A non-uniform along-channel velocity profile therefore produces a transverse circulation $v(z)$. The Coriolis acceleration has opposite sign for ebb and flood and is thus tide-averaged neutral. The centrifugal acceleration $\mathbb{C}u^2$ acts in the same direction irrespective of the direction of flow (ebb or flood). In estuaries, density differences may also play a role in the momentum balance: They influence the vertical structure of the along-channel flow, the magnitude and vertical structure of eddy-viscosity N and the magnitude and vertical structure of the transverse pressure gradient p_n. In case of a strong density stratification, the vertical structure of the centrifugal acceleration may be compensated by cross-channel inclination of isopycns and transverse circulation will be suppressed. In case of weak density stratification the transverse circulation will be enhanced due to eddy-viscosity damping.

We write $z^* = z/h$ and use the following assumptions:

- no density stratification (i.e., $p_n = g\rho\eta_n$);
- constant eddy-viscosity N;
- linear profile of along-channel flow ($u(z) = u_0(1 + z^*)$);
- no-slip, respectively slip conditions at bottom and surface ($v(z^* = -1) = 0$, $v_z(z^* = 0) = 0$). We then integrate the momentum balance twice over depth and obtain the following expression for the cross-circulation (see Fig. 5.10):

$$v(z) = \frac{h^2 u_0}{24N} \left[-\frac{u_0}{5R} \left(10z^{*4} + 40z^{*3} + 33z^{*2} - 3 \right) \right.$$
$$\left. + \frac{f}{2} \left(8z^{*3} + 9z^{*2} - 1 \right) \right]. \tag{5.40}$$

Averaged over the water column the centrifugal force is balanced by a pressure gradient corresponding to a transverse surface slope. From (5.39) and (5.40) we find

$$\eta_n = \frac{9}{20} \frac{u_0^2}{gR} - \frac{5}{8} \frac{f u_0}{g}. \tag{5.41}$$

Meander growth as a self-enhancing process

In the lower part of the water column the centrifugal force is too small for making balance with the surface-slope pressure gradient. Therefore the transverse flow component is directed near the bottom to the inner bend and carries sediment from the outer channel bank to the inner bank. The outer bank will erode, while the inner bank will accrete.

At the surface, the transverse flow component is directed to the outer bend. Due to the transverse circulation the flow in a channel bend will have a spiraling character. The longitudinal flow component is stronger near the outer bend than near the inner bend. The flow concentration at the outer bend is in general the major cause of channel bend erosion.

By strengthening bend erosion, the spiraling flow structure in a channel bend enhances meander growth. Conversely, meander growth enhances the spiraling flow structure. Due to this feedback, natural channel bends are inherently unstable; meander growth is a self-enhancing process.

Bend instability mechanism

We have shown that the genesis of alternating bars can trigger channel meandering. However, the inverse may also be true. In the previous section the genesis of alternating bars was explained as a positive feedback of the flow response to an alternating-bar perturbation of the channel bed. The channel banks were imposed to remain straight, however, preventing the development of true channel meanders. One may therefore ask: What would happen if the channel banks were allowed to shift by accretion or erosion?

Based on a linear stability analysis, Ikeda, Parker and Sawai showed mathematically in 1981 [440] that bank erosion at the outer channel band is another instability mechanism leading to channel meandering. A small spatially periodic shift of the channel bank produces a small channel curvature $\mathbb{C}'$, which induces a perturbation u'

of the tangential flow velocity at the outer bend of the curved channel. Ikeda *et al.* assumed that the initial rate of normal bank erosion at the outer channel bend is linearly related to the excess tangential flow velocity u', with a constant proportionality factor E (bank erosion coefficient).

The excess velocity u' can be estimated by solving the equations (5.38) to first order in the bank perturbation. Ikeda *et al.* therefore considered the depth-averaged version of these equations. The excess velocity u' produces further erosion of the initial channel bend, increasing the channel curvature $\mathbb{C}'$ and thus initiating an ongoing process of channel meandering.

The analysis shows that the wavelength λ of the initial periodic shift of the channel bank that experiences maximum growth does not depend on the bank erosion coefficient E. However, λ depends crucially on the transverse channel bed slope h'_n induced by the meandering flow (see Fig. 5.10). This transverse channel bed slope is not resolved in the model; a linear relationship of the transverse bed slope with the channel depth and the channel curvature ($h'_n \approx -3h\mathbb{C}'$) was assumed, based on field observations. The resulting wavelength of maximum growth is given by

$$\lambda \approx 4h_0/c_D \approx 1400 h_0. \tag{5.42}$$

This value is somewhat larger, but still of the same order of magnitude, as the wavelengths predicted for alternating bars in wide rivers if settling-resuspension lag effects are taken into account (Fig. 5.9).

The bend perturbation analysis was further refined by Blondeaux and Seminara [86]. They found smaller values for the preferred meander wavelength and pointed in addition to a phenomenon of 'resonance' between the bar and bed instability processes. Bend instability produces channel meanders which stimulate the development of point bars at the inner channel bend. These point bars correspond to alternating bars that can result from the bar instability mechanism in straight channels, but they do not correspond to the bar wavelength of fastest growth. However, in conjunction with the bend instability mechanism, the wavelength of fastest growth of the bar instability mechanism might shift to the wavelength of strongest bend instability and become 'resonant' with the bend instability. In this situation, alternating bars do not appear as free instabilities, but as instabilities forced by the meander pattern.

Development of river sinuosity

The bed topography observed in field situations (for example Fig. 5.2) and simulated in numerical process models (Fig. 5.12) is much less regular than the periodic patterns predicted by linear stability models. Linear stability analysis is restricted to the first stages of river meandering; it does not allow to investigate the high sinuosity characteristic of many rivers. Therefore one has to rely on numerical models, which can simulate long-term channel morphodynamics. Caution is required, however, because not all processes are sufficiently understood as yet for a reliable model representation. For instance, a reliable model representation of the process of channel bend scouring is not yet available.

The development of river sinuosity was tested with different process-based numerical models, starting from a straight uniform channel with erodible banks [756]. These model tests show that the initial emergence of free propagating bars is not sufficient to trigger the development of sinuosity. The channel remains almost straight with regularly spaced alternating bars; the wavelength remains stable and comparable to the wavelength predicted by the linear stability analysis. Inflow perturbation, generating a non-migrating bar near the mouth, stimulated the development of channel sinuosity, depending on the bank erosion formulation. High sinuosity only developed if the inner bend was converted to floodplain. This suggests that gradual consolidation of inner channel bends, due to vegetation encroachment for instance, is a necessary condition for the development of strong river sinuosity.

5.4. Tidal Channel Meanders

Applying river theory to tidal channels

The same feedback processes which are responsible for the emergence of alternating bars and meanders in rivers also operate under tidal conditions. Taking a simplistic view of tidal motion, in which flood and ebb flow are represented by successive periods of alternating steady flow, we might expect that we can apply almost without modification the results of the stability analyses for river flow. Applying this simplistic view to the Dutch tidal basins, we can compare the wavelength relationship for alternating bars in river flow (5.24) with the observed meander lengths in these tidal basins, indicated in Table 5.1. This comparison is shown in Fig. 5.11.

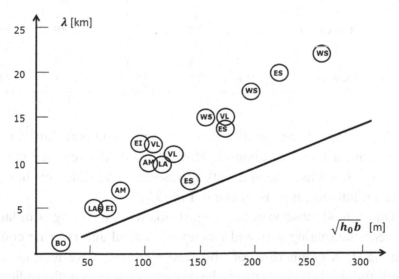

Fig. 5.11. Observed length scales λ of meanders in Dutch tidal basins plotted against $\sqrt{bh_0}$. The corresponding values are listed in Table (5.1). The solid line represents the theoretical wavelength for alternating bars in wide channels, $\lambda = 46\sqrt{bh_0}$. The observed meander lengths are almost a factor 2 larger than this theoretical wavelength, except for the Bornrif.

Table 5.1. Depth, width, tidal excursion and wavelength of meanders in some Dutch tidal basins. The meander wavelength is the distance between two successive shoals at the same side of the channel. The depth is the average channel depth and the width is the sum of the channel width and the bar width. The width-to-depth ratio is larger than 100 for all the basins.

Identifier	Tidal basin	Depth [m]	Width [km]	Tidal excursion [km]	λ [km]	$\lambda/\sqrt{hb}$
WS	Western Scheldt	20	3,5	22	22	83
WS		13	3	21	18	91
WS		11	2,2	20	15	96
ES	Eastern Scheldt	20	2,5	18	20	89
ES		12	2,5	14	14	81
ES		10	2	10	8	57
EI	Eijerland inlet	3,5	1,2	11	5	77
EI		5	2	13	12	120
VL	Vlie inlet	12	2,5	20	15	87
VL		8	2	15	11	87
VL		7	1,6	14	12	113
AM	Ameland inlet	4,5	2,5	16	10	94
AM		3	2	13	7	90
LA	Lauwers inlet	5	2,5	18	10	89
LA		4	1	12	5	79
LA		2,5	1,3	13	5	88
BO	Bornrif	1,8	0,3	1	1	43

The observed and predicted wavelengths compare fairly well, apart from a factor of about 2. However, this discrepancy can be removed, if we assume that settling-resuspension time lags have to be taken into account, as shown in Fig. 5.9.

Other field observations suggest likewise a strong similarity between alternating bars and meanders in tidal and riverine conditions [53]. A comparison of observed meander wave lengths λ in rivers and tidal channels over a large range of wavelengths indicates that the order of magnitude in both cases can be represented by the relationship $\lambda \approx 5b$, where b is the channel width [563]. This order of magnitude compares well with the linear stability analysis for the river case, see Fig. 5.9.

A numerical model of tidal channel meandering

We have considered so far analytical models of alternating bar development. In these models the channel geometry and the flow pattern are strongly simplified. The validity of the linear stability analysis is restricted to the initial stage of bar development.

With numerical models these restrictions can be partly overcome. Simplifying assumptions remain necessary in these models, because details cannot be reproduced down to the lowest relevant scales, but the restrictions are much less severe than for analytical models. It is possible, in particular, to simulate the development of alternating bars and channel meanders beyond the stage of initial emergence.

The first numerical simulation of channel and bar development under tidal conditions with a fully coupled 2D morphodynamic model was carried out by Hibma, De Vriend and Stive in 2003 [392]. The simulation started from a wide tidal basin with a uniform flat landward sloping sedimentary bed; the dimensions of the basin (width and length were fixed), the sediment grain size and the tidal forcing at the sea boundary corresponded roughly to those of the Western Scheldt estuary. Results of the simulation are shown in Fig. 5.12. The development of a pattern of flood and ebb channels meandering through a field of alternating bars is clearly visible. The overall picture shows strong resemblance with the actual morphology of the Western Scheldt, see Fig. 5.2.

The simulation demonstrates that the pattern which emerges initially (after a few decades) develops in the course of time into a pattern with larger spacing between the channel meanders. The authors attribute this increase of wavelength to a progressive increase of channel depths and tidal current velocities during the simulation. The simulated wavelengths are larger but still fairly close to the estimate of the linear alternating-bar stability analysis (5.24).

Simulations were carried out with two different sediment transport formulas: A formula assuming instantaneous coupling of the

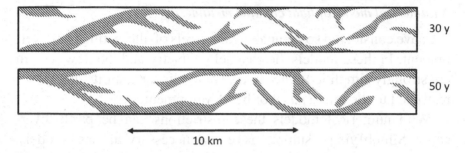

Fig. 5.12. Numerical simulation of the development of channel meanders in the middle section of a tidal basin; the seaward side is at the left and the landward side at the right. The channel has fixed lateral boundaries; the width is set at 2.5 km. The simulation started with a flat sloping bottom, with a channel depth decreasing from about 9 m at the left to about 4.5 m at the right. The top picture shows the main channel incisions (grey shaded) after 30 years of simulation; the bottom picture the channel configuration after 50 years. The shoal areas (in white) correspond to developing alternating bar structures. The wavelength of these bar structures is in the order of 8 km after 30 years. After 50 years the wavelength has increased to order 10 km, which is close to observed meander wavelength in tidal basins with similar dimensions. Equation (5.24) predicts an alternate bar wavelength of about 6 km. Figure redrawn from Hibma *et al.* [392].

transported load to the bed shear stress (as in the derivation of (5.24)) and a formula including suspended load transport. Because the results of both simulations were qualitatively similar, the authors concluded that the precise formulation of sediment transport is not crucially important. With the suspended load formulation a braided channel pattern emerged in the shallow landward part of the basin, similar to the results of the linear stability analysis. However, the braided pattern did not appear with instantaneously coupled sediment load.

Differences between riverine and tidal conditions

In spite of the correspondence between observed tidal meander wavelengths and river theory, there are several notable differences between the hydro-sedimentary dynamics in rivers and tidal channels.

- In rivers, bars and meanders migrate downstream, while in tidal basins almost no longitudinal migration occurs, in particular if flood and ebb flow are about equally strong.

- In rivers, the highest morphodynamic activity occurs at high river discharges, whereas morphodynamic activity in tidal channels is greatest during spring tides. During spring tides, the Froude number of the tidal flow remains in general much smaller than 1, while at high discharges the Froude number of the river flow may reach in some cases values close to 1.
- In many tidal environments, hydro-sedimentary dynamics are influenced by density currents, earth rotation (Coriolis), wave action and biological processes. These factors play hardly a role in riverine environments.
- In tidal channels, flood and ebb flow follow different pathways around alternating bars, because the inertial lag effect is opposite in flood flow and ebb flow. For this reason, the tide-averaged flow pattern exhibits strong horizontal circulations. Such circulations are absent in river flow.
- Much coarser bed material is transported in rivers during periods of high morphodynamic activity than in tidal channels. In rivers, sediment transport proceeds mainly through bed load, while in tidal channels suspended load transport generally dominates. Particle settling and resuspension in tidal channels is strongly influenced by the tidal phase. The influence of spatial gradients in shear stress related to presence of alternating bars is therefore less prominent in tidal flow than in river flow.

In the following paragraphs we will investigate in what way the difference between riverine and tidal conditions may influence the feedback mechanism responsible for the emergence of alternating bars and river meanders. We will focus on the latter two differences, related to horizontal circulation and suspended sediment transport.

Horizontal circulation in tidal channel bends

As shown in Fig. 5.10, the water level at the outer bank of a channel meander is higher than at the inner bank, both during flood and ebb. The corresponding transverse water level inclination thus causes an

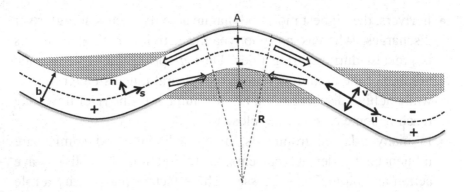

Fig. 5.13. Schematic representation of horizontal residual circulation in a bend of a mean-dering channel. The + sign at the outer channel bends indicates a water level maximum during both flood and ebb, and the − sign at the inner bends a water level minimum. The grey areas at the inner channel bends are tidal flats, possibly initiated as alternating bars in a straight channel. The cross-section at the channel bend (transect A–A') is schematically shown in Fig. 5.10. The horizontal circulation indicated by open arrows is a particularly effective mechanism for bar growth. The figure also shows notations and conventions used in the flow equations (5.38).

along-channel surface slope. The water level has a maximum at the outer bank and a minimum at the inner bank. The maximum and minimum levels occur in the transect where channel curvature is strongest, see Fig. 5.13. The corresponding along-channel pressure gradient, $\rho g \eta_s$, drives a horizontal circulation superimposed on the tidal flow, which is directed away from the bend along the outer bank and directed toward the bend along the inner bank. This circu-lation is enhanced by inertial acceleration, corresponding to the term $\rho u u_s = \rho (u^2)_s / 2$, in the along-channel momentum balance (5.38). This inertial term has the same sign as the along-channel pressure gradient, because u^2 has a minimum at the shallow inner channel bend and a maximum at the deep outer channel bend.

The resulting horizontal circulation is sometimes called 'headland eddy', because it is also present in coastal seas, at locations where the tidal flow bends around a headland. This residual eddy is related to the asymmetry of ebb and flood streamlines, caused by the opposite direction of inertial lag for flood and ebb at the outer channel bend, see Fig. 5.3.

The residual eddy has the same effect on sediment transport as the previously discussed transverse circulation: It drives a net flux of sediment from the deep outer bank to the shallow inner bank. The horizontal residual eddy contributes in general more to inner bank accretion than the transverse bend circulation.

In the northern hemisphere the Coriolis acceleration enhances transverse circulation in channel bends if the channel bend lies at the right hand of the main flow (opposite bends for ebb and flood). In wide tidal basins, earth rotation tends to concentrate the flood flow along the flood right-hand bank, whereas the ebb flow will be concentrated along the ebb right-hand bank.

Meander wavelength

Sediment is transported in tidal basins mainly as suspended load, due to the strength of the flow and the high fraction of fine sediment. In a numerical analysis, Fredsøe [315] investigated the influence of the ratio suspended-load to bed-load transport on the generation of alternating bars. This analysis shows that in situations where no alternating bars develop with bed-load transport alone, they may appear when suspended load is considered.

For tidal channels, the value of the settling-resuspension time lag T_{sed} and the lag distance l depend not only on spatial gradients in the shear stress, but also on the phase of the tide. For instance, there will be no settling at all during the tidal phases of accelerating flood or ebb currents. The process of bar formation extends over many tidal cycles and therefore integrates the effect of sediment settling for all tidal phases. Several stability analyses which take tide-induced settling and resuspension into account point to a significant influence of tidal phase effects on the wavelength of alternating bars [340,754,766].

This can be understood as follows. We first assume the initial presence of a pattern of alternating bars and associated channel meanders, generated by the mechanisms previously described, without considering tide-induced settling and resuspension. This channel pattern induces a field of residual eddies, as described in the paragraph above.

These residual eddies promote sediment transport towards the alternating bar crests at the inner channel bends.

We now consider sediment particles which are transported mainly in suspension and settle only around slack tide. Such a sediment particle, while being carried along with the tidal flow, will experience a net displacement in the direction of one of the alternating bar crests. However, if the meander wavelength λ is much smaller than the tidal excursion L, this net displacement is not necessarily directed to the same alternating bar for successive tides. It may therefore take many tides before the particle has reached an alternating bar crest at a moment coinciding with slack tide, i.e., during a tidal phase of current velocities which are sufficiently low for particle settling. This implies that an alternating-bar pattern with a wavelength λ much smaller than the tidal excursion will develop slowly. Inversely, if the meander wavelength λ is comparable to the tidal excursion λ, the net displacement of sediment particles during successive tides will always be directed to the same alternating bar crest. In this situation, the alternating-bar pattern will develop faster. We may therefore expect that in tidal basins with a high fraction of fine (non-cohesive) sediment, the meander pattern that grows fastest, will shift in the course of time towards a wavelength close to the tidal excursion.

This reasoning is similar to the mechanism proposed by Sleath [780] for the selection of a preferred wavelength of orbital ripples, see Sec. 4.3.2. These orbital ripples have a wavelength of approximately 2/3 of the wave-orbital excursion. By similarity, we may assume that a similar relationship holds for the meander wavelength in tidal basins with a high fraction of fine non-cohesive sediment.

Several semi-analytical stability analyses indicate that tide-induced settling and resuspension influence the preferred meander wavelength, such that meander wavelengths become comparable with the tidal excursion [754, 934]. These studies further show that the detailed formulation of the friction term (linear or quadratic) has no substantial influence on the results.

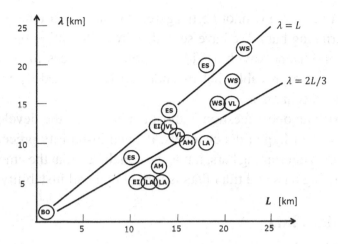

Fig. 5.14. Observed length scales λ of meanders in Dutch tidal basins plotted against the spring-tidal excursion $L = Tu_{max}\,\pi$. The corresponding values are listed in Table (5.1). The largest observed meander lengths are comparable to the spring-tidal excursion; the smaller observed wavelengths are closer to $\lambda = 2L/3$.

In Fig. 5.14 observed meander wavelengths λ in Dutch tidal basins are compared with the tidal excursion $L = Tu_{max}/\pi$, where u_{max} is the maximum spring tidal current velocity. The observed wavelengths are generally smaller than the tidal excursion, except for the largest basins. Although the comparison is not conclusive, the result is consistent with an increase of meander wavelength due to tide-induced settling and resuspension effects.

Bend instability as a leading process

The bend instability mechanism occurring in rivers also plays a role in tidal channels. A three-dimensional semi-analytical stability analysis of Solari *et al.* [790] reveals that the preferred wavelength of emerging tidal channel meanders by the bend instability mechanism is substantially smaller than the initial wavelength of alternating bars.

Meander lengths of channels in tidal flats and marshes observed in the Venice lagoon are in the order of ten times the channel width [563]. The ratio $\lambda/\sqrt{h_0 b}$ is in the order of 30–70 for the tidal flat channels $(b/h_0 \sim 8$–$50)$ and somewhat smaller (in the order of 25) for the marsh channels $(b/h_0 \sim 6)$. The strong sinuosity of tidal creeks observed in nature cannot be triggered by the initial emergence of free alternating bars. We have seen that free alternating bars do not develop in channels with a width-to-depth ratio less than 30. It is therefore more probable that meandering is triggered by the bend instability mechanism.

It is still an open question if this also holds for the development of meanders of large tidal channels. If bend instability precedes the formation of alternating bars, it may be concluded that the emergence of alternating bars and tidal flats is forced by bend instability.

5.5. Tidal Flats

5.5.1. *Tidal Flat Sedimentation*

Tide-induced circulations in meandering channels stimulate the growth of tidal flats by transferring sediment to the inner channel bends. Meanders hardly migrate in channels where tidal flow dominates. Therefore large banks can develop at the inner bends and become tidal flats, situated above the low water level. At this stage, accretion of tidal flats depend mostly on the flood flow, as explained in Sec. 5.2.2. Sediment supply to the tidal flat is stimulated by the flood dominance of tidal currents [682]. The highest tides determine the maximum height to which tidal flats can grow [832]. Typical planforms and profiles of tidal flats are shown in Figs. 5.15 and 5.16.

The finest sediment deposits are generally found in the upper landward part of the tidal flat, where wave action is generally less prominent than in the lower seaward part [327, 986]. However, for intertidal bars situated between tidal channels, the coarsest sediment is often found at the top, as shown by observations of grainsize distributions in the Wadden Sea [683].

The dynamics of muddy tidal flats strongly depends on biogeochemical processes [671], but also on tides and waves. Frequent small waves can be effective sediment resuspenders at the top of the tidal flat [358]. The tide determines the duration of exposure to wave action for different parts of a tidal flat. Therefore the gross morphologic

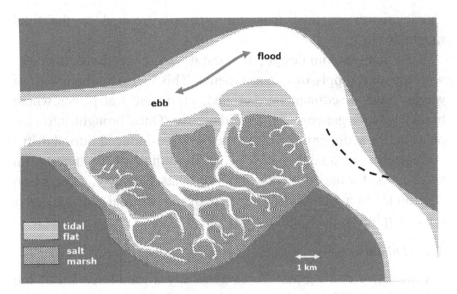

Fig. 5.15. Planform of a tidal flat marsh system at the inner bend of an estuarine channel: The Land van Saeftinghe in the Western Scheldt. The development of the tidal flat-marsh system is constrained by sea dikes. Several centuries ago, before the construction of the dikes, the tidal flats and marshes extended further landward. The largest marsh creek meanders have steep outer bends cutting into the marsh platform, and shallow inner banks with tidal flats. The number of creek branches increases in landward direction; the smallest dead-end creeks (not visible in the figure) have the highest sinuosity. The Western Scheldt is the sea entrance to the port of Antwerp, situated on the Scheldt river, at the right bottom of the figure. The entrance to the estuarine section of the Scheldt river is constricted by a submerged dike, indicated by the dotted line.

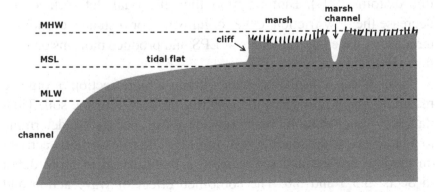

Fig. 5.16. Typical profile of a tidal-flat marsh system. The vertical scale is strongly exaggerated relative to the horizontal scale.

characteristics of tidal flats are primarily related to the tidal range [261].

Tidal flats seldom develop in coastal seas; exceptions are coasts with massive supply of fine sediments. This is due to the action of waves. Seabed sediments are eroded, particularly at places where breaking wave generate strong turbulence. Once brought into suspension, the sediments are dispersed by wave-induced littoral drift.

Estuaries and tidal basins are generally sufficiently sheltered from wave action for the development of tidal flats. However, wave action determines to a large extent the growth rate of tidal flats and their final morphological equilibrium.

Role of biota

The presence of seagrass causes generally a muddification of the intertidal soil [918], but in some cases (depending on type, density and location) vegetation may enhance turbulence and hamper fine sediment deposition [897]. Muddification is also stimulated by benthic diatoms; diatoms grow preferably on muddy soils and protect these soils from wave erosion by exuding extracellular polymeric substances (EPS). EPS bind sedimentary particles and form a biofilm at the tidal flat surface [338]. Therefore a positive feedback exist between mud accumulation and colonisation by benthic diatoms [804]. Biofilms smoothen the tidal flat surface and decrease the flow friction. Other organisms, for instance, cyanobacteria, also exude large amounts of EPS and produce biofilms on tidal flats [191].

In many cases, estuarine mud contains a high fraction of organic material, which influences the stability properties of the soil. Tidal flats provide a habitat for many species; these species also determine to a large extent the mudflat soil stability. The role of biotic activity on mudflat erosion and sedimentation is discussed in more detail in Sects. 3.5.3 and 3.6. The combined effect of wave action and biotic activity produces a strong seasonal variation in accretion and erosion.

Role of tidal currents and waves

We have seen that tidal currents and associated circulations are primary agents for creating the characteristic estuarine topography of channels with tidal flats at the inner meander bends.

Many observations confirm the existence of a sensitive balance between tide-induced accretion and wave-induced erosion of tidal flats. Tidal flats generally grow under moderate weather conditions, although suspended sediment concentrations are not as high as under strong wave action. Tidal flat accretion is observed even in situations where ebb velocities are higher than flood velocities [723]; this is illustrated by the example of Langstone Harbour (UK), see Fig. 6.28. The prominent role of tidal currents is illustrated by the observation that tidal flat accretion is substantially higher during spring tide than during neap tide [490,945]. The most significant bed level changes occur when water depths are shallow, at the beginning and the end of the inundation period [945]. This indicates that tidal flat accretion is more related to the increased transport capacity of the currents when tides are strong, than to the greater duration and depth of inundation.

Tidal flat erosion occurs during periods of increased wave action. Waves resuspend tidal flat sediments far more effectively than tidal currents, even if the strengths of tidal currents and wave orbital motion at the bottom are similar [223]. The wave influence on sediment resuspension is much stronger on the tidal flats than in the channel [452,514,723]. Under storm conditions, tide-induced sediment supply from the tidal channel is insufficient to counteract wave-induced tidal flat erosion and gravity-induced down-slope sediment transport. The height and extent of the tidal flats therefore results from a balance between tide-induced sediment supply and wave-induced erosion [491,694,916].

5.5.2. Tidal flat profile

The equilibrium depth profile of a tidal flat has a particular shape that depends on the prevailing hydrodynamic conditions. If the tidal flat is shaped mainly by waves, with only a minor role of tidal currents, the equilibrium profile is typically upward concave [987]. On the contrary, if tidal currents are the main morphodynamic agent, the equilibrium profile is typically upward convex [325]. This can be understood qualitatively by considering a theoretical tidal flat with a linear profile, see Fig. 5.17.

Wave-dominated tidal flats

We first assume that waves are the dominant factor. The amplitude of waves propagating onto a tidal flat with constant slope will decrease when the effect of shoaling becomes subordinate to the effect of friction and wave breaking. However, the near bottom orbital velocity is almost inversely proportional to the depth (according to linear wave theory). A linear depth profile thus implies that the wave-induced bottom shear stress increases in landward direction, in spite of frictional damping. The erosional action of waves will increase accordingly. If we assume that equilibrium requires uniform stress along the profile, then the linear profile must be modified in such a way that the effect of frictional damping is enhanced. This is achieved for a concave-up profile, where the depth decrease experienced by the wave is not as fast as for a linear profile. The above argument holds also in the presence of a varying tidal level, because the largest wave-induced bottom shear stress and the strongest sediment resuspension occur usually around mid-tide [357,902].

Tide-dominated tidal flats

It tides are the dominant factor, equilibrium requires that the maximum strength of the tidal current onto the tidal flat is approximately uniform along the profile. However, for a linear profile, the tidal current strength strongly decreases in landward direction. This is because flooding of the higher parts of the tidal flat occurs only near high water; during this tidal phase the water level variation is much slower than around mean water. The higher parts of the tidal flat thus

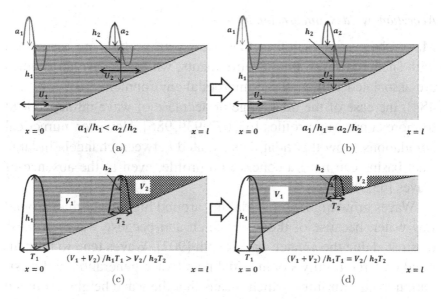

Fig. 5.17. Schematic representation of the cross-sectional profiles of a tidal flat. Top panels (a,b): Dominant wave conditions and small tide. The amplitude a of a wave propagating onto a tidal flat with a linear depth profile (Fig. (a)) is damped: $a_2 < a_1$. However, the damping is less than the decrease of depth h: $a_2/h_2 > a_1/h_1$. In shallow water, the wave orbital amplitude near the bottom is proportional to a/h. The wave-induced bottom stress thus increases in landward direction. For a concave-up profile (Fig. (b)), the wave damping is a bit less than for a linear profile, but the landward depth decrease is much less. Hence, the depth profile for which the wave-induced bottom stress is uniform, i.e., $a_2/h_2 = a_1/h_1$, must be upward concave. Lower panels (c,d): Dominant tide conditions and small waves; only the part of the tidal flat above mean sea level is shown. The curves represent the increasing and decreasing levels of the tide on the tidal flat. The average tidal current strength on the tidal flat is proportional to the flood volume ($V_1 + V_2$ at $x = 0$ and V_2 further landward) and inversely proportional to the flooding time (T_1, resp. T_2) and the local depth (h_1, resp. h_2). For a linear depth profile (Fig. (c)), the flooding time decreases in landward direction, but less than the depth. The tidal currents at $x = 0$ are therefore stronger than further landward. For a convex-up profile (Fig. (d)), the flooding time decreases in landward direction more than for the linear profile, but the depth decrease is much stronger. Hence, the depth profile for which the tidal current strength is uniform, must be upward convex.

will accrete. However, if the tidal flat profile is modified in such a way that the landward depth decrease is stronger than the landward decrease of the flood volume, the maximum current strength may remain approximately uniform along the profile. This corresponds to an convex-up tidal flat profile.

Accretion by tides and erosion by waves

Many observations confirm that convex-up profiles are associated with sheltered macrotidal environments, while concave-up profiles are associated with exposed microtidal environments [63, 236, 327, 484]. Increase of the tidal range or decrease of wave influence lead to more convex-up profiles [63, 672, 939, 986]. However, numerical simulations show that tidal flats situated between channels in large tidal basins can have a concave-up profile, even in the absence of waves [63, 980].

Waves grow higher on a tidal flat around high water than around low water, because of the larger fetch and because waves are less dissipated due the greater water depth [902]. Waves tend to be saturated on large mildly sloping tidal flats (wave generation and dissipation in equilibrium), which means that the wave height is limited by the water depth [514]. Observations of tidal flat hypsometries under non-equilibrium conditions show that retreating tidal flats by wave-induced erosion are associated with concave-up profiles and accreting tidal flats with convex-up profiles [485, 672].

5.5.3. *Feedback between Channel and Tidal-Flat Dynamics*

As tidal current strength and the associated sediment supply are a major driver of tidal flat development, a feedback exists between the morphodynamics of the tidal channel and the tidal flats.

This is illustrated by the evolution of tidal flats in the Eastern Scheldt basin, a large meso-tidal basin in the southwestern part of the Netherlands, see Fig. 5.2. During two successive periods of several decades each, the tidal currents in the Eastern Scheldt at first increased by about 10% and later decreased by 30–40%. The first intervention increased the area of tidal influence, producing an increase of tidal prism and tidal current strength. A significant increase of the tidal flat area was observed after this intervention and a corresponding increase of the channel depth, see Fig. 5.18. The second intervention, about 20 years later, decreased the tidal amplitude (decrease

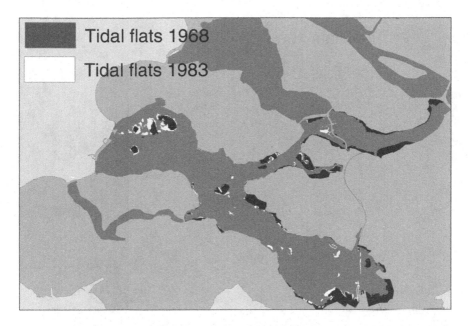

Fig. 5.18. Tidal flats in the Eastern Scheldt in 1968 (black) compared to tidal flats in 1983 (black and white), after [217]. In 1965 the tidal prism of the Eastern Scheldt was increased by about 10% due to the construction of a compartment dam between the Eastern Scheldt and the Grevelingen inlet further to the north. By 1983 the tidal flat area in the Eastern Scheldt had significantly increased compared to 1968. Most of the sediment was delivered by channel scour.

of about 10% by the construction of a storm surge barrier) and the area of tidal influence (decrease of about 30% by the construction of compartment dams). The resulting decrease of the tidal current strength was followed by a decrease of the tidal flat area and a corresponding decrease of the channel depth [545], see Fig. 5.19. During the second period, tidal flat erosion was mainly due to wave action and eroded sediments were deposited on the subtidal channel flanks. Before the first intervention, the Eastern Scheldt basin had evolved for many decades without major interventions; we may therefore assume that the basin morphology was close to equilibrium. The observed morphologic changes may therefore be interpreted as a natural morphologic response of the coupled channel-tidal flat system to the

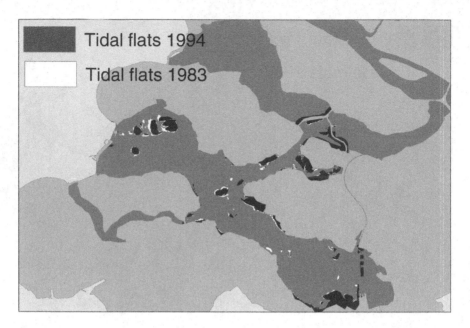

Fig. 5.19.   Tidal flats in the Eastern Scheldt in 1994 (black) compared to tidal flats in 1983 (black and white), after [217]. The construction of a storm surge barrier at the inlet and compartment dams in the seaward part of the basin in 1985 reduced the tidal prism of the Eastern Scheldt by more than 30%. By 1994 the tidal flat area had significantly diminished compared to 1983. Most of the sediment was deposited on the channel flanks below the LW line.

external interventions of first an increase and later a decrease of tidal prism and current strength.

When tidal currents are strong, sediment is transferred from the channel to the tidal flat. As a consequence, the channel cross-section increases. The inverse happens when tidal currents are weak; the tidal flat is eroded and the channel cross-section decreases. In the former situation tidal currents in the channel will progressively decrease, while in the latter situation the tidal currents will increase. Hence, a feedback exists between tidal channel development and tidal flat development. This feedback provides a control on the strength of tidal currents: Tidal currents are reduced while building up tidal flats and tidal currents are increased when tidal flats erode.

5.6. Salt Marshes

5.6.1. *Marsh Development*

Accretion

When the tide-induced sediment supply is sufficiently high, width and height of the tidal flat increase. The landward zone of the tidal flat rises above mean sea level and approaches the HW level. In this zone, the shear stresses exerted by currents and waves are relatively low. Plants can settle and the surface will become vegetated. This will happen in particular in regions with moderate climate conditions on the northern and southern hemispheres. The cohesion of the soil is strengthened by plant roots. The stems and leaves produce a further reduction of the shear stress exerted by waves and currents on the sediment bed and promote trapping of fine suspended sediments [163,626]. Waves propagating onto a vegetated marsh lose most of their energy over a distance on the order of 100 m or less [16].

Wave-induced resuspension of tidal flat sediments under storm surge conditions enhances vertical marsh accretion by inundating the marsh with sediment laden flood water [130,757]. Below-ground production of root material can also contribute significantly to raising the marsh platform [16,290]. Hence, vegetation accelerates further elevation of the marsh. The marsh platform finally reaches a level close to the highest astronomical tide level [832]. Accretion rates of young marshes are typically in the order of a few millimetres to a centimetre per year, but much larger accretion rates may occur in situations of high sediment supply [225]. In mature marshes, the platform elevation may decrease as a result of soil compaction [16].

Marsh creeks

Drainage channels (marsh creeks) develop during the marsh accretion process. Marsh creeks are stabilised by the vegetated marsh platform and incise deeply into the soft marsh soil [709,835]. In large marshes, the creeks are often strongly meandering, with smaller creeks draining into the large main creeks [874]. When the flood level exceeds the platform level, water spills from the marsh creeks over the marsh platform [833]. Small levees form near the marsh edges where the coarsest flood sediments are deposited. These levees generally exceed the level of the inner marsh platform by several decimetres. In some cases a major part of the marsh platform is flooded directly from the tidal flat [566]. The ebb flow passes mainly through the marsh creek; marsh creeks are typically ebb dominated and export sediment from the marsh to the tidal flat and the main tidal channel [566].

5.6.2. Interrelated Dynamics of Tidal Flats and Marshes

Marsh cliff formation

Waves dissipate their energy when propagating landward onto the tidal flat. At the level where most wave energy is dissipated, the tidal flat profile is locally steepened. Wave breaking on this steeper part of the profile enhances the steepness and a cliff is formed, see Fig. 5.16. The cliff location generally coincides with the transition between the bare tidal flat and the vegetated marsh. If the marsh cliff is stable, high wave-induced bottom stresses at the toe of the cliff prevent vertical accretion of the tidal flat zone fronting the marsh cliff. It can be shown that a landward marsh platform close to HW with step transition to a low seaward tidal flat (laying around and/or below mean water level) is the only stable configuration for the marsh-tidal flat system [289]. Only in cases of very high sediment supply a gradual transition from tidal flat to marsh may occur.

Marsh cliff retreat and progradation

However, a stable configuration does not mean that the cliff location is fixed. In the course of time, the marsh cliff base will be eroded by wave action until the cliff collapses. Tension cracks in the marsh platform, caused by tide-related pore pressure variations, can accelerate the collapse by undermining the cliff stability [313] (see Sec. 3.5.3 for a more detailed discussion). If the collapsed material remains close to the cliff toe and if sufficient sediment is supplied by the channel to the tidal flat, plant colonisation may start in front of the cliff, at the top of the tidal flat. In this case, a new cliff will be formed in front of the original cliff and the marsh will prograde [778,876]. However, if the collapsed material is washed away by waves and currents, a new cliff is formed landward of the original cliff and the marsh recedes. Observed marsh cliff recession rates are typically on the order of a few metres up to more than ten metres per year [16]. Meander evolution of the main channel bordering the tidal flat can also produce a shift of the boundaries of the tidal flat and the upper marsh.

5.7. Summary and Conclusions

Autogeneration of bedforms occurs in tidal channels likewise as in the sea. The lateral boundaries in tidal channels influence the flow structure such that other bedform types emerge. This holds for bedforms related to the horizontal flow structure. Bedforms related to the vertical flow structure, such as ripples and dunes, are basically the same in tidal channels as in the sea. Some differences may be expected, however, because of differences in sediment type (especially the occurrence of mud), because of the influence of salinity gradients on the flow structure and because of the minor influence of waves. We have not addressed these aspects in this chapter; the focus is on the interaction of bedforms with the horizontal flow structure.

The large-scale morphodynamics of tidal basins is discussed in the next chapter, which deals with the interaction of basin morphology with tidal wave propagation. Here we have looked at a lower scale level, the interaction of tidal currents with channel morphology. This interaction induces channel meandering and the development of shoals, tidal flats and marshes.

Bedforms in tidal channels have much in common with bedforms in rivers; they are generated by feedback mechanisms which are basically the same. We have discussed first bedform generation in river channels, because the physical processes are simpler and more transparent.

A striking feature of rivers and tidal channels is meandering and the occurrence of large bars. These bars are situated at the inner channel bends; they thus occur at alternating sides of the channel. Therefore they are called 'alternating bars'. Tidal channel meandering can be viewed as flow bending around alternating bars.

The question is: Are alternating bars a consequence of the initial emergence of channel meanders, or is channel meandering a consequence of the initial emergence of alternating bars? In this chapter the latter viewpoint is discussed more in detail, mainly because it is conceptually simpler — the lateral flow boundaries can be considered fixed.

The feedback mechanism is based on inertial delay in the horizontal flow structure, a phenomenon which is also responsible for the emergence of bedforms in the sea, as discussed in the previous chapter. In this case we consider an initial perturbation formed by a small bar at one side of the channel. The flow is diverted to the other side of the channel, where it accelerates. The flow over the bar decelerates due to increased friction. Inertial delay shifts the location of lowest current velocity downstream of the bar crest; the convergence of sediment transport at the bar crest implies accretion. The channel bed opposite to the bar deepens at the same time by erosion.

The linear stability analysis predicts the development of a sequence of alternating bars with a growth rate depending on the wavelength. The wavelength for which the growth rate is highest compares well with observed meander wavelengths in rivers. Meander wavelengths in tidal channels are generally larger, but the difference can be explained if the effect of settling and erosion lags is taken into account.

The other viewpoint — bend instability as leading meander mechanism — is also possible. A local outward shift of the channel bank induces a centrifugal force which concentrates the flow along the shifted bank. This flow concentration causes channel bank erosion and amplification of the initial outward bank shift. This feedback mechanism leads to channel meandering with a wavelength of maximum growth of similar magnitude as the wavelength of the alternating-bar instability. Comparison with observed meander wavelengths does not provide clear evidence for choosing between the two instability mechanisms. A stability analysis considering both mechanisms suggests that they interact, leading to a preferred 'resonant' wavelength of strongest positive interaction.

The tidal case is more complex than the river case. Flood flow and ebb flow tend to follow different pathways through a meandering channel due to inertial effects. Distinct flood-dominant and ebb-dominant channels can be observed in many wide tidal basins. The residual flow circulations induced by the flood-ebb channel pattern enhances accretion of the alternating bars, which may therefore develop into tidal flats. This accretion is strongest if the spacing between alternating bars is close to the tidal excursion length. Observed meander wavelengths in tidal basins are generally smaller than the tidal excursion length, but the order of magnitude is similar.

In summary, it seems most plausible that the formation of channel meanders, alternating bars and tidal flats proceeds through interaction of several instability mechanisms, which reinforce each other, because each mechanism promotes the development of a bar and meander pattern with similar wavelength.

The final sections of the chapter deal with the influence of waves on the development of tidal flats and marshes. Although most estuaries and tidal lagoons are sheltered from sea waves, small locally generated waves can increase considerably the bed shear stresses in shallow water. The channel-tidal flat-marsh profile is typically different in situations with low wave exposure compared to situations with strong wave exposure. As wave exposure depends on the fetch, and the fetch on the depth and width of the channel-tidal flat-marsh profile, only certain profiles will be stable under a given wind climate. In wide basins with a large wind fetch, the only stable profile corresponds to a high marsh platform separated from a low tidal flat by a steep marsh cliff. Vegetation dynamics determines to a large extent the marsh width.

Simulation of channel morphodynamics in numerical models remains a challenge. This is due in particular to the difficulty of modelling channel bank erosion. This process depends on soil mechanical factors, influenced by sediment properties and biota, which are not yet well understood. Better understanding this process still requires a major research effort.

Chapter 6

Tide-Topography Interaction

6.1. Introduction

This chapter deals with the physical principles of tidal inlet morphodynamics. In our terminology tidal inlets include tidal lagoons, estuaries and tidal rivers. Their development is related to sea-level rise; tidal lagoons are formed after breach of a littoral barrier and estuaries after marine transgression of a river plain. The morphology of tidal inlet systems is highly complex; yet the analysis of field data indicates that several gross features of tidal inlet morphology can be fairly well described by simple relationships. These relationships depend primarily on tidal characteristics, but also on river influences for estuaries and on wave influences for tidal lagoons. Before discussing the underlying tide-topography interaction we first present a general description of tidal inlet morphology and an idealised schematisation of the typical geometry of barrier tidal inlets and river tidal inlets. We review several empirical relationships; comparison with observations leads to the important concept of critical flow strength. We then discuss the characteristics of tidal propagation in shallow basins under different idealised conditions; a more general treatment is given in Appendix B.

Many tidal inlet systems have been strongly modified by human interventions during the past century. This raises questions about the impact of these interventions, the long-term resilience of tidal inlet systems and their capability to cope with an increased rate of sea-level rise. Therefore we investigate the tide-topography feedback

313

processes which are most relevant for morphological development. We derive relationships for the gross characteristics of the equilibrium morphology of estuaries and tidal lagoons, based on simple analytic models. These relationships are discussed with reference to the observed geometry of a large number of tidal inlet systems. Based on these relationships we discuss the response of tidal inlets to human interventions and to sea-level rise.

The simple morphodynamic equilibrium models ignore some important features of tide-induced transport of fine sediments. Therefore the final section of the chapter is devoted to tide-induced fine-sediment transport in tidal basins. The chapter ends with a summary and conclusions.

6.2. Tidal Inlets

6.2.1. *Tidal Inlet Characteristics*

Occurrence

Tidal inlets are ubiquitous along the world coastline. All present day tidal inlets are the result of marine transgression during the Holocene period of sea-level rise, see Fig. 6.1. On the time scale of geological evolution, tidal inlets can be characterised as very recent features. Their morphology has mainly been shaped during the past few thousand years of relatively minor sea-level variation.

The largest inlets are found along coasts where tectonic activity is weak, where the continental shelf is broad and where wide coastal plains have developed with a modest topographic relief. This is the case along the American Atlantic coast, the northwest coast of Europe, the southeast coast of Africa, the southeast coast of Asia and the northern Siberian coast, see Fig. 6.2 [89]. Strong tides can propagate far inland and shape the tidal inlet morphology by interaction with the sedimentary environment. In this chapter we will discuss this interaction and the resulting dynamic equilibrium morphology.

Tidal inlets also exist in coastal areas with high tectonic activity and pronounced topographic relief [773]. Along these coasts, tidal

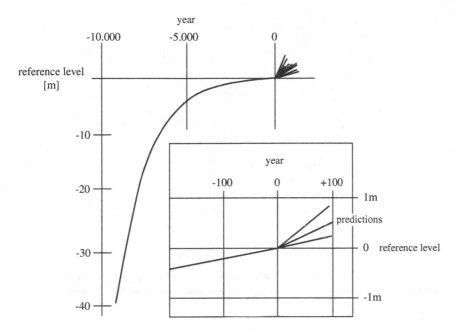

Fig. 6.1. Sea level curve for the southern North Sea during the past 10,000 years, derived from geological reconstructions. Expected sea level curves for the coming century according to IPCC [936].

inlet morphology is not only determined by tide-topography inter-action, but also by topographic constraints resulting from motions of the earth's crust or from glaciation processes. Fjords (canyons created by glacial erosion) and rias (submerged river valleys on a mountain coast) are examples of inlets where the morphology is to a large extent determined by pre-existing topography. Fjords and rias will not be considered in the following.

Some rivers with a high discharge of water and sediment build a delta that intrudes into the sea. The tidal influence in the delta branches is minor, even if the tidal range on the adjacent shelf is sub-stantial. Examples are the deltas of the Yellow River, the Yukon and the Mahakam. These deltas can hardly be considered 'tidal inlets'. Some aspects of the dynamics of delta river branches are discussed in Chapter 2. No further consideration will be given to these systems in this chapter.

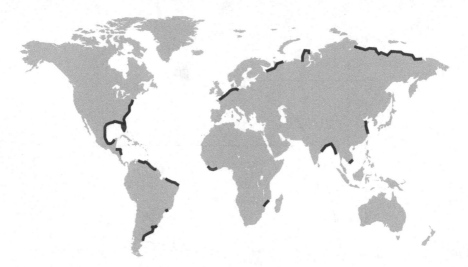

Fig. 6.2. Continental coastal plains where most of world's large tidal inlets are situated, after [89]. Most of the large continental plains are situated at subsiding coasts [80].

Barrier tidal inlets and river tidal inlets

The tidal inlets in low-lying coastal plains can be divided in two categories: barrier tidal inlets and river tidal inlets.

Barrier tidal inlets, which we will call tidal lagoons, are typical for coastal zones with important wave action, moderate to small tidal range and small river inflow (at least a few orders of magnitude smaller than the tidal exchange at the inlet). The inlet is restricted by a barrier (often in the form of a sand spit) or by barrier islands; these barriers result from wave-induced sand transport in longshore and cross-shore directions.

River tidal inlets, which are usually called estuaries, are typical for coastal zones with strong tides and substantial river discharge. The peak river discharge is comparable to the tidal discharge, except near the mouth, where tidal currents dominate. The width of the inlet is largest at the mouth; the width converges (more or less exponentially) in the upstream direction to the river width at the location where the tide has faded. Tides penetrate usually much further upstream than seawater does. The fresh water zone under tidal influence is called 'tidal river'.

The term estuary is used in the literature to designate semi-enclosed water bodies where marine and river waters mix; tidal lagoons with river inflow could be called estuaries, according to this definition. Here we will use the term lagoon for barrier tidal inlets and the term estuary for river tidal inlets, following the descriptions given above. The characteristics of tidal propagation in the two types of inlets are different. Therefore we will discuss tide-topography interaction separately for the two categories of tidal inlets. Some inlets have features of both categories. This is the case if wave action is substantial, while tide and river discharges are moderate.

Closure of tidal lagoons

A particular feature of tidal lagoons is the narrow entrance and the wide flood basin. The width of the entrance is a compromise between the scouring action of the tidal flow and the supply of sand along and towards the coast, which is mainly due to wave action. Both longshore and cross-shore wave-driven sand transport can play a role [676, 701]. If tidal currents are not strong enough, the inlet will become narrower and shallower. If tidal currents are increased, the inlet will widen and deepen. An increase of the inlet cross-section will induce a reduction of the maximum current strength (except in cases of an extremely narrow inlet). An equilibrium is established when the scouring action of the tidal currents is compensated by wave-induced sand supply. It appears (see next paragraph) that this equilibrium is generally achieved when the maximum ebb velocity at the inlet is of the order of 1 m/s.

If both the flood basin area and the tidal range are small, the equilibrium cross-section will also be small. However, there is a lower limit to the entrance cross-section, due to friction. For a very small cross-section, frictional losses at the inlet dampen the tide in the basin. This is due to frictional dissipation at the channel bed and channel walls, and also due to generation of turbulence when the flood and ebb jets from the entrance are slowed down in the much wider lagoon basin and in the sea. Tidal damping causes a further reduction of the currents and below a certain limit the tidal damping effect on the current strength is greater than the influence of flow concentration at the inlet. Hence, the inlet cannot stabilise and will finally close. This principle was shown by Escoffier in 1940 [287]; it has been further elaborated by Van de Kreeke and several other authors [117,340,877]. The ubiquitous closed-off lagoons along the world's coastline demonstrate that this is not just a theoretical possibility, but that tidal lagoon closure happens frequently when the tidal flow through the inlet is not strong enough. If tidal exchange is not strong enough to keep the inlet open, it may happen that opening of the inlet is forced during the rain season by high river runoff into the lagoon.

Critical flow strength at the inlet

O'Brien [640] was the first to notice that many tidal inlets follow a relationship of the type

$$A_{inlet} = C_A P^{n_A}, \qquad (6.1)$$

where A_{inlet} is the inlet cross-section area, P the tidal prism and where C_A, n_A are parameters. The tidal prism is the water volume stored by the tide in the lagoon; for lagoons much shorter than the tidal wavelength it is approximately given by

$$P \approx 2A_{inlet}U_{inlet}/\omega, \qquad (6.2)$$

where U_{inlet} is the maximum tidal velocity at the inlet and $\omega = 2\pi/T$ the semidiurnal tidal frequency.

The study of O'Brien was refined by Jarret [454] who investigated 162 inlets along the US coast(Atlantic, Gulf, Pacific). He found values of n_A ranging between 0.84 and 1.1, depending in particular on tidal range and wave climate. Other studies [335,427,625] confirmed the influence of these different factors on the relationship between inlet cross-section and tidal prism, including also the influence of longshore sand transport, flood-ebb asymmetry, river discharge and the presence and headlands.

Yet many inlets are fairly well represented by (6.1) as a long-term average, ignoring short-term fluctuations in the cross-sectional area of the channel. The observed maximum tidal velocity is typically of the order of $U_{inlet} \sim U_{inlet}^{crit} = 1 m/s$; this implies that the exponent n_A is close to 1. Taking $n_A = 1$, the value of the proportionality factor C_A must be close to $\omega/2U_{inlet}^{crit}$. The broad validity of the relationship (6.1) with these values of n_A and C_A suggests that tidal inlets have a natural tendency to evolve towards an equilibrium cross-section such that the peak ebb velocity U_{inlet} at the inlet reaches a critical value of the order of 1 m/s.

Human interventions

Many tidal inlets have been modified by human interventions. The most practiced interventions are inlet stabilisation, channel deepening and land reclamation.

Many inlets along the US coast have been artificially stabilised; Murrells Inlet (Fig. 6.7) is one of the jettied inlets. Jarret [454] observed that jetty construction modifies the equilibrium relationship (6.1) between the inlet cross-section and tidal prism. Jettied inlets have a smaller cross-section than non-jettied inlets and they are deeper.

Reclamation of tidal flats reduces the tidal prisms and perturbs the relationship between inlet cross-section and tidal prism. Observations in the western Wadden Sea (Fig. 6.3) show that natural sand transport processes respond to such a perturbation by reducing the inlet cross-section such that the equilibrium relationship is restored. The adaptation time scale depends on the size of the lagoon basin. The Frisian inlet in the Wadden Sea (east of the Ameland inlet) had a basin area of 200 km², which was reduced by 40% after the construction of a dam in 1969. Adaptation of the inlet cross-section to this intervention took about 30 years [879] (see also Sec. 6.5.3). Texel inlet was cut

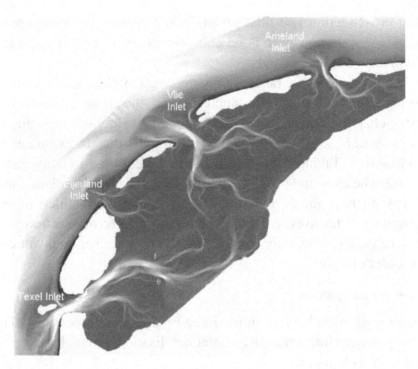

Fig. 6.3.   The tidal lagoons of the western Dutch Wadden Sea: Texel Inlet (length 45 km, mean tidal range 1.4 m), Eijerland Inlet (length 15 km, tidal range 1.7 m), Vlie Inlet (length 35 km, tidal range 1.9 m), Ameland Inlet (length 25 km, tidal range 2.1 m). Darkness indicates shallowness. The Wadden lagoons are sandy systems, with mostly fine sands (100–200 μ). The mud fraction ($d < 16\,\mu$) of the bed top layer (25 cm) is generally low (higher at the end of the summer than at the end of the winter). Even on the highest parts of the tidal flats the mud fraction almost never exceeds 10% and the average fraction is only a few %. Only the most landward part of the tidal flats is vegetated (*Salicornia dolichostachya, Spartina anglica*). The channel depth at the inlets is about 20 m and in the interior 5–10 m. The maximum spring tidal currents are of the order of 1.5 m/s at the inlet throat and of the order of 1 m/s in the interior channels. Cross-sectionally averaged velocities are about 30% lower. Peak currents are stronger during flood than during ebb. Sand from the North Sea enters the Wadden Sea inlets; the net yearly import is presently of the order of 5 Mm^3. This is a little more than needed for keeping pace with sea-level rise (15–20 cm over the past century). The channel system has a 'normal' configuration: Tidal flats are situated at the inner channel bends; channel branches shoot off at the outer channel bends; the strongest curvature occurs where the outer bend meets a boundary. Texel Inlet is the main inlet of the former Zuyderzee, which was dammed off in 1930. Bathymetric surveys indicate that the rate of infilling has strongly decreased during the pas decades, see Fig. 6.38. All the lagoons have pronounced ebb-tidal deltas. The sand volume stored in each ebb-tidal delta was comparable to the volume of the corresponding lagoon. After closure of the Zuyderzee the sand volumes of the Texel and Vlie ebb-tidal deltas have strongly decreased.

off from the Zuyderzee in 1930, which diminished the basin area of a few thousand km^2 to the present 700 km^2. The initial fast rate of infilling strongly decreased 60 years after this intervention [280] (see also Sec. 6.3.5). However, whether the inlet morphology is already close to a new dynamic equilibrium, is still under debate [931].

6.2.2. Ebb-Tidal Deltas

Obliquely incident waves generate a longshore current and an associated sand flux in the direction of the longshore wave vector component (also called littoral drift). The underlying processes are described more in detail in Chapter 7 on wave-topography interaction.

As discussed before, longshore sand transport closes tidal inlets, unless the ebb flow at the inlet is strong enough to resuspend the sand supplied to the inlet and to carry it offshore. The sand carried offshore by the ebb flow is deposited at some distance from the inlet, where the ebb jet is losing its momentum. At this location a shoal (or a series of shoals) is formed, called ebb tidal delta. The strongest accretion occurs at the tip of the ebb channel and along the sides of the ebb channel, producing the characteristic shield form of ebb tidal deltas. Flood channels develop along the shield at the updrift and downdrift margins. Figure 6.4 shows a sketch of a typical ebb tidal delta.

The morphodynamics of ebb-tidal deltas typically follows a cyclic process driven by tidal currents and waves [92, 303]. The ebb delta shoal can grow high because of permanent sand supply by the ebb flow. Waves break on the ebb-delta shoals, which may even become subject to swash (see Sec. 7.6). Swash-bars migrating over the shoal push the ebb channel in downdrift direction. The bending of the ebb flow is counteracted by inertia; part of the ebb flow spills over the shield at the outer channel bend and finally breaches the shield. A new ebb channel is formed in updrift direction. The old ebb channel is abandoned and the downdrift part of the shoal becomes attached to the downdrift inlet boundary. In this way the sand supply of the littoral drift crosses the inlet, preventing permanent sand accumulation at the

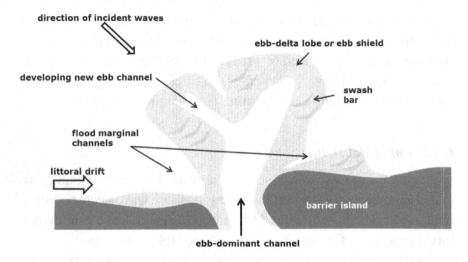

Fig. 6.4.   Schematic picture of an ebb tidal delta.

ebb tidal delta. The cycle starts again when a new ebb shoal is formed that migrates in downdrift direction by wave action.

Cyclic morphologic evolution has been observed in many ebb-tidal deltas [149, 371, 445, 645, 770, 787]. The period of the ebb delta cycle depends mainly on the size of the inlet; the time scale ranges typically between ten and hundred years.

Figure 6.6 shows the cyclic morphodynamics of the ebb tidal delta of Ameland Inlet. The Ameland Inlet differs slightly from the general picture, because the orientation of the main channel is strongly influenced in this case by the west-east propagation of tide along the coast.

The geometry of ebb-tidal deltas depends on tide and wave characteristics [771]. It appears however, that the gross characteristics of ebb-tidal deltas are mainly related to the tidal prism P. Walton and Adams [926] established an empirical relationship by comparing the tidal prism with the sand volume V_e stored in the outer tidal delta of inlet systems along the US Pacific, Atlantic and Gulf coasts. They found that these two quantities are almost linearly related,

$$V_e = c_e P^{n_e}, \tag{6.3}$$

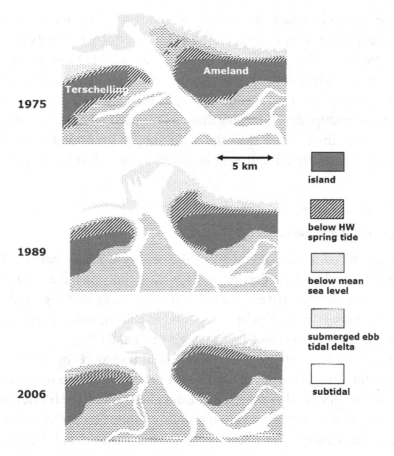

Fig. 6.6. The ebb tidal delta cycle of the Ameland inlet (western Wadden Sea). The tidal range is 2 m, the average wave height 1 m and the littoral drift is from west to east (left to right); longshore sand transport is of the order of 0.5–1 Mm^3/year. The lagoon area is 270 km^2, half of which is intertidal, see Fig. 6.3. The volume of the ebb tidal delta is 130 Mm^3. In 1975 the main channel was oriented west-east. A large shoal to the east was merging with Ameland Island. In 1989 a shoal at the west of the the main channel drove the ebb-dominant main channel into a north-south orientation. The shoal at the east merged with the island and developed into the Bornrif, see Figs. 2.21 and 2.22. In 2006 a new shoal developed at the east of the main channel. The Bornrif migrated eastward along Ameland Island and was separated from the new shoal by a marginal flood channel. The expected next stage of the ebb tidal delta, similar to the situation of 1975, is expected to occur around 2030 [280,931].

where c_e is a constant that mainly depends on wave exposure and n_e an exponent with average value 1.23. A similar relationship has been found for the outer deltas of the sandy tidal lagoons along the coast of Florida [686]; this study shows a considerable scatter in the constant c_e. Average values are $c_e = 0.2$ (S.I. units) and $n_e = 1$. The relationship (6.3) indicates that it is primarily the tidal exchange through the inlet (more than the sand supply by littoral drift) that determines how much sediment is stored in the ebb tidal delta.

6.2.3. Tidal Lagoons

Origin of tidal lagoons

Coastal plains are separated from the sea by a sandy barrier. The height and width of this barrier depends essentially on the existence of nearby sand sources (rivers, seabed, up-drift coast), on the granulometry and on sand transport by waves and wind. The morphodynamics of wave-dominated coasts is discussed in Chapter 7.

In the case of sufficient longshore sand transport (by waves and wind), the littoral barrier accretes at a rate comparable to or faster than sea-level rise. However, the coastal plain behind the barrier does not follow the rising sea level, unless it is flooded by sediment laden rivers. The existence of a low-lying coastal strip behind the littoral barrier is a common feature of coasts with high longshore sand transport. A breach of the littoral barrier during extreme storms allows tidal intrusion into this lowland. If the depression is sufficiently large and the tide sufficiently strong, the breach will be scoured by tidal currents. In this way a tidal lagoon develops.

Resilience of tidal lagoons

The morphology of sedimentary tidal inlets is shaped by processes of erosion, transport and deposition of sediments. Tidal currents are a major driver for these processes, which further depend on wave action and biotic activity. During the period of strong sea-level rise, tidal inlets drowned and migrated landwards. But when the rate of

5000 BP   1000 BP

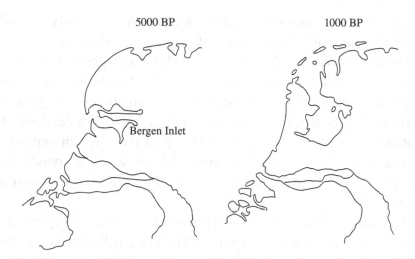

Fig. 6.5. The Dutch coastline 5000 BP and 1000 BP, according to geological reconstructions, redrawn after [982]. The large Bergen Inlet, located along the central coast of Holland, has completely filled in and the coastline was closed. Further to the north, the Texel inlet and Zuyderzee basin have developed during the past 1000 years.

sea-level rise decreased, some 5000 years ago (Fig. 6.1), the morphology of tidal inlets became more stable. Some inlets were filled in with sediments and closed off from the sea; this occurred in particular for inlets that were not connected to rivers. Many other inlets adopted a dynamic equilibrium morphology, even in the absence of substantial river inflow.

Along the coast of Holland several large barrier tidal inlets have disappeared by sediment infilling in excess of sea level rise (for example the Bergen inlet, see Fig. 6.5). This happened in the period between 7000 BP and 3500 BP, when the rate of sea level rise was higher than at the present [66]. The nearshore seabed was the major sand source [65]. The following sequence of infilling processes for these inlets was established, based on analysis of bottom cores [69, 882].

After the sea intruded into the low-lying coastal plain, the sediment bed was initially reworked by tidal currents into a pattern of channels and tidal flats. In the following stage, tidal flats became vegetated and transformed into marshes where fine sediments were easily trapped. While the marsh level was raised to the high water level, the tidal storage volume decreased. The decrease of tidal discharges was enhanced by wave-induced sediment supply to the inlet. The intrusion of saline water decreased and fresh water marshes with abundant vegetation developed at the landward boundary of the lagoons. Successive generations of marsh vegetation grew into peat bogs, which expanded vertically and horizontally. The tidal influence further decreased and finally the lagoons were closed by wave-induced sand transport at the inlet.

However, this sequence of infilling processes and subsequent closure did not take place everywhere. The eastern Wadden Sea tidal lagoons, for example, persisted without great alteration since at least a few thousand years.

Sea-level rise and morphodynamic equilibrium

Flood currents import sediment into the lagoon. The question therefore is why so many tidal lagoons with strong flood currents have survived.

Tidal currents generate morphological features, such as channels and tidal flats. These morphological features influence in turn the tidal motion in the lagoon. Tidal motion and lagoon morphology are continuously adapting to each other and evolve towards a morphodynamic equilibrium. A dynamic equilibrium corresponds to a situation where the (mainly tide-induced) residual sediment fluxes raise the bed level uniformly, such that depth profiles remain unchanged relative to the slowly raising sea level. 'Remaining unchanged' should be understood as 'remaining unchanged considering a sufficiently long period'. This sufficiently long period includes not only the periods of all significant tidal constituents, but also the period of long-term morphological fluctuations (for instance, cycles of channel avulsion at decadal timescales).

Observations of long-term average sedimentation in tidal inlets along the US Atlantic coast and Gulf coast and along the European

Atlantic coast and North Sea coast, point to a net accretion rate which is comparable with the rate of sea level rise [544, 631, 694]. The accretion is in some cases larger and in other cases smaller. Human interventions (sediment retention in upstream reservoirs, reclamation of tidal flats and marshes, dikes, dredging, etc.) are probably a major cause of discrepancies between accretion rates and sea-level rise.

In some microtidal lagoons, flood currents are not strong enough for carrying marine sediment sufficiently far inside the lagoon. These lagoons do not keep pace with the sea level and get progressively drowned, see Sec. 6.2.5. In all other cases, the import of sediment from the sea by wave action and flood currents is on average almost exactly balanced by export of sediment by ebb currents. If not, these tidal lagoons would have disappeared during the past millennia, during which the mean sea level hardly varied. Once a close-to-equilibrium situation was reached, the tidal storage volume of these lagoons did not significantly decrease and a sufficient tidal current strength was maintained at the inlet to flush the imported sediments.

Later in this chapter we will see how a close-to-equilibrium morphology can develop as a result of tide-topography interaction.

Morphology of tidal lagoons

Many characteristics of tidal lagoon morphology have already been described in the previous chapters and will not be repeated here: Small bedforms such as ripples and dunes (Sec. 4.2), alternating bars (Chapter 5), flood and ebb channels (Sec. 5.2.2), channel meanders (5.4), tidal flats (5.5) and marshes (5.6).

The detailed morphology of tidal lagoons is very complex. Typical examples of lagoon morphology are shown in the Figs. 6.3, 6.7 and 6.8. Yet, several general features of lagoon morphology are similar, meaning that they can be predicted with reasonable confidence. We will focus on these gross characteristics and establish some relationships that allow a comparison between tidal lagoons. Therefore we reduce the complexity to a very simple

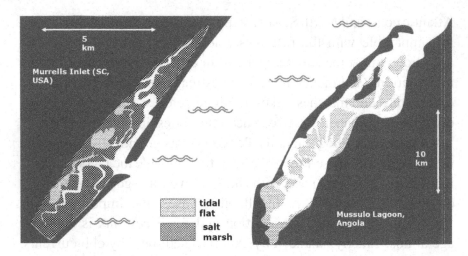

Fig. 6.7.   Two typical micro-tidal lagoons: Murrells Inlet (South Carolina, US) and Mussulo Lagoon (Angola). Land is dark and subtidal areas are white.

Murrells inlet is a small lagoon on the US Atlantic coast. The average tidal range is 1.4 m (maximum 2.2 m); flood and ebb flow are mainly concentrated in two marsh creeks, one to the north and one to the south. The depth of the largest creek is 3–4 m; there is little fresh water input. The tidal flats have almost entirely developed into salt marshes, which are only partially covered at high water. The marsh creeks meander, but the flow strength is insufficient for the development of distinct flood and ebb channels. The flood flow is stronger than the ebb flow [39]. Similar coastal systems exist lagoons the US Atlantic coast, along the Mexican Gulf of California coast and along the Portuguese Algarve coast. Other examples of lagoons with similar characteristics are Venice lagoon and Aveiro lagoon (Portugal).

Mussulo Lagoon is situated on the Atlantic coast of Africa, near Luanda. The average tidal range is 1.05 m (maximum 1.8 m). Sea breeze is the most common wind. Northward sand transport along the coast (in the order of 10^5 m^3/year, decreasing to the north) is driven by frequent swell waves from the SW with a significant height of 1–1.5 m [193]. The length of the sand barrier separating the lagoon from the ocean has decreased during the past century. A large shoal north of the entrance to the lagoon is a remainder of the previously longer sand spit. The lagoon is sandy without mudflats; it receives no fresh water runoff and little organic matter. Tidal flats are sparsely vegetated. Distinct ebb and flood channels are bending around tidal flats, which are intersected by meander cut-off channels. Other similar lagoons exist along the Persian Gulf coast (Al Jubail, Abu Dabi), and the coast of Tanzania (Lazy Lagoon).

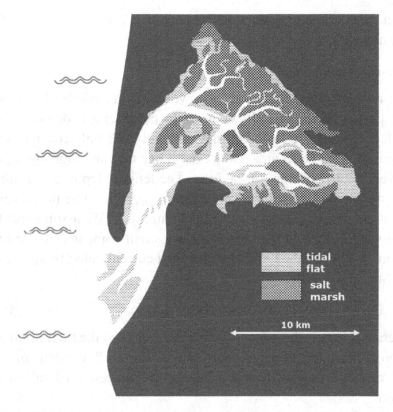

Fig. 6.8. Typical mesotidal lagoon: the Bassin d'Arcachon on the Atlantic French coast. Land is dark and subtidal areas are white.

The average tidal range is 3 m (maximum 4.8 m). The long sand spit at the entrance is maintained by a strong north-south (top–down) littoral sand transport of about 0.65 Mm^3/year. The basin substrate is sandy (grain sizes of 100–500 μ). The basin has been very stable over the past centuries. Sea-level rise over the past century is 15–20 cm; the overall increase in basin volume over this period is of the order of 10 Mm^3 [100]. Large entrance shoals cross the inlet with a periodicity of about 80 years [149]. Channel depths at the inlet are of the order of 15–20 m and in the interior of the order of 5–10 m. Maximum ebb currents are stronger than maximum flood currents (typically 1.6 m/s versus 1.1 m/s at the inlet, 0.85 m/s versus 0.6 m/s in the interior basin). The strongest flood currents occur 1–2 hours before HW, the strongest ebb currents at mid tide. Wave action in the basin is moderate (wave height 30 cm on average, 0.8 m maximum). The fresh water inflow is small (in the order of 30 m^3/s). The tidal flats are covered with muddy sands (mud percentage of the order of 10% or less). Marsh vegetation of the high parts consists of *Zostera Noltii*. The channel pattern inside the basin is complex, with meandering, branching and avulsing channels. Mesotidal lagoons with similar characteristics are found in the Wadden Sea and along the Georgia coast (US); Langstone Harbour (UK), Eastern Scheldt (Netherlands), Setubal Lagoon (Portugal), Willapa Bay (US) and Venus Bay (Australia) are other examples.

form, that crudely synthesises the main characteristics in a few parameters.

Schematisation of tidal lagoons

The inlet of tidal lagoons is narrower than the width of the interior lagoon. It is generally scoured to considerable depth, except for lagoons that are closed off intermittently. We only consider tidal lagoons with a basin area and a tidal range that are large enough to prevent natural closure of the inlet. The length, depth and width of the inlet are called respectively l_{inlet}, h_{inlet}, b_{inlet}. The tide level at sea is called $\eta_{inlet}(t)$ and a is the tidal amplitude. We assume that the freshwater inflow in the lagoon is very small compared to the tidal prism P. The cross-sectionally averaged current velocity $u_{inlet}(t)$ at the inlet is given by

$$u_{inlet} = Q/A_{inlet}, \qquad (6.4)$$

where $A_{inlet}(t)$ is the inlet cross-section and $Q(t)$ the tidal discharge through the inlet. The maximum inflow and outflow occur in general near the time of fastest tidal rise and fastest tidal fall at the mouth.

We adopt for the interior lagoon a one-dimensional description, shown in the Figs. 6.9 and 6.10. Therefore we project all the channels of the lagoon along a single x-axis; the lagoon cross-section $A(x, t)$ is the sum of all the projected channels, including neighbouring tidal flats. The corresponding wetted width of the lagoon is called $b_S(x, t)$.

In a crude model, which under certain conditions can be solved analytically, tidal flow is divided over a flow conveying part of the cross-section and a water storage part [248]. In the flow conveying part the tidal velocity is directed along the channel axis and has everywhere the same order of magnitude; in the water storage part the tidal velocity is substantially lower and the flow is mainly in cross-channel direction, see Appendix B and Fig. B.2.

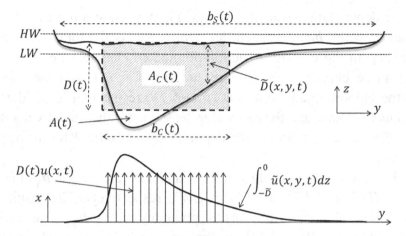

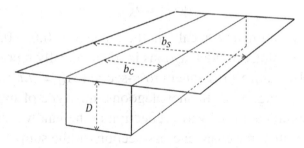

Fig. 6.9. Schematisation of a lagoon cross-section into a rectangular flow-conveying section and a storage section. The schematisation depends on the tidal phase. The transverse profiles of depth and depth-integrated current velocity are represented by $\tilde{D}$ and $\tilde{u}$; the schematised depth and current velocity by D and u.

Fig. 6.10. Sketch of the idealised geometry of a tidal lagoon, characterised by a non-converging, almost uniform basin width.

We call $u(x, t)$ the cross-sectionally averaged current velocity in the flow-conveying part of the channel. We assume that the current velocity in the remaining shallow part of the cross-section is so small that it can be ignored in the equations of motion. We will therefore use the term 'channel' only for the deepest part of the cross-section where the flow is concentrated.

The depth $h(x)$ is a representative depth of the channel cross-section relative to mean sea level and $D(x, t)$ is the instantaneous water depth, $D(x, t) = h(x) + \eta(x, t)$. We define a representative channel width $b_C(x, t)$ such that $Db_C u$ equals the total discharge $Q(x, t)$. In the case of a friction-dominated channel, the depth D should be chosen such that the integral of $\tilde{u}^2/\tilde{D}$ over the width of the lagoon, equals $b_C u^2/D$ ($\tilde{u}$ and $\tilde{D}$ are the real local along-channel velocity and the real water depth, and u the model velocity, see Fig. 6.9). In practice, the depth D can only be determined by iteration.

The total lagoon widths at HW and LW are called $b_S^+(x) = b_S(x, HW)$ and $b_S^-(x) = b_S(x, LW)$, respectively. The width of the intertidal area (tidal flat) is given by $b_S^+ - b_S^-$. The channel widths at HW and LW are called $b_C^+(x) = b_C(x, HW)$ and $b_C^-(x) = b_C(x, LW)$, respectively. In the case of steep channel walls we have $b_C^+ \approx b_C^- \approx b_S^-$. Otherwise we will assume that the average channel width b_C is proportional to the wetted width at LW, b_S^-,

$$b_C = \alpha b_S^-, \tag{6.5}$$

where α is a parameter typically in the range $\alpha \approx 0.6 - 0.9$.

The morphology of tidal lagoons is schematically represented by a more or less uniform channel cross-section $A_C = hb_C$ and basin width b_S, see Fig. 6.10. In short lagoons, this type of morphology produces significant tidal wave reflection at the landward boundary $x = l_A$. In reality, the channel cross-section (or the sum of the cross-sections of the different channel branches) decreases in landward direction. Nevertheless, we adopt the approximation of longitudinal uniformity, because it allows a simple analytical solution of the tidal equations.

Characteristic values for several typical tidal lagoons are given in Table 6.1.

Tidal motion in short lagoons

We consider now a short lagoon, with such a small length that spatial differences in the tide level η_{lagoon} within the lagoon can be neglected. The flow at the inlet channel is driven by the tide level difference

Table 6.1. Parameters of lagoons with characteristic barrier tidal inlet morphology: Lagoon length l_A, tidal range $2a$, ratio of wetted widths at HW and LW, b_S^+/b_S^-, average channel depth h and difference in duration of falling and rising tide at the inlet, Δ_{FR}^{inlet}. The length l_A is the length of the flood basin; the throat is excluded. The tidal range corresponds to mean spring tide at the inlet. The ratio b_S^+/b_S^- is estimated from the ratio of wetted lagoon surfaces S^+ at HW and S^- at LW. The channel depth h at the lagoon inlet is computed from Eq. (6.10), assuming $U_{inlet} = 1m/s$, $\alpha = 0.75$. The computed depths are consistent with published bathymetric data, but more representative for the seaward part than the landward part of the lagoons. See table 6.3 for references. In the analytic 1D model for tidal lagoons (see Appendix B) we use a linear friction coefficient $r = 0.003$ m/s for all the lagoons. The model results are consistent with $U_{inlet} = 1$ within 10%.

Identifier	Tidal lagoon	l_A [km]	$2a$ [m]	b_S^+/b_S^-	h [m]	Δ_{FR}^{inlet} [hour]
ES	Eastern Scheldt	40	3	1.3	13.	0.1
TE	Texel Inlet	50	1.5	1.2	7.7	0.6
EI	Eijerland Inlet	12	1.7	2.8	3.6	0.5
VL	Vlie Inlet	25	1.9	1.7	6.	0.4
AM	Ameland Inlet	22	2.1	2.	6.4	0.6
FR	Frisian Inlet	20	2.3	2.2	6.8	0.2
LA	Lauwers Inlet	17	2.3	2.5	6.4	0.4
ED	Ems-Dollard	20	3	2.2	8.9	0.3
OB	Otzumer Balje	10	2.8	3.	5.2	0.3
LD	Lister Dyb	20	1.8	2.	5.	1.2
LH	Langstone Harbour	5	3.25	4.	3.8	−1.4
BA	Bassin Arcachon	15	3.	1.5	5.2	−0.2
WA	Wachapreague	10	1.3	2.6	2.2	0.
MM	Murrells Main Creek	7	1.6	3.5	2.3	1.
MO	Murrells Oaks Creek	4	1.6	4	1.5	1.
NO	North inlet	6.5	1.5	3.5	2.0	0.
WB	Willapa Bay	32	3	1.8	12.5	−0.6
MU	Mussolo Bay	26	1.2	2.2	4.6	0.

between the sea and the lagoon and retarded by friction. Following these assumptions the tidal equations read

$$\frac{du_{inlet}}{dt} + \frac{g}{l_{inlet}}(\eta_{lagoon} - \eta_{inlet}) + c_D \frac{u_{inlet}^2}{D_{inlet}} = 0,$$

$$u_{inlet} = \frac{S_{lagoon}}{A_{inlet}} \frac{d\eta_{lagoon}}{dt}. \tag{6.6}$$

These equations can be solved to provide expressions for $u_{inlet}(t)$ and $\eta_{lagoon}(t)$; a general solution is given in [237, 591]. We will make further simplifying assumptions that yield first-order estimates. The surface area S_{lagoon} of the lagoon is taken independent of the tide level, the tidal variation of the inlet depth is neglected ($D_{inlet} = h_{inlet}$), the friction term is linearised as $r u_{inlet}/h_{inlet}$ with a constant friction coefficient r and the inertial term du_{inlet}/dt is neglected by assuming $r/h\omega \gg 1$. The solution of (6.6) for a sinusoidal marine tide $\eta_{inlet} = a\cos(\omega t)$ then reads

$$\eta_{lagoon} = a\cos\phi\cos(\omega t - \phi), \quad \tan\phi = \frac{r\omega S_{lagoon} l_{inlet}}{g h_{inlet} A_{inlet}}. \quad (6.7)$$

The amplitude of the tidal velocity at the inlet is given by

$$U_{inlet} = \frac{\omega a S_{lagoon}}{A_{inlet}}\cos\phi.$$

The velocity through the inlet is reduced if the inlet depth and inlet width are small, such that $\tan\phi > 1$. Lagoons with a reduced ebb velocity at the inlet are susceptible to natural closure; we will not consider such lagoons. We will only consider tidal lagoons with a sufficiently large inlet depth and width, such that the tidal range in the lagoon is comparable with the tide at sea.

Length of tidal lagoons

The characteristic geometry of tidal lagoons corresponds to an almost uniform basin width b_S that varies only with the tide level in the lagoon. We consider lagoons with a basin length l_A much smaller than the tidal wavelength L. We assume a sinusoidal tide, $\eta(x, t) = a\sin\omega t$, and a linear variation of channel cross-section A_C and the lagoon surface S with η. For short lagoons with a sufficiently wide inlet ($\phi \ll 1$) the tidal amplitude a is approximately equal to the tidal amplitude at the inlet.

The mass balance equation reads

$$Q_x + b_S\eta_t = 0, \quad (6.8)$$

where the subscripts x and t stand for partial derivation ($Q_x \equiv \partial Q/\partial x$, $\eta_t \equiv \partial\eta/\partial t$). Integration of the mass balance equation over the basin length and the flood period gives for the tidal prism

$$P = 2A_{inlet}U_{inlet}/\omega \approx 2aS_{lagoon} \approx a(S^+ + S^-), \qquad (6.9)$$

where A_{inlet} is the tidally averaged inlet cross-section, U_{inlet} the maximum tidal velocity at the inlet, $S^+ = l_A b_S^+$ is the lagoon surface at HW and $S^- = l_A b_S^-$ the lagoon surface at LW. We have neglected in (6.9) a term corresponding to the covariance of $A(t)$ and $u(t)$, assuming a phase difference close to $\pi/2$ and a tidal amplitude much smaller than the average channel depth h. The average channel depth h of the lagoon is approximated by $h = A_{inlet}/b_C$, with $b_C = \alpha b_S^-$, according to (6.5). From (6.9) we find a relationship between the lagoon length l_A and the average channel depth h and the ratio of the lagoon surfaces at HW and LW,

$$l_A \approx \frac{2\alpha h U_{inlet}}{a\omega}\left(1 + \frac{b_S^+}{b_S^-}\right)^{-1}. \qquad (6.10)$$

If no detailed bathymetric maps are available, the relation (6.10) can be used to estimate the average channel depth h. The basin length l_A and the ratio of the lagoon surfaces at HW and LW b_S^+/b_S^- can be estimated from LIDAR surveys or from satellite pictures. If no measurements are available for U_{inlet}, it will be assumed that $U_{inlet} \approx U_{inlet}^{crit} = 1m/s$, according to the empirical relationship (6.1) with $C_A = \omega/2U_{inlet}^{crit}$ and $n_A = 1$.

Unconstrained tidal lagoons

Tidal lagoons usually develop in a low-lying coastal plain behind a littoral barrier. If the plain is wide, the lagoon develops without boundary constraints. This is the case in the Wadden Sea, for example (see Fig. 6.3). The Wadden Sea consists of a series of inlets with tidal lagoons bounded by high tidal flats that act as natural boundaries. Hard boundaries at the landward side were built later, after initial lagoon development. The free development of the lagoons took place in two dimensions; the width b_S and the length l_A are therefore comparable. The basin surface S_{lagoon} is then proportional to l_A^2, the basin volume V_{lagoon} proportional to hl_A^2 and the tidal prism P proportional to al_A^2. According to 6.10, the basin length l_A is proportional to the ratio h/a. The values of the proportionality constants are approximately the same for all the Wadden lagoons, except the largest ones. For the smaller Wadden lagoons the inlet relationship (6.1) can thus be written

$$V_{lagoon} \propto A_{inlet}^{3/2}. \qquad (6.11)$$

Bathymetric studies confirm that this relationship between the lagoon volume V_{lagoon} and the inlet cross-section A_{inlet} is fairly well satisfied by the Waddden tidal lagoons and by other tidal lagoons that have developed under similar conditions [288,978].

Critical flow strength inside tidal lagoons

From several field and model studies it appears that the relationship (6.1) between cross-sectional area and tidal prism applies not only to the inlet, but also to the channels within tidal basins [122,156,194,222].

Friedrichs [324] compared the ratio of the peak spring tidal shear stress velocity and the critical shear stress velocity for channel bed scouring, U_*/u_*^{crit}, along the channels of several tidal lagoons and estuaries. When plotting this ratio as a function of the channel cross-section A_C within each basin, he found a good fit with a dependence according to $h^{1/6}$, corresponding to a constant Strickler–Manning coefficient (3.20). For sandy tidal lagoons and estuaries with large tidal channels, the observed ratios U_*/u_*^{crit} indicate peak tidal velocities of the order of 1 m/s. This value corresponds to the bottom shear stress required to flatten small bedforms (ripples, dunes) [117].

D'Alpaos *et al.* [194] simulated the development of tidal channels in idealised lagoons with characteristics similar to the Venice lagoon. They also found an approximately linear relationship between channel cross-section and upstream tidal prism, pointing to a uniform peak tidal velocity along the channels. According to this study the best fit is obtained with an exponent $n_A = 6/7$. The shallow tidal flat creeks did not follow the linear relationship between channel cross-section and the upstream tidal prism; this was imputed to drying of the creeks at low water.

Morphodynamic feedback for critical flow strength

The tendency of tidal lagoons to develop channels in which the peak current velocity is close to a 'universal' critical flow strength, points to the existence of a morphodynamic feedback mechanism. A plausible candidate for such a feedback mechanism for non-cohesive sandy lagoons is discussed in Sec. 5.5.3. When tidal currents are strong enough, the channel bed is scoured and sediment is transported towards the inner bend of channel meanders. Tidal flats are built up, the channel cross-section increases and the current velocities decrease. This process has to compete with the erosional action of waves on the tidal flats, which brings sediment back to the channel, reducing the channel cross-section and increasing the tidal current velocities. The critical flow strength is the result of these two competing processes. Its value depends on the bed grain size and on the wave climate at the tidal flats, which is a function of the tidal flat level, the wind strength, wind direction and fetch.

This feedback mechanism typically occurs in sandy tidal lagoons. It is less obvious in lagoons with a cohesive sediment bed or in areas with marsh vegetation. Fine sediments that are resuspended by wave action do not settle in the nearby channel, but are transported over larger distances. In marsh areas, the bed sediment is shielded from wave action by vegetation. In these cases, the peak value of the tidal currents is more likely determined by the minimum shear stress needed for channel bed scouring. The required peak current velocity for scouring consolidated cohesive sediment beds is higher than 0.5 m/s, see Fig. 3.17.

6.2.4. *Estuaries*

Morphologic characteristics

Estuaries have developed when the sea entered existing river plains and valleys. We only consider wide valleys that do not impose a strong topographic constraint on the morphological development of the inlet.

Estuaries are characterised by moderate to strong tides; sediment transport by tidal currents dominates over wave-induced transport. River-induced currents may dominate during peak flow events in the inner basin, but at the mouth, tidal currents are stronger.

Estuaries are typically funnel shaped, with a wide mouth and an upstream narrowing basin. In general, three zones can be distinguished [196, 198].

(1) The outer estuary, with ebb and flood channels and large submarine bars. At the intersection of ebb and flood channels, a bar is formed by the convergence of sand transport. These bars are dredged in many cases for navigation purposes. Wave-induced longshore and cross-shore transport is a major sand source for the bars in the outer estuary. The position and the shape of the bars is mainly determined by the tidal currents, see Sec. 5.2.2. In case of strong longshore sand transport a shoal or sand spit can be formed at the updrift inlet bank.

(2) The inner estuary, with meandering ebb and flood channels and tidal flats at the inner bends of the channel bends. These tidal flats have often been reclaimed and protected from flooding by the construction of levees.

(3) The upper estuary, with a single channel, which becomes a tidal river beyond the limit of seawater intrusion.

Characteristic parameters are presented in Table 6.2 for several large estuaries that share these features. The funnel-shaped inlet of the Elbe estuary is shown as an example in Fig. 6.11.

Table 6.2. Parameters of estuaries and tidal inlets with funnel morphology: convergence length l_b, average tidal range $2a$, ratio of basin surface areas b_S^+/b_S^- at HW and LW, maximum tidal current velocity U, average channel depth h and difference in duration of falling and rising tide at the inlet, Δ_{FR}^{inlet}. See Table 6.3 for references. The channel width b_C is related to the basin width b_S^- at LW by $b_C = \alpha b_S^-$, with $\alpha = 0.75$. The average channel depths are estimated from published bathymetric data. The width convergence of several estuaries (Elbe, Humber, in particular) cannot be well described with a single exponential function, as the convergence rate differs for different sections. This introduces an uncertainty in the estimate of l_b. The linear friction coefficient r is determined from the condition $\tau_{eq} = \rho r U$, with $\tau_{eq} = 2.5$ Pa.

Identifier	Tidal inlet	l_b [km]	$2a$ [m]	b_S^+/b_S^-	U [m/s]	h [m]	Δ_{FR}^{inlet} [hour]
WS	Western Scheldt	45	3.8	1.3	1.1	16	0.25
SC	Scheldt	21	5.3	1.1	1.1	10	0.75
TH	Thames	20	4.6	1.9	1	12	0.55
HB	Humber	30	5.6	1.4	1.6	12	0
DE	Dee	10	6	3	1.3	9	1.2
DY	Dyfi	6.5	3.6	4	1.	5.5	1.5
RI	Ribble	6	6	4	1	8	0.6
EL	Elbe	40	3.3	1.2	1.	12	0.9
WE	Weser	22	4	1.3	1	9	0.3
EM	Ems	22	3.3	1.	1.1	6.5	0.6
SE	Seine	25	5.5	1	1.6	8	2.4
LO	Loire	23	4.5	1.1	1.3	8	1.6
CH	Charente	10	5	1.1	0.8	6	1
GI	Gironde	40	4.2	1.2	1.6	9	1.4
SA	Satilla R.	18	2.7	1.8	0.8	7	0.5
OR	Ord	15	6	2	0.9	7	0
HO	Hooghly	36	4.2	1.1	1.2	11	0.65
FL	Fly	40	4	1.1	1.1	10	0.1
SO	Soirap	22	2.6	1.2	0.8	5	0.
GO	Gomso Bay	7.5	6	5	1.5	8	0.2
PU	Pungue	17	5.	2	1.1	11.5	0.8

Upstream sediment transport

Eroded terrigenous material carried downstream by the river is a major sediment source for estuaries. Gravelly and coarse sandy sediments are deposited in the river upstream of the estuary. Bed sediments in estuaries therefore consist predominantly of fine-grained material, such as silt and clay, especially in the inner tidal zone. At the

Table 6.3. Bathymetric references for tidal lagoons and estuaries.

Identifier	Region	Tidal inlet	References
TE	Wadden	Texel Inlet	[254]
EI	Sea	Eijerland Inlet	[254]
VL		Vlie Inlet	[254]
AM		Ameland Inlet	[254]
FR		Frisian Inlet	[254]
LA		Lauwers Inlet	[254]
ED		Ems-Dollard	[254]
OB		Otzumer Balje	[54, 155, 270]
LD		Lyster Dyb	[24, 120, 547]
ES	Netherlands	Eastern Scheldt	[254, 725]
WS		Western Scheldt	[254, 888, 912]
SC	Belgium	Scheldt	[875]
TH	England	Thames	[323, 587, 642]
HB		Humber	[467, 681, 848, 875]
LH		Langstone Harbour	[335, 639]
DE		Dee	[615]
SV		Severn	[379, 505, 860, 861, 935]
RI		Ribble	[7, 886]
DY	Wales	Dyfi	[112, 727]
EL	Germany	Elbe	[478, 505, 755, 875, 944]
WE		Weser	[350, 724, 875]
EM		Ems	[160]
SE	France	Seine	[43, 111, 311]
LO		Loire	[74, 793, 925]
CH		Charente	[847]
GI		Gironde	[13, 15, 140]
BA		Bassin Arcachon	[8, 100, 149]
SA	US Atlantic	Satilla R.	[85]
WA		Wachapreague	[321]
MM		Murrells Main Creek	[319, 320]
MO		Murrells Oaks Creek	
NO		North inlet	[321, 933]
WB	US Pacific	Willapa Bay	[521, 651]
CL		Columbia R.	[374, 420, 455, 614]
OR	Australia	Ord	[961, 964]
HO	India	Hooghly	[505, 587, 618, 776]
FL	New Guinea	Fly	[135, 380, 381, 959, 960]
SO	Vietnam	Soirap	[816]
GO	Korea	Gomso Bay	[154, 161, 162, 974]
MU	Angola	Mussolo Bay	[193]
PU	Mozambique	Pungue	[349, 548]

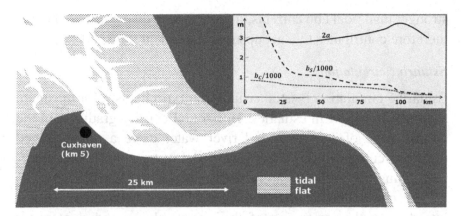

Fig. 6.11. Morphology of the entrance of the mesotidal Elbe estuary. Insert: The average tidal range $2a$, the overall width b_S and the channel width b_C along the estuary. The Elbe estuary was strongly modified by dredging and tidal flat reclamation during the past centuries. These interventions aimed primarily at enabling the passage of large ships to the harbour of Hamburg. The maximum channel depth at the inlet is 20–25 m and upstream about 15 m (maintained by regular dredging); the average channel depth h (channel cross-section divided by channel width) of about 12 m is fairly uniform along the estuary up to Hamburg (km 100). The strongly landward decreasing basin width at the mouth is characteristic for funnel shaped estuaries. The average tidal range at the entrance is in the order of 3 m; the increase of the tidal range near Hamburg is caused by the artificial deepening of the channel and the decrease of the channel width. In this part of the estuary the peak flood current is stronger than the peak ebb current, under conditions of average river discharge [944]. The estuary has a sandy bed (grain size typically 200–500 μ). The average river discharge is about 750 m^3/s, the peak discharge about 2500 m^3/s. The maximum salinity intrusion is 40 km at high river discharge and almost 80 km at low discharge. The turbidity is relatively low (usually below 250 ppm) in comparison to other meso/macrotidal estuaries (Weser, Ems, Charente, Loire, Humber, Gironde); the turbidity maximum is located in the zone of low low salinity.

outer estuary, which is subject to the combined action of currents and waves, the seabed is generally sandy. In some cases, the seabed may consist at certain locations of ancient coarse-grained river deposits (gravel — cobbles).

Several strong mechanisms operate to retain fine fluvial sediments within the estuary. The most important mechanism is tidal asymmetry: The distortion of the tidal wave when travelling up-estuary, causing a different duration and strength of flood and ebb currents (see Sec. 2.2 and later this chapter). Tidal flood currents are typically stronger than tidal ebb currents, if river flow is disregarded. The tide therefore contributes to a net upstream sediment transport.

Estuarine circulation

Estuarine circulation is another important mechanism for upstream sediment transport in estuaries. Horizontal density gradients in the mixing zone of sea water and river water drive a tide-averaged upstream current in the lower part of the water column and a corresponding net outflow near the water surface, see Fig. 6.12.

The circulation is sustained by exchanges between sea water intruding in the lower part of the water column and river water flowing in the upper part of the water column. The exchanges take place through turbulent mixing and entrainment. The landward near-bottom flow component of estuarine circulation contributes, together with the asymmetry of tidal currents, to the retention of fluvial sediments within the basin and to gradual infilling. The role of estuarine circulation is discussed further in Sec. 6.4.4.

Estuarine funnel shape

The upstream basin cross-section is narrowed as a consequence of these sediment import mechanisms. They contribute to the characteristic funnel shape of estuaries [470]. A more detailed description is presented in Sec. 6.4.2.

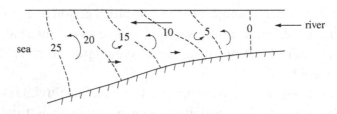

Fig. 6.12.   Schematic representation of estuarine circulation related to horizontal salinity-induced density gradients. Dotted lines: Salinity distribution (in ppt) in a longitudinal section of the estuary. Seawater of higher density drives a residual landward flow in the lower part of the water column, which is compensated by a net seaward flow near the surface.

The estuarine funnel shape does not develop if fluvial sediment supply is larger than the upstream sediment transport by tidal asymmetry and estuarine circulation. In this case the estuary will fill in and seaward progradation creates a subaerial river delta, from which the tide is excluded. Examples are the Yellow River, the Yukon and the Mahakam delta. Inlets on coasts with strong tides are funnel shaped, even if the fluvial sediment load is high. Examples are the estuaries of the Amazon, Fly, Ganges-Bramaputra and Hoogly.

Turbidity maximum

Tidal asymmetry and estuarine circulation contribute to trapping fine sediments in a zone near the limit of seawater intrusion (or even upstream of it, in the case of low river discharge [15]). In this zone, the turbidity reaches a maximum and a fluid mud layer may develop at the channel bed, in particular during neap tidal conditions (the Severn, Humber, Weser, Ems and Gironde are well documented examples [13, 181, 224, 350, 483, 868]). The turbidity maximum moves up and down the estuary with tide and river discharge, over distances comparable to the displacement of the seawater intrusion limit, see Fig. 6.13.

The turbidity maximum constitutes a pool of mobile sediment that, unlike sand, cannot easily develop into long-lasting bedforms. Most of the stable bedforms in the estuary consist of sand imported from the inlet region; these sand bodies may develop into sandbanks overlying the estuarine mud bed [196]. In a mature stage they may be capped with mud and become vegetated — a process which is presently observed in the Western Scheldt [231]. In some estuaries, for instance in the Gironde and in the Fly and Ord estuaries, sandbanks have grown into islands [380].

Mud is mainly deposited along the channel banks, where the shear stress exerted by currents is less strong than at the bottom along the channel axis. Mud deposits along the channel banks are often stabilised by marsh vegetation or by mangroves. In narrow tidal rivers

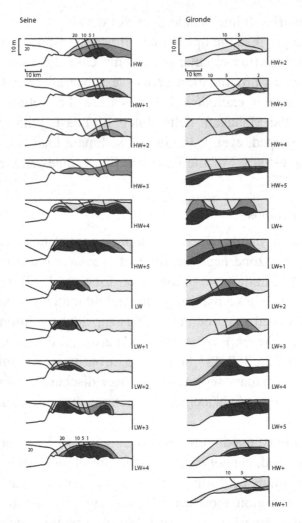

Fig. 6.13. Left: Longitudinal section of the Seine estuary showing the tidal evolution of a turbidity maximum at spring tide and low river discharge (October 1978). Numbers indicate hours after high or low water at Le Havre. Seawater intrusion is indicated by the salinity isolines for 20, 10, 5 and 1 ppt. Redrawn after [43]. Right: Longitudinal section of the south channel of the Gironde estuary showing the tidal evolution of a turbidity maximum at spring tide and high river discharge (May 1974). Seawater intrusion is indicated by the salinity isolines for 10, 5 and 2 ppt. Redrawn after [13] with permission from Elsevier. The turbidity maximum is best developed shortly before HW and LW, i.e., close to peak-flood and peak-ebb velocities. The turbidity maximum extends upstream from the limit of seawater intrusion.

the channel banks may become very steep; an example is the macrotidal King River (Northern Territory, Australia), with a 10 m deep channel and an average width-to-depth ratio of only 20 [964].

Mudflats along the channel banks have in many cases been reclaimed for human use. The resulting narrowing of the channel increases tidal velocities, because the presence of consolidated mud and underlying resistant substrate hampers channel scouring. In addition, the downstream velocity related to river flow will also increase. During high runoff events, part of the fine sediments accumulated in the turbidity maximum may be flushed temporarily out of the estuary [613] (see for example Fig. 6.42). Strong tidal currents may also prevent channel infill by dispersal of suspended sediment.

Equilibrium shear stress

Estuaries are efficient sediment traps, especially in periods of low river discharge. Tidal currents move large amounts of sediment up and down the estuary. Settling occurs around the high and low water slack tides. At spring tide freshly deposited sediment is readily resuspended, but at neap tide part of the deposits may resist resuspension. These deposits may consolidate and become resistant to erosion by the high spring tidal currents. Fine cohesive sediment deposits in particular (consolidating fluid mud) can strongly increase their resistance against erosion in the period between neap and spring. In this way an estuary gradually accretes; the cross-section is narrowed and tidal current velocities become stronger.

The accretion process stops when the shear stress of the spring tidal currents is high enough to resuspend all the sediment deposited during the neap-spring cycle. If we assume that the sediment composition is uniform along the estuary, the maximum spring tidal current shear stress should be the same everywhere in the estuary. This hypothesis was investigated by Friedrichs [324] for different estuaries. He found broad agreement when adopting for the equilibrium shear stress the formula

$$\tau_{eq} = \rho u_{*,eq}^2 = (1.65 \pm 0.5)(2a)^{0.79 \pm 0.35} \left[Nm^{-2} \right], \qquad (6.12)$$

where $2a$ ([m]) is the spring tidal range. The dependence on the spring tidal range reflects the fact that estuaries with a high spring tidal range generally have a more erosion resistant bottom than estuaries with low spring tidal range.

The existence of an equilibrium shear stress implies that, under the assumption of uniform sediment composition, estuaries tend to adapt their morphology in such way that the maximum tidal velocity is of the same order throughout the estuary. However, Friedrichs also noted deviations from the relationship (6.12). He suggested that this might be related to an along-channel variation in the residual sediment flux. We will come back to this issue in Sec. 6.4.

Schematisation

The gross morphologic characteristics of tidal lagoons and estuaries are basically different, see Figs. 6.10 and 6.14. The strong along-channel convergence of basin width b_S, which is characteristic for tide-dominated inlets (estuaries), hardly occurs for wave-dominated inlets (tidal lagoons). The basin width b_S, the channel width b_C, the channel depth h and instantaneous water depth D are defined for estuaries in the same way as for tidal lagoons, see Fig. 6.9.

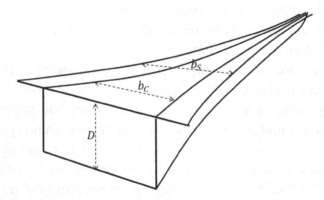

Fig. 6.14.   Sketch of the typical geometry of a river tidal basin characterised by a landward convergence of channel and basin width.

We assume again that the longitudinal flow is concentrated in the channel and that the tidal flats are lateral flood storage areas. We represent the channel flow by the cross-sectionally averaged velocity $u(x,t) = Q/A_C$, where Q is the total discharge and $A_C = Db_C$ is the channel cross-section, assuming that the flow velocity in the channel cross-section is sufficiently uniform.

Another related characteristic morphologic quantity is the propagation depth D_S, defined as $D_S = A_C/b_S = Db_C/b_S$; we call it 'propagation depth' because the tidal propagation speed c strongly depends on this quantity, as we will see later.

The morphology of an estuary (river tidal inlet) is typically funnel-shaped. The basin width b_S strongly decreases upstream and the channel cross-sectional area A_C decreases at approximately the same rate, see Fig. 6.14. The landward decrease of channel depth is much less, however [750]; we will assume that the channel depth can be considered constant along the channel, to a first approximation.

Characteristic values for several typical estuaries are given in Table 6.2. It appears that the maximum tidal velocity U is generally higher than 1 m/s in estuaries with strong tides.

Estuarine convergence

In many estuaries the longitudinal variation of the velocity $u(x,t)$ and the tidal amplitude $a(x)$ is much smaller than the longitudinal variation of the basin width $b_S(x,t)$ and the channel cross-section $A_C(x,t)$ [691]. In this case of strong convergence the mass balance equation

$$Q_x + b_S \eta_t = 0 \tag{6.13}$$

can be approximated by the simpler equation

$$U(A_C)_x + \omega a b_S = 0, \tag{6.14}$$

where we have assumed a sinusoidal tide with frequency $\omega = 2\pi/T$ and where U is the maximum cross-sectionally averaged tidal current velocity. We further assume that the tidal amplitude is much smaller than the channel depth, $a \ll h$. Most estuaries have relatively small tidal flats, compared to the channel width (this is often the result of channel embanking and reclamation). In this case we approximate $b_S^+ \approx b_S^- \approx b_C/\alpha$, with α constant. If the channel depth has a constant small slope $h_x = I$, we find from Eq. (6.14)

$$b_C = b_C(o)e^{-x/l_b}, \qquad (6.15)$$

where l_b is the convergence length of the estuary. The x-axis is in the up-estuary (landward) direction and $x = 0$ corresponds to the estuarine mouth. Uniformity of the maximum tidal current velocity along the estuary thus implies exponential convergence of the channel width (for strongly converging estuaries with small tidal flats).

With the above assumptions, we find for the convergence length

$$l_b \approx \frac{\alpha h U}{a\omega + \alpha U I}. \qquad (6.16)$$

Many estuaries are heavily dredged for ensuring a constant navigation depth, such that $I \approx 0$. In that case we have

$$l_b \approx \frac{h_S U}{a\omega}. \qquad (6.17)$$

The convergence length is proportional to the tidal velocity amplitude U and inversely proportional to the relative tidal amplitude a/h. The average propagation depth h_S is related to the average channel depth h by

$$h_S = h b_C/b_S \approx 2\alpha h(1 + b_S^+/b_S^-)^{-1}, \qquad (6.18)$$

where $b_S^\pm$ is the basin width at HW (LW). The channel width b_C depends on the shape of the channel cross-section. It is smaller than the LW basin width b_S^- and typically of the order of $\alpha = b_C/b_S^- \approx 0.6 - 0.9$.

Length of tidal rivers with triangular cross-section

Prandle [688] derived morphologic relationships for tidal rivers without a priori assumptions on the longitudinal variation of the channel depth $h(x)$ and the tidal velocity amplitude $U(x)$. These relationships are based on the assumption of a triangular channel cross-section with constant width-to-depth ratio and a uniform (x-independent) tidal amplitude. In the case of friction-dominated tidal flow he found for the tidal velocity amplitude $U(x)$ the expression

$$U(x) \approx 1.2 \, c_D^{-1/2} g^{1/4} \omega^{1/2} a^{1/2} h^{1/4}. \tag{6.19}$$

According to this expression the tidal velocity amplitude decreases slightly in landward direction with decreasing depth. The depth decreases from the seawards boundary $x = 0$ to the landward boundary $x = l_p$ according to $h(x) \approx 1.33 \, a\omega(l_p - x)/U(x) \propto (l_p - x)^{4/5}$.

The estuarine length l_p, defined by $h(0) = -\int_0^{l_p}(dh/dx)dx$, is given by

$$l_p \approx 0.7 \, \frac{h(0)U(0)}{a\omega}, \tag{6.20}$$

where $h(0)$ is the depth at the estuarine mouth. This expression is similar to (6.17), if we consider $0.7\,h(0)$ as a typical average depth of the triangular channel cross-section.

6.2.5. Other Inlet Types

Mixed inlet types

The previous description of tidal lagoons and estuaries is a crude simplification of reality. Most tidal basins deviate from these simple descriptions and possess characteristics belonging to both categories. Generally, both microtidal and macrotidal inlets have an ebb-tidal delta influenced by wave action. Macrotidal inlets have a much wider mouth than microtidal (wave-dominated) inlets. The inner basin of lagoons and estuaries is often tide-dominated with a secondary role of wave action and river run-off; the characteristic funnel shape occurs when tidal currents are strong enough. Typical examples of mixed inlets, with a tide-dominated inner basin and a strongly wave-influenced inlet zone are the Western Scheldt, Humber, Dyfi, Weser-Jade and Ems-Dollard.

Not all macrotidal estuaries have a wide funnel-shaped mouth. This is due to topographic constraints of geological origin. Examples are the Loire, the Tagus and the Mersey.

Tidal bays

In our terminology tidal bays are shallow wide open basins with strong tides and small river inflow. The largest tides occur in tidal bays with a converging geometry and a length close to the resonance length of the semidiurnal tidal wave (typically of the order of 100 km). Examples are the Bay of Fundy, the Bristol Channel, the Baie du Mont Saint Michel and Hangzou Bay, with a mean spring tidal range exceeding 10 m. The tide is strongly distorted when propagating into these bays; a tidal bore develops during spring tide. The dominance of flood currents over ebb currents promotes the import of marine sediment into these bays. Sediment import is limited if marine sediment sources are scarce, as in the case of the Bay of Fundy and the Bristol Channel.

Smaller bays in macrotidal coastal zones exhibit the characteristic funnel shape. Examples are the Baie de Somme, the Ribble estuary, the Dee estuary, and Gomso Bay; the morphology of Gomso Bay (South Korea) is shown as an example in Fig. 6.15. The strong flood and ebb currents carry large amounts of sediments. Large tidal flats are the result of marine sediment import.

The sediment import during flood must be compensated by export during ebb. If not, these small macrotidal bays would have filled in completely during the past period of slow sea-level rise. These small bays are very sensitive to waves incoming from the sea. These waves probably play an important role by stirring sediment deposited on the tidal flats and increasing the sediment load of the tidal ebb currents.

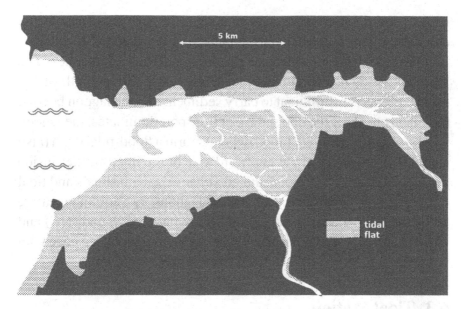

Fig. 6.15. Typical macrotidal basin: Gomso Bay (South Korea). Land is dark and subtidal areas are white.

The tide is semidiurnal; the average tidal range is 4.6 m (maximum 8 m). River inflow is very small. The basin has a funnel-shaped morphology, characteristic for macrotidal inlets. The water depth of the main channel reaches 20 m during spring tide. The maximum ebb current (about 1.5 m/s) is stronger than the maximum flood current. The inner basin is tide dominated, with large bare tidal flats. The tidal flats have typically an upward convex profile; they are muddy and strongly reworked by bioturbation. The outer bay is wave dominated with large muddy concave-up tidal flats. Significant offshore wave heights during winter are 2–4 m. The middle part of the basin is sandy. Macrotidal inlets with similar morphology include Baie de la Somme, Baie du Mont St. Michel (France), Morecambe Bay, the estuaries of the Ribble, Dee (UK), Haeju Bay (North Korea) and Baia de Sofala (Mozambique).

Tidal systems not in pace with sea level

At the opposite of macrotidal bays are the microtidal lagoons. We consider lagoons which are hardly influenced by import of river water and river sediment.

A lagoon is formed when, shortly after marine transgression, a coastal plain is separated from the sea by a wave-built sand barrier. It becomes a tidal lagoon if the tidal prism is large enough to prevent closure by wave-induced longshore transport. If flood currents inside the lagoon are strong enough to carry sediment and if the sediment load of the flood currents is higher than the sediment load of the ebb currents (or at least equal), a pattern of flood and ebb channels will develop, with intermediate shoals, tidal flats and marshes. The Wadden Sea is a typical example. However, if the strength of the flood currents is insufficient to carry sediment into the lagoon beyond the inlet zone, or if the ebb sediment transport dominates, the lagoon will be sediment starved. Flood deposits form a flood-tidal delta (also called 'flood shield') at the landward side of the inlet; further inside the lagoon there are no flood and ebb channels, with shoals and tidal flats. Such sediment-starved microtidal lagoons are a common feature along the US Atlantic coast; examples are Pamlico Sound (NC) and Great South Bay (Long Island). These lagoons cannot keep pace with sea level, even when the sea level rises slowly.

6.3. Tidal Motion

Ocean tides

Tidal motion is the oscillation of ocean waters under influence of the attractive gravitational forces of the moon and the sun. The response of the ocean to the gravitational forces follows a pattern of rotating ('amphidromic') systems, as a consequence of the earth's rotation, see Fig. 6.17. The frequencies are determined by the relative periodic motions of moon, sun and earth surface. The amplitude of the tidal oscillation is very small compared to ocean depths. The tidal oscillation in each point can therefore be represented by a linear superposition of sinusoidal tidal components. The most important tidal components have a periodicity which is close to semidiurnal or diurnal, due to earth's rotation. A more detailed discussion of ocean tides is given in Appendix C.1.

In this chapter we will generally assume that the semidiurnal tides are much stronger than the diurnal tides, unless stated otherwise. This assumption is reasonable for the Atlantic region; in some parts of the Pacific diurnal tides can be substantial. The semidiurnal lunar tide (called M_2, period ≈ 12.5 hour) dominates the other semidiurnal

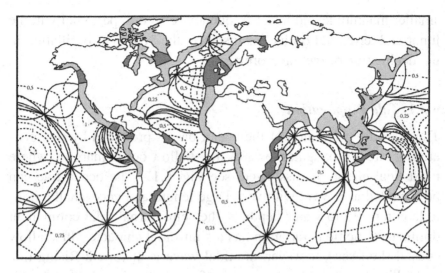

Fig. 6.17. System of semidiurnal lunar tidal waves (M2) in the oceans represented by lines of equal tidal phase (solid, intervals of 30°) and lines of equal tidal amplitude (dashed, intervals of 0.25 m). Three ranges are indicated for the tidal amplitude a on the continental shelf: Microtidal (white fringe, $a < 1$ m), mesotidal (light grey fringe, $1 < a < 2$ m) and macrotidal (dark grey fringe, $a > 2$ m). Redrawn after [62].

tidal components. The second strongest semidiurnal component is the semidiurnal solar tide (called S_2, period 12 hours). The superposition of these two semidiurnal tides has the character of a sinusoidal semidiurnal oscillation with an amplitude modulated at a frequency equal to the difference of the M_2 and S_2 tidal frequencies. This amplitude modulation corresponds to the fortnightly spring-neap tide cycle.

In regions where the diurnal tidal components are strong, the tide cannot be approximated by a single sinus curve. The main diurnal tidal components are the lunar diurnal component O_1 (period ≈ 26 h), the solar diurnal component P_1 (period ≈ 24 h) and the component K_1 (period ≈ 24 h), which is related to the declination of the lunar and solar orbits relative to the equatorial plane. In ocean regions with substantial diurnal tides, the tide curve exhibits a strong daily inequality. Interference of constituents (in particular M_2, K_1, O_1) can produce a systematic asymmetry between tidal rise (upward motion) and tidal fall (downward motion). There are many other tidal components of smaller magnitude and lower frequency related to periodicity in the lunar and terrestrial orbits, for instance, the 18.6 year oscillation in the declination of the lunar orbit.

6.3.1. *Tide Propagation*

The periodic oscillation of the ocean basins produced by tide generating forces propagates towards the shallow ocean margins, where most tidal energy is dissipated, see Fig. 6.17. The ocean tide can be regarded as a wave with very large length and very small amplitude compared to water depth. If the semidiurnal tidal component dominates, the tidal variation has a sinusoidal character. In shallow coastal shelf seas, however, the tidal wave is distorted: Rising tide and falling tide become asymmetric (for a discussion of coastal shelf tides, see Appendix section C.3). If the water depth at HW is much larger than the water depth at LW, the tidal wave becomes so strongly distorted that it evolves into a breaking tidal bore (see the example of Qiantang estuary, Sec. 2.2). The distortion of the tidal wave (leading to an asymmetric tide) is caused by a difference between the propagation speeds of the wave crest and the wave trough. The generation of tidal asymmetry can be understood by examining the propagation speeds of the high-water wave crest and the low-water wave trough.

In this section the tidal propagation process will be described for a few idealised situations. We first discuss tidal propagation for frictionless tidal flow in a rectangular channel of infinite length. This case is not representative for real tidal inlets. The tidal flow is in general strongly influenced by friction; the cross-section is non-rectangular and depends on the tidal phase. We discuss not only this second case but also a third case where the tide propagates in a strongly converging channel. The qualitative description highlights the fundamental differences between the propagation processes in these three cases.

Overtides and harmonic analysis

Tidal distortion in shallow water can also be described in terms of higher harmonic tidal components, the so-called overtides. If it is assumed that the leading ocean tide is the semidiurnal lunar M_2 tide, the higher harmonic tidal components M_4, M_6, M_8, etc. can be determined by Fourier development of the non-linear tidal equations. This alternative method (harmonic analysis) is practical from a computational point of view, if the higher harmonic components are small, to ensure rapid convergence of the Fourier development [248]. However, the generation mechanism of the overtides is not straightforward from a physical point of view. Strong tidal distortion, leading to tidal bore formation, cannot be described in this way. For this reason we prefer here to consider tidal distortion as a consequence of different propagation speeds of the high-water crest and the low-water trough of the tidal wave.

Frictionless tidal wave propagation

In the ocean, tidal wave propagation is almost frictionless. Frictionless wave propagation corresponds to a local departure of the sea surface from equilibrium, which propagates by converting continuously potential energy into kinetic energy and vice versa, without loss of energy. We consider frictionless tidal propagation in a rectilinear uniform channel of infinite length. This situation does not occur in nature, but it bears some resemblance with the propagation of the ocean tide along the shelf break, or with the propagation of the tide in an estuary with uniform cross-section, small amplitude-to-depth ratio and small bottom friction (due to the presence of a lutocline, for instance).

If the tidal wavelength λ is much larger than the average water depth h, we may assume that the pressure distribution over the vertical is hydrostatic. In this case the wave propagation in a single spatial dimension x can be described by the equations for conservation of mass and momentum (see Appendix B),

$$\eta_t + (Du)_x = 0, \quad u_t + uu_x + g\eta_x = 0. \qquad (6.21)$$

In these equation $\eta(x, t)$ is the departure of the sea level from equilibrium, u is the depth-averaged current velocity and $D = h + \eta$ is the local instantaneous depth. We assume a uniform (x-independent) average depth h. We use the convention that indices x, y, z, t

stand for partial differentiation to these variables; for instance, $\eta_t \equiv \partial\eta/\partial t$, etc.

Propagation speed of wave crest and wave trough

The tidal wave has a maximum (wave crest) and a minimum (wave trough) at locations where $\eta_x = 0$. These locations migrate while the wave propagates. We call c^+ the propagation speed of the wave crest and c^- the propagation speed of the wave trough. When following the wave crest on its path during propagation we have

$$\frac{d}{dt}\eta_x = \eta_{xt} + c^+\eta_{xx} = 0. \tag{6.22}$$

Because $u_x = 0$ coincides with $\eta_x = 0$, the propagation speed of the location of maximum current velocity equals the propagation speed of the wave crest,

$$\frac{d}{dt}u_x = u_{xt} + c^+u_{xx} = 0.$$

From (6.21) we substitute expressions for η_{xt} and η_{xx} and find at the locations where $\eta_x = 0$

$$\eta_{xt} = -(D^+u^+)_{xx} = -u^+\eta_{xx} - D^+u_{xx} = -c^+\eta_{xx},$$
$$g\eta_{xx} = -u_{tx} - u^+u_{xx} = u_{xx}(c^+ - u^+), \tag{6.23}$$

where $D^+ = h+a$ is the water depth at the wave crest and a the wave amplitude. In a frictionless propagating wave $u(x, t)$ is maximum at the wave crest ($u^+ = a\sqrt{g/h}$). Combination of the two equations (6.23) gives

$$(c^+ - u^+)^2 = gD^+. \tag{6.24}$$

After substitution of u^+ we find to order a/h,

$$c^+ = u^+ + \sqrt{gD^+} \approx \sqrt{g(h + 3a)}. \tag{6.25}$$

For the propagation speed of the wave trough we find similarly

$$c^- \approx \sqrt{g(h - 3a)}. \tag{6.26}$$

The wave crest and the wave trough propagate at different speeds. The wave crest propagates faster and therefore tends to catch up with the wave trough. In the ocean this effect is insignificant, as the water depth is much larger than the wave amplitude. But in shallow water, the tidal wave will be distorted, with a faster rise (as a function of time) and a slower fall.

Friction-dominated tidal wave propagation

If the water depth is small, the assumption of frictionless propagation does not hold. Frictional effects may even dominate inertia (see Appendix Sec. B.2). In the case of strong friction ($r \gg h\omega$, where $\omega = 2\pi/T$ is the tidal frequency), the tidal equations read

$$\eta_t + (Du)_x = 0, \quad g\eta_x + ru/D = 0, \tag{6.27}$$

where the friction term has been linearised with a friction coefficient $r \approx 0.85 c_D U$. Elimination of the velocity u from the two equations gives

$$\eta_t = \left(\frac{gD^2}{r} \eta_x \right)_x. \tag{6.28}$$

We solve this equation around the high water crest of the tidal wave ($\eta_x \approx 0$) under the assumption that the reflected tidal wave is strongly damped by friction. The tidal equation (6.28) becomes linear at the wave crest if we ignore the time dependence of η. The solution then reads

$$\eta = ae^{-k^+ x} \cos(k^+ x - \omega t), \quad k^+ = \frac{\sqrt{r\omega/2g}}{h+a}.$$

The propagation speed of the wave crest is thus given by

$$c^+ = \frac{\omega}{k^+} = (h+a)\sqrt{2g\omega/r}. \tag{6.29}$$

For the propagation speed of the wave trough we find similarly

$$c^- = \frac{\omega}{k^-} = (h-a)\sqrt{2g\omega/r}. \tag{6.30}$$

In shallow water, the propagation speed of the tidal wave crest can become substantially larger than the propagation speed of the wave trough. The tidal wave is distorted with a steeper rise and slower fall when propagating into shallow coastal waters.

However, in a friction-dominated tidal wave, HW (LW) does not coincide with the wave crest (trough). This implies that the tidal equation (6.28) is nonlinear around the wave crest (and wave trough). The expressions (6.29) and (6.30) for the propagation speed are therefore approximations. In Appendix Sec. B.3 it is shown that, for small amplitude $a \ll h$, the factors $h \pm a$ in the r.h.s. of (6.29) and (6.30) should be replaced by $h \pm (2 - \sqrt{2})a$.

Tidal diffusion

It is worthwhile noticing that the tidal equation (6.28) is a diffusion equation with diffusion coefficient gD^2/r. The diffusion character of tidal wave propagation implies damping of the tidal amplitude during propagation. The damping rate is inversely proportional to the wave length $\lambda = 2\pi/k$. A small diffusion coefficient (i.e., a large friction coefficient r) corresponds to strong tidal wave damping. Equation (6.28) illustrates the fundamentally different nature of tidal wave propagation in the frictional case compared to the frictionless case (6.21).

Tidal wave propagation in a channel with tidal flats

In the foregoing sections the width dependence of tidal flow has been ignored; the validity of the results is therefore restricted to tidal channels with rectangular cross-section and constant width. In reality, tidal flow is not restricted to the deepest parts of the channel where most of the tidal flow takes place. Some time before high water the flow spreads over tidal flats, while some time before low water the flow is confined to the deepest channel parts. Therefore a distinction has to be made between the channel width, b_C, representative of along-channel flow, and the surface width, b_S, representing the water-covered surface of the tidal basin. We will assume, as for the tidal lagoon schematisation, that they do not depend on x, see Fig. 6.10, the tidal Eqs. (6.21) and (6.27) then become

$$b_S \eta_t + (b_C D u)_x = 0, \quad u_t + u u_x + g \eta_x = 0 \qquad (6.31)$$

for frictionless propagation and

$$b_S \eta_t + (b_C D u)_x = 0, \quad g \eta_x + r u / D = 0 \qquad (6.32)$$

for friction-dominated propagation. The propagation speed of the tidal wave crest and wave trough can be derived in the same way as before. For frictionless tidal flow we have to replace (6.24) by

$$g D_S^{\pm} \approx (c^{\pm} - u^{\pm})(c^{\pm} - \frac{h_S}{h} u^{\pm}) \approx (c^{\pm} - \frac{h + h_S}{2h} u^{\pm})^2, \qquad (6.33)$$

where terms of second order in a/h are neglected. In the case of friction-dominated tidal flow, (6.29) and (6.30) are replaced by

$$c^{\pm} = \sqrt{2g D^{\pm} D_S^{\pm} \omega / r}. \qquad (6.34)$$

In these expressions

$$D^{\pm} = h \pm a$$

is the water depth at HW resp. LW and

$$D_S^{\pm} = (h \pm a) b_C^{\pm} / b_S^{\pm}$$

is the propagation depth at HW resp. LW.

In basins with large tidal flats, $b_S^+ \gg b_S^-$. If the amplitude-to-depth ratio a/h is small, we then have $D_S^+ < D_S^-$ and $D^+ D_S^+ < D^- D_S^-$. In this case the wave trough propagates faster than the wave crest. This can be understood by realising that the propagation of the tidal crest is retarded due to lateral diversion over the tidal flats. Analytical expressions for $c^{\pm}$ are derived in Appendix B.3 for the nonlinear tidal equations without friction (6.31) and with strong friction (6.32). This derivation shows that the dependence of the tidal propagation speed on the storage widths b_S^+ and b_S^- is correctly represented by the formulas (6.33) and (6.34).

Friction coefficient

The general form of the 1D tidal equations reads

$$b_S \eta_t + (Db_C u)_x = 0, \quad u_t + u u_x + g \eta_x + c_D \frac{|u|u}{D} = 0. \quad (6.35)$$

The inertial term $u u_x$ can in general be neglected, because for tidal flow the Froude number $F = |u|/\sqrt{gh}$ is small.

An analytic solution for the tidal equations only exists for the linearised form, where the local instantaneous depth D is replaced by a constant h. The friction term $c_D |u|u/D$ must be linearised too. We might expect that this affects the result, because the friction around HW and LW is much smaller in the non-linear case than in the linear case. However, according to numerical models it appears that linearisation of the friction term according to

$$c_D |u|u/D = ru/h, \quad r = \frac{8}{3\pi} c_D U \approx 0.85 c_D U \quad (6.36)$$

yields fairly good results for tidal wave propagation, if the semidiurnal tidal amplitude is strongly dominant [505]. The symbols c_D and U in this expression represent the friction coefficient (both grain and form drag, see Sec. 3.2.2) and the maximum tidal current velocity respectively.

For the estuaries of Table 6.2, we derive the value of the linear friction coefficient following the hypothesis of Friedrichs [324], discussed in Sec. 6.2.4. A simplified version of this hypothesis states that the maximum shear stress for average spring tidal conditions,

$$\tau_{max} = \rho r U, \quad (6.37)$$

should be everywhere the same in each estuary. For all the estuaries of Table 6.2 we use the value $\tau_{max} = 2.5$ Pa. The value of the linear friction coefficient can then be determined from the measured maximum cross-sectionally averaged tidal current velocities indicated in Table 6.2.

This prescription is not fully correct, for several reasons, pointed out by Friedrichs [324].

(1) The critical shear stress for erosion depends on the constitution of the channel bed; it is different for consolidated cohesive beds, for sand beds and for gravel beds. For beds consisting of consolidated cohesive mud or gravel beds it is higher than for sandy beds, see Sec. 3.5.3 and Fig. 3.17.

(2) The relationship between the shear stress acting on bed particles and the corresponding depth-averaged velocity depends on form drag (ripples, dunes) and on turbulence damping by density stratification, see section 3.2.2.

(3) The maximum tidal velocity in some estuaries is not only determined by the critical shear stress, but also by channel narrowing as a result of net upstream sediment transport.

In the following we will use very simplified hydro-sedimentary models in which these refinements are ignored. However, when interpreting the results, we have to be aware of these simplifications.

The tidal velocity amplitude U computed with this friction coefficient, using the 1D linear model (6.41), is generally in fair agreement (within 10%) with the observed tidal velocity amplitude (Table 6.4). The tidal velocity amplitude is underestimated for the Ems estuary; a lower friction would be needed. The computed tidal velocity amplitude is too large for the Weser, Ord and Fly estuaries; here a higher friction coefficient would be needed.

Table 6.4. Observed and modeled maximum tidal current velocities [m/s] for the estuaries of Table 6.2, following the 1D model (6.41), using a linear friction coefficient r determined from the condition $\tau_{eq} = \rho r U$, with $\tau_{eq} = 2.5\,\text{Pa}$. The bathymetric parameters for convergence length l_b, tidal amplitude a and ratio of basin surface areas b_S^+/b_S^- at HW and LW are taken from Table 6.2.

Identifier	WS	SC	TH	HB	DE	DY	RI	EL	WE	EM
u obs	1.1	1.1	1	1.6	1.3	1.	1.	1.	1.	1.1
u mod	1.1	1.14	1.06	1.76	1.36	1.0	1.0	0.98	1.16	0.92
Identifier	SE	LO	GI	SA	OR	HO	FL	SO	GO	PU
u obs	1.6	1.3	1.6	0.8	0.9	1.2	1.1	0.8	1.5	1.1
u mod	1.64	1.26	1.61	0.76	1.4	1.26	1.24	0.77	1.57	1.1

Strongly convergent basins

We have assumed so far that the channel geometry is uniform in the tidal propagation direction. This implies that a propagating tidal wave is damped by friction. However, when propagating into a progressively narrowing channel, frictional damping is compensated by concentration of the tidal energy flux. In the following we will assume an exponentially converging channel width (b_C, $b_S \sim \exp(-x/l_b)$), according to the schematisation of estuarine inlets discussed before. For strongly converging basins ($l_b \ll |u/u_x|$) and friction dominated tidal flow, the tidal equations read (see Appendix B.2)

$$b_S \eta_t + D u (b_C)_x = 0, \quad g \eta_x + r u / D = 0. \tag{6.38}$$

Eliminating u from these equations yields

$$\eta_t = -\frac{\omega}{k} \eta_x, \quad c = \frac{\omega}{k} = \frac{g D D_S}{r l_b}. \tag{6.39}$$

At the wave crest and wave trough, the x- and t-derivatives of η are zero. Therefore the tidal equation (6.39) is approximately linear around the wave crest and trough. For the HW (LW) propagation speed we thus have the solution (disregarding second-order terms $(a/h)^2 \ll 1$)

$$c^{\pm} = \frac{g D^{\pm} D_S^{\pm}}{r l_b}, \tag{6.40}$$

where r is the linearised friction coefficient, D the water depth and $D_S = D b_C / b_S$ the propagation depth.

From (6.38) we note that the phase difference between tidal velocity and tidal elevation in a strongly converging, friction-dominated tidal basin is equal to $\pi/2$ (up to order a/h). Yet, we do not have a standing wave, because the tide propagates. Tidal propagation characteristics are strongly dependent on basin geometry. The propagation velocity is inversely proportional to the linearised friction coefficient. From (6.39) it follows that the tidal wave amplitude is approximately constant along the basin, in spite of the fact that the flow is friction dominated. It appears that tidal propagation in a strongly converging, friction-dominated basin combines features of both a standing tidal wave (although there is no reflected wave) and a frictionless propagating wave.

Weakly convergent estuaries

The simplifying assumption of strong convergence are for many converging estuaries questionable. The product of wavenumber k and convergence length l_b is generally not much smaller than 1 and the acceleration term u_t has often the same order of magnitude as the (linearised) friction term ru/h. Because an exact analytical solution of the tidal equations is possible only if all terms are linear in η and u, we assume that the variation of the depth and width with x, t can be ignored. We therefore solve the equations for limited time intervals around HW and LW, where the variation of D and D_S is small. The tidal equations (6.35) then become

$$\eta_t + D_S u_x - D_S u/l_b = 0, \quad u_t + g\eta_x + ru/D = 0. \qquad (6.41)$$

The solution of these linear equations reads

$$\eta = ae^{-\mu x}\sin(kx - \omega t - \varphi), \quad u = -\frac{a\omega \sin\varphi}{kD_S}e^{-\mu x}\cos(kx - \omega t),$$

$$(6.42)$$

where

$$kl_b = \sqrt{\frac{1}{2}\left(p + \sqrt{p^2 + (k_c l_b)^2}\right)},$$

$$\mu l_b = -\frac{1}{2} + \sqrt{\frac{1}{2}\left(-p + \sqrt{p^2 + (k_c l_b)^2}\right)},$$

$$k_0 = \frac{\omega}{\sqrt{gD_S}}, \quad k_c = \frac{\omega r l_b}{gDD_S}, \quad p = (k_0 l_b)^2 - \frac{1}{4},$$

$$\tan\varphi = \frac{kl_b}{\mu l_b + 1}. \qquad (6.43)$$

Ignoring the time dependence of η at the wave crest (trough) we find for the HW and LW propagation speeds

$$c^{\pm} = \frac{\omega}{k^{\pm}}, \quad k^{\pm} = k(D^{\pm}, D_S^{\pm}). \tag{6.44}$$

It can easily be seen that (6.43) transforms into (6.33) for large $k_0 l_b$ and small $k_c l_b$, into (6.34) for large $k_c l_b$ and into (6.40) for small $k_0 l_b$. The phase delay φ between HW and HWS is equal to the phase delay between LW and LWS and approximately constant along the estuary. This is illustrated for the Loire estuary in Fig. 6.16.

The expressions (6.44) differ from the propagation speeds that would result from solving the tidal equations (6.41) with variable D and D_S. From the solution (6.42) it appears that the wave crest

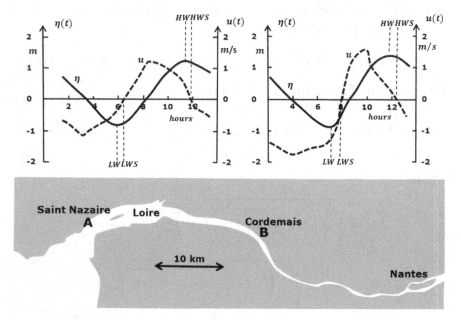

Fig. 6.16. Curves for tidal elevation and velocity in the Loire estuary, at stations A (Saint Nazaire) and B (Cordemais, 22 km upstream), during low river discharge, October 2000. The time delay between HW and HWS is almost equal to the time delay between LW and LWS. The time delays hardly vary throughout the estuary. (Measurements by HOCER, Projet Loire Grandeur Nature.)

(trough) does not coincide with HW (LW). We are in a situation intermediate between the cases of strong convergence, friction-dominated flow and frictionless flow. We therefore assume that (6.44) is a reasonable first approximation for the HW and LW propagation speeds. The inaccuracy in (6.44) is in most cases probably not greater than the inaccuracies inherent to the one-dimensional simplification of the lagoon and estuary bathymetries.

Comparison with data

Figure 6.18 shows a comparison of the HW and LW propagation speeds according to the expressions (6.40) and (6.44) with the propagation speeds derived from tide tables, see Table 6.5. At high water the tide is still flooding and during low water still ebbing, see for instance Fig. 6.22. Therefore a term $\pm U \sin \varphi$ has been added to $c^{\pm}$ in the expressions (6.40) and (6.44), to take into account the tidal velocity experienced by the tidal wave during propagation. A typical

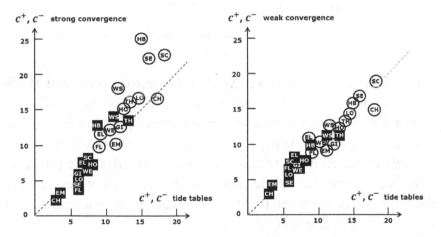

Fig. 6.18. Comparison of propagation speeds for HW and LW from tide tables with the propagation speeds computed from (6.40) and (6.44) for several estuaries listed (with acronyms) in Table 6.2. The black squares are for the LW propagation speed c^{-} and the white circles for the HW speed c^{+}. The formula for strongly converging estuaries overestimates in most cases the observed HW propagation speeds. The model (6.44) for weak convergence reproduces the observed propagation speeds fairly well, considering the uncertainties in the bathymetric parameters and in the tide table data.

Table 6.5. Propagation speed of HW and LW for several estuaries between $x = 0$ (mouth) and $x = l_b$, based on tide tables. The tide tables are based on harmonic analysis of data from tide gauges at different locations along these estuaries. The HW and LW propagation speed is not constant along the estuary (see, for instance, figure 6.19). Because tide stations generally do not coincide with $x = 0$ and $x = l_b$, the uncertainty in the values is substantial (up to 30%).

Identifier	Tidal inlet	c^+ [m/s]	c^- [m/s]
WS	Western Scheldt	11.7	11
SC	Scheldt	18.3	7.5
TH	Thames	13.5	13
HB	Humber	15	9
EL	Elbe	9	6
WE	Weser	10	7.2
EM	Ems	11.4	3.5
SE	Seine	16	6
LO	Loire	14	6
CH	Charente	17	3
GI	Gironde	12	6.1
HO	Hooghly	13	8
FL	Fly	9	6

value of the phase difference $\Delta_S \equiv \varphi/\omega$ between HW and HWS (or LW and LWS) is in the range 30–60 minutes. The more elaborate expression (6.44) represents the propagation speeds derived from tide tables in general better than the simpler expression (6.40) for strongly converging and friction-dominated estuaries. The propagation speeds derived from tide tables are also less well represented by the expressions (6.33) and (6.34) for non-converging frictionless and friction-dominated estuaries.

6.3.2. Tidal Asymmetry

Role of tidal asymmetry

Tidal asymmetry plays an essential role in producing large-scale residual sediment transport in tidal inlets. This tide-induced residual transport has important consequences for the equilibrium morphology of tidal inlets and for the morphologic response to changing external conditions. Tidal asymmetry depends on the characteristics of tidal propagation and these characteristics depend on the morphology of the tidal inlet, as shown in the previous section. In other words, there is a mutual dependency between tidal asymmetry and inlet morphology. In this section we will examine this mutual dependency and derive explicit expressions. These expressions will be used later to analyse the stability of tidal inlets and to derive relationships for the equilibrium morphology.

Inequality of flood and ebb duration

In the previous sections we have seen that the form of the tidal wave changes when the tide propagates into shallow water. The wave crest propagates at a different speed than the wave trough, due to differences in water depth and basin width. Figure 6.19 shows, as an

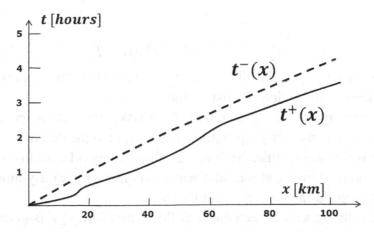

Fig. 6.19. Propagation of HW, $t^+(x)$, and LW, $t^-(x)$, in the Elbe estuary [102]. The times of HW and LW at the estuarine mouth at $x = 0$ (near Cuxhaven) are both set at t=0, to facilitate the comparison. In reality the time delay $t^-(0) - t^+(0) = (T + \Delta_{FR}^{inlet})/2$. The propagation speed of HW and LW is given by $c^\pm = (dt^\pm/dx)^{-1}$. The tidal wave crest propagates faster than the wave trough in the converging estuarine section (km 0–km 40; $c^+ \approx 9$ m/s, $c^- \approx 6$ m/s). The data show that the HW propagation speed is not constant along the estuary.

illustration, the propagation of the tidal wave crest (HW) and tidal wave trough (LW) in the Elbe estuary.

The propagation velocities of high water (the wave crest) and low water (the wave trough) depend on the basin geometry and on tidal propagation dynamics. For a few particular cases the propagation speeds can be approximated by simple analytical expressions given in (6.33), (6.34), (6.40) and (6.44).

Once the times of high water (HW) and low water (LW) at the seaward boundary of the tidal basin are known, the times of HW ($t^+(x)$) and LW ($t^-(x)$) within the basin can be derived from the high water propagation velocity c^+ and the low water propagation velocity c^-,

$$t^+(x) = t^+(0) + x/c^+, \quad t^-(x) = t^-(0) + x/c^-, \qquad (6.45)$$

where x is the distance to the seaward boundary, see Fig. 6.20. The difference between the periods of falling and rising water within the basin is given by

$$\Delta_{FR} = 2\left[t^-(x) - t^+(x)\right] - T = \Delta_{FR}^{inlet} + 2x\left[1/c^- - 1/c^+\right], \tag{6.46}$$

where

$$\Delta_{FR}^{inlet} = 2t^-(0) - 2t^+(0) - T$$

is the asymmetry of the offshore tide at the inlet (difference between duration of falling tide and rising tide).

For instance, if $\Delta_{FR}^{inlet} = 0$ and if the HW propagation speed c^+ is higher than the LW propagation speed c^-, then the duration of the rising tide is shorter than the duration of the falling tide, see Fig. 6.21. This figure shows that the tidal wave becomes (more) asymmetric when propagating through the tidal basin.

The difference between ebb and flood duration Δ_{EF} is given by a similar formula:

$$\Delta_{EF} = 2\left[t_S^-(x) - t_S^+(x)\right] - T, \tag{6.47}$$

where t_S^+ is the tidal phase of flow reversal from flood to ebb (high water slack tide HWS) and t_S^- is the tidal phase of flow reversal from ebb to flood (low water slack tide LWS).

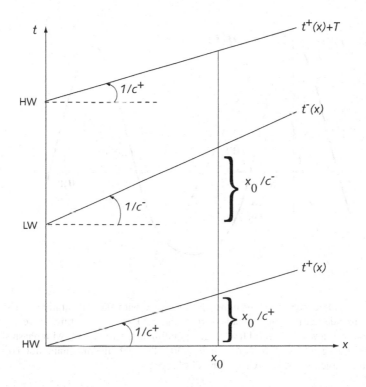

Fig. 6.20. Propagation of HW and LW in a tidal basin. The wave speed is assumed constant throughout the basin, but different for HW (c^+) and LW (c^-). In this example the propagation speed is greater for HW than for LW ($c^+ > c^-$); the period of rising tide in the basin will then be shorter than the period of falling tide. The corresponding tidal curve is depicted in Fig. 6.21.

For lagoons with uniform width the times of HW and HWS coincide at the head of the basin ($x = l_A$), where the tidal wave is reflected. The same holds for the times of LW and LWS: $t_S^+(l_A) = t^+(l_A)$ and $t_S^-(l_A) = t^-(l_A)$. This is illustrated in Fig. 6.22 for the Eastern Scheldt tidal lagoon.

In Appendix (B.39) it is shown that in this case Δ_{EF} can be approximated by

$$\Delta_{EF} \approx \Delta_{FR}^{inlet} + \frac{r}{3gh^2h_S^2}\left[3l_A^2 - (l_A - x)^2\right]\left[D^+D_S^+ - D^-D_S^-\right].$$

$$(6.48)$$

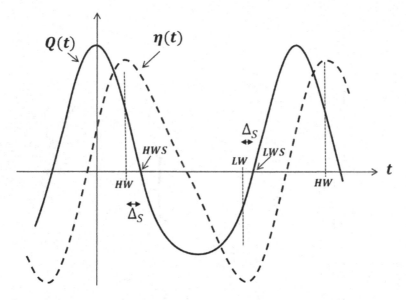

Fig. 6.21. Tide curve for an asymmetric tidal wave with shorter duration of rising tide than falling tide (dashed line), obtained from Fig. 6.20, by interpolating between times of high water and low water. Solid line: Corresponding discharge curve, with phase shift Δ_S relative to tidal elevation. Flood duration is shorter than ebb; the maximum flood discharge is therefore larger than the maximum ebb discharge.

For exponentially converging estuaries the phase differences $t_S^+(x) - t^+(x)$ and $t_S^-(x) - t^-(x)$ are approximately equal and independent of x. In this case

$$\Delta_{EF} = \Delta_{FR} = \Delta_{FR}^{inlet} + 2x \left[\frac{1}{c^-} - \frac{1}{c^+} \right]. \qquad (6.49)$$

The propagations speeds $c^\pm$ can be approximated by (6.44). However, according to these formulas, the dependence of the propagation speed on the depth D and the propagation depth $D_S = Db_C/b_S$ is more complex than for the case of strongly converging friction-dominated basins. For the interpretation of the results we use the simpler expression (6.40), as a first approximation. In this simpler model, the difference between the HW and LW propagation speeds is given by

$$\Delta_{EF} \approx \Delta_{FR}^{inlet} + \frac{2kx}{\omega h h_S} \left[D^+ D_S^+ - D^- D_S^- \right]. \qquad (6.50)$$

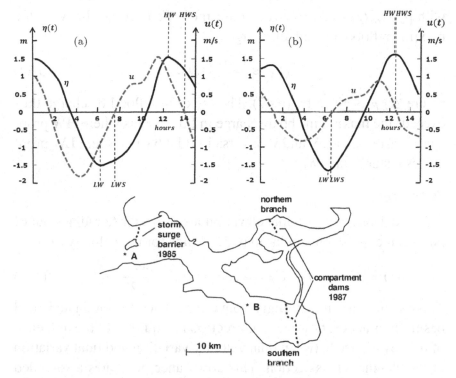

Fig. 6.22. Representative curves for tidal elevation and velocity in the Eastern Scheldt before 1985, at stations A (inlet) and B (30 km upstream). At the mouth there is a time delay of almost one hour between HW/LW and the corresponding slack water, characteristic of a partially standing tidal wave. Near the head of the basin there is no time delay, characteristic of a standing tidal wave. (Measurements by the Survey Department Rijkswaterstaat.)

Inequality of maximum flood and ebb discharge

When excluding the river runoff flow component, tidal inflow must on average equal tidal outflow. Therefore, if the flood duration is shorter, flood discharges must on the average be higher than ebb discharges; the opposite holds if ebb duration is shorter than flood duration, see Fig. 6.21.

If we assume that asymmetry is represented by a M4 tidal component,

$$Q(t) = Q_M \cos \omega t + Q_A \cos 2\omega t + Q_B \sin 2\omega t, \qquad (6.51)$$

with $|Q_A|, |Q_B| \ll Q_M$, we find for the difference Q_A between the maximum flood and ebb discharges

$$Q_A \approx \frac{\omega \Delta_{EF}}{4} Q_M, \tag{6.52}$$

where Δ_{EF} is given by (6.47). The maximum flood discharge thus exceeds the maximum ebb discharge in tidal basins where HW propagates faster than LW and vice versa in tidal basins where LW propagates faster than HW.

Stokes drift

If the tidal velocity and tidal elevation are almost $\pi/2$ radians out of phase, an expression similar to (6.52) holds for the velocity $u(t)$,

$$u(t) = U \cos \omega t + U_A \cos 2\omega t, \quad U_A \approx \frac{\pi \Delta_{EF}}{2T} U. \tag{6.53}$$

This is the case for strongly converging friction-dominated tidal basins. In practice, however, a correction is required for the influence of the covariance between tidal velocity variation and tidal variation of the channel cross-section. This covariance generates a so-called Stokes drift Q_S, given by

$$Q_S = \frac{1}{T} \int_0^T D(t) b_C(t) u(t) dt. \tag{6.54}$$

The Stokes drift is directed landward, as HW and LW precede HWS and LWS, respectively (delay $\Delta_S = \varphi/\omega$). This is illustrated in Figs. 6.22 and 6.23 for the Eastern Scheldt and Vlie tidal lagoon.

Mass conservation requires an increase of ebb velocities relative to flood velocities to compensate for the Stokes drift. Therefore an additional term has to be introduced in (6.53),

$$u(t) = U \cos \omega t + U_A \cos 2\omega t - U_S, \quad U_S \approx \frac{a}{2h} U \sin \varphi. \tag{6.55}$$

The difference between ebb and flood duration Δ_{EF} is also increased,

$$\Delta_{EF} = \frac{4}{\omega U}(U_A + U_S). \tag{6.56}$$

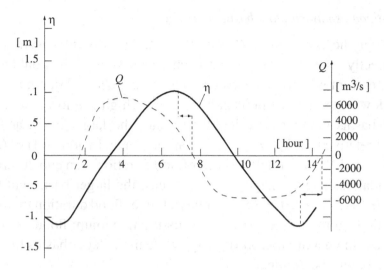

Fig. 6.23. Simultaneous tide curves of water elevation η and tidal discharge Q at the Vlie Inlet (Wadden Sea), near the entrance. The average slack water phase lag relative HW/LW is approximately one hour, indicative of a partially standing tidal wave and of an important Stokes drift contribution. (Survey Department Rijkswaterstaat)

In the derivation we have assumed b_C constant (t-independent) and $|U_A|, |U_S| \ll U$. Including a term $U_B \sin 2\omega t$ in (6.53) or (6.55) does not change the result. In the case of a symmetric sinusoidal tide ($U_A = 0$), the durations of ebb and flood are still different, because of the Stokes drift:

$$\Delta_{EF} = \frac{4U_S}{\omega U} \approx \frac{2a}{h}\Delta_S. \tag{6.57}$$

This term has to be included in the expressions for (6.48) and (6.49).

For exponentially converging basins an approximate expression for Δ_S is given by (6.43). For strongly converging friction-dominated estuaries $\Delta_S = 0$. For lagoons with uniform width one may use the expression (B.40) derived in Appendix B.4,

$$\Delta_S \approx \frac{r l_A^2}{3ghh_S}, \tag{6.58}$$

at the inlet $x = 0$.

Ebb-flood asymmetry and basin geometry

Through the expresssions (6.48), (6.49) and (6.56), tidal asymmetry is directly related to the geometric characteristics of the tidal basin. The relevant geometric characteristics are the channel depth at high and low water and the propagation depth at high and low water. For tidal basins without large intertidal areas, the larger depth at HW (D^+) relative to LW (D^-) and the larger channel width at HW (b_C^+) relative to LW (b_C^-) imply a shorter flood duration than ebb duration; for tidal basins with large intertidal areas, the larger basin width at HW (b_S^+) relative to LW (b_S^-) implies a longer flood duration than ebb duration. In the former type of basins the maximum flood velocity exceeds the maximum ebb velocity, while in the latter basins the ebb velocity will be dominant.

Ebb-flood asymmetry on tidal flats

When the tide rises above the tidal flat level, part of the flow will be diverted from the channel to the tidal flat. Flood flow runs over the tidal flat, while ebb flow tends to concentrate in runnels and creeks. The dominance of flood flow over ebb flow on tidal flats is a well established phenomenon [327, 514]. It explains why tidal flats can accrete vertically even if the flow in the main channel is ebb dominant [431]. This tidal flat building process complements the process of lateral tidal flat accretion, which is mainly due to secondary flow circulation at the inner channel bend (see Sec. 5.5.1).

Tidal wave asymmetry develops in a different way on tidal flats and in tidal channels. In the case of large tidal flats with a level well below HW, the part of the tidal wave around HW propagates faster than other parts of the tidal wave that propagate at lower water levels. The tidal wave is therefore distorted in such a way that flooding takes place in a shorter period than emptying. Hence, flood currents are on average stronger and carry more sediment than ebb currents, both due to nonlinear tidal propagation and due to different ebb and flood flow pathways.

However, the situation is reversed in the presence of waves. The ebb current will carry more sediment than the flood current when tidal flat sediments are resuspended by waves in the period around HW [514].

The Baie du Mont Saint Michel (France) is an example of a tidal basin with large tidal flats that are lying several metres below HW spring level. The basin geometry is shown in Fig. 6.24, with the measuring station on the tidal flat, a few kilometres east of the Mont Saint Michel. The observed tide level variation indicates a shorter duration of submergence compared to emergence. The flood velocity is higher than the ebb velocity. The highest velocity occurs shortly after the start of submergence. At that time a sharp rise of the suspended sediment concentration is observed. Wave activity was low during the survey. It appears that under this condition the tidal flat accretes.

6.3.3. *Slack-Water Asymmetry*

Asymmetry in the duration of high slack water (HWS) and low slack water (LWS) does not only arise from tidal wave propagation. This asymmetry is mainly related to the local channel-tidal flat geometry and to asymmetry in the offshore tide.

Slack water asymmetry can be demonstrated by relating the tidal discharge Q to a change in tidal volume in the landward part of the basin:

$$Q(x,t) = A_C(x,t)u(x,t) = \int_x^l b_S(x',t)\eta_t(x',t)dx', \quad (6.59)$$

where $\eta_t \equiv \partial\eta/\partial t$. We consider tidal basins where at HW and LW the tide level is almost the same throughout the basin (synchronous tide); this is the case if the length scale l (l_b for strongly converging basins, l_A for lagoons) is much smaller than the tidal wave length. If the basin cross-section has either a uniform width or an exponentially converging width, Eq. (6.59) is equivalent to

$$u(x,t) \propto l\, b_S(x,t)\eta_t(t)/A_C(x,t) = l\, \eta_t(t)/D_S(x,t). \quad (6.60)$$

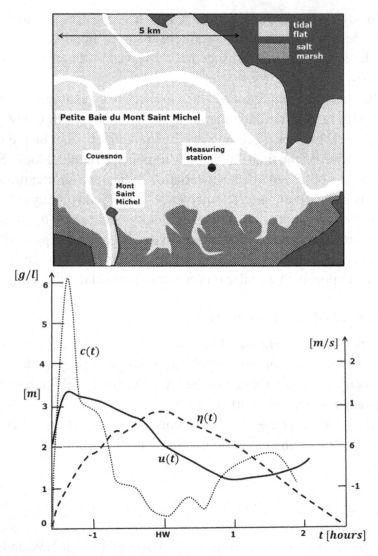

Fig. 6.24. Illustration of flood dominance and sediment transport on a large tidal flat, Baie du Mont Saint Michel (France, see also Fig. 5.5). The upper panel shows the geometry of the Petite Bay, near the Mont Saint Michel, with the measuring station on the tidal flat. Supratidal is dark and subtidal is white. The tide level η (relative to the tidal flat level), the velocity u and the near-bottom suspended sediment concentration c were measured during a mean spring tide (coefficient 86) in October 1975 [600]. The data illustrate shorter tidal rise than tidal fall and flood dominance of the tidal velocities. The suspended sediment concentration rises sharply as soon as the tide starts expanding over the tidal flat. The data indicate tidal flat accretion under the then occurring tidal and weather conditions.

This equation shows that the highest tidal velocities occur when b_S is large, i.e., just after the tidal flats become inundated. For low-lying tidal flats the highest tidal velocities occur close to LW, while in the case of high tidal flats the highest tidal velocities occur close to HW. This is consistent with many observations (Baie du Mont Saint Michel, Fig. 6.24; Satilla River [84]).

In the short basins considered here, the tidal variation of $\eta(t)$ is $\pi/2$ radians out of phase with $u(t)$; this implies $\eta_t \approx 0$ at HWS and LWS. By differentiating (6.60) we then obtain the following formula for the ratio of the velocity variation around HWS and the velocity variation around LWS:

$$\frac{|u_t|_{HWS}}{|u_t|_{LWS}} \approx \frac{D_S^-}{D_S^+} \frac{|\eta_{tt}|_{HWS}}{|\eta_{tt}|_{LWS}}. \tag{6.61}$$

The second ratio on the r.h.s. of (6.61) equals 1 if the offshore tide is symmetric. Equation (6.61) shows that in basins with large tidal flats and deep channels $(D_S^+ < D_S^-)$, the tidal velocity changes faster around HWS than around LWS: The period of slack water around HW is shorter than at LW. In basins with small tidal flats and shallow channels $(D_S^+ > D_S^-)$, the tidal velocity changes slowly around HWS compared to LWS: The period of slack water around HW is longer than at LW. In the former case sedimentation per unit area will be greater at LWS than at HWS and vice versa in the latter case.

Settling and resuspension lag

Sediment settling responds with some delay to a decrease of the current velocity. Freshly deposited sediment is not immediately resuspended when the current velocity increases again. These delays in deposition and resuspension are called settling lag and resuspension (or scour) lag. Settling and resuspension lag imply that sediment deposited around LWS will experience a net seaward displacement, while sediment deposited around HWS will experience a net landward displacement.

The process of settling and resuspension lag is most effective for sediment that settles around HWS on the high parts of a tidal flat. When the current velocity increases after HWS, the water level has already dropped below the zone of deposition. The resuspension lag is then at least a full tidal period, during which a sub-aerial deposit can consolidate (especially a deposit of fine cohesive sediment). This results in a large net landward displacement of (fine) sediment. A more detailed discussion of settling and resuspension lag effects is presented in Sec. 6.6.2.

6.3.4. *Asymmetry of Shelf Tides*

The tide becomes asymmetric when the tidal wave propagates into shallow water. The tide on broad continental shelves is therefore generally asymmetric, with a shorter duration of tidal rise compared to tidal fall, see Fig. 6.25. An important reason is the faster propagation of the HW crest of the tidal wave in the shallow continental zone compared to the propagation of the LW trough.

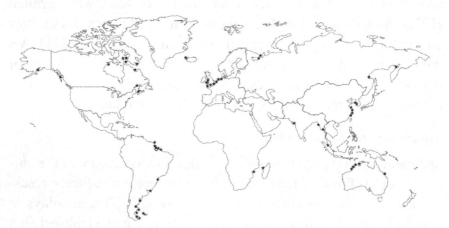

Fig. 6.25. Coastal zones with an important offshore quarter-diurnal M4 tide, indicative for tidal wave distortion with shorter duration of tidal rise compared to tidal fall. These zones coincide largely with zones where the continental shelf is broad and where the semidiurnal is tide strong, see Fig. 6.17. Many of the major tidal inlets are also located in these zones, see Fig. 6.2. Redrafted from [549].

In some cases, however, tidal fall is faster than tidal rise. This may happen for particular combinations of tidal constituents, especially in zones with relatively strong diurnal tidal components. At the US Pacific coast, for instance, the combination of O1, K1 and M2 tidal constituents produces an astronomical tide with faster tidal fall than tidal rise [632], while for south-eastern Australia these tidal constituents produce an astronomical tide with faster rise than fall [461]. A more general discussion of tidal asymmetry generation in the coastal zone is given in Appendix C.3.

Offshore tidal asymmetry influences tidal asymmetry within adjacent tidal basins. A faster tidal rise than fall at sea causes higher flood flow velocities than ebb flow velocities in the adjacent tidal basins, especially at the entrance. A slow tidal rise has the opposite effect. In long tidal basins offshore tidal asymmetry can be offset by the influence of basin geometry; in short tidal basins the influence of offshore tidal asymmetry dominates over internally generated asymmetry. Hence, the net sediment import or export and the morphologic development of lagoons and estuaries depend on offshore tidal wave asymmetry.

Figure 6.26 shows tide curves for the Dutch coast and its tidal basins; it appears that the shape of the offshore tide often persists within the adjacent basins.

The role of offshore tidal asymmetry in basin infill can be illustrated by the evolution of tidal inlets along the Dutch coast between 5,000 BP and 3,000 BP. When marine transgression reached the present Dutch coastline, some 6,500 BP, the sea was able to intrude into the low-lying plain of the rivers Rhine, Meuse and Scheldt. Large tidal inlets developed in the southern, central and northern parts of the Netherlands, scouring deep channels into the Pleistocene floor, see Fig. 6.5. The tidal inlets in the southern and northern parts of the Netherlands still exist, but those in the central part have disappeared, in particular the large Bergen Inlet.

What made the difference between the evolution of these tidal inlets? Figure 6.27 shows the distortion of the tidal wave along the

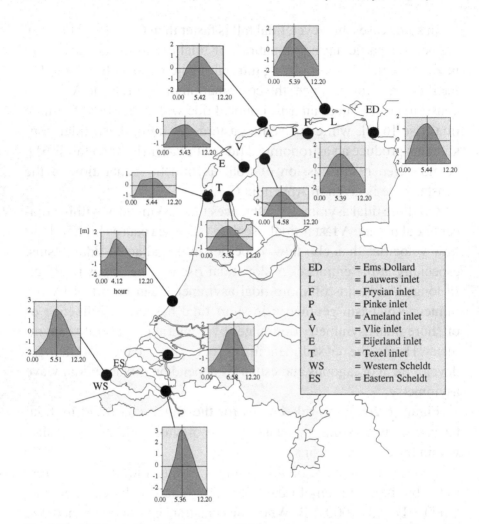

Fig. 6.26. Dutch coastline and tidal basins. Tidal curves averaged over a decade (for spring tide) give an impression of tidal amplitude and tidal asymmetry at various coastal and inland locations. Offshore tidal asymmetry persists at least partly in most tidal basins.

Dutch coast; tidal asymmetry is greatest at the coast of central Holland, where tidal rise is three hours shorter than tidal fall. The inlets on the central Dutch coast therefore experienced much stronger flood currents than ebb currents [882], causing a rapid infill of the basins and the development of large intertidal areas. The compensating

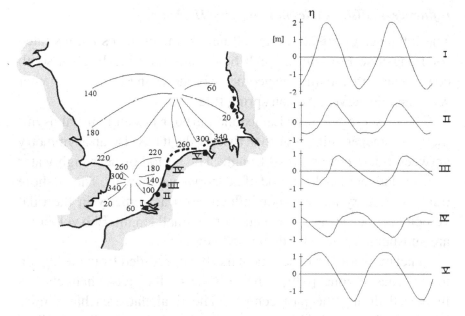

Fig. 6.27. While propagating along the Dutch coast from south to north the tidal wave becomes increasingly distorted. Asymmetry between tidal rise and fall is most pronounced along the coast of central and northern Holland (between III and IV). This asymmetry is mainly due to the relative shallowness of the inner shelf in this stretch of the coast, see Appendix C.3. Further to the north and along the Wadden Sea the tidal wave is part of the amphodromic system of the central North Sea, which develops less asymmetry than the amphidromic system of the Southern Bight.

effect of these intertidal areas on HW and LW propagation was probably insufficient to offset flood dominance imposed by the offshore tide. Around 3,500 BP the inlets at the central Dutch coastline were closed [982].

Asymmetry in the offshore tide does not pertain only to a difference in the duration of tidal rise and fall. It also happens that the tidal wave crest is longer or shorter than the tidal wave trough. A long period of HW (or LW) at sea causes a long period of slack water in the adjacent tidal basins, whereas a short period of HW (or LW) at sea causes a short period of slack water. In the following we discuss the consequences of this type of asymmetry for the net sediment fluxes in lagoons and estuaries.

Influence of offshore tides in Langstone Harbour

The influence of offshore tides on net sediment fluxes is illustrated for Langstone Harbour, a small tidal lagoon on the English south coast, near Portsmouth. The geometry is shown in Fig. 6.28, together with the tide curve for mean spring tide.

The offshore tide near Langstone harbour has an unusual asymmetry: A steeper falling tide than rising tide. It has also an asymmetry between tidal crest and tidal trough: The period of near-high water levels is longer than the period of near-low water levels. The offshore tidal asymmetry has a major influence on asymmetry in the tidal velocity inside the lagoon, because of its small length. Ebb velocities are substantially stronger than flood velocities.

A net outward flux of sediment has been recorded from the lagoon to the coastal zone [639]. The volume of the tidal channels has increased during the past century. The tidal flat area has diminished and about 80% of the salt marshes has disappeared since 1946. Channel dredging and the construction of sea defences may have contributed to this evolution.

However, it appears that in spite of this sediment loss, the tidal flat levels keep pace with sea level. This is probably related to the particular form of the tidal curve, which exhibits a long period of near-high water levels. The fine sediments deposited during this period on the high parts of the tidal flats is not resuspended when the ebb flow increases, because the tide level has dropped below before the critical value for resuspension is reached. This suggests that fine sediments are entering the lagoon, whereas coarser sediments are leaving.

6.3.5. *Influence of Human Interventions on Tidal Propagation*

Most estuaries in the world have been strongly modified by human interventions. The first interventions were dike constructions, to protect the hinterland against flooding. Dike construction was also

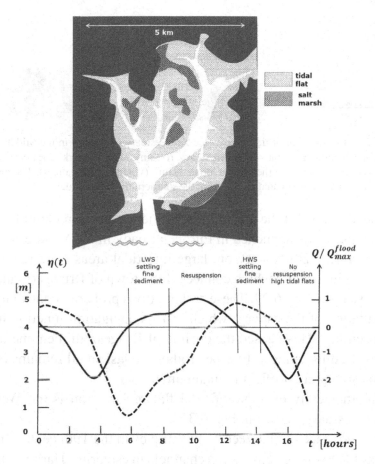

Fig. 6.28. Influence of tidal asymmetry on sediment transport in Langstone Harbour. The upper panel shows the present geometry of Langstone Harbour (subtidal = white, supratidal = dark); typical bathymetric dimensions are given in Table 6.1. The mean spring tidal curve $\eta(t)$ (dashed line) is shown in the panel below. The tidal discharge curve $Q(t)$ (solid line) is estimated by equating $Q \approx S(t)d\eta/dt$, where the wetted lagoon surface area S is approximated by $S/l = b_S^- + \left[b_S^+ - b_S^-\right]\left[(\eta - \eta_{LW})/2a\right]^{1/2}$, assuming a convex parabolic tidal flat profile (low wave activity, η_{LW} is the LW level, l is the distance from the landward boundary and $2a$ the tidal range). Tidal rise is slower than tidal fall; the maximum ebb discharge is therefore higher than the maximum flood discharge. The discharge curve also exhibits long periods of low velocity shortly after LW and HW. In the low-tide slack-water period sediment settles mainly in the channel; most of these deposits will be resuspended by subsequent strengthening of the flood currents. In the high-tide slack-water period, sediment settles on the higher parts of the tidal flats. These deposits are not resuspended when the ebb flow increases, because the water level has dropped below the high tidal-flat level before the ebb current has become sufficiently strong. One may thus expect net tidal-flat accretion, in spite of general ebb-current dominance in the tidal lagoon.

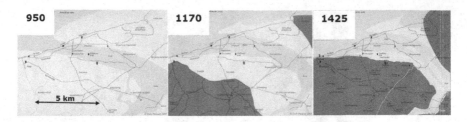

Fig. 6.29. Reclamation of tidal flats in the Zwin estuary (Belgium) in the Middle Ages. Subtidal = medium grey, intertidal = light grey, reclaimed areas = dark. The dots and lines in the background indicate the present infrastructure (villages, roads, dikes). The maps do not show Bruges; it is situated 10 km down the southern tip of the estuary.

intended to restrict the area of tidal influence and to reclaim land for human use. This happened in Europe already in the Middle Ages.

From the 12th century on, large intertidal areas were reclaimed in the Zwin estuary, which connected the town of Bruges (Flanders) to the sea, see Fig. 6.29. These reclamations probably contributed to the siltation of the estuary, which lost its navigation function in the 14th century. This caused the decline of Bruges, till then one of the richest trading cities in Europe. Today the last small remains of the Zwin estuary are artificially maintained.

Another early example of tidal flat reclamation is the Western Scheldt estuary, shown in Fig. 6.30.

The advance of engineering techniques in the 19th century made it possible to dredge navigation channels in estuaries. Harbour development boomed along the estuaries. Major harbours were situated at the estuarine head, beyond the salt intrusion zone, where the impact of storm waves was minor and where fresh river water was available. Maritime trade expanded and ship size increased. This led to the channelisation of estuaries. An extreme is the Seine estuary, shown in Fig. 6.31. The impact on the tidal wave propagation is shown in Fig. 6.32.

The coastal zone became increasingly urbanised and the dimensions of engineering structures increased. In the Netherlands, inland seas were completely dammed off from the sea during the past century.

Western Scheldt 1650

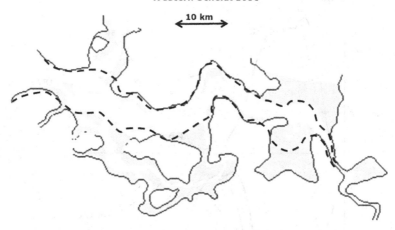

Fig. 6.30. Western Scheldt basin in the 17th century. The present embankments are indicated by the dotted line. Tidal flats are shaded. About 9×10^6 m^2 intertidal area is at present left of the former 3.4×10^7 m^2[883].

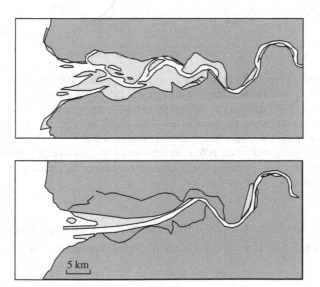

Fig. 6.31. The Seine estuary in the beginning of the 19th century (upper panel) and the estuary today (lower panel). The intertidal area is shaded light grey. Tidal flats have been reclaimed; the main channel is constrained by training walls.

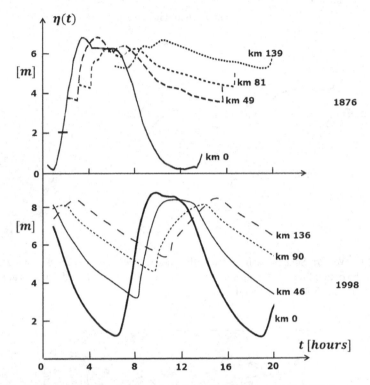

Fig. 6.32. Tide curves in the Seine estuary in 1876 [183] and 1998 [111] for mean spring tides (coefficient for 1876 is slightly lower than for 1998); km 0 corresponds to Honfleur, near the mouth. Tidal damping in the upstream tidal river is much less than in the past, mainly due to channel dredging. The HW levels have remained approximately the same; the LW levels are much lower now. The tidal bore, visible at km 81 in 1876, has disappeared. The upstream propagation of the HW wave crest and LW wave trough is faster than in the past. The increase of c^- is mainly due to channel deepening; the increase of c^+ is mainly favoured by embanking of tidal flats.

In this section we will discuss some consequences of these interventions for the hydro- and morphodynamics of lagoons and estuaries.

Embanking of tidal flats

When the flood wave spreads over the tidal flat, tidal rise is slowed down. Embankments constrict the tide to the channel and eliminate the delay in the progression of the tidal flood wave. The HW crest thus propagates faster upstream; the LW propagation is not much affected. Embankments therefore change the ebb-flood asymmetry. The strength of the flood flow increases relative to the ebb flow. This favours upstream sediment transport.

When the flood submerges the tidal flats, part of the sediment stays behind. The coarsest sediments are deposited near the channel; these deposits form small natural levees. The tidal flat area behind the levees acts as a sink for fine sediment, especially when tidal flats are sheltered from wave action. Embankments therefore diminish the sediment storage capacity of the estuary. The amount of sediment in suspension in the tidal channel increases. Suspended sediment is trapped in the estuary and accumulates at low river discharge. Embankments therefore favour the development of a turbidity maximum and the formation of mud layers. This is illustrated in Fig. 6.33 for the Loire estuary.

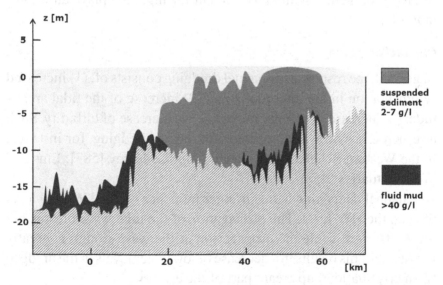

Fig. 6.33. Longitudinal section of the Loire estuary along the thalweg, showing the turbidity maximum and fluid mud layers. Horizontal coordinate: distance from Saint Nazaire (km 0); Nantes is at km 55–65. Vertical coordinate relative to mean sea level. Data for mean spring tide and low river discharge of $210\,\mathrm{m^3/s}$ (survey of 4 June 2011). Adapted from [74].

In turbid estuaries, the flow can become stratified with sharp interfaces between high and low turbidity layers (lutoclines). The tidal flow is shielded from bottom roughness; turbulence and related momentum dissipation are suppressed, as discussed in Sec. 3.5.2. A substantial suppression of turbulence occurs even at moderate stratification. Under such conditions the tidal wave propagates at larger speed (according to (6.34)); tidal current velocities are also increased. The high tidal velocities in the turbid Severn, Loire and Gironde estuaries are indicative for the presence of turbulence-suppressing lutoclines.

Upstream sediment transport contributes to a progressive narrowing of the tidal channel. Propagation of the tidal wave into a narrowing convergent channel induces tidal amplification. If the amplification is stronger than frictional damping, the tidal range increases. Embankments therefore contribute to a relative (or absolute) tidal amplification in the upstream part of the estuary. Channel deepening, which often goes together with embankment raising, also plays an important role.

Channel deepening

In general, the response to channel dredging consists of: (1) increased sedimentation in the intertidal zone; (2) increase of the tidal amplitude; (3) increase of salinity intrusion; (4) increase of turbidity. Such a response has been documented after heavy dredging, for instance, in the Western Scheldt [460], Ribble [886], Mersey [587], Ems and Weser estuaries [958].

Channel deepening has a stronger influence on the LW tidal level than on the HW level. The propagation of the tidal wave crest is not much affected, while the propagation of the wave trough is greatly facilitated. This results in a decrease of low water levels in the dredged (formerly shallow) upstream part of the estuary.

A greater part of the tidal flow is concentrated in the main channel, at the expense of secondary channels; flow reduction and rapid infilling of these secondary channels has been observed, for instance, in

the Ribble estuary [886]. Most of the sediment is delivered by intertidal sand banks around these channels; these sand banks experience
erosion. At long term, sand banks may recover by sediment import
through the inlet, see Sec. 6.5.3.

Flow concentration in the main channel is further enhanced by the
construction of training walls. Accelerated accretion of shallow subtidal and intertidal zones is observed as an immediate consequence
of training walls, in the Mersey, the Ribble, the Lune estuary [587]
and the Seine [43]. In the Ribble estuary, the outer delta has acted as
major sediment source [886].

Frequent dredging is required for adapting the Western Scheldt
navigation channel to the increasing draught of ships sailing to
Antwerp, see Fig. 6.34. Disposal of dredged material in the estuary
has accelerated the infilling of secondary flood channels. The loss
of tidal storage volume in the basin due to sedimentation is partly
compensated by an increase of the tidal amplitude [883]. No strong

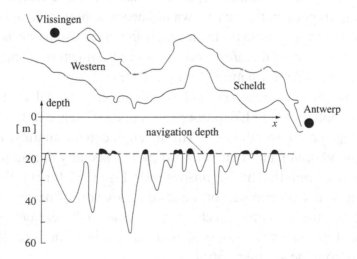

Fig. 6.34. Depth profile of the Western Scheldt basin along the channel axis. Dredging
in the Western Scheldt navigation channel is concentrated at the shoals at the junctions of
flood and ebb-dominated channels. The dredged material is not removed from the basin, but
mainly disposed in secondary flood channels and along the channel banks.

impact of channel deepening on the flow velocities has been observed in the Western Scheldt [460].

Channel deepening contributes to increasing the LW-propagation speed relative to the HW-propagation speed; shortening of the ebb duration relative to the flood duration may turn the basin from flood dominant into ebb dominant. In the Western Scheldt and the Ribble estuary the evolution towards ebb dominance is offset by a decrease of HW basin width; this decrease is due to accretion of intertidal areas above HW level and reclamation of intertidal areas. In the Western Scheldt the duration of rising tide relative to falling tide has not changed significantly during the past century, in spite of considerable channel deepening.

Channel deepening increases the tidal influence in the basin relative to the river influence. Saline marine water penetrates further into the estuary and the estuarine circulation, driven by the upstream pressure gradient of the saline wedge, is enhanced.

The dredging activities and the resulting change in the flow pattern increase sediment mobility, especially that of the fine fraction. The sediment trapped in the marine waters near the bottom is not easily expelled from the estuary, even when the river discharge is high. Channel deepening therefore contributes to sediment trapping in the estuary and to the formation of fluid mud layers.

Embanking of tidal flats and simultaneous channel deepening increase both HW and LW propagation. In the absence of tidal flats, HW propagates faster than LW. Highly engineered estuaries are typically flood dominant in the seaward part, especially during periods of low river runoff. This is illustrated in Fig. 6.35 for the Western Scheldt, where a large part of the mud deposits is of marine origin, over the entire estuarine reach. In the Seine, sediment deposits of marine origin have been observed far upstream, in the docks of Rouen, 100 km from the sea inlet [361].

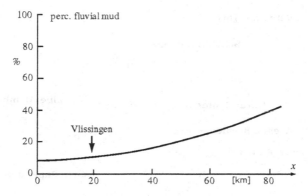

Fig. 6.35. In the estuarine part of the Western Scheldt fine organic sediment deposits are mainly of marine origin. Data are based on carbon isotope analysis, after [747, 901].

Embankment raising and channel deepening are together responsible for an increase of the tidal penetration in estuaries, by diminishing the relative influence of the river discharge, by constricting the tide into a converging channel and by reducing the friction experienced by the tidal flow. Figure 6.36 illustrates the increase in tidal range that has occurred in several European estuaries during the past century.

Closure of inland flood basins

In the Netherlands, large parts of tidal basins have been dammed during the past century. Primary motivation for the construction of these dams was the reduction of flooding risks. The largest closure was the damming of the Zuyderzee in 1930, see Fig. 6.37. The Zuyderzee was a large inland sea that was connected to the North Sea in the Middle Ages; it originated from a breach in the littoral barrier south of Texel. During several centuries Texel Inlet and the Zuyderzee formed the main shipping way to Amsterdam.

The closure of the Zuyderzee had several consequences for the tidal dynamics and morphodynamics of Texel Inlet, which are well

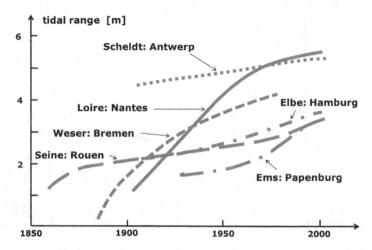

Fig. 6.36. Evolution of the tidal range in European estuaries resulting from human interventions. The tidal range has strongly increased in the upstream part of all these estuaries.
The increase is mainly related to a decrease of the LW level; the HW level has hardly
changed.

documented [201, 277]. Although the size of the flood basin strongly
decreased, the tidal exchange at Texel Inlet increased. The nature
of the tidal wave changed from a progressive damped wave into an
almost standing wave. The reflection of the tidal wave on the closure dam caused a substantial increase of the tidal range inside the
remaining lagoon.

Before closure, the tide level variation was in phase with the tidal
current. The resulting inland Stokes drift was compensated by tidal
ebb dominance. The Stokes drift strongly decreased when the tide
changed into an almost standing wave after the closure. Ebb dominance changed into flood dominance, as a result of the fast rising tide
at the inlet.

The flood dominance of the tidal currents caused high sediment
import from the North Sea. Most of this sediment was extracted from
the ebb-tidal delta; the volume of the ebb-tidal delta decreased about
as much as the volume of sediment imported into the lagoon, see
Fig. 6.38. Fifty years after closure of the Zuyderzee the infilling of
Texel Inlet was strongly reduced.

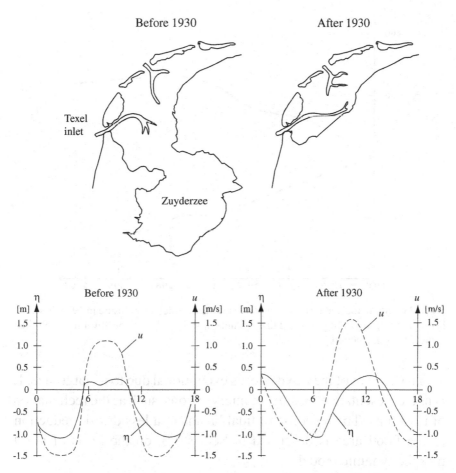

Fig. 6.37. Morphology, tide curves and velocity curves at the mouth of the Texel inlet, before and after closure of the Zuyderzee. The tidal curves are measured, the velocity curves are computed with a numerical two-dimensional tidal model. Before closure in 1930, the Texel Inlet was the main inlet channel of the Zuyderzee. Tidal energy was almost completely dissipated in the shallow Zuyderzee; the nature of the tide at Texel was a progressive damped wave, with almost no phase difference between tidal elevation and velocity. Texel Inlet was ebb-dominated due to Stokes drift. After closure of the Zuyderzee the tide was reflected at the closure dam. A phase shift of more than two hours between tidal elevation and velocity strongly decreased the Stokes drift; the flood duration became shorter than the ebb duration and the tide became flood-dominated.

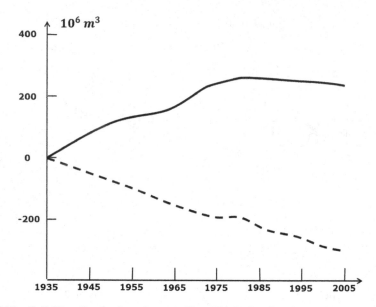

Fig. 6.38.   Solid line: Sand volume imported into Texel Inlet after closure in 1935, based on bathymetric surveys [280]. Dashed line: Volume of sand lost from the Texel Inlet ebb-tidal delta in the same period.

The closure of the Zuyderzee is exceptional due to its large size. It is not representative for smaller interventions, such as the reclamation of tidal flats. The response of tidal basins to a less drastic reduction of the flood area is described in Sec. 6.5.3, on the basis of a 1D morphodynamic model.

Change of the tidal prism

Human interventions in tidal lagoons may change the tidal prism (the water volume entering the lagoon during flood). An illustration of the consequences of tidal prism change is provided by the Delta Works in the Netherlands from 1960 to 1990. Several estuarine branches of the Rhine–Meuse–Scheldt delta were closed with dams at the inlet. At the inlet of the largest basin, the Eastern Scheldt, a storm surge barrier was constructed. The storm surge barrier is closed under storm surge conditions, but under normal conditions the barrier reduces tidal exchanges through the inlet by no more than about 10%.

During the closure of adjacent basins, the tidal prism of the Eastern Scheldt temporarily increased by about 10%, from 1968 to 1983. The storm surge barrier was completed in 1986. From then, the tidal prism was reduced by about 30%, due to the cumulative effect of the barrier and of compartment dams, which cut off the head of the basin, see Fig. 6.22.

Successive changes of the tidal prism had a significant influence on the channel-shoal relief in the basin, as described earlier in Sec. 5.5.3. In the period of enhanced tidal prism the topographic relief increased; channels deepened and the height and extent of tidal flats increased, see Fig. 5.18. The opposite process is taking place in the present situation of reduced tidal prism; tidal flats are eroding and channels are silting up, see Fig. 5.19. These observations illustrate the sensitivity of basin morphology to changes in the tidal prism and in tidal current strength.

6.4. Equilibrium of Estuaries

What is an estuarine equilibrium?

An estuary is in morphodynamic equilibrium if the net (tide-averaged) sediment flux is the same everywhere along the estuary and equal to the fluvial sediment flux upstream of the tidal limit. Such a condition can not be met at any timescale in the estuary. An equilibrium at the tidal timescale is not necessarily an equilibrium at the seasonal timescale or at the timescale of decades or centuries. An equilibrium at basin scale is not the same as an equilibrium at the local scale.

During a long period, a few thousand years, the fluvial and tidal regime have been fairly stable for many estuaries and the same holds for the sea level. We may reasonably assume that estuaries have adjusted during this period to a state close to morphological equilibrium, provided sediment availability and sediment transport capacity were sufficient. If this is the case, the net sediment flux through the estuary must have become close to the fluvial sediment input. Because the average flood discharge through the inlet is typically much larger than the fluvial discharge, a state close to equilibrium can exist only if the net sediment transport from or to the sea is much smaller than the total sediment import during flood or the total sediment export during ebb.

Estuaries have been strongly altered during the past centuries by human interventions. Therefore we may expect that the present morphology is not always in equilibrium with the existing fluvial and tidal regimes. It is generally not well known how far the estuary have shifted away from equilibrium.

Outline of the approach

We will investigate in this section if present estuarine morphologies are consistent with a long-term net sediment flux of the same order as the net fluvial sediment input. We will use very simple models; therefore we may not expect an accurate answer to the equilibrium question. The aim, however, is to get at a meaningful indication.

We will focus on the convergence length of estuaries. The convergence of the estuarine width (from a wide estuarine mouth to a much narrower tidal river) is a typical feature of estuaries with strong tides. In Sec. 6.5 on tidal lagoons, we will see that in the absence of river inflow, morphodynamic equilibrium of tidal basins is related to the extent of the intertidal area. This is not the case for estuaries with high river inflow, where the intertidal area is usually small, especially after the human interventions of the past centuries. We will show that morphodynamic equilibrium of estuaries is related to river inflow, even though river flow is much smaller than the tidal flow through the inlet.

In Sec. 6.2.4 another condition for morphodynamic equilibrium was introduced. This condition was based on the consideration that sedimentation might be expected where the peak tidal bed shear stress is lowest. This implies that in an equilibrium situation, the peak tidal bed shear stress should be approximately uniform along the estuary. We have assumed that this peak bed shear stress is the same for all estuaries. Based on this assumption we have determined friction coefficients for the estuaries of Table 6.2. Uniformity of peak tidal velocities requires that the estuarine width should be converging, approximately as an exponential function of the along-channel distance x for strongly converging basins. In this case the convergence length scale l_b is proportional to the tidal current amplitude and inversely proportional to the relative tidal amplitude (Eq. (6.17)). The consistency of the assumptions has been checked with field data of maximum tidal velocities and tide propagation speeds, using a simple 1D model.

With this model we will investigate if the observed convergence length of these estuaries is consistent with a morphological equilibrium state based on uniformity of the long-term average sediment flux. Comparison of the observed convergence length with the equilibrium convergence length following from the 1D morphodynamic model, provides an indication whether these estuaries are close to equilibrium or not.

Major assumptions

In order to investigate the long-term average sediment balance, we assume that it is possible to construct a theoretical cyclic semidiurnal tide at the estuarine inlet that generates the same sediment transport as the real long-term average sediment transport in the estuary. We do not know exactly what this tide looks like, but we assume that the tidal range is of the order of a mean spring tidal range. The form of the tidal curve depends on the offshore tide and the river discharge. It will appear, however, that the convergence length depends primarily on a few gross characteristics of the tide at the inlet, in particular the tidal amplitude and the tidal asymmetry (duration of tidal rise relative to tidal fall).

The river discharge has a major influence on the estuarine shape. The discharge is strongly variable, mainly due to seasonal fluctuations. Less frequent high floods can also have a substantial impact on the estuarine morphology. Glacier-fed rivers are generally less fluctuating than rain-fed rivers. River flow fluctuates less in temperate regions than in Mediterranean or subtropical regions. The long-term average river discharge representative for morphological evolution is therefore difficult to estimate. It can probably vary between less than two times the average discharge to several times the average discharge. We will discuss this later for a few particular estuaries.

We will use a sediment transport formula of the form (see Sec. 3.8)

$$q(x, t) = \alpha b_C(x)u(x, t)|u(x, t)|^n, \qquad (6.62)$$

where b_C is the channel width. The coefficient α is of the order of 4×10^{-4} (SI units), but it does not play a role in the following. We take $n = 2$, assuming that most of the sediment is sandy material transported as bed load. However, if we use a higher coefficient $n = 3$ or $n = 4$ to simulate total load transport (sand and fine sediment), the results are essentially the same.

The velocity $u(x, t)$ is the instantaneous current velocity, averaged over the channel cross-section. We should have included in (6.62) a threshold velocity for the uptake of sediment. We assume, however, that the velocities that produce really significant morphological change are much higher than this threshold. The transport formula overestimates the sediment flux due to the neglect of the threshold velocity; it underestimates the sediment flux because of the cross-sectional averaging of the velocity. We further assume that the estuarine bed consists of erodible material, with uniform grain size along the estuary.

The formula (6.62) considers only sediment transport by advection; diffusive transport is not included. Settling and resuspension time lags are also ignored. We assume that they do not play an essential role under the high energy regimes relevant for estuarine morphodynamics. For tidal lagoons with large intertidal areas the situation is different, as will be discussed later.

6.4.1. *Residual Sediment Transport*

Sediment load related to flow strength

Accepting these assumptions and simplifications, the cross-sectionally integrated and tide-averaged sediment flux is given by:

$$\langle q \rangle = \frac{\alpha b_C}{T} \int_0^T u^3(t)dt. \tag{6.63}$$

We consider a current velocity with an asymmetry between maximum flood flow (positive) and ebb flow (negative) of the form

$$u(t) = U \cos \omega t + U_A \cos 2\omega t + U_B \sin 2\omega t - U_S, \tag{6.64}$$

where U_S is the residual velocity due to the Stokes effect and the river discharge. We assume $|U_A|, |U_B|, |U_S| \ll U$. The Stokes effect is caused by a non-zero phase difference φ between high water HW and high slack HWS (and the same phase difference at low water). We therefore write the tide level variation $\eta(t)$ as

$$\eta(t) = a \sin(\omega t + \varphi),$$

where a is the tidal amplitude; we assume $a/h \ll 1$. The Stokes effect follows from the requirement

$$\frac{b_C}{T} \int_0^T (h + \eta(t))u(t)dt = -Q_R,$$

where Q_R is the river discharge. This requirement yields

$$U_S = U_R + \frac{aU}{2h} \sin \varphi, \tag{6.65}$$

where $U_R = Q_R/hb_C$.

We adopt the schematisation discussed in Sec. 6.2.4. The depth h is assumed uniform along the estuary; the width b_C is an exponentially decreasing function of x in upstream direction. In this part of the estuary (upstream converging and downstream widening), U_S, U_A, U_B are near the estuarine mouth always much smaller than the tidal velocity amplitude U. In this zone U_S and U_A are related to the difference Δ_{EF} in the durations of flood (positive flow) and ebb (negative flow) by the approximate expression (see also (6.56))

$$\omega \Delta_{EF} \approx \frac{4}{U}(U_A + U_S). \tag{6.66}$$

Substitution of (6.64) in (6.63) gives

$$\langle q \rangle = \frac{3\alpha b_C}{4} \left[U^2(U_A - 2U_S) - \frac{4}{3}U_S^3 \right.$$

$$\left. - 2(U_A^2 + U_B^2)U_S - U_A U_B^2 \right]. \tag{6.67}$$

Equilibrium requires that the long-term net sediment flux $\langle q \rangle$, the difference between flood and ebb fluxes, is uniform along the estuary and equal to the net fluvial sediment input. It does not imply uniformity of the separate velocity components U, U_A, U_B, U_S. This equilibrium requirement is therefore basically different from the requirement of uniform maximum tidal shear stress along the estuary.

Estuarine transport zones

According to the expression (6.67), we can distinguish different sediment transport zones in the estuary.

Most upstream, in the river, the tide has faded. Here the net sediment flux equals

$$\langle q \rangle = -\frac{\alpha Q_R^3}{h^3 b_R^2}, \tag{6.68}$$

where b_R is the river width.

Going downstream, we arrive in the tidal river, where the tidal current velocity U is comparable with the river flow velocity U_R. The width b_C of the tidal river is larger than the upstream river width b_R; the river flow velocity U_R is therefore smaller than Q_R/hb_R. The tide is strongly asymmetric; U_A is not much smaller than U. All the terms in the expression (6.67) are of comparable magnitude.

The net sediment flux results from a complex balance of these terms. Equilibrium requires uniformity of the sediment flux and equality to the upstream fluvial sediment influx. The channel cross-section will adapt by erosion or accretion until this equilibrium is achieved.

Further downstream we enter the wider estuary. Here the tidal velocity amplitude U dominates; U_R decreases further in downstream direction and becomes very small compared to U in the downstream widening (upstream converging) section close to the mouth. The net sediment flux then reads

$$\langle q \rangle \approx \frac{3}{4} \alpha b_C U^2 \left[U_A - 2U_R - \frac{aU}{h} \sin \varphi \right]. \qquad (6.69)$$

We have $b_C \gg b_R$ and consequently $b_C U^2 \gg Q_R^2 / h^2 b_R$. Equilibrium in the convergent zone near the mouth corresponds to an almost zero net transport; it thus requires

$$U_A \approx 2U_R + \frac{aU}{h} \sin \varphi. \qquad (6.70)$$

This condition only holds for estuaries where the river flow U_R at the mouth is much smaller than the maximum tidal velocity U. This is the case for all the estuaries listed in Table 6.6. Estuaries were peak river flow U_R at the mouth and maximum tidal current velocity are of comparable order of magnitude have to be excluded; examples are the Columbia River and Mekong River estuaries.

6.4.2. *Equilibrium Convergence Length*

The term $2U_R$ in (6.70) is smaller than the other terms, but increases when going upstream. The tidal velocity asymmetry U_A also increases with x, since HW propagates faster than LW in estuaries with small intertidal area ($c^+ > c^-$). We assume that the longitudinal variation of the Stokes term $aU \sin \varphi / h$ is small compared to the variation of the U_A and U_R. This is consistent with the linear 1D model (6.41) and consistent with observations in estuaries. Uniformity of the sediment flux along the estuary thus requires that the

Table 6.6.   Width at the mouth of several estuaries, long-term average river discharge [m³/s] and river flow [cm/s] at the mouth. The estuarine mouth is situated at the entrance of the estuarine section of the tidal inlet. For inlets with a lagoon type entrance section and multiple meandering channels, the estuarine section starts where a single channel remains. The location is indicated in the table.

Number	Tidal inlet	Inlet location	b_C [m]	$Q_R m^3/s$	$U_R cm/s$
SC	Scheldt	Belgian-Dutch border	1000	100	1.0
EL	Elbe	Cuxhaven	6000	750	1.0
WE	Weser	Bremerhaven	2000	324	1.8
EM	Ems	Emden	600	80	2.1
SE	Seine	Honfleur	900	440	6.1
LO	Loire	Paimboeuf	1200	825	8.6
GI	Gironde	20 km from mouth	7000	1000	1.6
SA	Satilla	river mouth	1700	70	0.6
OR	Ord	Pantin Island	3500	163	0.7
HO	Hoogly	Saugar	25000	7000	2.5
FL	Fly	Tirere	10000	6000	6.0
SO	Soirap	Vam Co river junction	1800	250	2.8
PU	Pungue	mouth	5000	137	0.2

increase of U_A equals the increase of $2U_R$:

$$\frac{dU_A}{dx} = 2\frac{dU_R}{dx}. \tag{6.71}$$

According to (6.66) and (6.49) we have

$$\frac{d}{dx}(U_A + U_R) = \frac{\omega U}{4}\frac{d\Delta_{EF}}{dx} = \frac{\omega U}{2}\left[\frac{1}{c^-} - \frac{1}{c^+}\right],$$

where $c^{\pm}$ is the propagation speed of HW (LW). The last equality is based on the 1D description of the estuary, which assumes uniform depth. We also neglect the along channel variation of the relative tidal amplitude a/h. Both from observations in the considered estuaries and from the model it appears that the variation of the tidal amplitude a with x is generally less than 10% over the distance l_b.

The convergence of the estuarine width was approximated by an exponential function $b_C = b_C(0) \exp(-x/l_b)$, primarily because it greatly simplifies the tidal equations. In reality, the convergence can often be approximated as well by a hyperbolic function $b_C = l_b b_C(o)/(x + l_b)$, which differs only slightly from the exponential function for $x < l_b/2$. The hyperbolic approximation of the width convergence yields $dU_R/dx = (Q_R/h)d/dx(1/b_C) = Q_R/(l_b h b_C(0))$. Substitution in (6.71) gives an estimate of the convergence length l_b consistent with equilibrium:

$$l_b = \frac{6Q_R}{\omega h U b_C(0)} \left[\frac{1}{c^-} - \frac{1}{c^+} \right]^{-1}. \tag{6.72}$$

If we had used an exponent $n = 4$ in the transport formula (6.62), instead of $n = 2$, then the factor 6 in (6.72) would change in a factor 5.

The expression (6.72) is evaluated from the model (6.41) for weakly converging estuaries, using the parameters of Table 6.2. Figure 6.18 shows that the HW and LW propagation velocities $c^{\pm}$ computed with the 1D model are in fair agreement with tide table values (Table 6.5). Table 6.6 gives values for the discharge Q_R and the width at the mouth $b_C(0)$. The results are compared with the observed convergence lengths in Fig. 6.39.

It should be emphasised that the expression (6.72) only holds in the convergent estuarine section where the influence of river discharge on sediment transport increases in upstream direction, such that the increase in river-induced velocity U_R matches the upstream increase of the tidal asymmetry velocity component U_A.

In the case of very weak (or zero) river discharge such a match is not possible. In that case morphodynamic equilibrium requires that the HW and LW propagation speeds are approximately equal. This can occur if large intertidal areas are adjacent to the estuarine channel, as will be discussed in Sec. 6.5. The expression (6.72) clearly makes no sense if c^+ and c^- are equal; the condition (6.71) is not relevant for the morphodynamic equilibrium of estuaries with large tidal flats and small river inflow; such estuaries are not considered here.

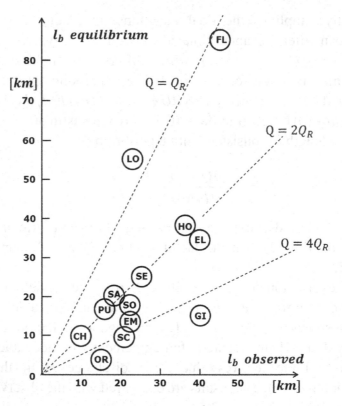

Fig. 6.39. Comparison of the actual convergence length with the equilibrium convergence length for several estuaries with small intertidal area and high peak river discharge. The equilibrium convergence lengths are derived from the condition (6.72) for uniformity of the sediment flux, using the 1D linear tidal model. The river discharge Q_{morph} representative for the long term average sediment flux has been set twice the long term average river runoff Q_R of Table 6.6. The estuarine acronyms are also indicated in this table. In estuaries situated below the line $Q_{morph} = 2Q_R$ sediment moves upstream when the river runoff is twice the average. For $Q = Q_R$ all estuaries have upstream sediment transport, except the Fly and the Loire. For $Q = 4Q_R$ all estuaries have seaward sediment transport, except the Gironde and the Ord.

Morphological river discharge

The river discharge Q_{morph} which is representative for the long term influence of river runoff on the estuarine morphology, is not well known. The impact of high discharges is greater than the impact of the average discharge Q_R. Therefore we have used in the comparison $Q_{morph} = 2Q_R$. This choice is somewhat arbitrary. For some estuaries it may be too large and for others too small, see Fig. 6.40. Therefore we have plotted in Fig. 6.39 also the theoretical equilibrium lines for $Q = Q_R$ and $Q = 4Q_R$.

The choice $Q_{morph} = 2Q_R$ seems reasonable for estuaries with modest seasonal and yearly fluctuations in river runoff; we consider as modest, a typical yearly maximum runoff not more than 4 times the average runoff and a once-in-10-year peak runoff not more than 6 times the average runoff (considering monthly averages).

Most of the estuaries fall into this category, with the exception of the Fly river (FL). The seasonal fluctuation of the Fly river runoff is very small [960] (see Fig. 6.40); therefore, a morphological discharge $Q_{morph} \approx Q_R$ is probably more appropriate for the Fly river.

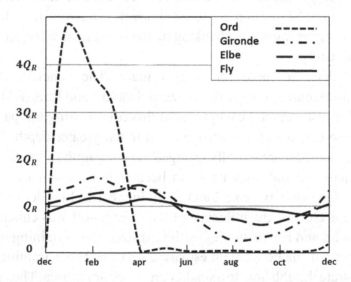

Fig. 6.40. Multi-annual average values of the monthly discharges of the rivers Ord, Gironde, Elbe and Fly, divided by the long-term mean discharge Q_R. The discharge variation of the Ord is very large (data before dam construction in 1969) and contrasts strongly with the small seasonal variation of the Fly river discharge. The Gironde and the Elbe are intermediate; the Gironde has a larger discharge variation than the Elbe. The multi-annual averaging masks part of the actual monthly variability, which is substantially larger. Discharge fluctuations on a daily basis are larger than fluctuations in the monthly averages.

According to Fig. 6.39, this would imply that the Fly river estuary could be close to morphological equilibrium. This is also a conclusion of several field studies of the Fly river estuary [381].

Comparison with data

Channel deepening and the related increase of tidal amplitude occurred in several other estuaries, such as the Scheldt, Elbe, Ems, Weser, Seine and Loire. Most of the intertidal areas were embanked when the estuarine channel was deepened. In spite of these interventions, the Elbe, Weser and Seine have a convergence length close to equilibrium, according to Eq. (6.72). Channel deepening and simultaneous embanking of intertidal areas produce an increase of both the HW and the LW propagation speed, as discussed earlier. The equilibrium convergence length may not be strongly affected by channel deepening and simultaneous embanking of intertidal zones. For the Elbe, Weser and Seine estuaries the antagonistic effects of channel deepening and embanking on the convergence length cancel approximately.

This does not hold for every estuary: The Scheldt (SC, the upstream estuarine part of the Western Scheldt) and Ems (EM) estuaries, for instance, are exceptions to this rule. During the past century these estuaries were dredged to a much greater depth, but the (small) intertidal area hardly changed. The strength of the tidal currents increased and flood transport became more dominant (see the large difference between c^+ and c^- in Fig. 6.18.) In order to reduce channel accretion, the influence of the river runoff was enhanced in the Scheldt and Ems estuaries by the construction of training walls at the mouth of the convergent estuarine section. These training walls concentrate the ebb flow in a smaller cross-sectional area. They mimic in a way the effect of stronger channel convergence, as required for equilibrium. The training wall at the entrance of the Scheldt estuary is shown in Fig. 5.15.

The Gironde estuary (GI) and the Ord (OR) fall far below the equilibrium line, while the Loire is situated far above. The Gironde estuary has developed by filling a wide pre-Holocene river valley. It is still importing sediment and has probably not yet reached its equilibrium configuration [14]. Besides, the choice $Q_{morph} = 2Q_R$ is probably too low for the Gironde, because of the large seasonal river discharge fluctuation (Fig. 6.40). The cases of the Loire and the Ord are discussed below.

Channelisation of the Loire estuary

The Loire estuary was much wider in the past, with large intertidal areas and a braiding meandering channel system. The width of the estuary was strongly reduced after channel reconfiguration and embanking in the past century, see Fig. 6.41. Table 6.6 shows that at present, the width-to-river discharge ratio of the channelised part of the estuary is small in comparison with other estuaries. In spite of strong flood-dominant tidal asymmetry, the estuary becomes ebb dominant when the river discharge is larger than average. In the channelised part of the estuary erosion dominates over accretion; the channel has deepened naturally over the past 50 years [601].

At low river discharge, flood currents in the Loire are stronger than ebb currents. The estuary becomes very turbid during such periods, and large fluid mud patches form on the channel bed, see Fig. 6.33. The same happens in other estuaries, in particular the Gironde estuary. There is an important difference, however. In the Loire the fluid mud

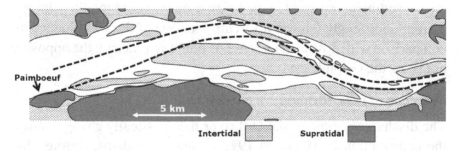

Fig. 6.41. The estuarine section of the Loire in the 19th century, characterised by large intertidal areas and a braiding meandering channel system. The dotted line is the contour of the estuary after channelisation in the 20th century.

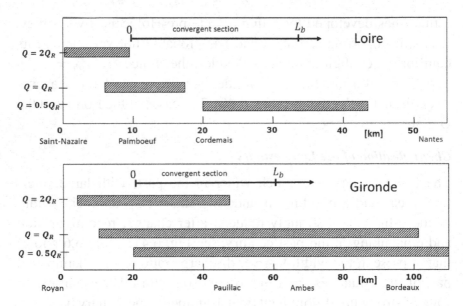

Fig. 6.42. The extent of fluid mud layers (gray bars) along the Loire and Gironde estuaries for small river discharge $Q = 0.5Q_R$, medium river discharge $Q = Q_R$ and high river discharge $Q = 2Q_R$. The Loire expels the fluid mud from the estuarine section already at medium discharge, while in the estuarine section of the Gironde the fluid mud is still present at high discharge. From [74, 792].

is expelled when the river discharge is above average, while in the Gironde a fluid mud layer persists, even when the river discharge is twice the annual mean, see Fig. 6.42. This is consistent with the model results for the equilibrium convergence length. Uniformity of the net sediment transport in the Loire estuary would require a larger convergence length (slower upstream increase of fluvial influence in comparison with upstream increase of tidal asymmetry); the opposite holds for the Gironde estuary.

Reduction of peak discharges in the Ord river

The discharge regime of the Ord river has drastically changed after the construction in 1969 and 1972 of upstream dams. Before, the Ord river discharge was highly variable, with peak discharges that could reach 100 times the average runoff (on a daily basis). It is probable that these peak discharges controlled to a large extent the width and depth of the river and the estuary. According to the previous analysis, the elimination of high discharges should induce a reduction of the convergence length of the estuary. This is consistent with the findings of Wolanski *et al.* [961], who studied the impact of the river dams on the morphodynamics of the Ord estuary. They observed a 50% reduction of the channel width at 20 km from the mouth, thirty years after the construction of the dams. The convergence length of the channel has decreased by a factor 2; the overall width has decreased less as the intertidal area has increased.

The agreement of the observed convergence length adaptation with the model prediction can only be qualitative. Because of the large relative tidal amplitude and the large intertidal area, the 1D estuary model is less suited for the description of the present hydrodynamics of the Ord estuary. This is illustrated by the fact that, in spite of the high turbidity, a high roughness coefficient is needed in the model to simulate the observed current velocities and tidal damping. The equilibrium condition is not fulfilled, because the estuary is still accreting, according to [961].

Equilibrium or not?

The equilibrium condition (6.70) predicts a value of l_b of the right order of magnitude. For estuaries with a small convergence length the predicted equilibrium value is small and for estuaries with a large convergence length the predicted value of l_b is large. The considered estuaries have apparently a tendency to evolve according to the equilibrium condition (6.70).

According to the simple 1D-model, morphodynamic equilibrium is the result of a balance between upstream sediment transport due to tidal asymmetry and downstream transport due to river runoff in combination with tidal flow. If the estuarine geometry is modified into a less converging shape, then the upstream increase in tidal asymmetry transport will prevail over the upstream increase in river-induced transport. This will induce upstream sediment transport, which will tend to reduce the upstream estuarine width and therefore reduce the convergence length. The inverse will happen if the the estuarine geometry is modified into a more strongly converging shape.

On the basis of the previous results it is difficult to say how close the estuaries are to equilibrium. The uncertainty is large, due to inaccuracy in the estuarine data (Tables 6.2, 6.5 and 6.6), as well as to the oversimplification of the model. This latter point will be addressed later more in detail.

The equilibrium condition gives an indication of the response of estuaries to interventions. Deepening of the estuarine channel leads to an increase of the convergence length, according to (6.72), because the difference $1/c^- - 1/c^+$ becomes smaller. Embanking and reclamation of tidal flats will reduce the equilibrium convergence length, because the difference $1/c^- - 1/c^+$ becomes larger. Before the interventions of the past centuries, which transformed several estuaries into shipping channels, the convergence length was longer than today. The seaward part of these estuaries had a weakly converging width and large intertidal area, similar to tidal lagoons. This is illustrated in Fig. 6.31 for the Seine estuary.

6.4.3. *Equilibrium at the Sea Boundary*

The condition (6.71) is a requirement for uniformity of the net sediment flux inside the funnel-shaped estuary. This condition does not exclude a net sediment import or export. Morphodynamic equilibrium therefore also requires continuity of the sediment flux at the sea boundary,

$$\langle q(x = 0) \rangle = q_{river}, \qquad (6.73)$$

where q_{river} is the average fluvial sediment import. The average sediment flux at the sea boundary depends on the tidal asymmetry Δ_{FR}^{inlet}, and on the residual velocity U_S due to Stokes transport and river runoff. Tidal asymmetry at the inlet corresponds usually to a shorter rise and longer fall of the tide, and therefore induces a net sediment import. The Stokes-related transport is ebb directed and contributes to sediment export, together with the (usually smaller) river-induced outflow.

Equality of these ingoing and outgoing sediment fluxes is not enough for equilibrium, however. The net tidal sediment flux should be sufficiently ebb dominant to evacuate the fluvial sediment input, to remove the sediment which is brought to the estuarine mouth by wave-driven longshore and cross-shore transport, and to neutralise the sediment import due to estuarine circulation.

Estuarine mouth bar

Table 6.2 shows that tidal asymmetry at the inlet of several estuaries is often quite strong, with a shorter duration of tidal rise compared to tidal fall. This causes a strengthening of flood currents relative to ebb currents. Sediment accumulates in the inlet region and produces a mouth bar. The presence of a mouth bar strengthens the ebb-directed flux due to river runoff and Stokes transport, because both are inversely proportional to depth. Hence, the development of a mouth bar is crucial for fulfilment of the equilibrium condition (6.73), in the case of strong flood-dominant offshore asymmetry.

Tidal propagation over the mouth bar enhances tidal asymmetry. The tide is damped, except when the estuarine mouth is strongly converging. In this case a tidal bore can develop (see also Sec. 2.2.2).

Strong tidal asymmetry ($\Delta_{FR}^{inlet} > 2$ hours) occurs in particular at the Seine estuary inlet, see Table 6.2. In the past, a large mouth bar did exist at the mouth of the Seine estuary. The funnel shape of the estuary, in conjunction with the mouth bar, produced a high tidal bore, see Fig. (2.6). After dredging of the mouth bar in the 1960s, the tidal bore disappeared (Fig. 6.32).

Strong tidal asymmetry ($\Delta_{FR}^{inlet} > 2$ hours) also occurs at Hangzhou Bay, the mouth region of Qiantang River. A large mouth bar is present in the strongly converging inlet of the Qiantang estuary. This produces the famous Qiantang River tidal bore (see Fig. 2.7 and the accompanying discussion).

The presence of a mouth bar is a feature that distinguishes natural estuaries from tidal lagoons. In the case of tidal lagoons, the water depth is greatest at the inlet. The water depth decreases inshore, and also offshore at the ebb-tidal delta. In the case of natural estuaries with a mouth bar, the water depth is greatest offshore and upstream in the tidal river.

In most estuaries the mouth bar has been dredged to facilitate navigation. As a consequence, the natural equilibrium state is disturbed. Fluvial sediment supply and net tidally induced sediment import produces accretion at the estuarine inlet. An artificial equilibrium is maintained by continuous dredging. An extreme example is the mouth bar of the Changjiang (Yangtze) estuary, where the amount of dredging for maintaining the navigation depth exceeds 20 million m^3/year [231].

6.4.4. *Other Processes that Influence Sediment Transport*

The strongly simplified 1D description of estuarine morphodynamics has many shortcomings, as important processes have been ignored. A qualitative discussion is presented below.

Role of estuarine circulation

The influence of estuarine circulation has been disregarded thus far. In some estuaries, sea water is almost completely flushed out at high river discharges. This is the case, for instance, in the Seine [43], the Weser [350], the Elbe [755] and the Loire [925]. However, in the zone close to the estuarine mouth, a well developed estuarine circulation is generally present, with a net upstream flow along the bottom, even at high river discharge, see Fig. 6.12. The residual velocities related to this circulation may exceed the velocity related to river runoff, especially in the deeper channel sections near the mouth. The neglect of estuarine circulation in the 1D model therefore leads to underestimation of sediment import from the sea.

At least two aspects of estuarine circulation influence tidal propagation [457]:

(1) Concentration of the ebb current in the upper layer;
(2) Density stratification and related turbulence damping, especially during ebb tide.

The effects (1) and (2) influence mainly the LW propagation speed, but in opposite ways (effect (1): Lower speed; effect (2) higher speed). This may explain why both the HW and LW propagation speeds are nevertheless fairly well reproduced by the 1D-model.

Comparison of depth-averaged and 3D numerical models for the Seine estuary shows that estuarine circulation increases the import of fine marine sediment; however, it hardly influences the formation and location of the turbidity maximum [111]. Model simulations in the Gironde, the Charente and the Tamar led to a similar conclusion [15,847,865]. This suggests that salinity-induced density effects have no strong impact on sediment transport gradients related to tidal asymmetry and channel width convergence.

Another argument was given by Prandle, who established predictive relationships for the depth and length of estuaries [689]; these relationships are essentially based on the empirical finding that the tidal amplitude is approximately uniform along estuaries with a length much smaller than the tidal wavelength (see also Sec. 6.2.4). He compared these relationships with observations for a large number of UK estuaries, and considered the good general agreement as evidence for the minor influence of salinity-induced processes on estuarine morphology.

Fine sediment and fluid mud

For the derivation of the equilibrium condition we have used a bedload sand transport formula (6.62). However, many estuaries are more muddy than sandy; the bottom substrate contains a high fraction of silt and mud. This is the case for very turbid estuaries such as the Gironde [13], the Hudson [850] and the Fly [380], but also for the less turbid estuaries of the Seine [111] and the Thames [642]. In all these estuaries fluid mud layers have been observed when the river discharge is low.

Fluid mud, and more generally, high concentration turbidity layers, influence estuarine hydrodynamics. The bed roughness is strongly reduced when fluid mud is present, as discussed in Sec. 3.5.2.

Lutoclines (sharp interfaces between layers of different suspended sediment concentration) dampen turbulence due to density stratification. Large, rapidly falling flocs may persist when turbulence is subdued. This favours settling of fine sediment, especially during the ebb phase [970]. Flocs are trapped in the near-bottom layer and transported upstream by the flood current. Upstream transport is further enhanced by estuarine circulation. Observations and model simulations show that mixing asymmetry enhances the formation of a turbidity maximum [119, 342, 764]. The underlying processes are quite complicated; even for sophisticated numerical models it is a challenge to represent these processes in a satisfactory way.

At high river discharge and spring tide only small fluid mud patches remain, at places sheltered by irregularities in the bottom profile. Under such conditions, bed roughness is less reduced than at low discharge. The reduction of bed roughness is accounted for in the 1D model by using a linear friction coefficient inversely proportional to the observed maximum tidal velocity, see Eq. (6.12).

One might expect that fluid mud is less relevant for estuarine morphodynamics, assuming that long-term morphological evolution occurs mostly at spring tide and high river discharges. However, this is not always true. During long periods of low river discharge fluid mud may consolidate at the bottom and become resistant against erosion under high river discharge. This is consistent with the presence of consolidated mud on the channel bed of highly turbid estuaries.

According to Winterwerp [954], transport of fine sediment is fairly well described by the formula (6.62) with $n = 3$, if easily erodible fine sediment is widely available (saturated mud transport capacity). Engelund and Hansen proposed a formula with $n = 4$ (3.79), for representing both bedload and suspended load transport. Using a higher exponent in the sediment flux formula has only a minor influence on the expression (6.72), as indicated earlier.

A major difference between sand transport and fine sediment transport is related to suspension and erosion lags, which are not included in the sediment transport formula (6.62).

Suspension and erosion time lags

The suspended load of fine sediments does not adjust instantaneously to changes in current strength. The delayed response of the suspended load to velocity variations induces a net transport of fine sediment which depends on tidal asymmetry in a different way than rapidly responding bedload or suspended load. In Sec. 6.6 this will be discussed in more detail; it will be shown that suspension and erosion lag effects are particularly important in tidal basins with large intertidal areas. It is therefore less relevant for the estuaries of Table 6.6.

Flood and ebb dominated channels

Flood and ebb currents are often concentrated in different parts of the channel cross-section. In some cases the inlet consists of two or more channels which convey different fractions of the tidal flow during ebb and flood. Examples are the Gironde [13] and the Fly [960], where the difference between flood and ebb discharge in each channel is greater than the river discharge. But single-channel inlets may also exhibit significant lateral differences in flood-ebb asymmetry, as observed, for instance, in the Hudson [106] and Tamar [862]. This may be related to channel meandering or to the influence of earth's rotation. In these estuaries fluvial sediment can escape to the sea via the ebb-dominated inlet channel (or via the ebb-dominated part of the inlet).

The Gironde is an example of wide two-channel estuary. The dredged southern inlet channel carries the major part of the flood flow; flood dominance is enhanced in this channel by estuarine circulation. A large part of the ebb flow is conveyed by the shallower northern channel; during high river discharge the tide-averaged near-bottom flow in this channel is directed downstream, allowing fluvial sediment to escape the estuary [13].

The 1D model is not adequate for describing morphodynamic equilibrium of estuaries with multiple inlet channels, such as the Gironde estuary. In the case of the Fly, we have therefore considered the tidal river upstream of the delta apex.

Wind influence

Suspension and subsequent export of fine sediment from mud flats is strongly stimulated by wind-induced waves, see Sec. 5.5.2. An example is the Seine estuary, with mudflats bordering the main channel near the mouth. The highest sediment export from the estuary occurs during storm events, when the mudflats are eroded by wind waves and highly turbid waters are dispersed over the outer bay [515]. Fluidisation of mud deposits by waves also plays a role [582]. Many other observations exist of sediment export caused by wave-induced mudflat erosion, for instance in the Wadden Sea [219].

Non-uniform tidal asymmetry

Some of the estuaries (Ord, Seine) have a large amplitude-to-depth ratio a/h. This invalidates approximations based on $a/h \ll 1$, made in the derivation of the model results. Moreover, in the case of a large amplitude-to-depth ratio, new phenomena may come into play. This is illustrated by the following examples.

Uncles *et al.* [863] have observed ebb-dominant flow over the intertidal areas in the otherwise strongly flood-dominant Tamar estuary. They explain this phenomenon by the inertial delay of flow reversal in the deeper parts of the channel relative to the shallow parts. After high water the ebb-flow therefore starts earlier at the shallow channel banks than at the centre of the main channel.

In estuaries with a large amplitude-to-depth ratio the contribution of increased ebb velocities due to Stokes drift is more important than in estuaries with a smaller amplitude-to-depth ratio. In the Tamar, model computations indicate that there is a small downstream residual flow near the bottom in the shallow upper part of the estuary, which can be explained in this way [864].

Model as a tool for diagnosis

The above discussion illustrates that simple flow models may give an inaccurate or even misleading picture of the net sediment transport in tidal inlets, and that interpreting the results requires additional field and model data. One may wonder what the use is of idealised models if there are so many limitations?

It is clear that the simple linear 1D-models (6.32) and (6.41) are not adequate for predictive purposes. The strength of idealised models is their transparency; therefore they should be considered and used as a tool for analysis, that helps define and understand phenomena observed in the field. They are particularly useful for a comparison of tidal inlets, as shown in the previous sections. Analytical models are also valuable in interpreting sensitivity tests in numerical models.

Although the models themselves do not contain much physics, they greatly contribute to better understanding of estuarine morpho-dynamics, if used in combination with field observations.

6.5. Equilibrium of Tidal Inlets without River Inflow

6.5.1. *Equilibrium Morphology*

In tidal lagoons and estuaries large quantities of sediment are transported every tide back and forth by flood and ebb currents. An imbalance between the flood transport and ebb transport results in erosion or accretion. An imbalance of 5 percent may raise or lower the bed level on average by a few metres, if it is persistent over several hundred years. This is significantly more than the average rate of sea-level rise during the past two or three thousand years. Most tidal lagoons would have disappeared if the flood transport persistently exceeded the ebb transport by more than 5 percent. The ubiquitous presence of tidal lagoons along the world's coastlines therefore indicates that flood and ebb transport are on average equal within a few percent. This means that most existing tidal lagoons should be in a situation close to morphological equilibrium.

This equilibrium is not arbitrary, however, because flood and ebb transport depend on morphology. The morphology must satisfy certain characteristics, such that flood and ebb transport are almost equal. For coastal lagoons, these characteristics depend mainly on tide and wind. For estuaries, the equilibrium morphology depends primarily on river discharge, as we have seen before.

In the following, we will derive morphological characteristics of tidal lagoons that are consistent with the balance of flood and ebb transport. We will also consider tidal bays and estuaries with very small river inflow. We will not look at the detailed morphological structure of these systems, but to the overall characteristics. Therefore we use simple one-dimensional models that can be solved analytically. The relationships we find in this way have more a qualitative than a quantitative character. They provide an indication how tidal lagoons and estuaries may change as a result of external changes, like sea-level rise, or as a result of human interventions.

When comparing model predictions of equilibrium morphology with observations, it should be realised that some lagoons and estuaries could have moved away in recent times from an equilibrium state. This might be due to natural causes — an extreme storm or an exceptional river discharge. Human interventions, such as dredging and reclamation, can also strongly disturb the equilibrium morphology. An example is previously discussed the closure of the Zuyderzee in 1930; it is not sure that the basins in the Western Wadden basins have already adapted to this intervention.

Sediment transport dominated by tidal flow alone

We restrict the analysis to tidal inlets where river flow is small. We only consider lagoons and estuaries where river flow is much smaller than tidal flow, even for peak runoff events. The equilibrium condition for estuaries with substantial river discharge were dealt with in the previous section. We assume that sediment transport is mainly dependent on tidal flow. Our analysis of morphological equilibrium is based on the sediment transport formula (3.79).

The influence of wave action on the net (tide-integrated) sediment transport is not taken into account. This implies that a tidally induced net transport gradient in the channel will produce local accretion or erosion, either in the channel or on the adjacent tidal flat. This simplification seems reasonable for medium to coarse sandy sediment, but it is questionable for fine sediment. Wave action on tidal flats can resuspend fine sediments that were imported by tidal currents. The suspended sediment concentration of the ebb flow is increased and the ebb sediment transport will be larger than the tide induced transport of formula (3.79). The consequence of neglecting wave action on tidal flats will be discussed qualitatively in the interpretation of the results of the 1D model (Sec. 6.5.2). Fine sediment transport is treated in more detail in Sec. 6.6.

We consider tidal basins shaped by the feedback between sediment transport and morphology, such that the requirement of net zero sediment transport imposes a condition on the bathymetry. Tidal basins that inherit their morphology for a substantial part from pre-holocene geological structures, such as drowned river valleys, do not fall into this category. The model therefore should not apply, for instance, to Chesapeake Bay and Delaware Bay, although these estuaries have a low river discharge.

Equilibrium relationship

Morphodynamic equilibrium requires uniformity of the long-term average sediment flux. In the absence of fluvial sediment import, the long-term average sediment transport must be equal to zero:

$$\langle q \rangle = 0. \tag{6.74}$$

A crucial assumption is the existence of a cyclic tide that represents the average sediment transport characteristics within the lagoon of all tides, including wind and wave effects, during a very long period of stable average sea level. One may argue that this is not a realistic assumption, because even if the tide is perfectly periodic, an equilibrium morphology will not be perfectly periodic at the time scale of a single tidal period. If a long-term equilibrium exists, it will probably evolve in cycles of different periodicity. We will assume, however, that the gross morphological features, such as the average length, depth and width, do not vary substantially during these cycles.

We further assume that the sediment transport depends on the tidal current velocity and that it is represented by a total load transport formula of the type (3.79):

$$q = \alpha b_C u(x, t)|u(x, t)|^n. \tag{6.75}$$

For the exponent n we take the value $n = 3$, but other values, $n = 2$ or $n = 4$ lead to almost identical expressions.

The propagation of the tide inside the lagoon changes the duration of the flood and ebb periods. We call Δ_{EF} the difference between the durations of the ebb period and the flood period. If $\Delta_{EF} > 0$, the maximum flood current will be stronger than the maximum ebb current and vice versa if $\Delta_{EF} < 0$. The flood and ebb velocities are related through the condition of zero net discharge,

$$\langle Q \rangle = \frac{1}{T} \int_0^T (h + \eta(t))u(t)dt = 0, \tag{6.76}$$

where T is the tidal period, h is the average channel depth and η the tide elevation relative to mean sea level. The tidal amplitude is a and $\varphi \equiv \omega \Delta_S$ is the phase difference between the HW tide level (LW tide level) relative to the time of high slack water HWS (time of low slack water LWS).

We assume again that the current velocity can be represented by a semidiurnal tide and a quarter-diurnal overtide generated by the nonlinear tidal propagation in the lagoon and by the offshore tide at the lagoon inlet. According to (6.55) and (6.56) we have

$$u(t) = U \cos \omega t + U_A \cos 2\omega t - U_S, \qquad (6.77)$$

with

$$U_S = \frac{aU}{2h} \sin \varphi \approx \frac{a\omega U \Delta_S}{2h}, \quad U_A = -U_S + \frac{\omega \Delta_{EF}}{4} U,$$

where Δ_{EF} is the difference in duration between ebb and flood. We assume $|U_A|$, $|U_S| \ll U$ and $\omega \Delta_{EF} \ll 1$. In (6.77) we have neglected a term $U_B \sin 2\omega t$, because it contributes only higher order terms in the small variables a/h and $|U_B|/U$.

Substitution of (6.77) in the tide-averaged sediment flux (6.75) gives

$$\langle q \rangle = \frac{\alpha bc}{T} \int_{t_1}^{t_1+T} u|u|^3 dt' = \frac{\alpha bc}{T} \int_{t_1}^{t_2} u^4 dt' - \frac{\alpha bc}{T} \int_{t_2}^{t_1+T} u^4 dt',$$

$$(6.78)$$

where $t_1 = (-T + \Delta_{EF})/4$, $t_2 = (T - \Delta_{EF})/4$. The integrals over the small intervals $[-T/4, t_1]$; $[t_2, T/4]$, etc. yield higher order contributions in a/h and $\omega \Delta_{EF}$ that can be ignored. Integration then gives

$$\langle q \rangle \approx \frac{16\alpha bc}{5\pi} U^3 \left(U_A - \frac{5}{3} U_S \right) \approx \frac{4\alpha bc\omega}{5\pi} U^4 \left(\Delta_{EF} - \frac{16a}{3h} \Delta_S \right).$$

$$(6.79)$$

This expression relates the net sediment flux to the asymmetry in ebb-flood duration Δ_{EF} and to the Stokes effect included in the phase shift Δ_S. It can be shown that the result is not essentially different if we had taken $n = 3$ or $n = 5$ in in the sediment transport formula (6.75).

Tidal lagoons with uniform width

For tidal lagoons with uniform width the expressions (6.48) and (6.58) relate Δ_{EF} and Δ_S to the morphologic characteristics. We substitute these expressions in (6.79), taking into account (6.57),

$$\langle q \rangle \propto \Delta_{FR}^{inlet} + \frac{r l_A^2}{3 g h h_S} \left[\frac{2a}{h} - \frac{16a}{3h} \right]$$

$$+ \frac{2 r l_A^2}{3g} \left[\frac{1}{D^- D_S^-} - \frac{1}{D^+ D_S^+} \right], \qquad (6.80)$$

where the HW (LW) depth $D^{\pm} = h \pm a$, $D_S^{\pm} = D^{\pm} b_C^{\pm}/b_S^{\pm}$, $h_S = (D_S^+ + D_S^-)/2$, l_A is the lagoon basin length and r the linearised friction coefficient. The first term in this expression is the contribution of the tidal asymmetry at the inlet, the second term is the contribution of the Stokes drift and the third term the contribution of the tidal asymmetry generated by nonlinear tidal propagation inside the lagoon. We simplify this formula by assuming that the channel widths at LW and HW are not very different: $b_C^- \approx b_C^+ \approx b_C$. We further assume that the basin widths $b_S^{\pm}$ at HW and LW satisfy the condition $b_S^+ - b_S^- \ll b_S^+ + b_S^-$ and that the tidal amplitude a is small compared to the average channel depth h. We may then approximate

$$\frac{1}{D^- D_S^-} - \frac{1}{D^+ D_S^+} \approx \frac{2}{h h_S} \left[\frac{2a}{h} - \frac{b_S^+ - b_S^-}{b_S^+ + b_S^-} \right].$$

Substitution in the condition for morphological equilibrium $\langle q \rangle = 0$ finally gives

$$\frac{b_S^+ - b_S^-}{b_S^+ + b_S^-} = \gamma \frac{a}{h}, \quad \gamma = \frac{7}{6} + \frac{h}{4a} \frac{\Delta_{FR}^{inlet}}{\Delta_S}. \qquad (6.81)$$

In this expression a/h is the relative tidal amplitude, b_S^+/b_S^- is the ratio of basin areas at HW and LW and Δ_{FR}^{inlet} is the difference in duration of falling and rising tide at the inlet. The time delay Δ_S

(average of HWS-HW delay and LWS-LW delay), is approximately given by (6.58)

$$\Delta_S \approx \frac{r l_A^2}{3 g h h_S}.$$

The condition (6.81) suggests that morphological equilibrium requires that the relative intertidal area $(b_S^+ - b_S^-)/(b_S^+ + b_S^-)$ for lagoons with a large relative tidal amplitude a/h should be larger than for lagoons with a small relative intertidal area. However, this also depends on the tidal asymmetry at the inlet, represented in the expression of γ by the difference in duration of tidal rise and tidal fall, Δ_{FR}^{inlet}.

Comparison with data

The relation between relative intertidal area and relative tidal amplitude is shown in Fig. 6.43 for the lagoons of Table 6.1. The figure also includes several estuaries of Table 6.2 for which the fluvial discharges are very small compared to the tidal discharges.

The overall picture supports the idea that the relative intertidal area for lagoons with a large relative tidal amplitude is generally larger than for lagoons with a small relative tidal amplitude. However, it also appears that some tidal lagoons with a rather large a/h have a smaller relative intertidal area than other lagoons with a smaller a/h. In the case of the Humber (HB) and the Baie d'Arcachon (BA) this can be related to zero or small negative offshore tidal asymmetry Δ_{FR}^{inlet}.

The factor γ is generally larger than 7/6, as may be expected from the fact that the rise of offshore tides is often shorter than the fall ($\Delta_{FR}^{inlet} > 0$). Most usual is a γ value between 1.5 and 2.

Figure 6.44 compares the ratio of the wetted lagoon areas at HW and LW, $S^+/S^- = b_S^+/b_S^-$, computed from the model (6.81), with the observed ratios indicated in Table 6.1. The agreement is reasonable, considering that Δ_{FR}^{inlet} is derived from observed HW and LW times at the inlet. This estimate of Δ_{FR}^{inlet} provides only a rough indication of

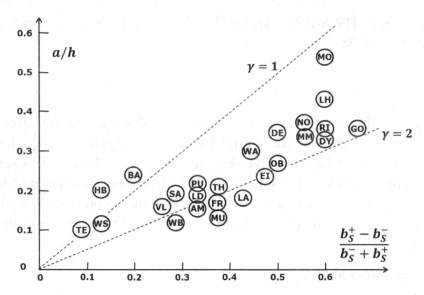

Fig. 6.43. Comparison of relative tidal amplitude a/h and relative intertidal area $(b_S^+ - b_S^-)/(b_S^+ + b_S^-)$ for tidal lagoons of Table 6.1 and for converging tidal basins of Table 6.2 with a small river inflow. These tables also explain the acronyms of the tidal inlets. The figure shows that tidal basins with a large relative tidal amplitude have in general a large relative intertidal area and vice versa.

the maximum steepness of the tidal rise and tidal fall, which is most relevant for the maximum flood and ebb velocities generated by the offshore tide. For this reason we may expect a rather large scatter in the computed values of b_S^+/b_S^-.

Nevertheless, a few lagoons do not match the observed values at all. For Murrells Inlet (MM) this may be due to the large value of the relative tidal amplitude; in this case the approximations in the model based on $a/h \ll 1$ are not valid. Therefore, we may not expect a good agreement between the model and the observations.

The case of Langstone Harbour (LH) has been discussed before, see Sec. 6.3.2. From observations [639] it appears that Langstone Harbour exports sand to the coastal zone; therefore the lagoon cannot be in morphological equilibrium. The presence of large muddy tidal flats points to the existence of other sediment transport processes not

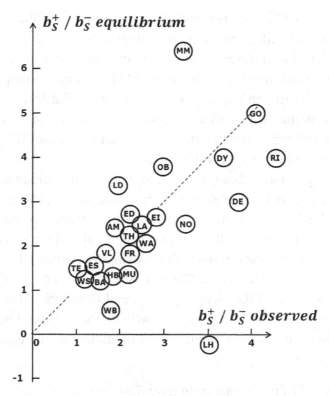

Fig. 6.44. Comparison of observed relative intertidal areas b_S^+/b_S^- with the relative intertidal areas that satisfy the condition $\langle q \rangle = 0$, according to the analytic 1D-models for tidal lagoons and for converging tidal inlets. The observed and the computed intertidal areas follow the same trend, with a few notable exceptions discussed in the text. One may conclude that most of the lagoons and converging basins are rather close to morphologic equilibrium. The computed intertidal areas are very sensitive to the input data, which are afflicted with substantial uncertainties. Part of the scatter in the computed intertidal areas is possibly caused by inaccuracies in the input data, especially regarding the offshore tidal asymmetry. The simplicity of the tidal and sand transport models also plays a role. The results should therefore be interpreted with care, as discussed in the text.

represented by a formula such as (6.75). This could be, in particular, an asymmetry between the HW and LW slack periods.

For Willapa Bay (WB), situated on the US Pacific coast, the discrepancy is large too. The offshore tidal asymmetry is negative (steeper tidal fall than rise), favouring sand export. This negative tidal asymmetry is due to interaction of the astronomical tides O1,

K1 and M2 [632], see also Appendix section C.1. Willapa Bay has strong (mixed) tides, but it is also subject to strong ocean storm waves [651]. From observations it is known that large sand volumes can be imported during a single storm [611], eventually compensating for the tidally driven export. The low intertidal flat level is indicative for important wave influence inside the Bay. The inlet is moving steadily northwards; it is therefore questionable whether Willapa Bay is close to a morphologic equilibrium.

Tidal lagoon morphology can be influenced by human activities. The impact of land reclamation, damming and partial closure has been discussed in Sec. 6.3.5. Another factor is fishery activity, which may enhance erosion, especially in shallow subtidal and intertidal zones. Intensive oyster culture exists in the Bassin d'Arcachon (BA); mussel fishery is practiced in several Wadden Sea basins in the Netherlands (TE, EI, VL, AM, FR), Germany (OB) and Denmark (LD). Bottom perturbation by fishery may explain why in some of these basins the present intertidal area is smaller than predicted by the equilibrium model (Fig. 6.44).

Converging tidal inlets with small river flow

We only consider converging inlets with substantial intertidal areas and with river discharges which are a few orders of magnitude smaller than the tidal discharges. Among the estuaries of Table 6.2, the Western Scheldt (WS), Thames (TH), Humber (HB), Dee (DE), Ribble (RI), Dyfi (DY), and Gomso (GO) satisfy this condition. In these estuaries, river flow is too small for providing a transport mechanism that can compete with tide-induced net sediment transport. Tide-induced net sediment transport should thus be small in a state close to morphological equilibrium. It will appear that this requires a relationship between the relative tidal amplitude and the relative intertidal area, as in the case of tidal lagoons.

The residual sediment flux (6.79) can be evaluated by using for Δ_{EF} the formula (6.49) and by computing $c^{\pm}$ and Δ_S from (6.42) and (6.44). After a few algebraic manipulations we find

$$\langle q \rangle \propto \Delta_{FR}^{inlet} - \frac{10a}{3h}\Delta s + \frac{2kx}{\omega p_2}\left[\frac{2ap_1}{h} - \left(p_2 + \frac{1}{4}\right)\frac{b_S^+ - b_S^-}{b_S^+ + b_S^-}\right],$$
(6.82)

with

$$p = (k_0 l_b)^2 - \frac{1}{4}, \quad p_1 = p_2 - \frac{1}{2}p + \frac{1}{8}, \quad p_2 = \sqrt{k_c^2 l_b^2 + p^2}$$

and

$$k_0 = \frac{\omega}{\sqrt{gh_S}}, \quad k_c = \frac{\omega r l_b}{ghh_S}, \quad \omega\Delta s = \varphi = \arctan\left(\frac{kl_b}{\mu l_b + 1}\right),$$

$$kl_b = \sqrt{\frac{1}{2}(p_2 + p)}, \quad \mu l_b = -\frac{1}{2} + \sqrt{\frac{1}{2}(p_2 - p)}.$$

The symbols in these equations have the following meaning: a is the mean spring-tidal amplitude, h is the average channel depth, Δ_{FR}^{inlet} is the difference between durations of rising and falling tide at the inlet, Δs is half the sum of the time delays HWS-HW and LWS-LW, k is the wave number, k_0 the wave number for frictionless flow, k_c the wave number for friction-dominated flow, r is the linearised friction coefficient, l_b is the estuary convergence length, $b_S^\pm$ the total wetted width at HW (LW), and h_S the ratio of average channel cross-section and average wetted width.

The equilibrium condition $\langle q \rangle = 0$ can only be satisfied for all x if the first two terms on the r.h.s. of (6.82) cancel and if the term between square brackets vanishes. For the selected estuaries with small river inflow, $\Delta_{FR}^{inlet} - 10a\Delta s/3a$ is about half an hour or less, with the exception of the Dee and the Dyfi. Most of the estuaries are close to a net zero sand transport at the mouth, considering uncertainty margins related to the simplicity of the model and to the inaccuracy of the data.

Vanishing of the term between square brackets (meaning $c^+ = c^-$) gives a model estimate of the b_S^+/b_S^- ratios,

$$\frac{b_S^+ - b_S^-}{b_S^+ + b_S^-} = \gamma \frac{a}{h}, \quad \gamma = \frac{2p_1}{h\left(p_2 + \frac{1}{4}\right)}. \tag{6.83}$$

These ratios are shown in Fig. 6.44, together with the ratios computed from the lagoon model. They compare fairly well with the observed ratios of the wetted areas at HW and LW. We thus conclude that the present morphology of most of these estuaries is consistent with a state close to morphological equilibrium.

The Dee and the Dyfi have a highly asymmetric offshore tide, with a much faster rise than fall. For these estuaries we should expect dominance of flood currents over ebb currents and a corresponding net sand import.

The fast tidal rise at the Dee inlet may explain why the actual intertidal area is larger than would be needed for morphological equilibrium in the case of a symmetrical offshore tide (Fig. 6.44). Besides, observations indicate ongoing accretion of the Dee estuary [615]. During the past century large intertidal flats were turned into agricultural land. This loss of intertidal area may still contribute to the present net import of sediment.

For the Dyfi estuary the situation is less clear. Field and model studies suggest that the Dyfi is at present in morphological equilibrium, with flood dominance over the tidal flats and ebb dominance in the inlet channel [112,673,727]. This can only be reconciled with the fast tidal rise indicated by the tide gauge (assuming that the data are reliable) if other competing sediment transport processes play a role. One may think of wave-induced sediment resuspension on tidal flats related to the basin orientation along the prevailing western wind direction, or residual circulation around the large sandbanks in the estuary.

6.5.2. *Other Processes that Influence Sediment Transport*

Limitations of the 1D model

The present analysis is not accurate; major shortcomings are related to (1) inaccuracy of the data describing the bathymetries, (2) an over-simplification of the bathymetric schematisations and tidal equations, (3) a relative tidal amplitude too large for linearisation of the equations, (4) the omission of other processes contributing to sediment transport (wave action, for instance), and (5) simplification of the tidal curve. The question therefore is, whether the analysis provides a meaningful evaluation of the equilibrium condition of tidal basins.

Several tidal basins are rather shallow in comparison with the tidal amplitude. For Murrells inlet and North inlet (NO, US Atlantic coast) and for Langstone Harbour the relative tidal amplitude a/h is close to 0.5. The condition $a/h \ll 1$ is clearly not satisfied; the model results for these inlets are therefore not very reliable. For other inlets, the Ribble, Dee, Dyfi and Gomso, the relative tidal amplitude is close to 0.3, which is not really a very small number.

We have represented asymmetry in the tidal current velocity by the expression (6.77). The M4 tidal component was estimated from the differences in the duration of tidal rise and tidal fall. In reality the form of the tidal velocity curve can deviate substantially from (6.77) and the M4 tidal component does not depend only on the difference in the duration of tidal rise and tidal fall. More realistic results would have been obtained by considering, in particular, the actual hypsometry of each tidal lagoon (i.e., the cross-sectional profile of channels and tidal flats). Some tidal lagoons have low lying tidal flats that inundate shortly after LW and other lagoons have high tidal flats that inundate only shortly before HW. It is clear that the tidal curves (and the tidal asymmetry) are not the same in these lagoons, but this difference does not appear if we consider only the ratio b_S^+/b_S^-. The scatter in

Fig. 6.44 may be partly due to the neglect of differences in tidal flat elevation.

Influence of wind waves

Wave action and storm surges have been ignored so far. This may lead to erroneous results, as illustrated above for Willapa Bay.

Several studies show that wave action contributes substantially to the export of fine sediment from tidal lagoons [514]. The influence of wave-induced sediment resuspension and corresponding sediment export has been well documented for the Wadden Sea tidal lagoons. Fine sediments deposited on the tidal flats during summer are removed during winter. After the storm surge of 1976, fine Wadden sands were found offshore even at 20 km from the coast [953]. Large sediment export during storms has been reported also for the Lister Dyb (LD) in the Danish Wadden Sea [547].

The inlet channels of Texel Inlet and Lister Dyb are oriented along the prevailing (western) wind direction. We may expect that tidal flats will develop less easily in these lagoons. The intertidal area is relatively small; favouring flood dominant sediment transport. The equilibrium intertidal area of these lagoons is over-predicted by the model (see Fig. 6.44). This discrepancy may be due to the absence of wave influences in the model.

Observations and model simulations show that tidal flat profiles tend to adjust to wave action by assuming an upward concave profile. For similar tidal conditions, the elevation of tidal flats under strong wave influence is lower than the elevation of tidal flats under weak wave influence. Wave influence depends not only on depth, but also on fetch, the distance over which the wave generating wind force acts. As pointed out by LeHir and others, wide tidal flats will not easily accrete, because of the large fetch [289,431,514,902]. Wave induced resuspension on a tidal flat during HW increases the sediment load of the ebb flow. This implies that in tidal lagoons with large tidal flats, compared to tidal lagoons with small tidal flats, the ebb flow from the tidal flat to the channel will carry more sediment (irrespective of the tidal asymmetry in the channel). This wave-induced transport is not included in the 1D model and in the sediment transport formula (6.75). We may thus expect that for tidal lagoons with large tidal flats, ebb transport is more effective compared to flood transport than the 1D model predicts. This should be the case in particular for the fine sediment fraction.

A more general discussion of the influence of wave action on sediment transport and sedimentation is given in Secs. 3.8.1, 5.5 and 6.6.

Local differences in flood and ebb dominance

In the simple one-dimensional representation of tidal inlets we have ignored cross-sectional differences in flow strength. Yet, such differences can be quite substantial. This is illustrated in Fig. 6.45, for the feeding channel of the Eastern Scheldt's inner flood delta. The

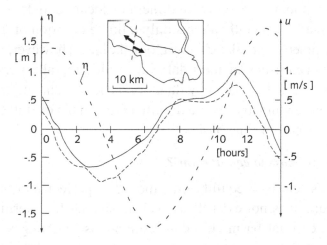

Fig. 6.45. Residual circulation in the inlet channel to the inner flood delta of the Eastern Scheldt (insert). The solid line is the depth-averaged velocity in a vertical near the southern channel bank and the fine-dotted line the depth-averaged velocity in a vertical near the northern channel bank; both are determined from simultaneous measurements. The southern part of the channel is flood dominated, while the northern part is ebb dominated. The dashed line indicates the tidal elevation. (Survey Department Rijkswaterstaat)

northern part of the channel is ebb-dominated, while the southern part is flood-dominated. Hence, even if the feeding channel is ebb-flood neutral as a whole, a net import or export of sediment will take place.

Local flood or ebb dominance is often stronger than the cross-sectionally integrated flood or ebb dominance due to tidal asymmetry. Therefore, the sediment balance of an inlet cannot be attributed solely to tidal asymmetry. If, for instance, the deepest parts of the channel are ebb-dominated and the shallower parts are flood-dominated, flood dominance will prevail over a larger cross-sectional width than ebb dominance. This may produce a net flood-dominated sediment transport, as observed, for instance, in the Dyfi estuary [112].

The meandering character of tidal channels sets a limit to the length of flood or ebb-dominated channels. At the end of a flood or ebb channel sediment is deposited due to flow divergence; in this way sills are built up at the transition between flood and ebb channels. Therefore, import or export of sediment by local flood dominance or ebb dominance generally affects only a limited portion of the basin. Tidal asymmetry may also change along the basin; however,the asymmetry characteristics of the tidal wave will generally persist over longer distances than the individual flood and ebb channels. For this reason tidal asymmetry plays a dominant role in the overall sediment import or export of tidal basins.

How close is 'close to equilibrium'?

What does 'close to equilibrium' mean? A perfect morphological equilibrium does not exist. Sea level trends and long term climate cycles force tidal basins to adapt continuously. Changes in basin morphology may become perceptible only over periods of several hundred years.

A major uncertainty is related to the estimation of the offshore tidal asymmetry Δ_{FR}^{inlet}. A fair estimate of the upper limit to this uncertainty is an error of 15 minutes in the difference between the durations of tidal rise and tidal fall. From (6.79) we can evaluate the influence of this error on the net sediment transport and the ensuing accretion or erosion in a tidal basin. The error $\delta\langle q \rangle$ produced by an error $\delta\Delta_{FR}^{inlet}$ is given by

$$\delta\langle q \rangle = \frac{4}{5\pi}\alpha b_C U^4 \omega \delta\Delta_{FR}^{inlet},$$

where $\alpha \approx 4 \times 10^{-4} m^{-2} s^3$, $U = 1 \, m/s$ is the maximum tidal velocity and ω the tidal frequency. An error of 15 minutes in Δ_{FR}^{inlet} therefore produces an error $\delta\langle q \rangle \sim 400 \, b_C \, m^3/year$ dry sediment. Assuming that this net sand transport is deposited with a density of $1.5 \times 10^3 \, kg/m^3$ in a basin with a length of 20 km and with a constant width b_S of three times the channel width b_C, we find an accretion of about 1 cm/year on average.

Because the inaccuracy in Δ_{FR}^{inlet} is probably not much smaller than 15', we have to conclude that the analysis cannot distinguish between basins in morphologic equilibrium and basins which accrete or erode about 1 cm per year on average. This is about five times the present rate of sea-level rise.

6.5.3. Morphodynamic Stability

In spite of the limitations of the 1D model of tidal lagoon morphodynamics, we will discuss here some consequences, assuming that it is a reasonable first approximation for the majority of tidal lagoons.

The model does not only specify the equilibrium condition (6.81), but it provides also insight in the response of a tidal lagoon when the equilibrium is disturbed. How does a tidal lagoon respond to interventions, such as channel dredging or tidal flat reclamation? Will it keep pace with sea level rise, or not? In the following we will consider these questions, which are relevant from a management perspective.

Equilibrium morphology

The equilibrium condition (6.81) shows that tidal basins tend to a morphodynamic equilibrium with a relative intertidal flat area that increases with increasing relative tidal amplitude. This can be explained by observing that flood dominance increases with increasing relative tidal amplitude (faster HW propagation than LW propagation, resulting in shorter flood and longer ebb duration), while flood dominance decreases with increasing intertidal flat area (slowing down of HW propagation relative to LW propagation). At equilibrium, deep tidal basins (small relative tidal amplitude) have relatively small intertidal area, while shallow basins have a relatively large intertidal area, see Fig. 6.46. This has the following consequences:

- Deep tidal basins with large tidal flats export sediment, except in case of a strong positive asymmetry Δ_{FR}^{inlet} at the mouth;
- Shallow tidal basins with small tidal flats import sediment, except in case of strong negative asymmetry at the mouth;

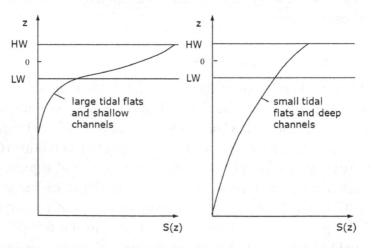

Fig. 6.46. Hypsometric curves for two different tidal basin geometries, which are both consistent with morphologic equilibrium, in the absence of tidal asymmetry at the inlet. $S(z)$ is the basin surface area as a function of the distance z from a reference level.

- Basins in long-term equilibrium import sediment during spring tide and export sediment during neap tide. Inversion of tidal asymmetry during the spring-neap tide cycle has been observed in the Charente estuary [847].

One may also expect an influence of the 18.6 year lunar nodal cycle on sediment import or export. During the period of increased M2 tidal amplitude, sediment import should be larger (or sediment export smaller) than during the period of decreased M2 amplitude. This is consistent with observations and modelling studies in the Wadden Sea [653] and the Humber estuary [932].

Response to tidal flat reclamation

The 1D morphodynamic model described before predicts the following scenario for the response of a tidal lagoon to tidal flat reclamation (see Fig. 6.47).

Tidal flat reclamation reduces the flood storage volume of the tidal lagoon, and causes a corresponding decrease of the tidal current velocities. We assume that reclamation affects only the HW basin width, and not the subtidal zone. After reclamation, the HW-propagation is less delayed when filling the remaining intertidal storage area; the HW-propagation speed is increased and the flood duration is shortened. The flood currents are thus less reduced than the ebb currents and the basin shifts to an infilling mode.

However, initial infill is not caused only by tidal asymmetry, but also by settling lag. The flood current is loaded with sediment in the

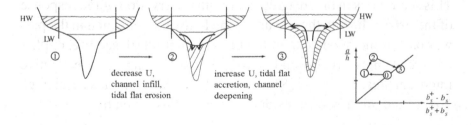

Fig. 6.47. Different stages of the morphodynamic response to tidal flat reclamation.

ebb-tidal delta zone, where sediment uptake is less reduced than in the lagoon. The fine sediment fraction does not settle immediately when the current speed decreases, but settles further inside the lagoon. The ebb current inside the lagoon is not strong enough to resuspend and transport all the imported sediment. This is another important mode of net sediment import after tidal flat reclamation.

The strength of the tidal currents after reclamation is not sufficient to carry sand to the tidal flats and to maintain the tidal flats against erosional wave action. Channel infill is, therefore, not only caused by import of marine sediment, but also by erosion of the remaining tidal flats. This infill reduces the channel cross-section; it increases the tidal velocities and finally restores the capability of tidal currents to counteract wave induced tidal flat erosion.

At this stage the basin is still flood dominant, because erosion has caused an additional reduction of the intertidal area. Tidal asymmetry now strengthens the sediment import. This import serves to rebuild tidal flats and goes on until the HW-propagation speed has become similar to the LW-propagation speed. At that stage a new morphologic equilibrium is achieved. The process is schematically depicted in Fig. 6.47.

The channel cross-section in the new equilibrium situation is smaller than in the former situation. The basin has imported sediment, which has mainly been supplied by the ebb-tidal delta; the volume of the ebb-tidal delta has decreased.

Tidal flat reduction at the Frisian Inlet

This scenario is in broad agreement with observation of the response of the Frisian Inlet to the closure of the Lauwerszee, which in the past was part of the flood delta of the Frisian Inlet [880], see Fig. 6.48.

According to (6.79) and (6.81), a small change in the relative intertidal area, $\delta S_R = \delta\left[(S^+ - S^-)/(S^+ + S^-)\right]$, induces a change $\delta\langle q \rangle$ of the initially zero residual sediment flux given by

$$\delta\langle q \rangle = -\frac{16\alpha b_C U^4 \omega \Delta_S}{5\pi}\delta S_R.$$

The closure of the Lauwerszee has decreased the ratio S^+/S^- from about 3 to 2. Other values for the Frisian inlet are: $b_C = 1.5$ km, $\omega\Delta_S = 0.25$, $U = 1$ m/s and $\alpha \approx 4 \times 10^{-4}m^{-2}s^3$. This yields an annual sediment import by tidal asymmetry in the order of 0.8 million m^3 sediment. Tidal asymmetry is further strengthened as a consequence of the reduced channel depth, but this is not included in the above formula. Assuming a bed density of about 1.5 10^3 kg/m^3 [306,665], the observed annual accretion is of the order of 1.5 million m^3 sediment in the initial stage of infill and of the order of 0.5 million m^3 in the final stage, see Fig. 6.48.

The case of the Frisian Inlet shows qualitative agreement between model results and observations, but for quantitative results the model is not reliable. Morphologic adaptation involves processes that are not accounted for in the model. For instance, some parts of the original channel system are more strongly affected by tidal flat reclamation than others. Therefore the original channel system will be reshaped; this process involves significant sediment displacements. The adaptation time scale of the basin therefore depends not only on sediment transport through tidal asymmetry, but also on other processes, such as the reconfiguration of the main channel system.

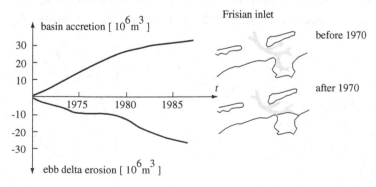

Fig. 6.48. Sedimentation in the Frisian Inlet and simultaneous erosion of the ebb-tidal delta after closure of the Lauwerszee flood delta, after [880]. The accretion volume includes a porosity factor; the corresponding sediment volume is 1.4–1.7 times smaller.

Response to channel deepening in tidal lagoons

The response to channel deepening in estuaries has been discussed already in Secs. 6.3.5 and 6.4.2. Here we concentrate on channel deepening in tidal lagoons which are initially in a state of equilibrium. The 1D morphodynamic model predicts the following scenario (see Fig. 6.49).

An increase of the channel depth produces in general a decrease of the relative tidal amplitude a/h and the tidal velocity amplitude U, with the exception of very shallow tidal lagoons where the tide strongly depends on friction. The equilibrium between tide-induced accretion of tidal flats and wave induced erosion is disturbed in favour of the latter. Tidal flats will therefore erode; the width and height of the tidal flats decreases and the eroded material will settle in the channel. By this process the channel depth decreases until return to its initial value. This channel infilling process is also simulated by settling lag, in the same way as described above for tidal flat reclamation. The amplitude of the tidal velocity is restored at its critical value, where accretion and erosion of tidal flats are in balance.

However, the tidal flat area is decreased. This decrease of tidal flat area influences tidal asymmetry such that the duration of rising tide is decreased relative to falling tide. The net sediment flux therefore becomes flood dominated. The channel depth is further decreased, the tidal current strength increases, the channel bed is scoured and the tidal flat accretes. The channel erosion is compensated by sand

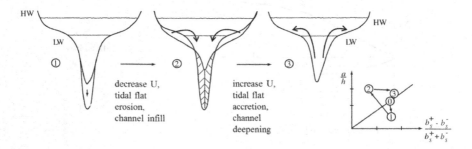

Fig. 6.49.   Different stages of the morphodynamic response to channel dredging.

import from the ebb tidal delta. If marine sand import is insufficient, the basin may go through a cyclic process of increasing and decreasing channel depth and corresponding decreases and increases of the tidal flat area. If such a cyclic adaptation process exists, it will be probably masked by natural fluctuations caused by storms, by long-term periodicity in the tide and by other (man induced) disturbances.

6.5.4. *Response to Mean Sea-Level Rise*

The response to sea-level rise is more complex than the previous two cases. We will assume, for simplicity, that the wave climate is not affected and that the tidal amplitude and tidal asymmetry at the inlet also remain unchanged. We further assume that the tidal lagoon is initially in a state of morphodynamic equilibrium.

A small sea-level rise δz increases the average total basin width at HW by a quantity δb_S^+ (if it is not bounded by dikes), the average basin width at low water by a quantity δb_S^-, and the average channel depth δh by a quantity δz. Sea-level rise also increases the flood storage volume of the basin, implying an increase δU of the tidal current amplitude. The response $\delta \langle q \rangle$ of the residual sediment flux to a small change δz of the sea level follows from (6.79) and (6.81),

$$\frac{\delta \langle q \rangle}{\delta z} = \frac{8 \alpha b_C U^4 \omega \Delta_S}{5 \pi b_S} \left(\frac{\delta b_S^- - \delta b_S^+}{\delta z} - \frac{7 a b_S}{3 h^2} \right). \qquad (6.84)$$

The first term between brackets is related to the influence of sea-level rise on the intertidal area, and the second term to the influence on the channel depth.

Embanked basins

In this case $\delta b_S^+ = 0$. The increase of the LW basin width δb_S^- with sea-level rise leads to a decrease of the intertidal area, an increase of the flood propagation speed and to a relative increase of the strength of the flood currents. The increase of the channel depth has the opposite effect, as discussed before. The net effect of these two contributions on the residual sediment flux is generally small, however. Typical values of the difference $\delta b_S^- / \delta z - 7 a b_S / 3 h^2$ are in the range 100–1000.

The sediment volume V_{sed} needed for a δz raise of the lagoon level equals $V_{sed} = (1 - p_b) S \delta z$, where $S = l_A b_S^+$ is the lagoon surface area and $p_b \approx 0.4$ the porosity. We call T_{SLR} the time needed for importing a sediment volume equal to V_{sed} and δz the sea-level rise during this time. We assume that $\delta \langle q \rangle$ following (6.84) is approximately a linear function of δz. Keeping pace with sea-level rise, on average over the period T_{SLR}, then requires

$$T_{SLR} = \frac{2(1 - p_b) S}{\delta \langle q \rangle / \delta z}.$$

We consider, as an example, a lagoon of length $l_A = 20$ km, total width $b_S^+ = 3$ km and channel width $b_C = 1$ km. We use typical values: $\delta b_S^- / \delta z - 7 a b_S / 3 h^2 = 500$, $\Delta_S =$ half an hour (average delay between HW and HWS), $U = 1 m/s$ and $\alpha = 4 \times 10^{-4} m^{-2} s^3$. This yields a time T_{SLR} of about two hundred years (independent of δz). At shorter time scales accretion lags behind sea-level rise. One may conclude that, in this case, sediment import trough tidal asymmetry will compensate for sea-level rise only after a long period. This time lag is hardly perceivable if the sea level rises less than 1 mm/year. However, if sea-level rise exceeds 1 cm/year, a substantial portion of the the intertidal area will be lost.

Sea-level rise influences sediment dynamics not only through tidal asymmetry, but also through other processes. Due to the increase of the tidal prism, tidal flats become more tide-dominated than wave-dominated. Sediment will be transferred from the channel to the tidal flats, which will accrete, while the channel depth increases. If sea-level rise is not too fast, the tidal-flat accretion rate may equal the rate of sea-level rise and channel deepening may be compensated by accretion through settling lag. In this case, the lagoon morphology shifts entirely upward with sea-level rise, without loss of intertidal flat

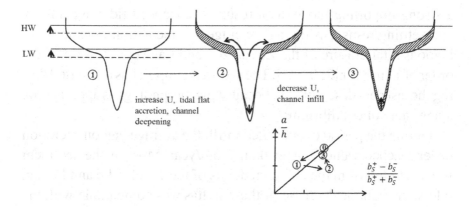

Fig. 6.50. Different stages of the morphodynamic response to sea level rise.

area; this is schematically depicted in Fig. 6.50. Sediment is delivered by the ebb-tidal delta, which decreases.

If the accretion rate is smaller than the rate of sea-level rise and if channel deepening is not (completely) compensated by accretion trough settling lag, the lagoon will evolve to an equilibrium morphology with deeper channels and smaller intertidal area. The loss of intertidal area with increasing sea level is referred to as 'coastal squeeze' [671].

Response to sea-level rise in the Wadden Sea

The eastern part of the Wadden Sea tidal lagoon system has been more or less stable with respect to the sea level for over 5000 years [280, 544, 665]. During the past century, the observed accretion was of the same order as required to keep pace with the mean sea-level rise (approximately 1.5–2 mm/year). The sediment import was probably even higher; this might be due to an increase of the tidal amplitude in the period 1960–1990 (increase of about 10 cm [399, 617]). The closure of the Zuyderzee and the Lauwerszee had a major impact on several lagoons in the Western Wadden Sea, especially Texel Inlet, Vlie Inlet and Frisian Inlet. These lagoons were shifted far away from their equilibrium morphology and turned into a strong importing mode, as a result of enhanced tidal asymmetry and settling/resuspension lag. Average sedimentation rates in these lagoons after closure of the Zuyderzee and Lauwerszee were of the order of 1 cm/year [280,544]. The sediment import has decreased during the last decades, probably because the morphology is approaching a new state of equilibrium.

During the period of enhanced infill, the average lagoon accretion never reached values higher than 2 cm/year. Most of the sediment was extracted from the ebb tidal deltas of the Texel, Vlie and Frisian inlets; the volume decrease of these deltas was comparable with the sediment volume imported in the lagoons, see Figs. 6.38, 6.48. The barrier islands near Texel Inlet and Frisian Inlet also experienced strong erosion, see Fig. 7.48. Further shrinking of the ebb-tidal deltas would imply that less source material will be available for keeping pace with sea level. These observations and arguments suggest that the delay in morphological adaptation may become substantial if the rate of sea-level rise exceeds 1 cm/year.

Non-embanked basins

If the tidal flats can extend landward, such that the intertidal area $S^+ - S^-$ increases with rising sea level, tidal asymmetry will induce sediment export and a corresponding increase of the channel cross-section. Due to the increased tidal velocities, the tidal flats will be more tide dominated than wave dominated. Sediment is therefore exported from the channel not only to the ebb-tidal delta, but also to the tidal flats, which may grow in height even faster than sea-level rise. Ebb dominance is strengthened by channel deepening; it can be neutralised by widening of the channels. This implies that the lagoon will evolve to an equilibrium with wider channels and higher tidal flats.

The removal of embankments and the abandonment of polders adjacent to tidal flat areas may prevent the loss of intertidal area with rising sea level. This 'managed realignment strategy' against coastal squeeze is already put into practice in the UK.

Similarity with dike breach

From a morphological point of view, managed realignment is comparable to non-managed realignment. In the past, polder dikes in the Rhine–Meuse–Scheldt delta were breached by storm surges at several occasions. One of these events, the St. Elisabeth Flood in 1421 (AD) described in Sec. 2.5.1, invaded a large polder area. The tidal prism of the Haringvliet Inlet increased and large channels were scoured to accommodate the increased tidal flow. The inundated land turned into a new flood delta with tidal flats and marshes. A substantial part of the eroded sediment was exported and deposited in the ebb tidal delta; the volume of the ebb tidal delta increased. It may be expected that a similar process will take place when the intertidal area increases as a consequence of sea-level rise, albeit at a much slower rate.

6.5.5. Other Models for Lagoon Response to Sea-Level Rise

Several other models are currently used to simulate tidal lagoon morphodynamics. We will shortly describe below two types of models, that follow different approaches.

The ASMITA model

The ASMITA model (Aggregated Scale Morphological Interaction between Inlets and Adjacent coast) is based on empirical relationships between tidal and morphological parameters [229, 238, 890]. The most important relationship is between channel cross-section and tidal prism, already discussed in Sec. 6.2.3. The existence of a linear relationship between the channel cross-section and tidal prism is broadly verified with observations in many tidal lagoons. It is assumed that this relationship holds when the lagoon is in morphodynamic equilibrium. Another basic ingredient of the ASMITA model is the assumption that the long-term mean sediment flux is proportional to the along-channel gradient of the suspended sediment concentration. This implies that morphodynamic equilibrium requires a zero gradient of the equilibrium suspended sediment concentration. The equilibrium suspended sediment concentration is related to the tidal current strength, which follows from the ratio of local tidal prism $P(x)$ and channel cross-section $A_C(x)$. A departure from the equilibrium morphology (change of the ratio P/A_C) induces a gradient in the suspended sediment concentration. This gradient is reduced to zero by (quasi-diffusive) sediment transport, settling and erosion. In this way the evolution to a new equilibrium morphology can be simulated.

The equilibrium requirement of a zero gradient in the suspended sediment concentration is equivalent to the requirement of equal strength of turbulent sediment suspension throughout the basin (assuming that sediment availability does not play a role). This requirement of the ASMITA model is similar to the equilibrium condition of uniform maximum tidal shear stress, proposed by Friedrichs [324] (see Sec. 6.2.4).

The ASMITA model was specifically designed to include the combined lagoon-ebb-tidal delta dynamics [807]. Predictions with this model indicated that tidal flats in the Wadden Sea will probably not keep pace with a rate of sea-level rise higher than about 1.5–2 cm/year [895].

Physics-based 2D simulation models

As a result of increased computing power it has become possible to employ 2D numerical models for the simulation of lagoon morphodynamics. Such models provide a far more realistic representation of the physical processes involved in morphological evolution than the analytic 1D models described previously. They provide the opportunity to asses the influence of different processes and parameters, which cannot be included in simple models. A drawback of numerical 2D models is their complexity; it is often not easy to pinpoint the interactions that determine the overall morphodynamic behaviour. The outcome of the models is sensitive to choices regarding the initial geometry, the grid size and the sediment transport formula [176].

Numerical 2D models have been used to simulate the development of tidal lagoon morphology, starting from an initially flat sloping seabed [564, 900]. For sediment transport a simple formula of the type (3.82) was used; wind effects were ignored. The initial flat seabed was randomly disturbed to trigger morphological development. A remarkable result of such models is the generation of a network of channels, steered by morphodynamic feedback. Stability analysis of the underlying morphodynamic equations indicates that the morphology is unstable for certain wavelengths of a lateral bed perturbation. It appears that the number and spacing of the channels generated by the numerical model is related to the wavelength of strongest instability [927]. Scouring of channels at mutual distances corresponding to the most unstable perturbation is auto-enhanced. The wavelength of strongest instability depends primarily on seabed slope and water depth; at the initial stage development, secondary flow circulations do not play a role [564].

The tidal lagoons simulated by the numerical 2D morphodynamic models evolve toward an equilibrium morphology consistent with the outcome (6.81) of the 1D analytical model [900],

$$(S^+ - S^-)/S^+ \approx a/h,$$

for values of a/h between 0.25 and 0.75.

Another relationship investigated with 2D models is the dependence of channel cross-section on tidal prism. Simulations with tide alone indicate a linear relationship between inlet channel cross-section and tidal prism, similar to the empirical relationship (6.2) established from data for a large number of inlets. Further numerical model experiments suggest that the relationship between channel cross-section and tidal prism also holds in the sheltered channels inside the lagoon [194]. However, it appears that in the absence of wave driven sediment supply, the relationship is not stable at the long term [887]. The numerical model experiments indicate that the relationship between cross-section and tidal prism depends on lagoon geometry and sediment grain size.

The long-term response of a schematised tidal inlet to sea-level rise, was tested in a 2D numerical model [241]. The model included the adjacent coastal zone, but considered only tide forcing (no waves) and a single sediment fraction (200 μm). Sediment transport was described by an advection-diffusion equation for suspended load and by a bedload transport formula for bedload. The dimensions of the inlet and the tidal conditions were similar to the Ameland Inlet in the Wadden Sea. The simulation started with an initially flat bottom; sea-level rise was imposed after the establishment of a morphology close to equilibrium. The model tests indicate that sediment import into the basin is enhanced when the sea level rises. However, the enhanced import is insufficient for compensating a rate of sea-level rise higher than 5 mm/year. This contrasts with the result of the ASMITA model. Simulations with more complete numerical models, including wind forcing and different sediment fractions, have not yet been reported.

6.6. Transport of Fine Sediment

6.6.1. *Transport Processes*

Fine sediment deposition areas

Tidal lagoons and estuaries are deposition areas for fine sediment. This is primarily due to their wave-sheltered location and low tidal current velocities. Fine sediment deposits do not occur at places with high wave activity. The grainsize of sediment deposits typically decreases with decreasing wave activity and increases with increasing wave activity. Exceptions are coasts with massive mud supply; examples are the coasts of Guiana and Surinam.

Fine sediments settle at places where currents are strongly reduced. Dredged trenches, channels and harbour basins are notorious traps of fine sediment. Fine sediments also accumulate in abandoned channels, often after engineering interventions. In natural estuaries and lagoons, fine sediments are mostly deposited on tidal flats.

Non-saturated transport

The transport formulas used in the morphodynamic analysis of estuaries and tidal lagoons, (6.62) and (6.75), assume that settling and resuspension processes adjust the transported sediment load instantaneously to changes in the current velocity. The transported sediment load is thus permanently in equilibrium with the transport capacity of the current (saturated transport). This is a reasonable approximation for non-cohesive sediment particles with a fall velocity exceeding a few cm/s (medium sand and coarser), considering the large distances over which particles are transported.

For particles with a smaller fall velocity and for particles that stick to the bottom by cohesive forces, the validity of this approximation is questionable, see Sec. 3.7.2. It can be argued that the approximation still holds for fine sediments when they move up and down the channel, because the shear stresses in the channel are high at conditions which are most pertinent for morphological evolution. However, shear stresses on tidal flats are much lower. For accretion and erosion of tidal flats, the approximation of instantaneous adaptation of the transported sediment load to changes in current velocity is generally not justified. The transported sediment load can be higher or lower than the equilibrium load. The delay in the response of the transported sediment load to the equilibrium load is called settling lag when the current strength decreases, and scour lag or resuspension lag when the current strength increases. Settling and resuspension lag induce a net sediment transport, which is particularly important for fine sediment accretion in regions where tidal current velocities are locally decreased.

Other transport processes

Settling lag can be very long for fine sediments with a low fall velocity. Fall velocities are low when floc formation is inhibited by strong current shear stresses; the presence or absence of organic binding substances plays also a role (see Sec. 3.5). In such cases fine suspended sediments are transported in a way similar to dissolved substances.

Tidal dispersion is the primary transport mechanism for dissolved substances in tidal lagoons, by inducing large random net displacements of water parcels [250, 991].

Estuarine circulation also contributes to net random displacements of water parcels. In well mixed estuaries, transport by estuarine circulation can be described as a dispersion process, which enhances tidal dispersion. If the concentration of suspended sediment is much higher near the bottom than further up in the water column, estuarine circulation contributes to upstream sediment transport. It appears that the influence of estuarine circulation on upstream sediment transport is minor compared to the influence of tidal asymmetry, under conditions of high morphodynamic activity. This has been discussed in Sec. 6.4.4.

Tidal asymmetry influences fine sediment transport not only through a difference between flood and ebb current strength, but also through asymmetry between the durations of high and low water slack tides. This mechanism has been discussed and illustrated for a few tidal lagoons in Sec. 6.3.3.

At very high suspended concentration (typically larger than $10\,g/l$), sediment transport may take place through density-induced turbidity currents or mud avalanching. Such conditions will not be considered in the following.

In previous sections we have seen that tidal flats are generally flood dominant. There are several causes: Asymmetry in the offshore tide, asymmetry generated by tidal propagation over wide tidal flats (for instance, Baie du Mont Saint Michel), and different pathways of flood and ebb flow (flood flow preferentially over the tidal flat, ebb flow preferentially through runnels and channels). These different mechanisms have been discussed in Secs. 6.3.2 and 6.5. Flood dominance is an important transport mechanism for fine sediment, just as for medium and coarse sediment. Simulations with a numerical Delft3D model suggest that its effect overrides the effect of other transport mechanisms for tidal flat accretion [902]. We will focus in the following on the settling and resuspension lag mechanism and the tidal dispersion mechanism and demonstrate that they can also contribute substantially to fine sediment transport.

Morphodynamic feedback

Net sediment transport through tidal asymmetry yields a feedback to the strength of tidal asymmetry. In this way estuaries and tidal lagoons can evolve to a morphological equilibrium with zero net sediment transport, as discussed in Secs. 6.4 and 6.5. Such a feedback between the mechanism of net sediment transport and morphological evolution is less direct for the other transport processes.

Estuarine circulation and tidal dispersion still exist at morphological equilibrium. Their contribution to net sediment transport depends on gradients in the suspended sediment concentration; these gradients depend on turbulent water motions and are therefore influenced by basin morphology. According to the ASMITA hypothesis, the basin morphology should evolve to an equilibrium such that suspended sediment gradients have disappeared.

Net transport of fine sediments induced by settling and resuspension lags also continues when equilibrium for the coarser sediment fraction is achieved. In this situation accretion through settling and resuspension lags is counteracted by erosional processes, such as wave-induced erosion. In the absence of wave action, tidal flats grow to an equilibrium level corresponding to the maximum high water level.

6.6.2. Settling and Resuspension Lag

Moving reference frame

Due to settling and resuspension lag, the suspended sediment concentration depends on its past trajectory. A water column moving over a fresh mud patch will keep a significant amount of mud in suspension a long distance away from the patch. Data of fine suspended sediment sampled at a fixed location in a tidal basin cannot, in general, be interpreted in terms of local erosion and deposition.

A way to handle this difficulty is by considering a water column moving with the tidal current. This eliminates partly the influence of concentration fluctuations advected from other locations, but it does not rule out advection completely; the water column moves at different speeds near the bottom and near the surface. We use this moving-frame approach, as it makes the influence of tidal asymmetry more transparent than the fixed-frame approach. Tidal variation of the suspended sediment concentration measured from a vessel floating with the mean tidal current is shown in Fig. 6.51. The peak concentration at maximum flood is mainly due to suspension of fine sand and silt. The variation of the suspended sediment concentration exhibits the expected behaviour: the degree of settling is related to the duration of the slack water period (currents below the critical strength for erosion) and the overall variation is delayed relative to the tidal variation of current strength by one hour or even more.

Net tidal sediment flux

For investigating net tidal fluxes we will assume that the moving reference frame returns to its initial position after a tidal period. This assumption is reasonable only if the tidal discharge is several orders of magnitude larger than the river runoff and if estuarine circulation is small; for tidal lagoons these assumptions generally hold. We also ignore horizontal residual circulation (e.g. circulations between ebb and flood channels); its influence needs to be considered separately. Residual circulation contributes mainly to the net sediment flux through dispersion; this contribution can be estimated by means of dispersion coefficients, see Sec. 6.6.3.

The velocity of the moving reference frame is taken as $u = Q/A_C$, where Q is the instantaneous tidal discharge and A_C is the channel cross-sectional area. $X(t)$ is the path of the water column moving with velocity $dX/dt = u$. If we consider cyclic tides and assume that there is no phase difference between tidal channel flow and water exchange with tidal flats, then the path $X(t)$ is cyclic too,

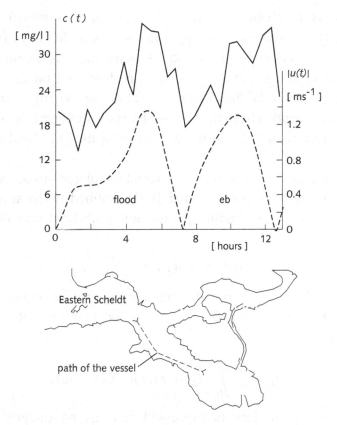

Fig. 6.51. Current velocity (dashed line) and depth-averaged suspended sediment concentration (solid line) measured from a vessel that followed a buoy floating with the mean tidal current [252]. The experiment was carried out in the Eastern Scheldt in 1984. Most of the suspended material is fine sediment ($< 63\,\mu$m), except at maximum flood. The flood peak velocity is slightly higher than the ebb peak velocity. The slack water asymmetry is the opposite of that in the Wadden Sea: The period of low slack water is longer than the period of high slack water. The net fine sediment flux in the Eastern Scheldt is on average directed seaward [837]. The tidal variation of the suspended sediment concentration exhibits an overall time lag relative to the tidal current of about one hour. (Survey Department Rijkswaterstaat).

$X(t + T) = X(t)$. In that case the tidally averaged sediment flux through the cross-section $x_0 = X(0)$ can be written

$$\langle q \rangle = \frac{1}{T} \int_0^T \int_{x_0}^{X(t)} \left[b_C (De - Er) + q_{flat} \right] dx dt, \qquad (6.85)$$

where b_C is the channel width, De the deposition (settling) rate, Er the suspension (erosion) rate, q_{flat} the sediment flux to or from the tidal flats. This integral simply expresses the fact that sediment only passes through the moving plane if it is deposited on the channel bed or stored on tidal flats before arrival of the moving plane $X(t)$; sediment suspended from the channel bed or inflowing from the tidal flats before arrival of the plane will not pass the plane and must be subtracted.

Now we assume that the time variation of the suspended load $A_C c$ in the moving frame is mainly determined by local erosion-sedimentation and by sediment exchange with tidal flats (see also Sec. 3.7.2),

$$d(A_C c)/dt \approx b_C(Er - De) - q_{flat}, \qquad (6.86)$$

where c is the cross-sectionally averaged suspended sediment concentration. The integral (6.85) can thus be rewritten, after partial integration, as

$$\langle q \rangle = \frac{1}{T} \int_0^T Q(X(t), t) c(X(t), t) dt. \qquad (6.87)$$

The net sediment flux is expressed now as an integral of the instantaneous flux through the moving plane. Terms related to the cross-sectional variation of flow velocity and suspended sediment concentration are left out of consideration. The expression (6.87) shows that the net import or export of fine sediment can be determined by measuring the suspended concentration in a frame moving along the channel with the average tidal flow.

HW-LW velocity asymmetry in a moving frame

Postma, in his influential analysis of sedimentation in the Wadden Sea, first pointed to the occurrence of velocity asymmetry in tidal basins and its effect on net import of fine sediment [684]. He analysed the tidal variation of the currents in several shallow tidal channels in the landward part of the Ameland Inlet. He noted a swift reversal from a strong ebb current to a strong flood current around low slack water. The reversal from flood to ebb current around high slack water was quite slow, on the contrary. He attributed this difference to the small wetted channel cross-section at low water slack, compared to the much larger cross-section at high water slack. In fact, this is not the only reason; the offshore tide at Ameland also varies more slowly around HW than around LW (see Fig. 6.52). The sediment suspended by the strong ebb current has not enough time to settle in the period around low slack water (LWS); most of the suspended sediment is immediately carried landward by the flood current. The result is a substantial landward displacement of sediment, see Fig. 6.52.

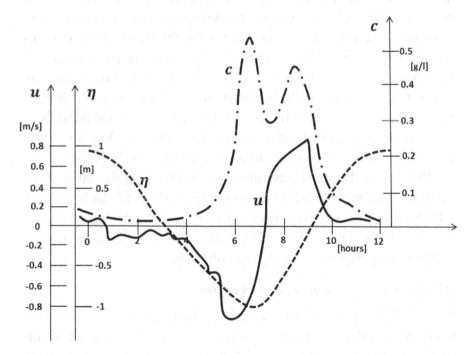

Fig. 6.52. Current velocity $u(t)$ (solid line) and fine suspended sediment concentration $c(t)$ ($d \leq 63 \, \mu$m, dashed-dotted line) in a shallow tidal channel close to the head of the Ameland tidal lagoon, redrawn after [684]. The dashed line is the offshore tide $\eta(t)$ at the inlet. All the suspended sediment settles from the water column around HW slack tide, but the reversal of ebb to flood current at LW slack tide is too short for substantial deposition, in spite of the small channel depth (less than 2 m).

A second source of velocity asymmetry becomes apparent when following the moving frame; even if there is no tidal velocity asymmetry at a fixed location, the velocity variation experienced by suspended sediment particles can be asymmetric in a moving frame. Around high water slack tide (HWS) many sediment particles move over tidal flats, where the current velocity is low; around low water slack tide (LWS) sediment particles are moving through channels, where velocities are much higher. The two sources of velocity asymmetry add up; hence, for a suspended particle moving with the current, the period of low current velocities is shorter around LWS than around HWS. Therefore, more particles will settle to the seabed at HWS than at LWS. This implies that at the beginning of the flood period most particles, which have been suspended during ebb, are still in suspension, while at the beginning of the ebb period most particles are deposited. As it takes some time for sediment particles to get resuspended, the suspended sediment concentration at the beginning of flood flow is substantially higher than at the beginning of ebb flow. Even if the strength of flood and ebb currents is equal, the flood flow, on average, will carry more sediment than the ebb flow and a net landward displacement of sediment will result, see Fig. 6.54.

This net landward displacement is further strengthened on high tidal flats, where the water level drops below the tidal flat level soon after high water slack. If the resuspension lag is greater than this time interval, the sediment deposited around HW becomes subaerial and will not be resuspended during the ebb period.

HW-LW depth asymmetry in a moving frame

When following a sediment particle on its tidal trajectory we have to consider both the velocity asymmetry and the bathymetric asymmetry. Around low water most sediment particles move in the main channels, where the water depth is large, even at LWS. Shortly before HW, many of these particles will move onto a tidal flat, where the water depth is small, even at HWS. Sediment particles therefore have a higher probability of reaching the bottom at HWS than at LWS, even

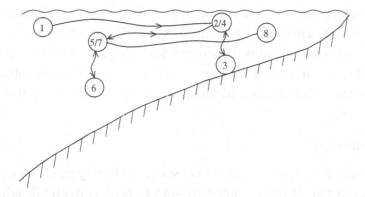

Fig. 6.53. Net tidal displacement of a sediment particle due to settling-erosion lag. The particle starts at 1 near the inlet at the onset of flood. Due to settling lag it is transported in suspension over almost the entire flood excursion. At high slack water the particle has arrived in shallow water, where it settles on the bottom at 3. When the ebb flow starts, the current is at first too weak to resuspend the particle from the bottom. It is resuspended later, but the remaining ebb excursion is shorter than the previous flood excursion. Towards low slack water the particle starts falling at 6, but the duration of the low slack water period is too short and the distance to the bottom too long for deposition. The particle is picked up by the flood flow immediately after slack water and travels landward again over the entire flood excursion. Towards high slack water it starts settling, at a location 8 landward of the previous settling location 3.

if the periods of low velocities are equal. This induces a greater resuspension lag at HWS than at LWS. The net landward displacement of a sediment particle due to these different mechanisms is depicted in Fig. 6.53.

Wave influence

Wind waves have an important influence on tidal flat sedimentation. Fine sediment will not accumulate on exposed tidal flats, except in periods of calm weather. In the Wadden Sea, for example, mud deposits are found on many tidal flats during summer, but on an annual basis mud accumulates only on the most sheltered tidal flats. Fine sediment deposited on the tidal flats in calm periods is resuspended during storms and carried seaward by the ebb flow. The erosion-sedimentation asymmetry is reversed; erosion-sedimentation lag at low water dominates over erosion-sedimentation lag at high water. Under storm conditions much sediment remains in suspension; dispersive transport may then play an important role in equalising sediment concentrations within the tidal basin and exchanging sediment with the near coastal zone.

A simple model

The integral expression (6.87) may serve as starting point for a semi-quantitative analysis of suspension-lag effects on net tidal sediment transport. Following Groen [359], we assume that the settling and erosion time lags are comparable and given by T_{sed}; expressed in radians we call the time lag $\phi_{sed} = \omega T_{sed}$. This time lag is defined as the time scale at which the depth-averaged suspended sediment concentration $c(t)$ adapts to an equilibrium concentration c_{eq}. In fact, a distinction should be made between high water slack tide and low water slack tide. In the moving frame, more sediment remains in suspension during LWS than during HWS, as explained before. The adaptation time scale should therefore be taken longer at HWS than at LWS. This effect is ignored in the model, for simplicity. The settling-erosion lag relation then reads, in a frame moving with the average tidal flow,

$$c_t = \frac{1}{T_{sed}} (c_{eq} - c). \tag{6.88}$$

The solution of this equation reads

$$c(t) = c_0 + \frac{e^{-t/T_{sed}}}{T_{sed}} \int_0^t e^{t'/T_{sed}} (c_{eq}(t') - c_0) dt', \tag{6.89}$$

where the initial concentration $c_0 \equiv c(0)$.

We will assume that the equilibrium concentration c_{eq} depends quadratically on the tidal flow velocity u, with constant proportionality factor α, $c_{eq} = \alpha u^2$. We will also assume that the velocity u in a moving frame exhibits slack water asymmetry, such that the

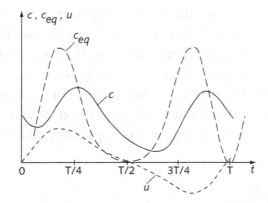

Fig. 6.54. Current velocity $u(t)$ (6.90) with asymmetry parameter $\epsilon = 1/3$ (short-dashed curve), equilibrium suspended sediment concentration $c_{eq} = \alpha u^2$ (long-dashed curve), and delayed suspended sediment concentration $c(t)$ (solid curve) according to the model (6.88). The high slack water period at $t = T/2$ is much longer than the low slack water period. The erosion-sedimentation delay timescale T_{sed} has been taken as 2 hours. This delay, together with the slack water asymmetry, results in a suspended sediment concentration which is lower on average during ebb than during flood.

duration of high slack water is longer than the duration of low slack water,

$$Q(t) = A_C(t)u(t), \quad u(t) = U(\sin \omega t + \epsilon \sin 2\omega t), \qquad (6.90)$$

where the relative strength of the second harmonic tidal component is much smaller than the main component, $\epsilon^2 \ll 1$. The velocity $u(t)$, the equilibrium concentration $c_{eq}(t)$ and the delayed suspended sediment concentration $c(t)$, according to (6.89), are shown in Fig. 6.54 for $\epsilon = 1/3$. The net sediment transport $\langle q \rangle$ given by (6.87) is found by substitution of (6.90) and (6.89) with $c_{eq} = \alpha u^2$. Assuming a constant cross-section A_C the result is

$$\langle q \rangle = \frac{A_C}{T} \int_0^T u(t)c(t)dt = \epsilon \alpha A_C U^3 \frac{\pi \phi_{sed}^3}{2(1 + \phi_{sed}^2)(1 + 4\phi_{sed}^2)}. \tag{6.91}$$

Numerical example: for a suspension time lag of 2 hours ($\phi_{sed} \approx 1$) and asymmetry parameter $\epsilon = 1/3$, the net sediment import is 25%

of the total flood import (the integral of q over the period $[0, T/2]$). This illustrates that settling and resuspension lag can be a powerful mechanism for transporting fine sediment from areas with high tidal current velocity to areas where the velocity is low.

Many effects have been ignored in this simple model. It is not adequate for reliable estimates but provides qualitative insight into the influence of settling and erosion lag on net tidal import of fine sediment.

6.6.3. *Tidal Dispersion*

Similarity with dissolved substances

Fine sediments can be kept in suspension, even when the current velocity is temporarily low. Once suspended, fine particles move with the tidal current during most of the tidal cycle. At spring-tide, a substantial fraction of the fine particles may remain suspended during the entire tidal cycle. In the Eastern Scheldt, typically half of the suspended sediment remains in suspension at slack water, see Fig. 6.51. The displacement of these particles is similar to the displacement of dissolved substances. This does not hold for particles which are deposited during part of the tidal cycle. Settling and resuspension may produce a strong and consistent net displacement of fine sediment, as discussed in the previous section. Here we focus on the net displacement of sedimentary particles that remain in suspension for a long time — longer than the tidal period.

The tidal dispersion mechanism refers to residual transport related to non-uniformity of the spatial flow distribution. It tends to smooth the distribution of suspended sediment by dispersing particles away from regions of high concentration, for instance the zone of maximum turbidity.

Scale dependency of dispersion processes

Dispersion is the general designation of transport processes which smooth gradients in the concentration of dissolved or suspended matter. These transport processes correspond to the mixing of water masses, i.e., to the relative displacement of water parcels without net displacement of water volume. The dispersion coefficient is a measure of the efficiency of mixing processes; it can be expressed as the ratio of a squared length scale X^2 and a time scale T. The length scale X is a measure of the distance over which water parcels are displaced relative to each and the time scale T is a measure of the time involved. Dispersion processes occur at many different spatial and temporal scales; the definition of concentration is scale dependent accordingly. The term diffusion is normally used for any dispersion process that takes place at sub-model scales. Brownian motion is the dispersion process at the smallest temporal and spatial scale (molecular scale). Turbulence is the most powerful mechanism for dispersal of dissolved or suspended matter at the scale of water depth. Morphologic time scales are generally quite long; we therefore focus on dispersion processes at temporal and spatial scales which are much longer than the scales of Brownian motion or turbulence. We are mainly interested in transport along the tidal basin in longitudinal direction; the corresponding large scale mixing processes are known as longitudinal dispersion.

Random walk

Smoothing of concentration peaks will not occur if water parcels move all together in a coordinated way. If, for instance, all individual water parcels return to their initial position after a tidal cycle, there will be no dispersion; the same is true if all water parcels are displaced over an identical distance. On the other hand, if initially close water parcels travel over very different distances in a tidal cycle (illustrated in Fig. 6.55), an initial concentration peak will be smoothed. If these individual parcel displacements after a tidal cycle are randomly distributed, there is practically no chance that a concentration peak is created from an initially smooth concentration distribution. This illustrates that dispersion can be described as a random walk process

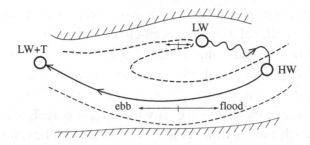

Fig. 6.55. Schematic representation of dispersion by the combined action of lateral tidal velocity gradients and cross-sectional mixing. A water parcel at low water in a shallow secondary channel moves landward with low flood velocity. Transverse currents and horizontal eddies take the water parcel towards the main channel, where it arrives at around HW. Subsequently it travels with the strong ebb current in the main channel far seaward from its initial position at low water.

at the time scale of a tidal cycle or several tidal cycles. The distance over which water parcels move away from each other depends on the strength of their respective net tidal displacements and the time scale T_A at which these net displacements become statistically uncorrelated. The time scale T_A is defined as the correlation time scale of water parcels in the same cross-section and we call X_A^2 the average quadratic net displacement of these water parcels away from the original cross-section over the period T_A. We assume that the semidiurnal tide has a cyclic character and that the net discharge is zero; the sum of all individual displacements thus equals zero. Random walk theory then states that the average quadratic displacement X^2 of water parcels after a time $t \gg T_A$ increases proportionally with time t [302].

Dispersion equation

An equivalent statement is that the tidally and cross-sectionally averaged concentration $c_0(x, t) \equiv \langle \bar{\bar{c}} \rangle$ satisfies the dispersion equation

$$c_{0t} = D_L c_{0xx}, \tag{6.92}$$

with a dispersion coefficient D_L given by [251]

$$D_L = X_A^2 / 2T_A. \tag{6.93}$$

In Eq. (6.92) it has been assumed, for simplicity, that the dispersion coefficient D_L and the channel cross-section A are independent of x. We will use the same simplification in the following. For the inclusion of non-dispersive sediment transport processes (transport induced by tidal asymmetry and by settling and suspension lag) we consider the general sediment balance Eq. (3.56). When integrating this equation over a tidal cycle, we assume that the net tidal effect of non-dispersive transport can be added to tidally induced dispersive transport. This is probably a reasonable assumption for the fine sediment fraction that is not strongly affected by non-dispersive transport. It is, in fact, the assumption underlying the ASMITA model (see Sec. 6.5.5). In the presence of a residual discharge Q_R and taking into account other sediment transport processes, the dispersion equation reads

$$c_{0t} - U_R c_{0x} + \langle q'_x \rangle / A_0 = D_L c_{0xx} + \langle Er - De \rangle / h_0. \qquad (6.94)$$

In this equation U_R is the river discharge velocity $U_R = Q_R/A_0$; $\langle q' \rangle$ is the net tidal sediment flux due to non-dispersive sediment transport; A_0 is the tidally averaged cross-sectional area and Er, De are the local erosion and deposition rates per unit area.

Dispersion coefficient

It is important to note that the dispersion coefficient (6.93) depends on the flow field in the entire basin during the tidal cycle. However, D_L is independent of the concentration distribution $c_0(x, t)$, if it is assumed that the cross-sectional mixing time scale T_A is much smaller than the time scale at which sediment particles travel in the longitudinal direction through the basin [251]. In that case D_L can be derived experimentally from the observed tidally averaged salinity distribution $\langle S(x, t) \rangle$ in a tidal basin. If the river discharge Q_R is constant during a time long enough for the establishment of an equilibrium salinity distribution $S_{eq}(x)$, then the tidally averaged salt flux through any estuarine cross-section equals zero,

$$U_R S_{eq} - D_L S_{eq\,x} = 0. \qquad (6.95)$$

The cross-sectionally averaged salinity distribution S_{eq} is often a smoothly varying function of x. In that case the sea water intrusion length L_S can be approximated by $L_S \approx S_{eq}/|S_{eq\,x}|$, giving

$$D_L \approx U_R L_S. \tag{6.96}$$

Typical values of the dispersion coefficient D_L are in the range $100\text{--}300\,\mathrm{m^2/s}$, as shown in Fig. 6.56 for the Eastern Scheldt and

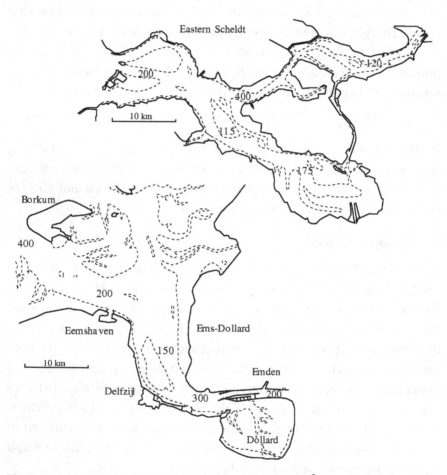

Fig. 6.56. Values of the longitudinal dispersion coefficient $[\mathrm{m^2/s}]$ in the Eastern Scheldt and the Ems-Dollard tidal basins, determined from the observed salinity distribution at constant river runoff.

the Ems-Dollard tidal basins. Higher values up to 1000 m^2/s have been found in tidal basins with a very large width-to-depth ratio ($\geq$ 1000) and complex geometry (meandering and braiding channels, tidal flats), like the Wadden Sea basins [988] and Chesapeake Bay [40]. In such basins tidal dispersion counteracts the formation of a strong turbidity maximum. Strong turbidity maxima are seldom observed in wide tidal basins with complex geometry.

Numerical examples

From the observed values of the dispersion coefficients we can estimate the contribution of dispersion processes to the net transport of fine sediment. Therefore we assume, as a rough approximation, that the dispersion coefficients for fine sediment and dissolved substances have comparable values. We call c the tidally and cross-sectionally averaged value of the suspended sediment concentration; we only consider the fine sediment fraction. Typical values for the Wadden Sea are of the order of $\rho_{sed}c = 2.5 \ 10^{-2}$ kg m^{-3}. Gradients are of the order of $\rho_{sed}dc/dx = 10^{-6}$ kg m^{-4}. The ratio of dispersive transport during a tidal cycle and the total flood transport can be approximated by

$$\frac{Uc}{\pi D dc/dx},$$

where $U \approx 1$ m/s is the cross-sectionally averaged tidal velocity amplitude and $D \approx 400$ m^2/s is a representative estimate of the dispersion coefficient in the Wadden Sea. In this numerical example dispersive transport of fine sediment is of the order of 5% of the total flood transport of fine sediment. This transport is in the ebb direction, if, as usual, the suspended sediment concentration in the lagoon is higher than offshore.

The importance of tidal dispersion can also be illustrated in another way. The net random displacement Δx of a water parcel in a time interval Δt by tidal dispersion is given by $\Delta x = \sqrt{2D\Delta t}$ [302]. A dispersion coefficient $D = 400$ m^2/s implies that sedimentary particles carried by water parcels are randomly displaced with respect to the mean tidal flow in one tidal cycle over a distance of about 6 km. In a neap-spring tidal cycle this distance is more than 30 km, comparable to the length of many lagoons. For an accurate description of fine sediment dynamics in tidal lagoons, dispersion processes have thus to be taken into consideration.

Dispersion processes

Processes responsible for tidal dispersion are related to the spatial flow structure. Spatial inhomogeneity is generated by vertical shear, lateral depth variations, dead zones and residual circulation. The greatest contribution to dispersion is provided by cross-sectional fluid exchange between zones of high and low tidal velocity at the timescale of the tidal period [251, 302, 991]. An example is shown in Fig. 6.55. In shallow well mixed estuaries vertical mixing takes place over a timescale much shorter than the tidal period; differences in flow velocity over the water column do not strongly contribute to longitudinal dispersion in the absence of stratification. Lateral mixing over the entire basin width often takes place at a timescale substantially greater than the tidal period, but local mixing over lateral tidal flow gradients takes place at the tidal timescale and therefore causes strong tidal dispersion.

Residual circulation contributes most to tidal dispersion if the timescale for mixing across the circulation is long [250]. Vertical circulation is most effective in the presence of density stratification; otherwise lateral circulation yields a greater contribution to longitudinal dispersion. High dispersion coefficients are often found near the estuarine mouth. This strong dispersion is related to "tidal pumping", the flood inflow of 'new' seawater after dispersion of estuarine ebb water in the coastal zone. Analytical estimates of dispersion coefficients are at best qualitative, due to the complexity of the flow structure in most tidal basins. The best estimates of dispersion coefficients are derived either from observations or from numerical models [722].

6.7. Summary and Conclusions

Different inlet types

Tides, waves, fluvial discharge, sediment supply and sea-level rise steer the development of sedimentary tidal inlet systems.

We have made a distinction between two types of inlets: River tidal inlets, which we called estuaries, and barrier tidal inlets, which we called tidal lagoons. The existence of estuaries is related to their river discharge function; river discharge plays an important role in the morphological development, even if the tidal current is much stronger than the river-induced current. Tidal lagoons owe their existence to sea-level rise and sediment supply by wave-driven longshore transport. Depending on the rate of sea-level rise and their capability to trap sediments, they can disappear after some time or they can get drowned. They can also evolve towards a morphodynamic equilibrium. Such an equilibrium results from mutual adjustment of morphology, hydrodynamics and ecodynamics to minimise gradients in the residual sediment transport. Such a morphydynamic equilibrium is not a static equilibrium, firstly because steering agents are highly variable (storms, river runoff) and secondly because there are cycles in the bio-geomorphological evolution (e.g. channel meandering and avulsion, ecological cycles).

There are also mixed inlet types with characteristics of both estuaries and tidal lagoons: Estuaries with a littoral barrier (estuaries partly closed by a sand spit or estuaries debouching into a tidal lagoon) and tidal inlets without a littoral barrier and without significant river inflow.

Characteristic tidal current strength

Each tidal inlet system is unique, as it has developed under a unique combination of steering conditions. However, the physical and biotic processes underlying morphodynamic evolution are universal. Tidal inlet systems therefore share a number of common properties.

A striking common property of close-to-equilibrium estuaries and tidal lagoons is the concentration of tidal flow in channels. In tidal inlet systems with predominantly non-cohesive sediments, the maximum depth-averaged current velocity is typically of the order of 1 m/s (seldom above 1.5 m/s or below 0.5 m/s). This property can be related to several feedback processes:

(1) Tide-induced secondary flow moves sediment from deeper parts of the channel (the outer channel bends) to shallower parts (alternating bars, tidal flats); underlying processes are instability of channel-flow interaction and meander development.

(2) Wave-induced resuspension counteracts tidal flat building by tide-induced secondary flow. Tidal flat erosion reduces the channel cross-section and increases tidal current velocities.

(3) The channel scouring capacity strongly decreases when the channel cross-section increases and the current velocity drops. Inversely, an increase of tidal current velocity due to a decrease of channel cross-section induces a strong increase of the channel scouring capacity.

(4) The strong dependence of sediment transport on current velocity implies that channel-tidal flat morphology can be in equilibrium only in a narrow range of peak tidal velocities.

The peak tidal current velocity at the inlet of tidal lagoons is also of the order of 1 m/s (cross-sectionally averaged); it hardly depends on the tidal amplitude or on the sediment supply by longshore wave-driven transport. The strong dependence of scouring capacity and sediment transport on velocity is probably a major cause, together with the requirement of the continuity of net sediment transport throughout the tidal lagoon main channel.

Role of waves, currents and biota

Waves are essential for tidal flat stability. Without wave-induced resuspension, tidal flats grow up till the high water level and transform into marshes. The ensuing loss of flood storage area stimulates flood dominance and lagoon infill. Biota can stabilise (algae mats) or destabilise (filter feeders) tidal flat deposits. Vegetation is the most important stabilising factor. However, wave action counteracts vegetation development. The growth and expansion of tidal flats is also limited by channel meandering and avulsion processes. At present, only qualitative descriptions exist of tidal flat dynamics taking into account tides, waves, biota and their interactions.

Turbid estuaries

In tidal inlet systems (estuaries, in particular) with predominantly fine cohesive sediments, maximum tidal current velocities can be higher than 1.5 m/s. This is probably related to the high shear stress needed to scour consolidated mud beds. Fluid mud is removed by high river runoff or it is trapped in dredged trenches and harbour docks. Fine suspended sediments are deposited on sheltered parts of tidal flats. If no deposition areas are available (for instance, due to reclamation of tidal flats), the suspended concentration of fine sediments can become very high (turbidity maximum). The presence of high concentration fine sediment layers near the bottom decreases the bottom-induced flow resistance. In this situation the velocities needed to produce a given bottom shear stress are very high. According to observations and models the maximum tidal bottom shear stress in estuaries and tidal lagoons is generally of the order of 2.5 ± 1 Pa.

Channel convergence

Estuaries with small tidal flats have a landward converging width. The width convergence is often represented by an exponential function. This is is very practical, as it simplifies strongly the tidal equations and provides analytic expressions for the description of tidal propagation. For strongly converging estuaries, an exponentially decreasing width is consistent with a uniform tidal velocity amplitude along the estuary.

Nonlinear tidal propagation

Tidal propagation in estuaries and tidal lagoons is highly nonlinear. This is manifest in the difference between the propagation speeds of the crest and the trough of the tidal wave. In most estuaries the tidal amplitude is approximately constant along the estuary (not strongly damped nor strongly amplified). In this case the tidal wave crest and tidal wave trough correspond to HW and LW, respectively; the tidal equations are approximately linear during short periods around HW and LW. The difference between HW and LW propagation speeds increases with decreasing relative tidal amplitude; the tidal wave crest propagates at greater water depth than the tidal wave trough and the speed is therefore higher. The difference between HW and LW propagation speeds decreases with increasing relative intertidal area; the speed of the tidal wave crest is reduced by flood water spilling over the tidal flats; the speed of the wave trough is less affected as it propagates in the channel. When HW propagates faster than LW, the flood period is shortened relative to the ebb period and the flood current strength is increased relative to the ebb current strength. This tidal asymmetry in current speed induces an even stronger asymmetry between flood and ebb sediment transport. Dominant flood transport implies sediment import and accretion of a tidal inlet system, dominant ebb transport implies export and erosion.

Morphodynamic equilibrium

Morphodynamic equilibrium requires axial uniformity of long-term average sediment transport. The assumption that tidal inlet basins adapt their morphology such that flood and ebb sediment transport become equal, has the following morphological implications.

For estuaries: The width convergence length is adapted such that the gradient in the upstream sediment transport due to increasing tidal asymmetry equals the gradient in the downstream sediment transport related to tidal flow and river discharge. The convergence length typically decreases when the channel is embanked and when tidal flats are reclaimed; the convergence length typically increases when the channel is deepened. Strong tidal asymmetry at the inlet of an estuary promotes the development of a mouth bar. Propagation of the tidal wave over the mouth bar can generate a tidal bore.

For tidal lagoons: The channel depth and the intertidal area are adapted such that the ratio of the relative intertidal area and the relative tidal amplitude is of the order of 1 — in the absence of asymmetry in the offshore tide. The ratio is greater than 1 (typically of order 2) if the period of rising tide at the inlet is shorter than the period of falling tide; the ratio is smaller than 1 in the opposite case. This is consistent with observational and theoretical evidence.

Fine sediment dynamics

The net transport of the fine sediment fraction is not only due to asymmetry in the strength of ebb and flood currents. Other important mechanisms for fine sediment transport are related to settling and resuspension lag and to tidal dispersion. Settling and resuspension lag promotes net landward transport if the HWS period is longer than the LWS period and net seaward transport in the opposite case. Settling and resuspension lags also promote transport from regions where the tidal velocities are high to regions where they are low. Tidal dispersion is an important transport mechanisms for particles that remain a long time in suspension; it also contributes to the diffusion of turbidity maxima.

Turbidity maxima in estuaries are primarily caused by tidal asymmetry (greater strength of flood currents compared to ebb currents) and estuarine circulation. Under the highly energetic conditions of significant morphological development, estuarine circulation is subordinate to tidal asymmetry for upstream sediment transport.

Field observations

Morphologic adaptation is a slow process. Yearly average erosion or accretion rates at basin scale are of the order of 1 cm/year or less. Subaerial accretion or erosion can be measured with good precision. However, the inaccuracy of subtidal bottom soundings is much larger than 1 cm. For a quantitative assessment of morphological change at basin scale, detailed survey programmes are required, covering periods of several decades, because of strong spatial and temporal variability in erosion and accretion. Such programmes are scarce. Probably the best surveyed estuary is the Western Scheldt and the best surveyed tidal lagoons are the inlets of the Western Wadden Sea. These survey data have been used in several studies for the calibration and validation of morphodynamic models.

Modelling

Thanks to increased computing power it has become possible to develop and employ numerical models for the simulation of morphodynamic evolution at the scale of estuaries and lagoons. In several model studies the initial development of an estuary or tidal lagoon has been simulated. The results show fair qualitative agreement with observed morphologies. They are also consistent with the results obtained with the simple analytical 1D models discussed in this chapter. No model is capable yet to make reliable simulations for the long term morphodynamic evolution of estuaries and tidal lagoons. For instance, there is still quite some uncertainty regarding the response to sea-level rise. The 1D models used in this chapter are too simple for such predictions. Further development of 2D and 3D numerical models is required. For qualitative analyses, for identifying major morphodynamic processes and for intercomparison of different lagoons and estuaries the 1D models are quite suitable, if they are used in combination with observations.

Chapter 7

Wave-Topography Interaction

7.1. Introduction

Interaction of the sandy ocean shores ...

The land-sea transition zone consists of various materials, such as rock, boulders, shingles, gravel, sand, silt and mud. Sand is the most common type of substrate of sedimentary coasts which are exposed to moderate wave action. At sandy coasts, the transition from sea to land is formed by a narrow coastal fringe with a rather steep inclination, of the order of 1/10–1/100. This coastal fringe is called 'shoreface'. Further off-shore, where water depth exceeds 10 m, the average slopes are gentler, of the order of 1/1000. This zone is referred to as the inner shelf. Locally, slopes may be steeper, because of seabed structures such as dunes and ripples.

with incoming breaking waves ...

The dynamics of the shoreface is largely determined by energy dissipation from wind waves propagating onto the coast. Part of these waves is generated by local wind fields on the continental shelf; this is called 'sea'. Another part, called 'swell', is generated by remote storms on the ocean and has substantially greater characteristic wave period and wavelength. Before reaching the coast, sea and swell wave fields have often gathered wind energy over hundreds of kilometres. The breaking of waves on the shoreface converts this wind energy partly to heat and partly to forces acting on the seabed. These forces are generally larger than forces from other water motions acting on the shoreface, such as tides and wind-driven or density-driven circulations. Sand coast dynamics is therefore mainly determined by wave action. Large quantities of sediment are mobilized and set in motion, particularly in the surf zone, where waves are breaking.

produces a variety of seabed patterns ...

This motion is not just a symmetric oscillation. Due to the nonlinear nature of wave motion in shallow water on the one hand, and the nonlinear nature of sediment transport on the other hand, sediment particles experience a net displacement. This net displacement varies in space and time, especially due to seabed topography. As a result of the interaction between gradients in sediment transport and gradients in topography, morphologic structures emerge on various scales, such as longshore and transverse bars, rip cells, cusps and ripples. These structures often have a temporary character. Their development may be 'forced', for instance, through resonance with incoming waves or interaction with landscape structures. They may also emerge 'spontaneously' as free instabilities. Coastal morphology is therefore continuously changing. Complexity and variability are a natural characteristic of coastal morphology. Waves are the dominant morphodynamic agent. However, sea level change, sediment supply and sediment type also play an essential role in coastal evolution at long time scales.

which are studied in this chapter.

This chapter aims to explain the origin and evolution of coastal morphology. The focus is on underlying processes more than on accurate methods for predicting coastal evolution. It is organised in six sections. First we present a qualitative description of major morphological characteristics of the coastal zone under different wave and tide conditions. Then an introduction is given to the nature of wind waves in the coastal zone and their role in coastal morphodynamics; basic notions of the underlying theory are presented in Appendix D. In the next section some aspects of large-scale coastal behaviour are discussed with emphasis on the role of sea level rise and the issue of long-term forecasting. This section ends with an introduction to feedback processes in the coastal zone, causing instability and pattern formation. The fourth section starts with large-scale coastline adaptation to wave forcing and continues with morphodynamic feedback processes leading to different types of shoreline patterns, such as shoreline cusps, shore-normal or shore-oblique bars and rip cells. A separate subsection is dedicated to the beach cusp phenomenon, which was described already in the introduction to Chapter 2. The fifth section deals with coastal morphodynamics in the cross-shore plane. The concept of equilibrium profile is discussed; different models are introduced, based on feedback of the coastal profile to nonlinear wave transformation processes in the nearshore zone. A qualitative discussion is presented of the generation of longshore breaker bars, as a response to wave breaking processes in the surf zone. The final section is devoted to coastal erosion. The first subsection deals with temporary erosion by fluctuations in wave conditions and with prediction methods. Then we discuss structural coastal erosion, its causes and strategies for dealing with it. In the summary we reflect on the present state of our understanding of wave-topography interaction.

7.2. Morphology of Sandy Shores

This section is intended to familiarise the reader with some morphological characteristics of sandy ocean shores; a more quantitative discussion of the physical processes is postponed till later. First a qualitative description is given of typical morphologic features of sandy coasts and conditions are indicated under which these features are observed. An introduction is given to the major physical processes, which are closely related to shallow water wave dynamics.

Throughout this chapter we use the convention that the positive x-axis points in the offshore direction, with the exception of the sections dealing with swash flow. At the water line $x = 0$. For an alongshore uniform coast, the y-axis follows the shoreline. The z-axis is upward positive.

7.2.1. *The Coastal Profile*

A cross-shore section of the coast is called a coastal profile. Coastal profiles have typically a small slope of the order of 1/1000 at the inner shelf zone, about one kilometre up to some ten or twenty kilometres offshore, with water depths typically ranging from about 10 to 50 metres. Landward of the inner shelf the seabed slope becomes much steeper, typically between 1/100 and 1/10; this portion of the coastal profile is called the 'shoreface', see Fig. 7.1.

The seaward part of the shoreface is called 'shoaling zone'. In this zone incident waves 'feel' the seabed; they lose their sinusoidal character and become skewed (or 'peaked', in another terminology).

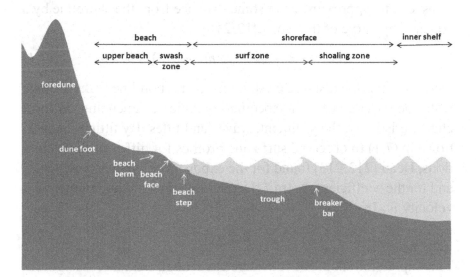

Fig. 7.1. Different zones of a microtidal coastal profile. Each of these zones is shaped according to a particular type of morphodynamic feedback.

We call $x = X_{br}$ the offshore distance at which waves start breaking (the breaker line). The position of the breaker line depends on the characteristics of the incoming waves, in particular the wave height H. The region landward of X_{br}, with water depths roughly between one and eight metres, is called 'surf zone'. In this zone waves dissipate energy by breaking; this is either a continuous process ('spilling waves') or a succession of breaking events ('plunging waves'). The former case is typical for breaking of locally generated waves on a gently sloping seabed and the latter case for high swell waves breaking on a steep shoreface or a pronounced longshore bar, also called breaker bar. The location and pattern of longshore bars is generally irregular and variable over time (see also next paragraph). A surf zone may be absent in the case of a steep shoreface dominated by low-energy swell waves.

When averaging coastal profiles over a long time interval, individual bedforms such as longshore bars are generally smoothed. The shape of the resulting profile exhibits a gradually increasing landward slope. Many coastal profiles, in particular under microtidal conditions, can be approximated at some distance from the shoreline by a concave-up curve of the type [212,215]

$$h(x) = A_b x^{m_b}, \quad m_b \approx 2/3, \qquad (7.1)$$

where x [m] is the distance offshore from the shoreline, h the average water depth and A_b a site-specific coefficient, depending on local characteristics of the sediment, waves and tides. By fitting the relationship (7.1) to observed surf zone profiles for different field situations, Dean [212,213] found for the exponent m_b values close to 2/3 and for the coefficient A_b [m$^{1/3}$] a relationship with the sediment fall velocity w_s [m/s],

$$A_b = 0.5 \, w_s^{0.44}. \qquad (7.2)$$

This expression should be considered as a rough estimate, as it ignores grainsize variation over the coastal profile. Dean [212] also showed that the exponent $m_b = 2/3$ is consistent with the assumption of uniform wave-energy dissipation per unit volume in the surf zone (to be discussed later in Sec. 7.6.1).

The 2/3 exponent in the power law (7.1) is not representative for all coastal profiles. Sometimes a lower exponent yields a better fit [442]. Laboratory experiments and field observations indicate that retreating coasts have a steeper profile, corresponding to a lower exponent m_b of about 0.3 [228].

The shoreface

Waves lose their sinusoidal character when propagating from deep water onshore. The wave crest is peaked and the wave trough is flattened. The orbital motion becomes asymmetric, with higher peak orbital velocities in the onshore direction than in the offshore direction. This generates a net onshore sediment transport, which in an equilibrium situation is mainly balanced by offshore transport induced by a seaward sloping bottom (assuming that cross-shore currents are small). In Sec. 7.6.1 it will be shown that the concavity of the long-term average coastal profile according to (7.1) can be explained by such a balance.

This picture holds under average conditions, but not in particular cases, as discussed in Sec. 3.8.2. Net onshore sediment transport is reduced or even reversed, when part of the sediment suspended by strong onshore orbital velocities remains in suspension during offshore orbital motion [243,356,717,734]. This lag effect may occur when the seabed is rippled, when sediment settling is slow or when incident waves are highly energetic.

When the depth decreases further along the propagation direction, wind-generated water waves cannot be considered any more as short waves. They become more similar to long waves, with a saw-toothed shape, due to faster propagation of the wave crest relative to the wave trough. Strong onshore wave orbital velocities and accelerations promote sediment transport in the onshore direction. The wave front steepens; the waves start breaking and lose gradually their energy upon approaching the shoreline.

Wave breaking modifies drastically the wave orbital velocity profile. A strong net onshore flow is concentrated in the upper part of the vertical (wave trough-to-crest part), and a net offshore flow occurs near the bed. This offshore flow, which is called undertow, drives a net seaward sediment transport. The coastal profile adapts to this change in sediment transport direction. The profile flattens and often a shore-parallel bar ('breaker bar') is formed in the convergence zone just seaward of the location where the waves break.

Undertow — a net seaward flow below the wave trough, compensating for the net shoreward transport in the trough-crest part of the vertical - is already present before wave breaking. However, the near-bed strength is less than in the breaker zone.

Waves approaching the shore under an angle, generate a longshore current when breaking. This current, which is strongest along the breaker bar at its landward flank, can exceed 1 m/s for high waves. It carries sediment suspended by breaking waves along the coast; this longshore sediment transport is called 'littoral drift'.

In general, waves are only partially broken on the bar and reform landward in the bar trough. They propagate further landward with a steep wave front and break a second or a third time when propagating into gradually shallower water. Several parallel breaker bars may be present in the surf zone, especially if the average seabed slope is small [893]. If the location of wave breaking is close to the shore (steep coastal profile, long swell, small tide), waves plunge directly on the beach. In this case, the undertow can be so strong that bar development is inhibited. A more detailed discussion of breaker bar formation is presented in Sec. 7.6.2.

Breaker bars have a typical height of a few decimetres up to a few metres; they can extend alongshore over distances of many kilometres. The cross-shore spacing of multiple bars is of the order of a few hundred metres. Breaker bars move usually offshore in periods of high wave activity (storms) and are stable or moving onshore under less energetic conditions.

Under such calmer conditions with less oblique wave incidence the longshore linear bar breaks up in smaller segments with a crescentic form [529]. Onshore motion may result in shore-attachment of the inside horns of the crescentic bars, see Fig. 7.32. Rip channels develop at the interruptions between the crescentic bars. The formation of crescentic bars and ripp cells is discussed in Sec. 7.5.5.

Coastal equilibrium profile

The concept of equilibrium beach profile is related to the self-organised response of the beach and the shoreface to wave forcing; under similar conditions (wave climate, sediment characteristics and availability), beaches adopt a similar profile. The coastal profile determines the shape of the waves in the shoreface zone, while being shaped itself by wave-induced transport.

The equilibrium profile concept holds only in a statistical sense. Beach profiles can go through a variety of transient states, depending on actual wave conditions and previous beach states. High energetic storm waves erode the beach and transport beach sand to the shoreface; under fair weather conditions (low-energy swell) this sand is gradually transported back to the beach, see Fig. 7.2. For complete recovery the time interval between successive storms should be sufficiently long.

In many cases the beach profile is not solely controlled by morphodynamic feedback; other factors such as vegetation or local geological setting (presence of hard outcrops and headlands or consolidated substrate) play a role as well. Such outcrops should be taken into account when applying the equilibrium concept in practical situations [543].

A true equilibrium does not exist in practice, for two reasons: (1) the time scale of wave climate fluctuations is shorter than the morphological adaptation time scale; (2) the adaptation of the beach state to sea-level rise is a permanent process.

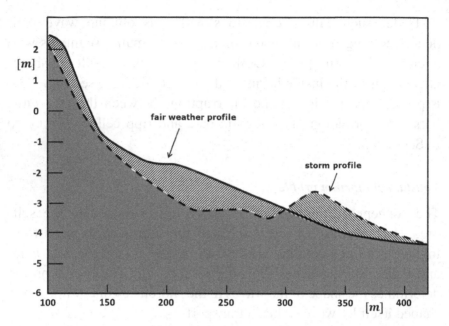

Fig. 7.2.   Coastal profiles near Duck (US, N-Carolina) under fair weather conditions (solid line, 14 August 1994) and storm conditions (dashed line, 26 October 1994). Sand eroded from the beach during storms by highly energetic waves is deposited in the surf zone and on the lower shoreface. During fair weather conditions these storm deposits are transported back to the beach by low-energy swell waves. The long-term averaged profile is the mean of these different conditions. The beach is in equilibrium if the long-term average profile does not change. Redrawn from [330].

Beach classification

Sediment grainsize (or equivalently, fall velocity) is an important parameter for the characterisation of beach profiles. Besides, beach profiles also differ according to wave properties and tidal range. Steep coastal profiles are generally associated with low-energy swell waves and coarse sediment (high fall velocity), while gentle coastal profiles are associated with dominance of energetic short-crested waves and fine sediment [492]. Therefore coastal profiles can be characterised by the non-dimensional settling time (also called 'Dean parameter') [211, 348]

$$\Omega = H_{br}/(w_s T), \tag{7.3}$$

where H_{br} is the wave height at which waves start breaking and T is the peak spectral wave period.

Wright and Short [966] developed a beach typology based on the parameter Ω, by analysing coastal profiles in Australia covering a broad range of different wave and sediment conditions.

They distinguished a category of 'reflective' beaches, corresponding to $\Omega \leq 1$, a category of 'intermediate' beaches, corresponding to $1 < \Omega < 6$, and a category of 'dissipative' beaches, corresponding to $\Omega > 6$. This classification implicitly assumes that coastal profiles tend to an equilibrium profile associated with the prevailing hydrodynamic forcing and sedimentary characteristics [115, 118].

A field survey at the microtidal Mediterranean coast indicates that beach states can be classified according to the Dean parameter when considering for H_{br}/T the average wave climate of the previous month and for w_s the average fall velocity of subtidal sediments [6]. However, it appears that beach states are also strongly influenced by the local geological setting (presence of headlands, river mouth, rocky outcrops).

The reflectiveness of coastal profiles can be related to the Irribarren number (or 'dynamic steepness') ξ [57],

$$\xi = \beta/\sqrt{H/L}, \tag{7.4}$$

where H is a representative amplitude of the incoming wave field (root-mean-square of the wave spectrum) and L the deep-water wavelength, $L \approx gT^2/2\pi$ and β the profile slope. Reflective beaches are characterised by $\xi > 1$ and dissipative beaches by $\xi < 0.25$. Instead of the Irribarren number, the surf similarity parameter ϵ_s is also used, defined as

$$\epsilon_s = H_\infty \omega^2/2g\beta^2 \approx \pi/\xi^2, \tag{7.5}$$

where H_∞ is the deep-water wave height and $\omega = 2\pi/T$ the radial wave frequency. Because the beach slope $\beta \ll 1$, we equate $\sin \beta \approx \tan \beta \approx \beta$. Reflective beaches are characterised by $\epsilon_s < 3$.

Although waves are the most influential agent at the seabed, tides also play a role, especially by shifting forth and back the zones of strongest wave influence. Coastal profiles in regions with strong tides differ from the profiles in tideless regions; this holds in particular for the upper part of the coastal profile. Several formulations have been proposed to generalise the coastal profile model of Dean (7.1) to the tidal case [70, 570], by considering the tidal range $H_{tide} \equiv 2a$ or the relative tidal range $RTR = 2a/h_{br}$ as an additional parameter (h_{br} is the water depth at the breaker line — the location where wave start breaking).

An analysis of beaches around the UK shows that the parameters Ω, ξ and RTR are not sufficient for classifying the different beach types, see Fig. 7.3. For differentiating between beach profiles with and without bars one should also consider the wave energy flux $F = \rho g H^2 c_g/8$, where c_g is the wave group velocity. Bars and rip cells occur typically for $F > 3kW/m$ [762].

Parametric profile expressions

Field surveys and laboratory experiments show that the concave profile shape (7.1) does not hold in the zone where waves start breaking [929]. Around the breaker line $x = X_{br}$, often an inflection point is present in the coastal profile, or even a terrace [83, 965]. Seaward of X_{br}, in the shoaling zone where incident waves become asymmetric, the coastal profile is again typically concave. Analysis of observed profiles show that this part of the shoreface can be represented by a similar power law as for the surf zone [70, 442], see Fig. 7.4,

$$h(x) = A_{nb}(x - x_0)^{m_{nb}}, \quad x > X_{br}. \qquad (7.6)$$

The subscript nb distinguishes the shoaling zone (non-breaking waves) from the surf zone (subscript b). At the breaker line the expressions (7.1) and (7.6) should match the breaker depth h_{br}, yielding $x_0 = X_{br} - (A_b/A_{nb})^{1/m_{nb}} X_{br}^{m_b/m_{nb}}$.

Shoreface profiles at the US Pacific coast are best fitted with the same exponents $m_b = m_{nb} = 2/5$ in the shoaling zone and in the

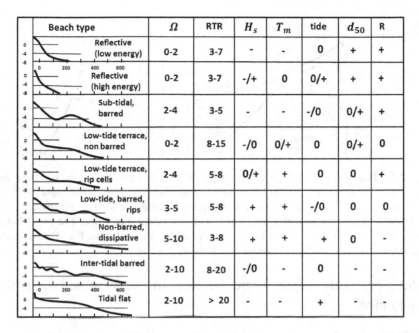

Beach type		Ω	RTR	H_s	T_m	tide	d_{50}	R
	Reflective (low energy)	0-2	3-7	-	-	0	+	+
	Reflective (high energy)	0-2	3-7	-/+	0	0/+	+	+
	Sub-tidal, barred	2-4	3-5	-	-	-/0	0/+	+
	Low-tide terrace, non barred	0-2	8-15	-/0	0/+	0	0/+	0
	Low-tide terrace, rip cells	2-4	5-8	0/+	+	0	0	+
	Low-tide, barred, rips	3-5	5-8	+	+	-/0	0	0
	Non-barred, dissipative	5-10	3-8	+	+	+	0	-
	Inter-tidal barred	2-10	8-20	-/0	-	0	-	-
	Tidal flat	2-10	> 20	-	-	+	-	-

Fig. 7.3. Beach types of the UK coast, adapted from [762]. Left column: Beach profiles, depth in [m] from mean sea level, distance from the shoreline in [m]. Other columns: Non-dimensional fall velocity Ω, relative tidal range RTR, significant wave height $H_s = \sqrt{2}H$ (+ > 1 m, 0 = 0.6 − 1 m, − < 0.6 m), mean wave period T_m (+ > 6 s, 0 = 4 − 6 s, − < 4 s), tidal range (+ > 6 m, 0 = 2 − 6 m, − < 2 m), medium grain size d_{50} (+ = gravel, coarse sand, 0 = medium, coarse sand, − = medium, fine sand), reflectiveness R (+ = high, 0 = medium, − = low). The beach parameters Ω and RTR alone are not sufficient to characterise the different beach profiles.

surf zone. The coefficients A_b and A_{nb} in the expressions (7.1) and (7.6) are found to be of order 1 [m$^{3/5}$] [442].

Field observations of tidal beaches in northern Spain yield exponents $m_b = m_{nb} = 2/3$ in the shoaling zone and in the surf zone. For the coefficients A_b and A_{nb} a linear dependence is found on Ω [70],

$$A_b \approx 0.21 - 0.02\,\Omega \quad \left[\mathrm{m}^{1/3}\right], \quad A_{nb} \approx 0.06 + 0.04\,\Omega \quad \left[\mathrm{m}^{1/3}\right].$$
$$(7.7)$$

For tidal beaches, the surf zone extends seaward to the breaker line at low water and landward to the HW mark. According to (7.1), the width X_{br} of the surf zone is then given by $X_{br}^m = (h_{br} + 2a)/A_b$,

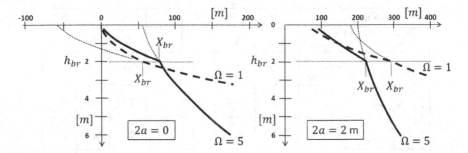

Fig. 7.4. Typical coastal profiles represented by a 2/3 power law, both in the surf zone and the shoaling zone, after [70]. The left panel is for the non-tidal case and the right panel for the tidal case (tidal range $2a = 2$ m). The dissipative profiles ($\Omega = 5$) are represented by a bold solid line and the reflective profiles ($\Omega = 1$) by a bold dotted line. The small dotted lines indicate the extrapolation of the shoaling profile in the surf zone. For all profiles the depth where waves start breaking is taken here as $h_{br} = 2$ m.

where h_{br} is the depth where waves start breaking and $2a$ is the tidal range.

Reflective beaches

At reflective beaches ($\Omega \leq 1$), only part of the wave energy is dissipated by breaking; the remaining part is reflected. In this case the steepness of the coastal profile close to the shoreline can be greater than suggested by the 2/3 power law. This is not necessarily in contradiction with uniform wave-energy dissipation per unit volume, because the energy of reflected waves has to be subtracted from the energy of incident waves. This yields a profile parametrisation of the form [52, 70]

$$A_b^{3/2}\, x = (h(x) + 2a)^{3/2} + B_b\,(h(x) + 2a)^3, \quad x < X_{br};$$

$$A_{nb}^{3/2}\,(x - x_0) = h^{3/2}(x) + B_{nb}\,h^3(x), \quad x > X_{br}, \quad (7.8)$$

where A_b, A_{nb} are given by (7.7) and B_b, B_{nb} by [70]

$$B_b \approx 0.89\exp(-1.24\,\Omega), \quad B_{nb} \approx 0.22\exp(-0.83\,\Omega).$$

The intercept x_0 follows from equating both expressions at $x = X_{br}$. Figure 7.4 shows typical average coastal profiles for the surf zone and the shoaling zone for non-tidal and tidal dissipative ($\Omega = 5$) and reflective ($\Omega = 1$) beaches. For dissipative beaches, a break in the shoreface slope is clearly visible around the breaker line X_{br}. The slope of the shoaling zone is typically steeper than for reflective beaches. In the tidal case, the width of the surf zone is substantially larger than for the non-tidal case.

7.2.2. Swash

Beach face

The beach face is situated between the surf zone and the dry beach. In this zone incident waves collapse and rush up the beach, a phenomenon called swash. It is also influenced by the exfiltration of groundwater, and by infiltration in the sediment bed, mainly during the uprush [410]. The beach face has typically a steep slope, especially in the case of coarse grained sediment.

Swash, the uprush and downrush of incident waves over the lower beach, plays a crucial role in the transfer of sediment from the surf zone to the beach. We therefore devote a more extensive discussion to this topic.

In this section we change our convention for the x-direction; the onshore motion is now positive and the offshore motion negative.

Uprush and downrush

In the case of single frequency, non-breaking incident swell waves, the frictionless cross-shore runup X_s on a beach with constant slope β can be determined analytically [137,364]. The result is

$$X_s = \sqrt{\frac{\pi}{2\beta^3}} H,$$

where H is the deep water wave height. Such non-broken waves are reflected at the shoreline and form with the incident waves a pattern of standing waves.

Here we consider the more usual case of partially reflective beaches, where incident waves collapse on the beach face and rush

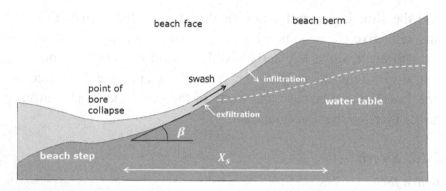

Fig. 7.5. Schematic representation of swash uprush on the beach face. The vertical scale is strongly exaggerated. The swash excursion X_s extends landward from the location where the incident wave bore collapses. The rapid uprush on the beach face slope causes the collapse of the incident wave bore.

up as a bore, see Fig. 7.5. The velocity V of a collapsing wave front can be estimated from the bore formula $V = 2\sqrt{gD_0}$ [726], where D_0 is the height of the collapsing wave (typically some decimeters).

The uprush loses energy by turbulent overturning at the bore front and by bed friction behind the front [698]. The uprush produces high bed shear stresses; the uprush velocity can reach values of 1–2 m/s and sediment concentrations may exceed $100 \, \text{kg/m}^3$.

The runup (swash excursion) X_s is related to exhaustion of kinetic energy. If we neglect friction and infiltration, the trajectory $x(t)$ of the bore front follows from $d^2x/dt^2 = -g\beta$ and $dx/dt = V \approx 2\sqrt{gD_0}$ at $t = 0$, where β is the slope of the beach face ($\beta \ll 1$). The uprush duration T_u is given by $T_u \approx V/g\beta$. With the above assumptions, the uprush and backwash durations are equal; the swash period T_s is thus about twice the uprush duration. The runup is given by $X_s = g\beta T_u^2/2 = V^2/2g\beta = 2D_0/\beta$. The velocities in the boundary layer of uprush and backwash have approximately a logarithmic profile. The friction coefficient is of the order of $c_D \approx 0.02 - 0.025$ [441, 481,706]. By including a friction term $c_D(dx/dt)^2/D$ in the equation for d^2x/dt^2 a more accurate expression is found for the runup. This nonlinear equation has an analytical solution, if the thickness D of the bore is assumed constant. The solutions for the uprush period T_u

and the runup X_s are [421]

$$T_u = \sqrt{\frac{D}{g\beta c_D}} \arctan K, \quad X_s = \frac{D}{2c_D} \ln(1 + K^2),$$

$$K = 2\sqrt{\frac{c_D D_0}{\beta D}}. \tag{7.9}$$

With $c_D = 0.025$, a beach slope $\beta = 0.1$, and with initial and average bore heights $D_0 = 0.5\,\text{m}$ and $D = 0.1\,\text{m}$, the friction reduces the uprush period T_u from 4.5 s to 2.3 s and the runup X_s from 10 m to 3.6 m. For strong friction and gentle slopes, K is generally much larger than 1 and we can approximate arctan $K \approx \pi/2$. The uprush period T_u and excursion X_s are then related by

$$\beta T_u^2 / X_s \approx \pi^2 [4g \ln K]^{-1}. \tag{7.10}$$

This relationship is almost the same as for the frictionless case, apart from the factor $\ln K$, with values typically between 1 and 1.5. Experiments show that the duration of backwash T_b is longer than the duration T_u of uprush by 20–40% [401,422].

The runup (7.9) is related to the bore thickness estimates D_0 and D, which depend on the height and period of the incident waves and on wave dissipation in the surf zone. Because no broadly generally expression is available, empirical relationships for X_s are used in practice. Observed maximum runup values are reasonably well represented by the relationship [811]

$$X_s \approx 0.4\sqrt{HL(1 + 0.007\beta^{-2})} \approx 0.16T\sqrt{gH(1 + 0.007\beta^{-2})}, \tag{7.11}$$

where H and $L \approx gT^2/2\pi$ are the wave height and wavelength at deep water, respectively. Another similar expression involves the sediment grain size [429,819],

$$X_s = 0.4T \exp(-10\,d^{0.55})\sqrt{gH}, \tag{7.12}$$

where d is the grain size [m]. We will see later that for dissipative beaches another relationship should be used.

The maximum runup depends on the mean water level at the shoreline. This water level is increased by wave-induced setup η_0, see Appendix D.3.1. The following empirical relationship has been derived for the sum of maximum wave setup and swash runup [811]:

$$\eta_0 + \beta X_s \approx 0.73\, \beta\, \sqrt{HL}.$$

Swash sedimentation and erosion

The sediment suspended in the surf zone and eroded from the lower beach face is deposited at the higher beach face, where the uprush velocity is small. Deposition is enhanced by infiltration, in particular for coarse sediment beaches with average grain size d_{50} larger than 1 mm. Infiltration has two opposite effects on the bed shear stress: A decrease due to reduction of the flow volume and an increase due to thinning of the boundary layer.

After reaching the highest point on the beach face, the water retreats along the beach slope in a very thin layer (called downrush or backwash). The backwash period is generally longer than the uprush period and the velocity is lower. However, the bed shear stress is initially high due to the small backwash depth. Exfiltration near the shoreline has two opposite effects: Thickening of the boundary layer and related decrease of the bed shear stress versus bed destabilisation [121]. Laboratory experiments show that bed shear stresses are similar for impermeable and permeable sandy beaches, whereas higher bed shear stresses are observed for permeable gravel beaches [481]. Field observations suggest that the net effect of infiltration/exfiltration depends on grain size: For medium sediment, it decreases the uprush sediment flux ($\approx -10\%$) and increases the downrush sediment flux ($\approx 5\%$); for coarse sediment the effect is opposite.

Entrainment of the bed top-layer (1–2 cm) contributes significantly to the uprush and downrush sediment fluxes [578, 603]. Most of the sediment deposited by the uprush is remobilized by the back-wash. The highest deposits may remain unaffected, however, because of scour lag; the backwash has already retreated below these deposits when it reaches velocities that are high enough for remobilization. The highest uprush deposits form a beach berm, see Fig. 7.5. Under extreme storms the beach berm evolves into a ridge on the upper beach [68].

Field experiments by Masselink *et al.* [578] showed that net accretion or erosion can be very different for successive individual swash events. In situations where several waves contributed to the same swash event, they observed net accretion if the first wave dominated the following waves and erosion otherwise. They also found that one single swash event could contribute significantly to the net accretion or erosion occurring over a complete tidal cycle. This illustrates the difficulty of simulating swash-induced accretion or sedimentation with process-based numerical models.

Holland and Puleo presented field evidence for the following scenario of negative morphodynamic feedback to the swash process [401]. If the swash period T_s is shorter than the average period T of the incoming waves, the backwash transports beach sediment to the surf zone. In such a situation the net result of swash is beach erosion. The beach face is lowered and the slope decreases. However, according to (7.9), a decrease of the beach slope induces an increase of the swash period. If the swash period T_s is longer than the average wave period T, the backwash collides with the next uprush. This precludes offshore transport; the net result of swash is now beach accretion. When the beach face accretes, the slope increases and the swash period decreases. Equilibrium is thus established for a beach face slope such that $T_s = T$.

Observations show that swash periods are not very different from the average period of incident waves at many reflective beaches. However, as mentioned before, long-term equilibrium does not only depend on swash processes, but also on surf processes (wave breaking on the beach) during storm periods [576].

The above feedback mechanism does not explain the dependence of beach slope on grain size. However, such a dependence may be expected, because swash on coarse-grained beaches is more strongly influenced by infiltration and exfiltration. Coarse sediment settles more easily at the end of the uprush than fine sediment; equilibrium with downslope transport by backwash therefore requires a steep beach slope. Hence, coarse-grained beaches (gravel beaches, for instance) typically have a steeper slope than fine- or medium-grained beaches.

The formation of beach cusps is intimately linked to the swash process. This will be dealt with in Sec. 7.5.7.

Infragravity swash

On dissipative beaches incident short waves lose most of their energy by breaking in the surf zone [734]. Near the shoreline only the longest waves survive, the so-called infragravity waves. Infragravity waves arise mainly from nonlinear interactions between wind waves with different wavelengths and frequencies [383]. They carry only a small part of the wave energy on the inner shelf (typically of the order of 1% or less), but their relative importance increases strongly in the surf zone. A more detailed discussion of infragravity waves is presented at the end of section 7.3.2.

Laboratory and field experiments suggest that the vertical component of the infragravity swash runup, βX_s, does not depend on the beach slope. The runup can be represented by a linear relationship with the deep-water wave height H [734, 744],

$$\beta X_s = aH + b, \quad H \geq 2\text{m}.$$

For the parameters a and b values are found ranging between 0.2 and 0.4. Other observations [769, 811] show a better correspondence with

$$\beta X_s = (0.05 - 0.06) \sqrt{HL},$$

where L is the deep-water wavelength. The runup X_s is bound to a maximum value for very high wave heights (H of the order of 5 m or more) due to breaking of infragravity waves.

While there is strong evidence that short-wave swash stimulates beach accretion, this is less clear for infragravity swash. Observations point to a net offshore directed transport by infragravity swash, especially in the seaward part of the swash zone. However, reliable models for simulating the complicated infragravity morphodynamics in the surf zone are not yet available [272].

7.2.3. Tidal Beaches

Beaches situated in regions with strong tides have a broad intertidal area. The swash zone moves with the tide up and down the intertidal area; there is no well defined beach face.

Tidal sweep

The morphology of the intertidal beach (also called foreshore) is primarily shaped by wave action. However, wave activity is strongly modulated by the tide. The major tidal effect is an alternating onshore and offshore shift of the breaker line and the surf zone ('tidal sweep'); the tide therefore stretches the width of the surf zone, see Fig. 7.6. Around low water, the surf zone is located further offshore, in a flatter portion of the coastal profile, where shoreface dynamics has a dissipative character. Around high water the surf zone is located onshore, in a steeper portion of the coastal profile, where shoreface dynamics has a more reflective character, especially if the tidal range is large [138, 775]. For moderate wave conditions and tidal range between 2 and 5 metres, the intertidal beach has a typical average slope of $\beta \approx 0.01$.

The duration of wave-induced morphodynamics in different zones of the intertidal beach is much shorter than for microtidal conditions, due to the tidal sweep effect. Morphological changes therefore unfold more slowly on tidal beaches [387, 522]. Dissipative tidal beaches

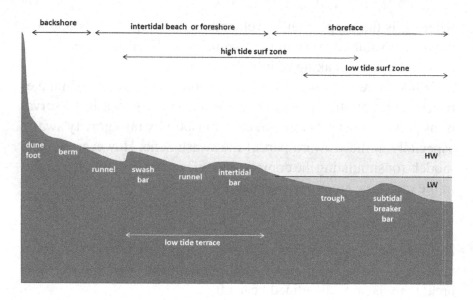

Fig. 7.6. Different zones of a mesotidal coastal profile.

are therefore less sensitive to seasonal changes in wave climate than reflective beaches [700].

Small-scale bedforms

The intertidal beach is a place of delectation for coastal morphologists. It is covered with structures shaped by morphodynamic feedback processes [604]. At low water, the tide provides free access to this morphological treasure house.

Small scale structures are current and wave ripples (typical height of a few centimetres, typical wavelength of a few decimetres), which can be found in zones where wave activity is low. Intermediate structures are megaripples (typical height of a few decimetres and typical wavelength of a few metres), in zones with low and intermediate wave activity. The morphodynamic processes that generate current and wave ripples and mega ripples are discussed in Secs. 4.3.2 and 4.2.

Intertidal bars

The largest structures on the intertidal beach are bars and troughs (height generally less than 1 metre and typical wavelength of order 100 metres), often called ridges and runnels. Most ridges are aligned with the shore, with a length of a few hundred metres. Transverse bars also occur, see Fig. 7.7.

Low-amplitude alongshore bars in the lower intertidal zone are similar to the breaker bars of the subtidal surf zone and mainly related to the same surf processes [498]. These bars are not very sensitive to changing wave conditions; they migrate slowly, on average in onshore direction. Wave breaking over the bars generates water-level setup

Fig. 7.7. Sequence of beach ridges and runnels at the meso-tidal coast of Goeree (The Netherlands). Rip channels and transverse bars are visible in the background. The beach ridge in the foreground is covered with megaripples. [Photo https://beeldbank.rws.nl, Rijkswaterstaat].

and onshore water transport concentrated near the water surface. This transport is partly compensated by a return current lower in the vertical (undertow), causing seaward sand transport. The water-level setup induces also a lateral circulation in the trough landward of the bar. It feeds a seaward flow concentrated in rip channels which intersect the bar [42]. The morphodynamics of rip cell generation is discussed in Sec. 7.5.5.

In the upper part of the intertidal zone, bars are mainly subject to swash. These bars are called slip-face bars or swash bars. They generally occur on gently sloping micro or mesotidal beaches. They are typically higher than the intertidal bars and are characterised by a steep landward slope (the slip face). Waves generate onshore swash transport when collapsing on the seaward bar face. The bar migrates onshore due to overwash by overtopping waves, in a way similar to the onshore migration of transgressive barrier systems. Under moderate wave conditions the swash bar merges with the berm in the upper swash zone. Under storm conditions the swash bar is subject to surf conditions [577]; the beach is flattened by infill of the landward runnel leading to rapid destruction of the slip-face bar morphology [579].

Megatidal beaches

For megatidal conditions (for instance, beaches of northern France with a tidal range of the order of ten metres), the surf zone exhibits three distinct zones [522] with different slopes: A steep, coarse-grained, reflective high tidal zone with slope $\beta \approx 0.1$; A mid-tidal zone with fine to medium sand and moderate slope $\beta \approx 0.02$ and a very dissipative, featureless low tidal zone with fine sediments and a weak slope $\beta \approx 0.005$. The low tidal zone is characterised by shoaling wave conditions and rapid breaker migration.

Longshore currents along macro/megatidal beaches are often more strongly driven by tides than by waves [390]. The alternating direction of tide-driven longshore currents produces longshore dispersal of sediment brought in suspension by waves; tidal asymmetry determines the net longshore transport direction. There is also evidence that strong longshore tidal currents tend to inhibit cross-shore sediment transport [390].

7.2.4. *Upper Beach*

Aeolian sand transport

The most shoreward part of the coastal profile is formed by the upper beach, also called supra-tidal or sub-aerial beach. It is nourished with sand from the lower (intertidal) beach, mainly by aeolian processes. The finer sand fraction is more easily blown by the wind to the upper beach than the coarse fraction, which means that in summer the upper beach usually consists of finer sand.

The upper beach has generally a steeper equilibrium slope than the lower beach; down-slope sand transport by gravity and wind is significant only for very steep beaches. Most down-slope transport occurs when the upper beach becomes part of the surf zone, during exceptional storm events with very high water levels. It is mainly the fine sand fraction that is removed. On the contrary, coarse sediment — gravel, for instance — can be deposited by energetic swash during storms at the top of the upper beach. So, the upper beach is not always sandy.

Aeolian sand transport proceeds mainly through saltation of sand grains. Fine sands are also transported in suspension [31]. The sediment flux depends not only on the wind speed, but also on the fetch (the distance traveled by the wind over the dry beach). Due to the shorter fetch, onshore winds are sometimes less efficient for transport from the lower to the upper beach than winds with a substantial along-shore component, especially in the case of narrow beaches [61]. When sand is blown over the beach, a pattern of streamers often appears: Streaks of high sand concentration of some 10 cm wide oriented in the wind direction. A satisfactory explanation of this phenomenon is not yet available [210, 232].

There are no easily applicable formulas that provide reliable estimates of aeolian sediment transport on beaches [60]. The wind stress required for picking up sand grains depends strongly on the moisture content of the beach surface. The moisture content is highly variable, in time and space. Several processes play a role, such as: Wetting of the beach by swash, spray, groundwater exfiltration and capilarity, precipitation, sheltering from solar radiation, vegetation [210]. Other factors also limit the uptake of sand, for instance, bed armouring by sediment sorting at the surface, salt crust formation and vegetation [232]. Wind stress itself is also highly variable. Formulas based on sand transport in deserts are therefore of little use for estimating aeolian sand transport over a beach; sand fluxes in the former case are generally much larger than in the latter case, for similar winds.

Storm conditions

Under storm conditions the beach is entirely covered with water and becomes part of the surf zone. Breaking waves induce a net seaward transport; during such periods large amounts of sediment migrate offshore. The beach is lowered and the slope is decreased. Under a calm or average wave climate the beach accretes and the beach slope increases. The beach state fluctuates continuously between these different conditions.

Beach erosion under storm conditions and post-storm recovery under calm to moderate weather conditions take place at very different time scales [808]. The time interval between storms is often too short for complete recovery. In such cases it is difficult to define the equilibrium beach state. The presence of subtidal bars reduces the impact of storms on dissipative beaches.

Dune growth

Sandy upper beaches may develop into a dune area; sometimes the transition is gradual with low foredunes, but a more abrupt transition, marked by a steep dunefoot, may also occur. Sand is blown from the beach into the dunes; in this way considerable amounts of sand can migrate landward, from a few m^3/m up to a few tens of m^3/m on a yearly basis. The average aeolian sand supply to the dunes along the Dutch coast is estimated at about 10 m^3/my [885]; it can contribute significantly to coastline retreat.

Dune formation is to a large extent controlled by sand availability and vegetation. On wide dissipative beaches more sand is available for aeolian transport than on narrow reflective beaches. This is not only due to the larger fetch, but also to the smaller grain size at dissipative beaches. Dune areas are better developed in the former situation than in the latter [61, 391].

Vegetation is crucial for seaward accretion of dunes [31]. Sand blown inland is captured near the vegetated dunefoot; only very strong onshore winds carry sand to the dune top. Bare dunes grow more easily in height, especially under shore normal winds; absence of vegetation also stimulates landward extension of the littoral dune belt.

Dune erosion occurs during storms, when the swash uprush reaches the dune foot. Erosion of the dune foot undermines the dune stability, leading to collapse of the dune slope. Undercutting of the dune foot by swash with subsequent sliding of dune mass is the most common dune failure mechanism [286].

7.2.5. Shoreline Patterns

Spatial scales

Sandy shorelines look fairly smooth when considered from a large-scale point of view, and this also holds for the bathymetric contour lines. This smoothness reflects the high mobility of the seabed which adapts to gradients in large-scale forcing by waves, wind and tides; these gradients are generally small. Large scale patterns, at the scale of 10 km and more (sand spits, delta lobes, headland shoals, for example), are often related to landscape topography, to coastal inlets or deltas and to man-made structures. Changes in these large-scale

Fig. 7.8.   Seen from far, sedimentary coasts look smooth; but a closer look reveals a variety of structures. (Coast of Goeree, Holland).

patterns are slow, exceeding several decades or even centuries [808]. However, a closer look at the coastal morphology yields a very different impression, see for example Fig. 7.8. At smaller scales (one km and less) a great variety of patterns shows up, both in alongshore and in cross-shore direction. These patterns can change quite rapidly, over periods of a day up to a few years. The smallest patterns, bed ripples, evolve on the timescale of minutes.

Many of these patterns emerge by morphodynamic feedback, even when the shoreline is straight and the incident wavefield uniform. Whether certain patterns emerge depends on different factors: Wave climate, sediment composition, seabed slope, tide and storms. At places where these conditions are similar, the same kind of patterns appear. A more detailed discussion, including theories of the generation processes, is presented in Sec. 7.5. Below we give a short overview, starting from the smallest scale alongshore patterns.

Shoreline cusps and undulations

- Beach cusps are familiar bedforms on reflective beaches, with medium-coarse-grained sand or gravel. The shoreline is structured as a series of prominent horns alternating with embayments which are typically 10–30 meters wide. Beach cusps appear when energetic waves approach the shore at an approximately right angle; they disappear under oblique wave incidence. Beach cusps are discussed in Sec. 7.5.7.

- Transverse bars occur at gently sloping intertidal or subtidal zones of dissipative beaches composed of fine-medium-grained sand. Transverse bars are observed with different orientations: Shore-perpendicular, down-current oblique or up-current oblique. In the the latter two cases wave-induced longshore currents play a role in their formation; these bars are generated when medium waves approach the shore at low-oblique angles. The shore-perpendicular bars, also called 'finger bars', owe their existence mainly to refraction of perpendicularly incident waves. The bar spacing of the oblique bars is typically a few tens to about 100 m and their crest length typically hundred or a few hundred metres. Finger bars at low wave energy have a wavelength of a few tens of metres and a crest length of a few hundred metres; finger bars at medium-high wave energy have a wavelength of the order of 100 m and a somewhat larger crest length. The generation mechanisms of these bars are discussed in Sec. 7.5.4.

- Shoreline cusps emerge when a longshore bar breaks up in crescentic bars, a phenomenon frequently observed after a storm period. These shoreline cusps with intermediate embayments are related to the development of a rip-cell circulation, see Sec. 7.5.5. Crescentic bars with rip cells are observed at coasts which are dissipative or slightly reflective. Wave breaking at the crescentic bars is a major driving mechanism for rip-cell development [42]; favourable conditions correspond to medium wave energy and small (close to shore-normal) incident angles. The cusps have an amplitude of a few tens of metres; their spacing varies from a few hundred metres to more than thousand metres. Widely spaced cusps associated with rip cell embayments are, for example, observed along the US Atlantic barrier coast [244, 492] and along the Dutch coast (see Fig. 7.31).

- Shoreline cusps have also been observed at larger spatial scales (several kilometres), without associated rip-cell morphology. They occur in particular at coasts with frequent highly shore-oblique waves (Namibia, Gulf of Finland). They can be explained by an instability related to positive feedback from wave refraction over a long-shore undulating surf-zone bathymetry, see Sec. 7.5.2. More subtle shoreline undulations at scales of several hundred to several thousand metres are observed at storm-dominated coasts (Holland, Denmark, Poland) [499]. Several hypotheses have been put forward for their formation; no firm single theory has emerged so far.

- Shoreline megacusps or megaspits exist at scales of many tens of kilometres up to several hundred kilometres. Examples are the spits in the Sea of Azov (Fig. 7.9) and the capes along the USA Atlantic coast. High-angle wave instability (HAWI) provides an explanation for their emergence [33], as discussed in Sec. 7.5.2.

7.3. Wind Waves Over a Shallow Seabed

7.3.1. *Wave Types*

Swell waves

The transition zone between land and sea receives a high energy input from waves which deliver their energy when breaking on the coast. This energy has been accumulated in the wave field by wind forcing on the ocean and the continental shelf. The amount of energy that can be absorbed in the wave field is limited, however, by white-capping: Waves break when they become too steep. Wave energy can travel over large distances compared to the wavelength, without much energy loss due to bottom dissipation.

Fig. 7.9. The northern coast of the Sea of Azov is characterised by a sequence of large-scale sand spits. The development of these spits can be explained by the growth of shoreline instabilities through positive feedback from longshore currents, driven by almost coast-parallel winds. See also Fig. 7.24. Image from Google Earth.

Incident waves generated far offshore in the ocean are called swell waves. The period of incident swell waves is relatively long — typically of the order of 10–15 s. They have long crests and travel in a well defined direction. Waves generated far away with strong directional spread and short wavelength are dispersed and dissipated before arriving at the coast. The strength of the longest swell waves increases due to energy transfer from shorter waves by nonlinear interaction [383].

Swell waves are almost perfectly symmetric at deep water. However, upon entering the shallow coastal zone (depths of about 10 m or less), their propagation characteristics become similar to those of long waves (e.g. tidal waves) with weak friction. This results in strong asymmetry, with higher orbital velocities and stronger acceleration in the propagation direction than in the opposite direction.

Locally generated waves

Waves generated by local wind fields have smaller wavelengths. They propagate shoreward from different directions; the wave field consists of short-crested random waves peaked in all directions.

Wave-skewing is mainly related to nonlinear interaction with the curvature of the wave surface, see Fig. 7.10. It hardly occurs for long waves, which have an almost flat wave surface.

Wave-skewing influences the wave-orbital water motion; the orbital motion in the wave propagation direction is of shorter duration, but stronger than the orbital motion in the opposite direction. This follows from a second-order approximation of nonlinear wave propagation, see Appendix D.

Close to the shore, the wave crests become more or less linear, due to the decrease of directional wave spreading by refraction. These sea waves are superimposed on the longer swell waves; together they form a complicated wave spectrum.

Single sine representation of an irregular wavefield

Nevertheless, in the simple models we will use for illustration purposes, the incident wave field is represented by a monochromatic sine shaped wave. This wave has the frequency of the most energetic part of the spectrum and a wave height H equal to the root-mean-square wave height, H_{rms} [318, 796]; $E = \rho g H^2/8$ represents the average energy density of the wave field. The wave direction is chosen equal to the average direction of the most energetic part of the incident wave field. Instead of H_{rms}, the significant wave height H_s (mean of the 33% highest waves) is often used in practice; they are related by $H_s = \sqrt{2} H_{rms}$.

Linear wave theory, based on a monochromatic wave representation, can be used to understand certain statistical properties of wave interaction with the seabed, but it should be kept in mind that it is a very crude simplification of reality. A more realistic representation includes the modulation of the wave amplitude (wave group envelope), due to interference of waves with slightly different periods

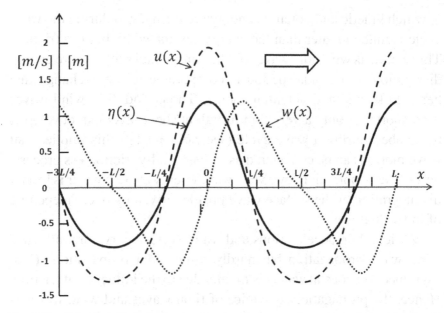

Fig. 7.10. Second-order Stokes approximation for a monochromatic wave propagating in positive x-direction at a water depth of $h = 6\,$m with amplitude $a = 1\,$m and period $T = 6\,$s, see Appendix (D.18) and (D.20). The wavelength $L \approx 40\,$m. Solid line: Surface elevation $\eta(x)$ at a given time t, dashed line and dotted lines: Orbital velocity $u(x)$ and vertical velocity $w(x)$ at the water surface, respectively. The wave crest is peaked and the wave trough is flattened. The wave-orbital velocity is higher in the propagation direction than in the opposite direction. The skewing results from the nonlinearity of the boundary conditions at the wave surface, related to its curvature. In deep water, wave skewing follows mainly from the kinematic boundary condition $\partial\eta/\partial t = w - u\partial\eta/\partial x$. The last term accelerates the upward and downward wave motion around the crest and decelerates the downward and upward wave motion around the trough, because of the $\pi/2$ phase difference between u and $\partial\eta/\partial x$. The second nonlinear boundary condition, which imposes the wave surface as a streamline, enhances wave skewing in shallow water. For long waves, tidal waves for example, the wave-skewing process is not significant because of the great wavelength and small water surface slope.

within the wave spectrum. An introduction to the theory is presented in Appendix D.

Comparison of wind waves and tidal waves

Wind wave propagation is much less influenced by frictional dissipation than tidal propagation. Due to the small wave period, the rate at which kinetic and potential energy are exchanged during the wave cycle is much greater than the energy dissipated by bottom friction. This contrasts with tidal waves on the continental shelf, where energy dissipation over a tidal period is comparable to the exchange rate between kinetic and potential energy. The period T of wind waves is so short that only a very thin turbulent shear layer can develop at the seabed during a wave cycle (see Sec. 3.3.1). This implies that wave motion can be considered as an essentially frictionless process, described by potential flow. Most wave energy at deep water is lost by disintegration of the surface wave profile, driven by over-steepening of individual waves.

Vertical fluid accelerations and water surface curvature influence wind wave propagation but hardly affect tidal propagation. Tidal asymmetry in coastal waters is mainly due to the influence of friction. Hence, the propagation dynamics of tidal waves and wind waves is basically different.

7.3.2. *Wave Transformation in the Surf Zone*

Wave breaking

When a wave travels from offshore into shallow water, the wave height and wave skewness increase. When the water depth gets smaller than about one to two times the wave height, the wave starts breaking [59]. Wave breaking does not depend only on the ratio of wave height to water depth, but also on the wave shape [695]. Physics-based breaking criteria involve the ratio of particle speed to wave speed at the crest or the particle acceleration at the crest [597, 699].

Wave skewness and wave breaking have important consequences for the net displacement of seabed particles under the influence of wave motion during a wave cycle. Below we will discuss qualitatively some important underlying processes; a more detailed mathematical treatment is presented in Appendix D.

Spilling or plunging

In the case of a gentle shoreface slope and strongly skewed waves, breaking starts at the wave crest. This breaking process is called 'wave spill'; air is entrained into a roller, which is pushed forward at the wave front. When the spilling wave propagates into gradually shallower water, the breaking process continues with a decreasing wave height and a growing roller volume, see Figs. 7.11 and 7.12. The wave height H decreases almost linearly with depth; the proportionality factor γ_{br} is typically of the order of 0.4–0.8. When reaching the shoreline, only a thin wave bore is left, that does not produce energetic swash. A simple theory for wave dissipation in the surf zone by spilling breakers is presented in Appendix D.2.1.

In the case of a steep shoreface slope and/or less peaked waves, wave breaking has a plunging character (Figs. 7.12 and 7.13), with

Fig. 7.11. Spilling and plunging breakers on a gently sloping beach. Waves start breaking when the wave height becomes comparable to water depth, often at the outer longshore bar. At the landward slope of the bar the ratio of wave height to the water depth is decreased. Waves may reform in the trough between the outer and inner bar and break a second time at the inner bar, close to the beach.

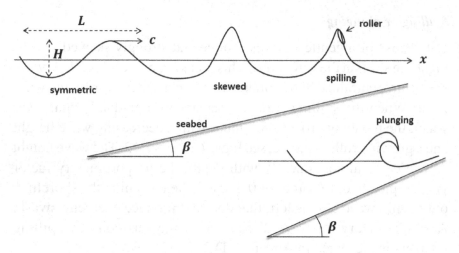

Fig. 7.12. Wave transformation on the shoreface. In shallow water the wavelength decreases and the wave surface slope becomes steeper. The wave then loses symmetry and the wave surface becomes skewed, with higher shoreward than seaward orbital velocities. The wave starts breaking when the wave height becomes comparable to water depth. Wave spilling is the most common breaking mode on gently sloping beaches; plunging breakers will occur more frequently on steep beaches.

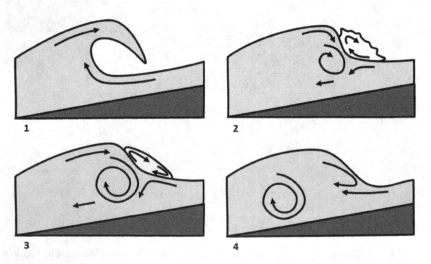

Fig. 7.13. Different phases of wave breaking. 1. Overturning at the wave crest. 2. Formation of a roller with air entrainment at the wave front. 3. Formation of a vortex behind the wave front. 4. The wave reforms; the vortex moves offshore and downward with the undertow. Redrawn after [55].

production of strong vortex motions and sediment uplift from the seabed. Sediment suspended by spilling breakers remains close to the bottom and is entrained offshore by undertow; sediment suspended by plunging breakers moves up higher in the vertical where the velocity is onshore. Plunging breakers therefore produce more onshore transport than spilling breakers [4]. After initial breaking, plunging waves may reform and break a second time. Upon collapsing on the beach face, the plunging wave produces an energetic swash uprush.

The ratio $\xi = \beta/\sqrt{H/L}$ of bottom slope β and quare-root wave steepness (H, L are height and wavelength of the incident wave) yields a practical criterium for the type of wave breaking [57,59]. We have spilling waves for $\xi < 0.45$ and plunging waves for $0.45 < \xi < 3.3$. For larger values of ξ waves do not break but surge onto the beach.

Undertow

Wave breaking is a complicated process. It is still challenging mathematical modelers, although much progress has been made [474,786]. Just before breaking, the speed of water particles at the crest increases and approaches the wave speed, while a strong return current develops at the bottom. This return current, called undertow, opposes shoreward transport caused by wave skewness. This may result in a net seaward sediment transport.

A return flow is present both in breaking and non-breaking waves; it balances the net onshore mass transport (Stokes drift) due to the covariance of water level and wave-orbital motion. In breaking waves the return flow is concentrated close to the bottom, while in non-breaking waves the return flow is distributed over the water column below the wave trough [710]. Offshore-directed undertow velocities are typically of the order of 0.2–0.4 m/s [220], see also Appendix D.1.2.

Radiation stress

Wave breaking in the surf zone entails not only a shoreward decrease of potential wave energy and wave height, but also a shoreward decrease of kinetic wave energy and wave-orbital momentum. Only a minor part of the momentum decrease in the surf zone is due to frictional losses. During shoreward orbital motion, more shoreward momentum enters the surf zone from offshore than is transferred from the surf zone to the beach. During seaward orbital motion more seaward momentum is transferred from the surf zone to offshore than enters the surf zone from the beach. The result is a net gain of shoreward momentum in the surf zone or, equivalently, a net loss of seaward momentum. This generates, in the case of shore-normal wave incidence, a net cross-shore stress gradient and in the case of shore-oblique wave incidence, an additional net alongshore stress gradient. This phenomenon is described in mathematical terms by means of 'radiation stresses' [539].

Breaking-induced gradients in the radiation stress terms produce a shoreward increase of the average sea level and a longshore current in the surf zone. This longshore current generates a substantial longshore sediment flux, in particular during storms, when wave breaking causes intense sediment suspension in the surf zone. Currents produced by radiation stresses play an important role in the morphology of the nearshore zone.

The radiation stresses themselves are very sensitive to seabed disturbances, because such disturbances affect locally the intensity of wave breaking. Therefore the interaction between seabed morphology and radiation stresses has a rich potential of generating seabed instability and morphologic patterns, see Sec. 7.5.3. An example is the development of rip channels, which results from seabed instability inherent to the mutual interaction of seabed morphology and radiation stresses.

Simple analytical expressions can be derived for the radiation stresses if incident waves are monochromatic and sinusoidal, see Appendix D.2.2. These analytical expressions will be used even in cases where (1) waves are not sinusoidal (i.e., in the breaker zone), (2) waves are not monochromatic (wave spectrum) and (3) the wave incidence angle is variable (wave directional spreading). Ignoring these effects leads in general to overestimating the radiation stresses [297]. However, the qualitative features of the radiation stresses are not strongly affected by these simplifications.

Wave refraction

In shallow water (water depth much smaller than wavelength) the wave propagation speed depends on depth; the smaller the depth, the slower waves propagate. A wave field approaching the coast under a non-zero angle propagates at different velocities according to the local depth pattern. In Fig. 7.14 the wave field is represented by hypothetical crestlines, perpendicular to the propagation direction. The part of the crestline which is closest to the shore (shallowest water) propagates more slowly than the part of the crest further offshore. Hence, further away from the coast the crestline moves faster shoreward than closer to the coast. This implies that the crestline rotates towards a shore-parallel orientation, as indicated in Fig. 7.14. This phenomenon is called refraction and is described by Snell's law

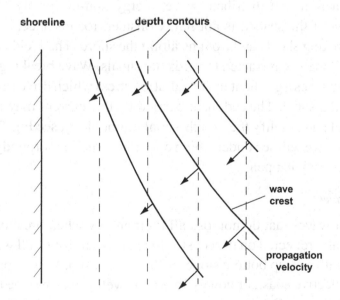

Fig. 7.14. The wave propagation speed increases with depth; an obliquely incident wavefront is therefore refracted towards the shoreline.

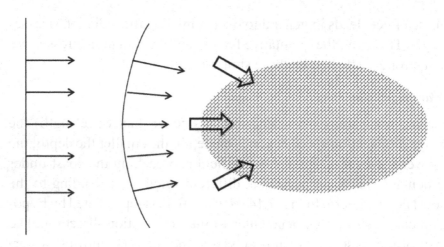

Fig. 7.15. Wave refraction concentrates wave breaking at a shoal and increases locally the radiation stresses. This drives sediment-laden flow onto the shoal and contributes to shoal growth.

($c/\sin\theta$ constant along wave rays, where θ is the angle between propagation and bottom-slope directions).

Refraction will distribute wave energy non-uniformly over the surf zone if the seabed is not flat. Consider, for instance, a pattern of alternating shoals and troughs along the shore. The incident wave field will then be refracted towards the shoals. Wave breaking on the shoal produces a gradient in the radiation stress which drives a current toward the shoal. The sediment carried by this current may nourish the shoal and amplify the trough-shoal morphology, see Fig. 7.15. In Sec. 7.5.3 we will see under which conditions such a morphodynamic feedback may happen.

Edge waves

Incident waves that do not lose all their energy when breaking, will be partially reflected on the coast. This may occur for swell waves on coasts with a steep bottom slope (reflective coasts). When approaching a reflective coast, obliquely incident swell waves will be trapped along the coast. Due to refraction, the reflected wave will turn back to the shore. Such coastally trapped waves are called 'edge waves'.

Incident and reflected edge waves may form a standing wave and have been hypothesised to generate a pattern of rhythmic bedforms structured to the nodes and antinodes [103]. Edge waves may also play a role in the formation of beach cusps, see Sec. 7.5.7.

Wave groups and infragravity waves

Incident waves with slightly different wavelengths interfere. The interference produces a modulation of the amplitude of the incident waves, see Appendix D.1.1. The wave envelope has a much larger wavelength than the individual incident waves. For an observer it seems that the waves arrive in groups, see Fig. 7.16.

The group wavenumber is approximately equal to the average width of the wavenumber spectrum of the incident waves. The group propagates with a velocity c_g given by (D.14), which is smaller than the propagation velocity c of the incident waves.

The amplitude modulation within a wave group induces a spatial modulation in the radiation stress, which has a maximum where the wave group amplitude is largest. The gradient in the radiation stress terms produces a small set-down of the mean water level where the wave group amplitude is largest and a small set-up where the wave group amplitude is smallest [539], see Fig. 7.16 and Appendix D.2.2. This long wave associated with the wave group is called 'bound infragravity wave'. In deep water, the height of infragravity waves is small (less than 1 cm), but close to the shore their magnitude increases and can reach even 1 m during storm conditions.

Infragravity waves modulate the propagation speed of incoming short waves in the nearshore zone by varying the water depth; successive wave bores in the surf zone therefore propagate at different speeds [840]. A wave bore travelling at a higher water depth (near the infragravity wave crest) can overtake an earlier wave bore travelling at a lower water depth (near the infragravity wave trough); merging leads to a decrease of the number of wave bores in the surf zone and to a decrease of observed wave frequency [844].

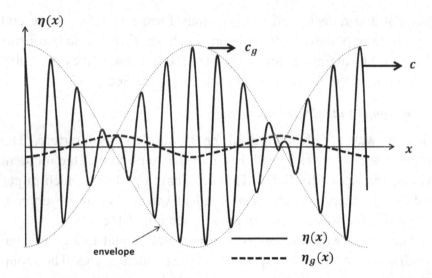

Fig. 7.16. Wavegroup, resulting from the interference of two incident waves with slightly different wavelengths; $\eta(x)$ is the water level variation of the combined incident waves. The infragravity wave $\eta_g(x)$ associated with the wave group and the wave group envelope propagate both at the group velocity c_g, which is lower than the propagation speed c of the incident waves. Incident waves thus travel through the envelope.

Incident waves dissipate in the surf zone due to wave breaking. In the surf zone, the infragravity wave then becomes a free wave, which is not bound any more to the incident waves. Part of the infragravity energy is dissipated in the surf zone (mainly through breaking [218]), but there is also generation of infragravity energy, because of energy transfer from short waves and interaction with topography [49]. Free infragravity waves can be partially reflected at the beach, even at dissipative coasts; the ratio of offshore and onshore propagating infragravity energy can be even larger than 1. Reflected infragravity waves can propagate to deep water (so-called "leaky waves") but often they are trapped in the nearshore by refraction and become edge waves. Observations show that edge waves at dissipative coasts are related to the occurrence of reflected free infragravity waves [329];

Assuming small frictional wave dissipation, the orbital velocity associated with the bound infragravity wave is directed offshore where the wave group amplitude is largest and onshore where the group wave amplitude is smallest. Suspended sediment concentrations are relatively larger where the wave group amplitude is large and relatively smaller where the wave group amplitude is small. The bound infragravity wave therefore enhances offshore sediment transport [734].

Infragravity sediment transport can be directed offshore as well as onshore in the surf zone, because the free infragravity wave is no longer related to the maxima and minima within the wave group [3]. This implies that, on average, infragravity waves carry sediment away from resuspension maxima at the breaker bars. Laboratory experiments indicate an overall tendency of free infragravity waves to promote onshore transport relative to offshore transport [51].

The slow wave modulation in the nearshore zone related to the arrival of wave groups and infragravity waves is called surfbeat. Swash motion is also modulated by infragravity waves, as discussed in section 7.2.2. The infragravity component is even dominant in the swash motion on dissipative beaches [734].

7.3.3. *Wave-Induced Cross-Shore Sediment Transport*

Here we summarize major mechanisms by which waves influence cross-shore sediment transport. A more detailed discussion is presented in Sec. 7.6.

On the inner shelf, wave orbital motion hardly reaches the seabed. This changes when incident waves enter the shoreface zone. The water depth at which the seabed is significantly disturbed by wave action is called 'closure depth' [369]. The closure depth is typically situated between 10 and 30 metres. Under storm conditions or under conditions of strong swell, it is situated further offshore.

Before arriving at the shoreface zone, waves have often lost their sinusoidal shape already. The wave front of long swell waves steepens when the product of wave number and depth, kh, becomes smaller than 1. Wave orbital velocities then become stronger in the onshore direction than in the offshore direction. This results usually in an onshore directed residual sediment transport.

For waves with a steep front the onshore orbital acceleration is stronger than the offshore acceleration. This acceleration asymmetry strengthens onshore sediment transport, see Sec. 3.3.1. Onshore transport is further enhanced by streaming in the wave boundary layer.

However, the residual transport direction can be reversed in the presence of steep vortex ripples, as discussed in Sec. 3.8.2. In this case the sediment resuspended by the strong onshore motion in the wave boundary layer is transported offshore in suspension after reversal of the orbital motion. Bound infragravity waves, associated with wave groups, also contribute to offshore sediment transport.

When entering the surf zone, waves start breaking. If breaker bars are present, wave breaking will be concentrated at the bars. Wave breaking strengthens the undertow current, which transports sediment offshore. This offshore transport and the resulting convergence of sediment transport induces bar growth and bar migration. If no bars are present, breaking of energetic waves can trigger bar formation.

After breaking at the offshore bar, waves may reform and assume again a skewed or saw-toothed shape with strong onshore orbital velocities producing onshore sediment transport.

The phase relationship between bound infragravity waves and wave groups is lost in the surf zone upon dissipation of incident waves by breaking. The residual sediment transport induced by free infragravity waves is directed onshore, except at locations just seaward of suspended sediment maxima. In the case of shore-perpendicular incidence, reflected infragravity waves may form a standing wave pattern. Such a standing wave can produce a cross-shore pattern of sediment transport convergence and divergence zones.

Wave dissipation induces an effective stress gradient in the direction of the incident waves. The cross-shore component of this stress gradient drives an onshore current and onshore sediment transport over shoals and indirectly an offshore current and offshore sediment transport through rip channels.

When arriving at the shoreline, incident waves collapse on the beach face, producing a swash bore. The net result of swash uprush and downrush contributes in general to beach accretion. On gently sloping beaches, the incident waves are almost completely dissipated when arriving at the shoreline. In this case, swash motion is mainly due to infragravity waves.

Under storm conditions the beach becomes part of the surf zone. Wave breaking on the beach produces offshore sediment transport and beach erosion.

7.4. Coastal Genesis

Sedimentary coastal systems consist mainly of material released by weathering processes on the continents. Cliff erosion and submarine abrasion can also deliver substantial amounts of sediment, but reliable estimates are scarce. In some regions, marine biological processes are a major source of coastal sediments. These sediments are reworked by waves and currents into the present submarine and subaerial coastal landscape. Sea-level rise plays a major role in the large-scale long-term evolution of sedimentary coasts. At smaller scales, the morphology of sedimentary coasts is highly variable. Much of this variability can be attributed to instabilities and symmetry breaking inherent to wave-topography interaction.

7.4.1. *Response to Sea-Level Rise*

Balance of large-scale longshore and cross-shore transport

Gradients in longshore and cross-shore sediment transport produce change in coastline position and coastal profile. Gradients in longshore transport are related to longshore variations in sediment availability and to longshore variations in wave energy, shoreline orientation and wave incidence direction. Cross-shore transport to or from the shoreface depends on the balance between onshore and offshore transport. Onshore transport is mainly wave-induced and occurs under moderate wave conditions. Offshore transport occurs under storm conditions; long-term offshore transport is enhanced by gravity effects. Cross-shore transport at the top of the coastal profile, the subareal beach, depends on the balance of aeolian inland transport and storm-induced erosion of the upper beach and the foredune [638]. Coastline position and coastal profile adjust in such a way that, when averaged over a sufficiently long time, the gradients in longshore and cross-shore transport are minimised. This adjustment is a continuous process, because of sea-level rise. During the last centuries, profile adjustment is strongly influenced by the response to human interventions.

Marine transgression

Since the last ice age, the sea level has risen more than a hundred metres in less than fifteen thousand years. In the initial phase, sea-level rise was fast, up to about one metre per century. During the last millennia sea-level rise has slowed down, to an average of about one decimetre or less; in low-latitude regions the sea level has even dropped [608, 923]. In the past century average sea-level rise was of the order of 20 cm, local uplift or subsidence set aside. Sea-level rise has produced great changes in the coastline of low lying coastal zones. Former coastal plains were drowned and at many places the coastline shifted landward over great distances, a process called 'marine transgression'.

Barrier formation

Marine transgression of low-lying coastal plains brought large areas under the influence of wave action. Where the drowned land consisted of unconsolidated sandy material, waves and currents eroded the new seabed and carried sediments onshore. On exposed coast, onshore sediment transport was mainly produced by wave action. Marine transgression was limited by accretion of beach berms, which were subsequently transformed into coastal barriers by aeolian processes [108, 732], see Fig. 7.17. When sea-level rise slowed down, the barrier could keep pace with sea level if sediment supply remained

Fig. 7.17. Beach berms are the embryos of new coastal barriers (prograding island of Goeree along the Dutch coast; sluices of the Rhine at the background).[Photo https://beeldbank.rws.nl, Rijkswaterstaat/Rens Jacobs].

sufficiently abundant; the barrier then acted as sediment source for dune formation.

Barriers were not necessarily attached to the shore but could enclose coastal lagoons. Lagoon development behind coastal barriers took place in low-lying areas where infilling and scouring processes were in close balance (see Sec. 6.5). Back-barrier lagoons also formed by down-drift elongation of sand spits at shoreline breakpoints or by onshore migration of sandbanks (see the Bornrif example, Fig. 2.21). Coastal barriers migrated landward by storm-driven sediment wash-overs [415]. The ensuing decrease of tidal volume resulted in many cases in the closure of inlets by littoral drift.

Not all the eroded seabed sediment was stored in the barrier. Fine sediments were washed away and deposited offshore or onshore in backbarrier basins. Sandy sediments got lost to the deeper shelf

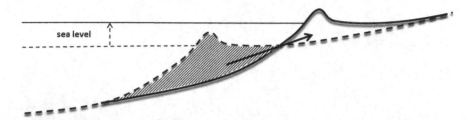

Fig. 7.18. Marine transgression of a barrier coast under slow sea-level rise. Former coastal profile (dashed line) and new profile (solid line) after sea-level rise. The former profile is eroded; part of the eroded sand volume is deposited onshore, mainly by wave-induced transport. The cross-shore profile of the new coast, relative to the new sea level, is similar to the profile of the former coast.

in situations where rocky headlands or submarine canyons were nearby [631].

Large amounts of beach sands are transported offshore under conditions of highly energetic storm waves. However, observations and model simulations indicate that at least part of this sand loss is recovered by onshore wave-driven transport under less energetic conditions, though at a much slower pace [580].

With ongoing sea-level rise, barriers will naturally migrate landward following a continuous process of overwash and inland aeolian transport, see Fig. 7.18. In the case of low overwash rate and fast sea-level rise, the barrier may be overstepped and decay; in this case a new barrier will form further landward [569]. Present coastal development and hard protection works prohibit at many places this natural coastal retreat.

Remains of ancient coastal barriers have been found on the inner shelf, in the core of offshore sandbanks [412]. This provides evidence that ancient coastlines may be remodelled as submarine ridges in interaction with the prevailing current and wave conditions [824] (see also Sec. 4.5).

Human interventions can strongly accelerate the process of wave-driven onshore sandbank migration. An example is the onshore migration of sandbanks after closure of the tidal inlet Grevelingen in the Rhine–Meuse–Scheldt delta, see Fig. 7.19. The ebb-tidal delta

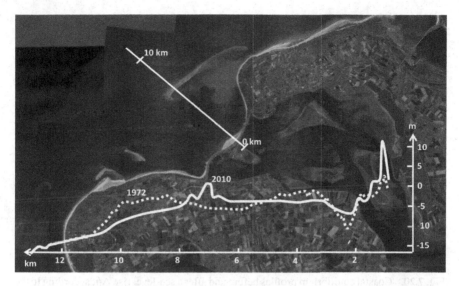

Fig. 7.19. Onshore migration of ebb-tidal delta banks after closure of the tidal inlet Grev-elingen (Netherlands) in 1970. Solid line: Depth profile in 2010; dashed line: Depth profile in 1972. The profiles correspond to the transect indicated in the figure. The image (from Google Earth) shows the situation in 2013. After [281]

banks moved several kilometres onshore in a period of 40 years. The shape of the banks was also changed; the broad subtidal banks of the ebb-tidal delta were transformed into narrow intertidal and supratidal banks.

7.4.2. *Coastline Retreat*

Climate change, and sea-level rise in particular, is the major cause of coastline retreat in the absence of human interventions. The beach slope, the erodibility of the beach material and the fate of the eroded material determine the rate of coastline retreat. We focus here on sandy or gravelly beaches, which adapt to changes at time scales much shorter than the time scale of sea-level rise.

In previous sections we have seen that beaches and shorefaces with sufficient longshore uniformity have a characteristic cross-shore profile. This profile depends on several factors, in particular sediment properties, wave conditions and tide. It has been suggested, since a

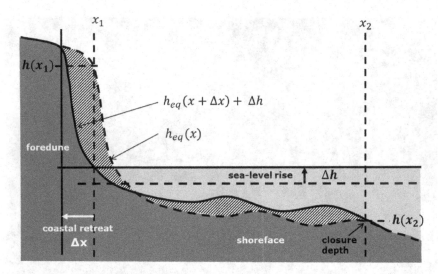

Fig. 7.20.   Coastal equilibrium profiles before and after a sea-level rise Δh, according to the Bruun hypothesis. The initial equilibrium profile $h_{eq}(x)$ (dashed line) is shifted landward over a distance Δx and upward by the same amount as the sea-level; the final equilibrium profile (solid line) is identical to the initial profile relative to sea level, landward from the closure depth.

long time already, that the long-term average coastal profile (not the position) approaches an equilibrium shape if these factors are sufficiently stable [186, 187, 465]. In this case, the coastal profile follows the long-term average average sea level [104,113,211]. When the sea level rises, sediment is eroded from the beach and the foredune and redistributed in such a way that the coastal profile is maintained relative to sea level. The rate of coastline retreat then depends not only on the coastal equilibrium profile, but also on the width of the cross-shore area subject to erosion and redistribution processes, see Fig. 7.20.

Bruun rule

Based on this insight, Bruun [113, 115] formulated in the fifties of the past century a simple rule for estimating the retreat of sandy or gravelly coasts resulting from sea-level rise. Bruun considered coastal sections that satisfy the following conditions, when averaging over a sufficiently long time interval:

(1) No exchange of sediment with the offshore shelf;
(2) No net loss or gain of sediment due to gradients in longshore transport;
(3) No inland loss of sediment by wind transport or overwash;
(4) No trend in wave climate.

The Bruun rule states that under these conditions the average coastal profile keeps its original form with respect to sea level, by redistributing sediment from the beach and the foredune over the coastal profile, see Fig. 7.20. The coastal retreat Δx corresponding to the redistribution of beach and dune sediment over the profile follows from equating the sediment volumes in the original coastal profile $h_{eq}(x)$ and the coastal profile $h_{eq}(x + \Delta x) + \Delta h$ after a sea-level rise Δh,

$$\int_{x_1}^{x_2} h_{eq}(x)dx = \int_{x_1}^{x_2} \left[\Delta h + h_{eq}(x + \Delta x)\right]dx, \qquad (7.13)$$

where x_1 and x_2 are the landward and seaward boundaries of the 'active' coastal profile, i.e., the part of the coastal profile subject to wave-induced cross-shore transport, see Fig. 7.20. Sea-level rise is a slow process, even with the present and projected increase rates (see Fig. 6.5). Considering sea-level rise over a period of, say, one century, Δx is a small quantity. We then neglect gradient contributions of $h_{eq}(x)$ in the intervals $[x_1, x_1 + \Delta x]$ and $[x_2, x_2 + \Delta x]$. The expression (7.13) then simplifies to the well-known Bruun rule

$$\Delta x \approx \Delta h \frac{x_2 - x_1}{h(x_1) - h(x_2)}. \qquad (7.14)$$

The Bruun rule has been applied in many studies of climate change impacts, but it is also much criticised [185]. Comparison of observed rates of coastal retreat with the rule (7.14) often shows large discrepancies. In fact, the conditions (1)–(4) are seldom satisfied, which limits the practical usefulness of the rule. A few examples given below illustrate reasonable agreement with the Bruun rule in cases where the conditions are approximately satisfied.

Closure depth

Condition (1) assumes that at some offshore depth h_{cl}, the influence of waves and currents on the seabed is not strong enough to move sediment in the cross-shore direction. The depth h_{cl} is called the closure depth. Under exceptional circumstances this depth can be very large. Several estimates have been proposed for the closure depth relevant for sea-level rise impact studies, depending on the time scale under consideration. For large time scales (centuries) one may use [806]

$$h_{cl} = 0.014(H_s - 0.3\sigma)T_p\sqrt{g/d}, \qquad (7.15)$$

where σ is the standard deviation of the significant wave height H_s (for Rayleigh-distributed waves $\sigma \approx H_s/2$), T_p the peak wave period and d [m] the mean seabed grain size. Values of h_{cl} typically range between 20 and 30 m. At smaller time scales (years) a smaller value, $h_{cl} = 2H_s + 11\sigma$, is more appropriate [369, 627].

During periods of fast marine transgression, the offshore distance x_2 to the closure depth can become very large. In this situation, sediments from far offshore can be transported onshore by wave action. Marine transgression is then followed by shoreface deepening and coastal progradation. At present, the slope of the shoreface is generally too large for substantial net onshore transport by waves. This could be one of the reasons for the worldwide dominance of coastal retreat over coastal accretion [81], together with the impact of human interventions, discussed in Sec. 7.7.

Longshore transport gradients

The Bruun rule assumes that longshore transport gradients can be ignored, but this is generally not the case for local profiles. Gradients in littoral drift occur for many reasons, such as coastline curvature, wave refraction at offshore sandbanks, presence of inlets and river mouths, welding of sandbanks to the coast, etc. Convergence of littoral drift contributes to advance of the coastal profile, while divergence contributes to retreat. However, the influence of longshore transport gradients on average coastline change becomes small when long coastal stretches are considered.

The shoreline shift along the 120 km long Holland coast over the past 130 years exhibits considerable longshore variation (from -0.4 m/year to $+1$ m/year). The overall average shift corresponds to a retreat of 0.18 m/year for a sea-level rise of 1.5 mm/year, which is close to the estimate obtained with (7.14). A study of coastal retreat in Chesapeake bay arrived at a similar conclusion; the overall average shoreline retreat is much better predicted by the Bruun rule than the retreat along shorter sections [731]. The current loss of coastal area along the 50,000 km sandy European coastline has been estimated at 15 km^2/year [246]. Assuming an average slope $(h_{eq}(x_1) - h_{eq}(x_2))/(x_2 - x_1)$ of $1/100$ and a sea-level rise of 2 mm/year, the Bruun rule gives a reasonably close estimate of 10 km^2/year. But for separate European coastal regions the observed coastline shifts are generally very different from the estimates according to the Bruun rule. These examples, and others [984] suggest that eliminating the influence of local gradients in longshore sediment transport by considering long coastal stretches greatly improves coastal retreat estimates with the Bruun rule.

Inland loss of sediment

The Bruun rule assumes that sand from the beach and the foredune is redistributed offshore over the active coastal profile such that the profile is shifted upward with the rising sea level. In many cases, however, the beach and the foredune lose sand also by inshore transport. Wind blows sand from the beach to the foredune and from the foredune to inland dune ridges. If the foredune is low, sand can be transported inland by overwash. The volume of inland sand transport is often considerable, exceeding 10 m^3/m/year on average.

If there is no long-term trend in the average sediment volume transported inland from the beach and the foredune, the Bruun rule can be extended to include this inland loss of sediment. If we assume that a sediment volume ΔV is transported inland per metre coastline during a time Δt, the Bruun hypothesis of coastal profile maintenance relative to the sea level is expressed as [730]

$$\int_{x_1}^{x_2} h_{eq}(x)dx = \Delta V + \int_{x_1}^{x_2} \left[\Delta h + h_{eq}(x + \Delta x)\right] dx, \quad (7.16)$$

where Δh is the sea-level rise during the time Δt and Δx the corresponding profile shift. It follows that the profile retreat Δx can be approximated by the expression

$$\Delta x \approx \frac{\Delta V}{h(x_1) - h(x_2)} + \Delta h \frac{x_2 - x_1}{h(x_1) - h(x_2)}. \quad (7.17)$$

The simplicity of this expression is a bit misleading, in the sense that it is generally difficult to obtain reliable estimates for the volume ΔV.

Cliff erosion

A major portion of the world's coast is formed by cliffs. Cliffs are subject to erosion both at the slope and at the base. Erosion processes are similar for rocky cliffs and soft cliffs (including bluffs), but the erosion rates are much higher for soft cliffs [853]. Recession rates may exceed a few metres per year.

Slope erosion involves processes such as surface ravinement, weathering, seepage and landsliding. For exposed cliffs, wave attack at the cliff base is the primary agent of cliff recession. The presence of a shore platform or a beach protecting the cliff base determines to a large extent the cliff recession rate [818]. Abrasion and removal of these protecting elements by wave action strongly accelerate cliff erosion. Unlike sandy beaches, removal of debris from the cliff base is not recovered by natural processes; cliff erosion is an irreversible process.

Cliff coasts recede even under a stable sea level. However, cliff recession is accelerated when the sea-level rises. By assuming that cliff recession depends linearly on the slope of the shore platform and the fronting beach, a simple model predicts that the cliff recession rate increases with the square-root of the rate of sea-level rise [36].

7.4.3. *Long-Term Large-Scale Coastal Behaviour*

The interdependent evolution of coastal subsystems

In the previous section we have discussed a long-term process — the response of a coastal barrier to sea-level rise — from the viewpoint of a local isolated system. By doing so we ran into a problem. It appeared that the long-term dynamics of coastal barrier retreat is influenced by the dynamics of nearby systems. We therefore had to make assumptions about sand losses by overwash and aeolian processes, assumptions about the closure depth and assumptions about sand supply or withdrawal by longshore transport. The evolution of nearby systems influences barrier retreat by affecting the sand budget of the coastal barrier system and the wave climate. For a proper description of barrier retreat we should have extended the model by including the dynamics of nearby systems. However, we then encounter new obstacles. The dynamics of nearby systems cannot be treated in isolation, because their dynamics are mutually coupled and possibly depending on the dynamics of other larger coastal systems. Moreover, the dynamics of each nearby system depends on sub-scale processes at smaller timescales; dealing simultaneously with sub-scale processes at smaller timescales in all coupled nearby systems is in general an unfeasible task.

In this book we have dealt so far with separate coastal subsystems. Therefore we have assumed steady boundary conditions with predefined values. Then we could determine equilibrium states for these subsystems: the wavelength of ripple and dunes, the wavelength of alternating bars and the meander length of tidal channels, the profile of tidal flats and marshes, the cross-section of tidal inlets, the intertidal area of lagoons and the convergence length of estuaries. However, the boundary conditions for these coastal subsystems are not fixed in reality. They evolve in mutual interdependence. At the longest timescale this evolution is steered by trends in sea level and wave climate. For understanding and simulating long-term coastal behaviour the interdependent evolution of different coastal subsystems has to be addressed.

The coastal tract

A way to deal with the interdependent evolution of coastal subsystems on large time scales is the coastal tract or cascade approach [230]. The coastal tract is defined as 'a spatially contiguous set of morphological units representative of a sediment-sharing coastal cell' [188].

The crucial step in this approach is the definition a coastal cell. The coastal cell is a morphological system that responds to trends in sea level and wave climate without being significantly influenced by other nearby systems. Determination of the boundaries of a coastal cell is primarily a matter of expert judgment, backed by field data and model estimates. The morphological units are first-order coastal subsystems with particular internal dynamics, which interact trough boundary conditions: In particular sediment fluxes, but also water levels, tide and wave parameters.

An example of a coastal cell is the set of morphological units: Tidal lagoon, ebb-tidal delta, barrier island and shoreface, bounded longitudinally by headlands, a submarine canyon or a divergence zone of longshore transport. The dynamics of each of these first-order subsystems involves second-order subsystems: Channels and tidal flats for the tidal lagoon, channels and shoals for the ebb-tidal delta, beach and dunes for the barrier island and slope, breaker bars and rip channels for the shoreface. Within each second-order subsystem one may distinguish third and fourth order subsystems such as bars, creeks, berms, dunes and ripples.

The response time scale of the higher-order subsystems to changing boundary conditions is smaller than the response time scale of the large first-order subsystems. At the time scale of sea-level rise, it may be assumed that the smaller subsystems are almost in equilibrium with their boundary conditions. By adopting equilibrium relationships for these smaller subsystems, the modelling of the coastal tract is greatly simplified.

The long-term coastal behaviour is simulated through the interactions between the first-order morphological units. These interactions are basically nonlinear and may therefore lead to complex dynamics, involving quasi-cyclic behaviour and disappearance of subsystems (for instance, barrier destruction or inlet closure). Realistic simulations require a precise specification of the initial state, in particular the stratigraphy of the substrate.

Models based on the coastal tract approach have been developed till present only for rather simple situations. An example is the ASMITA model for the impact of sea-level rise on coastal lagoons, discussed in Sec. 6.5.5. Other examples are models of shoreface evolution [189] and barrier island migration [569,813].

7.4.4. *Pattern Formation*
Inherent limitation to detailed predictions

There is no unique relationship between change in external conditions and coastal response [404]. The wave field and wave-driven currents interact with coastal morphology in a nonlinear way. From this interaction morphologic patterns arise on spatial and temporal scales which have no direct relationship with patterns in the original topography or in the offshore wave field. The first suggestions that these patterns might originate from instability of the local coastal morphology relative to hydrodynamic conditions date from the late sixties [791]. At present it is broadly accepted that many patterns in the sedimentary coastal environment arise in this way. The instability of the system involves a great sensitivity to small perturbations, which cannot be captured in deterministic models. Patterns that develop from similar initial states may be quite different, even though sharing some general characteristics. This imposes an inherent limitation on detailed predictions of the small-scale coastal evolution at longer time scales. However, statistical properties of these patterns can often be derived from the basic physical properties of the morphodynamic interaction process.

Perturbation growth and symmetry breaking

The process of pattern formation due to wave-topography interaction is in several respects similar to current-topography interaction. Through this process the system evolves without external intervention from a state with little or no structure and high symmetry to a more structured state with less symmetry. Pattern formation is inherent to the dynamics of the system and it has an irreversible character as long as external forcing conditions remain the same. The trigger for this evolution can, in principle, be infinitesimally small if the initial symmetric (i.e., unstructured) state is unstable against perturbation. The development toward a more structured and less symmetric state is therefore inevitable. If patterns are present in nature, they will automatically reappear after being destroyed by an external intervention. Inversely, if no patterns are present or emerging in the field, an initial perturbation will decay and will not start off pattern formation. Pattern formation results from the nonlinear nature of the interaction between morphology and hydrodynamics and becomes manifest as soon as an infinitesimal perturbation is applied to an unstable symmetric initial state. Analysis of initial perturbation growth is therefore an adequate method for revealing the basic nature of pattern-generating feedback mechanisms and will be used in the next sections.

Instability mechanisms

Wave-topography interaction produces seabed instability through several mechanisms. The relevant mechanisms have been mentioned in previous sections and are summarised below:

- Instability related to frictional delay of the wave-orbital momentum response in the wave boundary layer to seabed perturbations; this instability is responsible for the generation of wave-induced ripples, as described in Sec. 4.3.2.
- Instability related to the interaction between littoral drift and shoreline orientation. Under high-angle wave incidence, small perturbations of the shoreline orientation may be amplified by positive feedback from convergence/divergence of the littoral drift, leading to shoreline sandwaves or megacusps.
- Instability related to radiation stresses in the surf zone. Perturbations of the seabed topography produce gradients in radiation stresses, related to wave breaking and wave refraction; perturbation growth/decay depends on feedback from residual flow patterns produced by the radiation stresses. Positive feedback results in seabed patterns such as rip channels, crescentic bars and transverse bars.
- Instability related to swash flow on a sloping beach. Diversion of swash flow by cuspate beach structures produces a circulation pattern which provides a positive feedback to the growth of these cuspate structures.

The latter three processes will be discussed more in detail in the next section.

What is the practical use of understanding pattern formation?

Wave-topography interaction may produce more instability mechanisms, which have not yet been fully explored. Predictive modelling of these processes is only possible to a limited extent, even if the basic physics is perfectly understood. For example, it is often not possible to predict if a shoreline perturbation at a certain location and at a certain time will be positive (accretion) or negative (erosion). This raises the question: What is the practical use of better understanding pattern formation on short to medium spatial and temporal scales?

Perhaps the most important reason for studying pattern formation processes is not in forecasting, but in hindcasting, analysing and understanding coastal evolution. Trends in coastal morphology on large spatial and temporal scales generally proceed at a much slower pace than short-term local fluctuations; the former are easily masked by the latter. Interpretation of the short-term local phenomena is necessary for the detection of long-term trends from field observations. In addition, many short-term local fluctuations are indirect expressions of the large-scale morphology of the coastal system and its response to trends in forcing. For instance, the presence or absence of beach cusps provides information about wave climate and coastal profile. The observation of spatial and temporal patterns of breaker bars and rip currents provides information relevant for the optimisation of shoreline maintenance, design of more efficient nourishment schemes, safety and rescue of swimmers, etc.

Misinterpretation of coastal field observations may impair the effectiveness of coastal management strategies. A skilled coastal manager is capable to 'read' the beach and to derive qualitative expectations regarding the resilience of the coastal system and the natural response to eventual interventions.

7.5. Shoreline Dynamics

7.5.1. *Large-Scale Shoreline Stability*

Oblique wave incidence and littoral drift

In this section we will investigate how an equilibrium shoreline evolves after perturbation, when a constant uniform wave field is approaching an initially straight coast at a given non-zero angle θ_∞ at deep water.

Waves propagating obliquely onshore induce a longshore current in the surf zone, by the mechanism discussed qualitatively in Sec. 7.3. This longshore current produces a longshore sediment flux q_{br}, the littoral drift. The littoral drift depends on the angle θ between the crest lines of the incident wave in the surf zone and the depth contours following the expression,

$$q_{br} = q_0 \sin(2\theta). \tag{7.18}$$

A derivation is given in the Appendix D.3.2, together with empirical expressions for q_0. These expressions involve the wave height H_{br} at the breaker line, the average seabed slope β of the surf zone, the friction coefficient c_D and the sediment grainsize d. In the following we use x as cross-shore coordinate, positive in offshore direction, and y as longshore coordinate.

Shoreline perturbation

We assume here that the unperturbed shoreface is in equilibrium (net cross-shore transport equal to zero), with depth contours parallel to the initially straight shoreline. Due to the assumed longshore uniformity there is no transport gradient that causes erosion or accretion.

Then we perturb the shoreline with a small undulation $X(y)$; the shoreline is shifted locally a little forward or backward. The shoreface depth contours until the closure depth h_{cl} are shifted similarly, such that the cross-shore equilibrium profile is maintained, see Fig. 7.21. The angle of wave incidence relative to the depth contours now varies as a function of the longshore location. This causes a longshore variation of the littoral drift and an adjustment of the shoreline through deposition/erosion. How will the shoreline perturbation evolve; will it remain stable, will it grow or will it decay? We will examine this question in the following.

The shoreline evolution model of Pelnard-Considère

We consider the evolution of the shoreface between the shoreline and the closure depth h_{cl} (the active profile), resulting from the

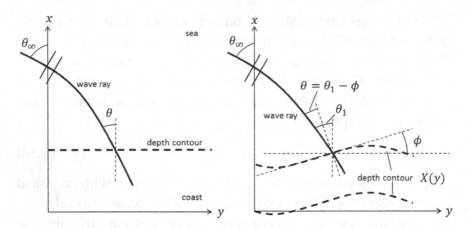

Fig. 7.21. Definition of the angles of wave incidence on a straight and on a perturbed undulating shoreline, which is rotated locally by an angle $\phi(y)$ relative to the unperturbed shoreline.

longshore variation of the littoral drift. The littoral drift varies in cross-shore direction and therefore modifies the cross-shore profile. A crucial assumption made by Pelnard-Considère is that profile adaptation by cross-shore transport takes place on a much shorter time scale than adaptation of the shoreline orientation by longshore transport. In this case the cross-shore variation of the littoral drift over the active profile can be eliminated from the sediment balance equation by integration over an appropriate timescale.

We use the notation $X(y, t)$ for the shoreline position at location y along the coast at time t; the rate of local shoreline advance or retreat, X_t, is given by the sediment balance [667]:

$$\tilde{h}_{cl} X_t + (q_{br})_y = 0. \tag{7.19}$$

Indexes t and y denote partial derivatives. The second term, representing the longshore gradient of the wave-induced littoral drift, includes a factor that takes into the account the pore-water fraction of the deposited or eroded sediment. The depth $\tilde{h}_{cl}$ is defined such that $\tilde{h}_{cl} X$ represents the volume change of the active profile (extending from $x = 0$ to $x = x_{cl}$) due to the shoreline perturbation X:

$$\tilde{h}_{cl} X = \int_0^{x_{cl}} (h_0 - h) dx,$$

where $h_0(x)$, $h(x)$ are the unperturbed and the perturbed still water depths, respectively, and x_{cl} is the location where $h_0 = h_{cl}$ [295]. Equation (7.19) is often called a 'one-line model'; it ignores cross-shore transport. More refined 'n-line' models take into account cross-shore transport by considering a number n of contour-parallel shoreface strips. The strips have given depths (h_1, h_2, ...) and widths (X_1, X_2, ...) which evolve as a result of gradients in longshore and cross-shore fluxes [47]. In the one-line model it is assumed that the cross-shore coastal profile maintains its equilibrium shape while retreating or accreting.

When approaching the coastal zone, the angle of wave incidence at deep water, θ_∞, changes by refraction to $\theta_1(y)$ in the surf zone, see Fig. 7.21. We call $\phi(y)$ the angle of the depth contour at location y with respect to the y-axis (the original shoreline) in the surf zone; $\tan \phi = X_y$. The angle of incidence θ relative to the depth contour normal is

$$\theta(y) = \theta_1(y) - \phi(y). \tag{7.20}$$

Rotating the depth contour by an angle ϕ is equivalent to rotating the deep water incident wave angle θ_∞ by an angle $-\phi$. This implies that θ decreases for increasing angles ϕ (we only consider small shoreline undulations of great wavelength, such that $\theta_\infty > |\phi|$), because waves refract more closely towards the contour normal if the angle $\theta_\infty - \phi$ is small.

Alongshore variation of the coastline orientation ϕ also affects the wave height H_{br} at the breaker line. More wave energy is propagating to portions of the breaker line making a small angle to the incident wave front than coastline portions making a high angle, see Fig. 7.24. If the depth contours in the shoreface zone (down to the line of closure depth x_{cl}) run approximately parallel to the shoreline undulation, the wave height H_{br} and the sediment flux q_0 in the breaker zone are functions of θ. The wave-induced longshore transport (7.18) takes the form [294]

$$q_{br} = q_0(\theta)\sin(2\theta). \tag{7.21}$$

Equation (7.21) can be simplified if it is assumed that the alongshore wavelength of the perturbation is much larger than the alongshore distance of significant wave ray curvature; in this case the refraction of each wave ray only depends on the local shoreline orientation ϕ, i.e., [294]

$$\theta = \theta(\phi). \tag{7.22}$$

This approximation is better satisfied for incident short sea waves than for long swell waves [295]. We further assume that the amplitude of the coastline perturbation is much smaller than its wavelength,

$$\phi \approx \tan\phi = X_y. \tag{7.23}$$

Substitution of the y-derivative of (7.21) in the sediment balance equation (7.19) then yields

$$\tilde{h}_{cl}X_t = -(q_{br})_y = -(q_{br})_\phi\phi_y = \tilde{h}_{cl}D_{PC}X_{yy}, \tag{7.24}$$

with

$$D_{PC} = -q_0\theta_\phi\chi/\tilde{h}_{cl}, \quad \chi = 2\cos(2\theta) + \frac{(q_0)_\theta}{q_0}\sin(2\theta), \tag{7.25}$$

where $(q_0)_\theta \equiv \partial q_0/\partial\theta$, $(q_{br})_\phi \equiv \partial q_{br}/\partial\phi$. With the above assumptions (validity of one-line approximation and conditions 7.22, 7.23), shoreline evolution can be described by a diffusion equation (7.24). The expression (7.25) for the diffusion coefficient D_{PC} incorporates morphodynamic coupling between the incident wave field and shoreline morphology. We recall that we only consider small shoreline undulations $X(y, t)$ such that $\theta_\phi < 0$; the sign of the diffusion coefficient D_{PC} thus depends primarily on the sign of the factor χ.

In fact, (7.25) is not a pure diffusion equation. It also contains an advection component, because D_{PC} depends on X_y. Hence, the maximum shoreline displacement (positive or negative) generally does not coincide with the perturbation crest $X_y = 0$.

Fig. 7.22. Decay of a shoreline irregularity by wave-driven longshore transport. A local shoreline perturbation is diffused along the coast if the angle of wave incidence at the breaker line is much smaller than 45°.

Decay of shoreline irregularities

For a wave field approaching the shoreline at a non-zero moderate angle ($0 < \theta_\infty < \pi/4$), the factor χ in (7.25) is close to 1 for sufficiently strong longshore transport q_0. The diffusion coefficient D_{PC} is positive, which implies that a shoreline perturbation will be damped by longshore transport gradients, see Fig. 7.22. A forward or backward shoreline displacement of length l will decay within a period of approximately $\Delta t \approx l^2/D_{PC}$.

During periods of intense wave action (storm events) the littoral drift is very strong. Under such conditions it is the dominant transport mechanism in the coastal zone, as assumed in the one-line model. The model predicts that shoreline irregularities present before the storm are straightened during a storm with oblique wave incidence. This is consistent with field observations [529, 966]. Wave-induced longshore sediment transport under moderate wave-incidence angles may thus be considered an important shoreline stabilising mechanism. However, if the wave incidence angle is very high, a different shoreline response will ensue, as we will see later.

Pocket beach

An illustration of the model is the shoreline morphology of a sandy beach confined between seaward extending headlands (cliffs or piers). As longshore transport is interrupted at the headlands, morphologic equilibrium requires zero average longshore transport everywhere along the shoreline. This implies that the shoreline will tend to adjust its orientation such that the average wave incidence after refraction is always normal to the shoreline.

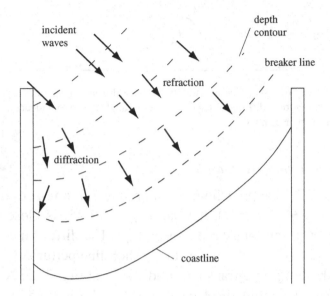

Fig. 7.23. Equilibrium shoreline for constant wave-incidence angle of a 'pocket-beach' enclosed between headlands. The shoreline is perpendicular to the direction of wave incidence. In the shadow of the headlands the shoreline is perpendicular to the locally diffracted wave rays.

Suppose we have an offshore wave field approaching the coast at a fixed angle θ relative to the headlands. The shoreline section which does not lie in the 'shadow' of the headlands will then take an orientation perpendicular to the angle of wave incidence at the breaker line (this angle is less oblique than further at sea, due to refraction). The shoreline orientation in the shadow of the headlands is determined by wave diffraction around the headlands, see Fig. 7.23. Theoretical shoreline curves of pocket-beaches based on refraction-diffraction models are generally in good agreement with observations [416,492].

7.5.2. *High-Angle Wave Instability (HAWI)*

Wave-induced longshore sediment transport has a completely different impact on the shoreline when waves approach the coast at a large angle, such that θ is close to $\pi/2$. This was first demonstrated by Ashton and Murray [33–35], who simulated the emergence of shoreline patterns under high-angle wave incidence with a cellular automata model. Here we we follow the analytical approach developed by Falqués [294].

When the angle of wave incidence is very large, the factor χ in the expression (7.25) becomes negative if $(q_0)_\theta \equiv \partial q_0/\partial \theta < 0$. This implies a negative diffusion coefficient D_{PC}. Figure 7.24 illustrates

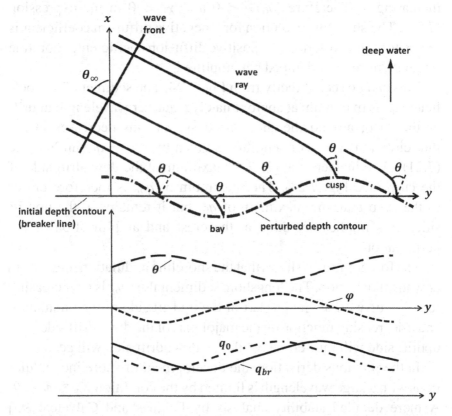

Fig. 7.24. Refraction of wave rays towards a perturbed undulating breaker line in the case of high-angle wave incidence. The strongest convergence of wave rays occurs on the updrift flank of the breaker line undulation. At this location wave energy is highest; q_0 varies in phase with ϕ. The wave incidence angle θ at the surf zone depth contour increases with decreasing contour angle ϕ and q_0 decreases with increasing θ. The diffusion coefficient D_{PC} (according to 7.25) is therefore negative for values of θ close to $\pi/4$: The perturbation is then unstable. The sediment flux $q_{br} = q_0 \sin 2\theta$ is maximum at the updrift side of the cusp. This implies also convergence of the sediment flux and accretion at the cusp.

that this is indeed the case when the shoreline is perturbed by a small undulation. Most of the incident wave energy reaches the surf zone in the interval between the bay and the cusp of the undulating shoreline. The cusp-bay interval lays in the 'shadow' of the bay-cusp interval and receives less incident wave energy [33]. This implies that q_0 is larger in the bay-cusp interval than in the cusp-bay interval. The largest value of q_0 occurs where θ is smallest; q_0 thus decreases with increasing θ. Therefore $(q_0)_\theta < 0$ and $\chi < 0$ in the expression (7.25). The shoreline evolution for a negative diffusion coefficient is opposite to the evolution of a positive diffusion coefficient: Shoreline undulations are not damped but amplified.

This can be seen directly from Fig. 7.24. The sediment flux coefficient q_0 is maximum at approximately a quarter wavelength updrift of the cusp; around the shoreline cusp, q_0 thus decreases in the flux direction. The sediment flux is given by $q_{br} = q_0 \sin(2\theta)$ (see (7.21)). The incidence angle θ is maximum at the downdrift side of the cusp (see Fig. 7.24). The maximum of q_{br} is therefore closer to the cusp than the maximum of q_0, but it remains at the updrift side. This implies accretion at the crest and amplification of the perturbation.

The foregoing also shows that the shoreline undulation migrates in downdrift direction. The longshore sediment flux q_{br} is an increasing function of y on the major part of the updrift side of the undulation and a decreasing function on the major part of the downdrift side. The updrift side will thus erode, while the downdrift side will accrete.

In the previous derivation, the development of shoreline undulations with a large wavelength is limited by the condition $X_y \approx \phi < \theta$. A more detailed stability analysis by Falques and Calvete [295] shows that negative diffusion coefficients only occur for shoreline undulations of great wavelength (of the order of 50–100 times the surf zone width or even more) and for deep-water wave incidence angles θ_∞ larger than about 45°. These authors further showed that shoreline instability is inhibited for gently sloping coasts and long swell waves. Long waves on a gently sloping coast refract almost completely to shore normal incidence, reducing the incidence angle θ to a value close to zero.

As shown in Appendix D.3.2, littoral drift also depends on alongshore gradients in wave-induced setup. This introduces a second term in the expression (7.18) for the alongshore transport, proportional to the alongshore wave-height gradient, $(H_{br})_y$ (see Eq. (D.52)). This term has a negative sign, because wave-setup in the surf zone is positively correlated with wave height. Therefore it has a damping affect on shoreline instability by wave focussing on the updrift shoreline undulation. However, the analysis of Falques and Calvete [295] shows that this additional term is in most cases small compared to the destabilising influence of high-angle wave incidence in the first term.

Simulations with a numerical morphodynamic model reproduced shoreline instability, in agreement with the previously discussed simpler diagnostic models [873]. The simulations also show that instability is reduced when taking varying wave incidence angles into account; when high wave angles occur less than about 80% of the time, no instability develops.

Spit growth

When the amplitude of the shoreline undulation grows under influence of high-angle waves ($\theta_\infty > \pi/4$), the angle θ will become at some moment equal to $\pi/4$ at the updrift flank of the undulation. The maximum of the sediment flux q_{br} then shifts in updrift direction. The downdrift migration of the cusp stops and the downdrift flank of the undulation becomes completely shielded from incident waves. Sediment trapped in the shadow zone develops into a spit at the downdrift side of the cusp. The spit grows initially in the direction of maximum sediment transport, i.e., $\theta = \pi/4$. For very oblique waves, the direction of the spit will therefore have an offshore component ('flying spit') [34]. A detailed numerical model study of spit formation under high-angle waves shows that spit growth will always occur in the unstable regime, if incident waves come from a single

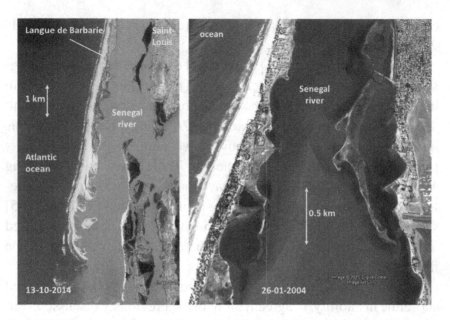

Fig. 7.25. High-angle wave instability at the mouth of the Senegal river, south of the town of Saint-Louis. Left panel: Expansion of a new spit in southward direction in 2014, eleven years after artificial breach of the Langue de Barbarie in October 2003. The spit progresses by a succession of small spits, each growing from a previous spit when it curbs onshore. Right panel: Close up of the Senegal river at the southern tip of the town of Saint-Louis in January 2004. Persistent waves from northern direction (top of the image) produce HAWI of the river shoreline. At the left river bank (at the right side on the image) two smaller cusps with flying spits are superposed on a larger moon-shaped spit of several km length. Flying spits are also visible at the opposite river bank. Images from Google Earth.

direction [474]. This study also shows that, after initial formation, the tip of the spit curves towards the trough of the undulation. When this happens, a new spit emerges from the curvature of the initial spit. At the long term, the initial spit is transformed into a succession of smaller spits, all tending towards the next trough. Some of the features described above are illustrated in Fig. 7.25.

Mega-scale shoreline features

Mega-scale shoreline cusps are remarkable features that have attracted much attention of the research community. However, not all large-scale shoreline irregularities are due to high-angle wave instability. Most shoreline cusps occurring along the world's coastline are related to rocky outcrops. Mega-scale shoreline cusps should not be confused with smaller scale shoreline irregularities, such as beach cusps and rip cells, which are related to completely different physical processes (see next section).

Three types of shoreline irregularities have been associated to high-angle wave instability (HAWI): Mega-scale capes and spits, large-scale shoreline cusps and spits and longshore sandwaves. Series of mega-scale capes and spits with wavelength of several tens to more than a hundred kilometres exist along the US Atlantic Coast (Carolina) and along the coast of the Sea of Azov (see Fig. 7.9). Ashton and Murray [33,34] demonstrated in a numerical model that the wavelength of shoreline undulations can steadily increase through the HAWI mechanism, due to absorption of smaller undulations into larger ones. This is a very slow process, but the timescale could match the assumed age of the mega-capes on the Carolina coast [35]. The same model showed that spits can grow on the down-drift side of these mega-capes under persistent almost shore-parallel incident waves, which may explain the large-scale spits along the Sea of Azov.

Large-scale shoreline cusps and spits associated with HAWI exist along the East Russian coast and in the Gulf of Finland. One of the best examples of large-scale shoreline cusps probably due to HAWI is found along the coast of Namibia [475,873], see Fig. 7.26.

Longshore sandwaves have been observed along the coasts of Holland, Denmark, Poland and along several other coasts, see Fig. 7.27 [114, 475, 499, 606, 737]. These sandwaves appear as fairly subtle shoreline undulations with wavelengths ranging from less than 1 km to several kilometres, width of several tens of metres and migration rates of a few hundred m/year to more than 1 km/year. It has been suggested that these sandwaves are triggered by HAWI. Numerical simulations based on the HAWI assumption indicate that sand wave perturbations have a maximum growth rate for wavelengths of a few kilometres or more [295]. However, other explanations for large-scale

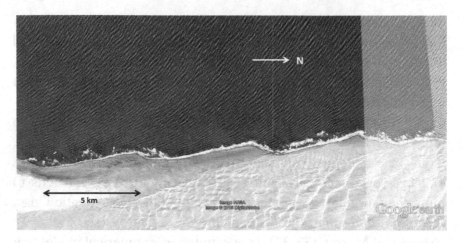

Fig. 7.26. Large-scale shoreline cusps along the coast of Namibia, Conception Bay. The prevailing wind direction is from the south. The wave crest lines illustrate the high angle wave incidence. The dune ridges on land are oriented in the same direction. Image from Google Earth.

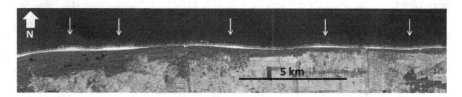

Fig. 7.27. Shoreline undulations along the Polish Baltic Sea coast near Lubiatowo, possibly related to instability caused by predominant shore-parallel westerly winds. The amplitude of the beach width undulation is about 50 m. The wavelength varies between 3 and 5 km. Image from Google Earth.

shoreline undulations have also been put forward. Shoreline undulations may result from intermittent sand supply by offshore banks welding to the coast, for instance, banks migrating over ebb-tidal deltas [202, 808]. Other possible sources are shoreface connected ridges, which are often present along storm-dominated coasts (see Sec. 4.5.5). Shore attachment of such ridges has been observed at the northern French coast [28].

Although HAWI provides a plausible explanations for certain large-scale shoreline irregularities observed in the field, direct observations of the HAWI process are still lacking.

7.5.3. *Medium-Scale Shoreline Patterns*

Shoreline straightening

In the previous section we have discussed large-scale long-term shoreline development under the influence of wave-induced longshore transport. Strong littoral drift by waves propagating onshore at moderate oblique angles, smoothes shoreline irregularities and straightens the shoreline. This happens especially under storm conditions. Under such conditions, the beach is eroded and sand is transported offshore. A shore-parallel, almost rectilinear sand bank is formed in the surf zone, where incident waves start breaking.

We have also seen that in the case of sandy shores bounded by headlands or by other obstacles to littoral drift, the shoreline will tend to an orientation approximately perpendicular to the average incident wave field, if local sediment sources are absent. Observations show that this is indeed the most common shoreline orientation [207].

Temporary patterns

However, under milder wave conditions with less oblique incidence angles the picture is very different [146, 208, 404, 492, 702, 966]. A great variety of shoreline structures develops, consisting of sequences of cusps, bars and channels. Structures with different spacing, ranging from around ten metres up to a few kilometres, may coexist and different bar orientations relative to the shoreline are possible. They are often associated with 'rip cells', circulations in the surf zone with spatially alternating shoreward and seaward currents. These features strongly depend on wave conditions and they are therefore quite variable and generally short-lived. Their emergence is often a matter of only days, or even less, and they may disappear as quickly. An illustration of a rhythmic transverse bar pattern is shown in Fig. 7.28.

Fig. 7.28. Transverse bars along the shoreline of Goeree (Dutch coast), at the tip of an accreting spit near the Haringvliet mouth, also shown in Fig. 7.17. Such bars are also called 'finger bars'. Small obliquely incident waves refracting close to the shoreline are visible near the bottom of the picture. However, it is not clear whether these conditions prevailed during the formation of the transverse bars. Photo taken at low water (mean sea level — 1 m). https://beeldbank.rws.nl, Rijkswaterstaat/Rens Jacobs.

The high longshore variability is characteristic of low-energy conditions; in the storm season most of the longshore variability is removed and the continuity of shore-parallel bars (breaker bars) is restored [18,905].

Segregation of shoreward and seaward water motions

Several mechanisms have been identified which may be responsible for the formation of alongshore cusp, bar and channel patterns. Some of these mechanisms will be discussed below. A common denominator of these mechanisms is their tendency to separate, along the coastline, zones of net shoreward water motion from zones of net seaward water motion. The cross-shore seabed profile of these zones is different. The adjustment of the seabed profile is enhanced by feedback mechanisms: The bed profile is shaped by asymmetry between shoreward-seaward motions such that spatial separation of shoreward-seaward motions is subsequently strengthened.

Shoreline instability

The observed regularity of shoreline patterns suggests that resonant wave conditions are responsible for their generation. For a long time, explanations of shoreline patterns were based on this assumption [103, 492]. It has become increasingly clear, however, that the major reason for the emergence of shoreline patterns is to be found in spontaneous symmetry breaking, caused by instability of the straight unperturbed shoreline. This means that even an infinitesimal initial disturbance may experience exponential growth through positive feedback with the flow pattern. Three types of shoreline instabilities will be discussed. First we will look at instabilities arising from interaction between the seabed in the surf zone and the incident wave field, then we will consider shoreline instabilities caused by interaction between the seabed and the longshore current and finally we will discuss bed-flow interaction in the swash zone.

Our understanding of shoreline pattern generation has been greatly improved during the past decade by the theoretical work of the Catalan morphodynamics school of Falqués and co-workers; many of the theoretical considerations exposed in the following pages are based on their work.

Radiation stresses

We have shown earlier that onshore propagating waves are strongly altered when entering the surf zone. Here we will examine the consequences of two wave modification processes: Wave breaking and wave refraction. Breaking waves lose most of their momentum in the surf zone. Part of the momentum is dissipated by friction at the seabed. Another part generates a radiation stress gradient, with a cross-shore component and a longshore component depending on the wave-incidence angle with the depth contour lines.

The wave direction in the surf zone depends on wave refraction over the seabed topography; the direction of refracted waves is given by Snell's law. In the following we will focus on situations with small wave incidence angles. Expressions for the radiation stress are derived in the Appendix, based on linear wave theory — see Sec. D.2.2. The validity of linear wave theory in the shallow surf zone may be questioned; it appears, however, that it gives a reasonable approximation from a qualitative point of view.

7.5.4. Shore-Normal Bars

In the case of perfect alongshore uniformity, the radiation stress gradient produced by breaking waves that enter the surf zone at shore-normal angles, is balanced by water-level setup at the shoreline, see Appendix D.3.1. However, a local disturbance of alongshore uniformity will immediately perturb this force balance. The perturbed force balance drives a flow which affects sediment transport, leading to adjustment of the seabed bathymetry. However, such adjustment will modify the radiation stress gradients, because wave height, wave orbital velocities and wave direction depend on bathymetry. A key question is, whether the modification of the radiation stress by bathymetric adjustment opposes or reinforces the perturbation of the force balance. In the first case the initial bathymetric disturbance will disappear; in the second case it will be amplified.

Morphodynamic feedback to a surf zone shoal

The feedback process is sketched in Fig. 7.29. It appears that the response of the seabed to a disturbance can yield a positive feedback in the case of normal wave incidence [293]. This can be understood as follows. A submerged low-crested initial shoal in the surf zone situated on an otherwise uniform sloping shoreface will enhance wave breaking and therefore strengthen locally the shoreward radiation stress gradient $S_x^{(xx)}$. Wave refraction plays an important role, because it enhances the radiation stress gradient by orienting waves towards

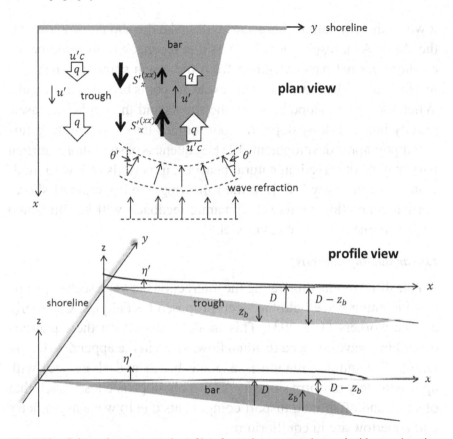

Fig. 7.29. Schematic representation of bar formation at normal wave incidence, plan view (top) and profile view (bottom). Wave breaking enhances the radiation stress $S^{(xx)}$ at the seaward edge of the bar, inducing a shoreward current u' over the bar and a shoreward sediment flux q. The suspended load is highest near the breaker line and decreases in landward direction. The resulting convergence of the shoreward sediment flux produces accretion of the bar. The opposite occurs in the trough neighbouring the bar. Wave breaking on the bar is enhanced by refraction of incident waves towards the bar.

the shoal [123, 633]. The excess radiation stress gradient will drive a current u' shoreward over the shoal; a compensating seaward return current flows along both sides of the shoal. The shoreward current over the shoal carries sediment from the outer surf zone, where the suspended sediment concentration c is high due to intense wave breaking, to the inner surf zone, where sediment concentrations are lower. The difference between high sediment input from offshore and lower sediment output shoreward will make the shoal grow towards the shore. As a result, there is a positive feedback to the emergence of shore-normal bars extending from the beach some distance into the surf zone. The return currents excavate pools between the shoals. When the seabed slope between the shoals and the pools increases, gravity induced downslope transport comes into play, limiting further topographic development [123]. Sequences of such shore-normal bars may be observed on natural beaches, in periods of low to moderate wave intensity [702]. In the following, the principle of shore-normal generation by morphodynamic feedback will be illustrated with a simple mathematical model.

Linear stability analysis

An idealised model illustrating the interaction between seabed profile and incoming wave field has been proposed by Falqués, Caballeria and co-workers [123, 293]. This model is based on the equations describing wave-induced residual flow, given in the appendix (D.34–D.36). The initial situation is a plane beach and shoreface, with approximately uniform slope $h_x \approx \beta$. This slope is chosen such that onshore and offshore transport components due to wave asymmetry and undertow are in equilibrium.

The equilibrium morphology is disturbed by a sequence of bars and troughs with longitudinal spacing $\lambda = 2\pi/k$ and amplitude f of order ϵh, $\epsilon \ll 1$. The bottom perturbation is represented by

$$z_b(x, y, t) = f(x, t) \cos ky. \qquad (7.26)$$

It is assumed, for simplicity, that the balance between cross-shore transport components due to wave asymmetry and undertow is not significantly disturbed by this small seabed perturbation.

The depth-averaged residual flow induced by the perturbation is represented by u', v' and the induced perturbation of the surface inclination by η'_x, η'_y. All quantities indicated by an acute accent refer to perturbations of the initial state. The total wave-averaged water depth $D = h + \langle \eta \rangle$ is modified by

$D' = \eta' - z_b$. Momentum dissipation of residual flow circulations induced by the perturbed radiation stresses $S_x'^{(xx)}$, $S_y'^{(xy)}$ occurs mainly through bottom friction; lateral mixing is not considered. The bottom shear stress $\vec{\tau}_b = \rho c_D \langle |\vec{u}|\vec{u} \rangle$ is linearised, $\vec{\tau}_b' \approx \rho r (2u', v')$, assuming that the residual velocities u', v' are much smaller than the maximum wave-orbital velocity U. The linear friction coefficient r depends on the amplitude of the wave-orbital velocity, $r \approx f_w U/2$, but is taken constant for simplicity. The perturbed flow is then to first order in ϵ described by the equations

$$g\eta'_x + 2ru'/D + \left(S^{(xx)}_x + S^{(xy)}_y/\rho D\right)' = 0, \qquad (7.27)$$

$$g\eta'_y + rv'/D + \left(S^{(yy)}_y + S^{(xy)}_x/\rho D\right)' = 0 \qquad (7.28)$$

$$(Du')_x + (Dv')_y = 0. \qquad (7.29)$$

Residual flow over the shoal

The modification of the radiation stress S' is due to the modification of water depth D' and to modification of the angle of wave incidence by refraction. Refraction of the incoming waves causes a deviation θ' from shore-normal incidence towards the shoals. The first order estimates follow from the expressions (D.40) given in the Appendix. After substitution of wave saturation in the surf zone ($H = \gamma_{br} D$, see (D.41)), we have

$$\left(\frac{S^{(xx)}_x}{\rho D}\right)' = \frac{g\gamma_{br}^2}{8} \left(2D'_x \left(\frac{1}{2} + \cos^2 \theta'\right) + D(\cos^2 \theta')_x\right)$$

$$\approx \frac{3g\gamma_{br}^2}{8} (\eta'_x - z_{bx}),$$

$$\left(\frac{S^{(xy)}_{y}}{\rho D}\right)' = \frac{g\gamma^2_{br}}{8}\left(2D'_y \sin\theta' \cos\theta' + D(\sin\theta'\cos\theta')_y\right)$$

$$\approx \frac{g\gamma^2_{br}}{8}D\theta'_y,$$

where we have assumed $\theta' \ll 1$. The shoal dips at the seaward flank, i.e., $z_{bx} < 0$. We may assume that the surface perturbation η' is smaller than the bottom perturbation z_b; it then follows that at the shoal $S'^{(xx)}_x > 0$. Due to refraction of the incident waves towards the shoal, $\theta'_y > 0$, or $S'^{(xy)}_y > 0$ at both sides of the shoal. It then follows from (7.27) that on the shoal $u' < 0$; hence, the flow on the shoal is shoreward.

Growth of the shoal

The implication of shoreward flow over the shoal for accretion or erosion can be derived from the sediment balance equation (3.57),

$$z_{bt} + \left\langle q^{(x)}_x \right\rangle + \left\langle q^{(y)}_y \right\rangle = 0, \tag{7.30}$$

where $\vec{q}$ is the sediment flux related to the perturbation and where $\langle \ldots \rangle$ stands for averaging over the wave period. We will assume here for the sake of simplicity that the gradient in the alongshore flux $q^{(y)}$ is much smaller than in the cross-shore flux $q^{(x)}$. The wavelength $\lambda = 2\pi/k$ of fastest bar growth depends crucially on the alongshore flux gradient; this wavelength cannot be determined in the following analysis where we only consider cross-shore velocities and fluxes. We use the formula of Bailard for suspended-load transport in a wave-dominated environment (3.84) and drop the superscript $^{(x)}$,

$$q = \alpha_s \langle |u^3|u\rangle, \quad u \approx U\cos(\omega t) + u', \tag{7.31}$$

where $U = Hc/2D$ is the wave-orbital velocity amplitude. In the idealised model we ignore, for simplicity, sand flux contributions from bottom-slope effects, wave asymmetry and undertow. We assume that the wave-orbital velocity amplitude U is much larger than the residual velocity u' of order ϵ. After averaging over the wave period we find, to first order in ϵ,

$$\langle q \rangle = 4\alpha_0 u', \quad \alpha_0 = (4/3\pi)\alpha_s U^3, \tag{7.32}$$

where α_0 stands for the amount of sediment brought in suspension by wave stirring. The bed evolution equation (7.30) can now be written

$$z_{bt} + 4(\alpha_0 u')_x = z_{bt} + 4\alpha_0 u'(\ln(\alpha_0/D))_x = 0. \tag{7.33}$$

For the second equality we have used the continuity equation, which to first order in ϵ reads: $(Du')_x = 0$.

In the case of wave saturation we have $U \propto \gamma_{br}\sqrt{D}$ and $\alpha_0/D \propto \sqrt{D}$. The offshore increase of the wave-stirring factor α_0 is therefore such that the derivative $(\ln(\alpha_0/D))_x$ is positive. Because at the shoal crest $u' < 0$ the shoal will grow $((z_b)_t > 0)$, according to (7.33). This confirms the result anticipated from the previous qualitative discussion.

Shore-normal bars

Gravity-induced down-slope transport introduces an additional term in the sediment flux $\langle \vec{q} \rangle$. This term is generally assumed proportional to the slope of the perturbation $\vec{\nabla} z_b$; it has a negative sign and, according to the Bailard transport formula (3.84), it decreases with increasing grainsize and increases strongly with increasing wave amplitude. The growth of seabed structures with a small wavelength (steep relief) is favoured by wave refraction, but counteracted by gravity-induced down-slope transport. Maximum growth of seabed perturbations occurs for two different types of bars, as shown by numerical experiments by Caballeria *et al.* [123]: Shore-normal bars and crescentic bars.

At low wave intensity and medium-coarse sand, the influence of gravity-induced down-slope transport is relatively small. The fastest growing perturbations in the linear stability analysis then correspond to shore-normal bars with a relatively small lateral spacing, less than the width of the surf zone. These bars extend over some distance from the beach into the surf zone, see Fig. 7.29. The bars are attached to the shore, due to strong wave-refraction, which induces shoreward flow over the bar (and thus accretion) from the breaker line up to the beach.

Shore normal bars may also develop at higher wave intensity. Under such conditions the strongest radiation stresses occur further offshore. Down-slope transport in this case is more important than for low wave intensity; it inhibits the development of closely spaced shore normal bars. The shore normal bars have greater lateral extension and go further offshore. Their characteristics are similar to crescentic bars; they are closely related to crescentic bars, from a morphodynamical point of view (see Figs. 7.32 and 7.33).

7.5.5. Crescentic Bars and Rip Channels

Field observations

Observations show that crescentic bars usually emerge by breaking-up of a shore parallel bar, formed after a period of strong littoral drift. Wave breaking at the bar generates water-level setup in the pool shoreward of the bar. This water-level setup is not everywhere in balance with the onshore radiation stress gradient because longshore bars are not perfectly homogeneous in longitudinal direction. A rip cell circulation develops, with onshore flow over higher parts of the bar and seaward flow at the lowest parts, where rip channels are subsequently scoured. A longshore radiation stress gradient drives a feeder current from the pool toward the rip channels. This results in a pattern of crescentic bars, approximately parallel to the shoreline. The alongshore undulating pattern of the crescentic bars is mirrored by the shoreline; shoreline bays are generally opposite the rip channels and shoreline cusps are opposite the inside horns of the crescentic bars, as in Figs. 7.32 and 7.33.

Rip-head bars

The lateral constraint on the rip channel vanishes seaward of the bar, leading to a drop of the rip current velocity [952]. Besides, wave-induced seabed stirring decreases as a function of offshore

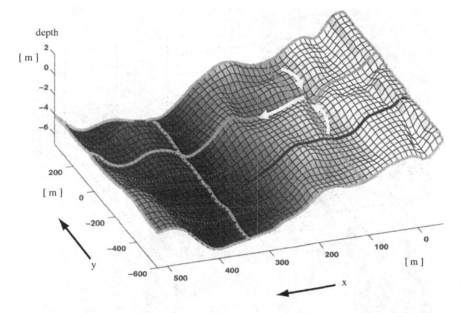

Fig. 7.30. Nearshore seabed morphology at Egmond (The Netherlands) in summer (17 August 1999), from [905]; notice the different cross-shore and longshore scales. The white arrows indicate a rip channel with feeder currents. The rip channel ends in a shoal (just landward of the longshore trough), formed by rip-head deposits. During the storm season the longshore bars are straight with little longshore structure; after the storm they are broken up in a sequence of crescentic bars.

distance. Therefore, the sediment load transported by rip currents is deposited, producing of a rip-head bar seaward of the rip channel, see Fig. 7.30. The onshore sand flux over the crescentic bar is fed with sand mobilized by increased wave stirring when approaching the bar; producing a trough at the seaward flank of the bar. Hence, a mirror bar pattern is formed opposite to the crescentic bar pattern. Observations show that over time the rip-head bar migrates shoreward and contributes to constriction of the rip channel [110]. The bathymetry shown in Fig. 7.32 corresponds to this stage of crescentic bar development.

The morphologic characteristics resulting from initial instability are qualitatively preserved upon further development of the bar-pattern (wave conditions remaining the same). This follows from

Fig. 7.31. Rip cells along the micro-tidal coast of North-Holland at low water (tidal range less than 2 m). Both the outer longshore bar and the inner intertidal bar are interrupted by rip channels at regular intervals. The rip spacing is of the order of 1 km. Wave breaking at the bars and the resulting gradients in water level drive the rip current circulation. Waves travelling shoreward through the rip channel of the outer bar break near the shoreline; the resulting water level elevation drives a counter circulation which forms a shoreline bay at the rip location and a shoreline cusp behind the outer crescentic breaker bar, see also Figs. 7.32 and 7.33.

laboratory experiments [367] and from numerical model investigations [221, 337] of rip channel development in a coastal system with pre-existing longshore bar.

Rip cell spacing

The strength of rip currents depends on the volume of water feeding the rip currents. Numerical model simulations indicate that the rip

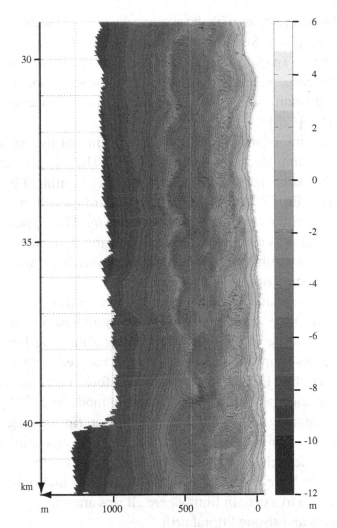

Fig. 7.32. Bathymetric map of the shoreface along the coast of North-Holland (at the right), viewed from south (bottom) to north (top) [801]. It is based on bottom soundings in cross-shore transects with 200 m alongshore spacing. An aerial photograph is shown in Fig. 7.31. The crescentic bar exhibits a pattern of shoreward pointing horns opposite seaward pointing horns of the inner intertidal bar. The cusp spacing is of the order of one to two kilometres. The width of the trough varies around a few hundred metres. At some locations the shoreward horns of the bar have almost welded to the inner bar. The alongshore resolution of the bathymetric survey is not sufficient to visualise the narrow rip channels. They are located at both sides of the shoreward horns of the crescentic bar.

channel spacing scales with the average distance of the crescentic bar to the shore [221]. Stability models for idealised situations indicate an average rip spacing $\lambda \sim 2X_{br}$ for low-energy waves and normal wave incidence, where X_{br} is the width of the surf zone. For higher waves and small oblique incidence angels the rip spacing is a few times larger [128].

The spacing of the rip channels is constrained by two opposite morphodynamic feedback processes [221]. The current velocity in the feeding area (the trough behind the bar) is limited by bottom friction (see Eq. (7.28)). Therefore the feeder flow will not increase indefinitely with increasing rip channel spacing. This imposes, on the one hand, an upper limit on the spacing of the rip channels. Strong wave setup (i.e., high waves) is needed to overcome frictional limitation in the case of large rip spacing.

On the other hand, a small spacing between rip channels causes a strong alongshore variability of the cross-shore flow, just seaward of the bar. In the case of oblique wave incidence, the longshore current advects cross-shore momentum along the bar, reducing the alongshore variability. Longshore currents therefore counteract small rip channel spacing, as indicated by numerical modelling [221].

The foregoing considerations imply that rip spacing will be relatively small in situations with small waves propagating onshore at small angles (close to shore-normal incidence). Rip spacing will be relatively large in situations with high waves and oblique wave incidence, up to a certain limit where all rips are wiped out by energetic waves and strong littoral drift.

Observed rip patterns are highly irregular in the alongshore direction and variable in time [551]. No significant correlation of rip spacing with wave height is observed in the field [334, 405, 858]. The largest spacing, up to about $10X_{br}$, occurs shortly after storm conditions, when antecedent rips are wiped out and waves are still high [894]. Later on, rip spacings become smaller, which is in line with results based on stability analysis [843].

The poor correlation of observed rip spacings with wave conditions suggests that the morphodynamic feedbacks controlling rip spacing do not impose very strong constraints. Another explanation is the time lag of morphodynamic feedback; this time lag is in many cases of the same order or longer than the time scale of wave variability [334]. This may explain the dependence of observed rip morphologies on antecedent conditions. Alongshore variability in rip spacings may be induced by alongshore variations in the initial bar height and also by alongshore variations in the cross-shore profile slope [129].

Directional spreading of incident waves

Numerical model experiments indicate that rip spacing can be influenced by directional spreading of incident waves [710]. Directional spreading in the wave spectrum produces spatial interference between incident waves from different directions. The resulting interference pattern is irregular and variable in time. The average alongshore scale of the pattern depends on the wave directional spreading and can be very large when the spread is small (several hundreds of metres for a spread of order $10°$). The pattern migrates alongshore with speed proportional to the spread in wave frequencies (slow migration requires a very small directional spread). It has been suggested that the circulation produced by the interference pattern forces the self-organised rip pattern towards these larger, irregular and variable spatial scales [710].

Rip current dynamics

The momentum advected by the feeder current toward the rip channel and vortex stretching in deep water produce a strong acceleration of the rip current [952], see Fig. 7.33. Recent rip current observations with marine radar indicate that rip currents can extent offshore over a distance of several surfzone widths [368]. This finding is in contradiction with drifter and dye release experiments in field and laboratory settings, which show that rip currents are contained within gyres

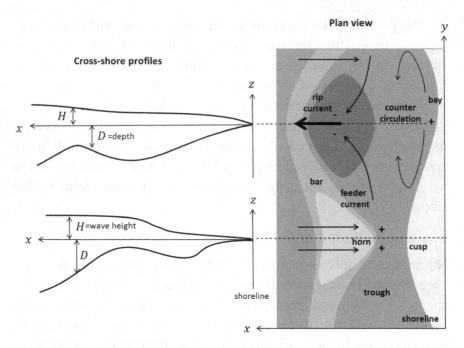

Fig. 7.33.   Schematic representation of rip circulation over a crescentic bar. Right panel: Plan view; increasing darkness = increasing depth. Left panel: Depth and wave height along two transects. Wave breaking over the horn of the crescentic bar produces a water level setup (+) in the trough behind the bar which is greater than the water level setup (−) in the rip channel. The resulting water level slope drives a circulation which feeds the rip current. Wave breaking at the shoreline behind the rip channel yields a local water level setup (+) that drives a counter circulation. This counter circulation produces a bay-cusp structure along the shoreline which mirrors the crescentic bar shape.

and do not extend far offshore [143]. Several reasons have been put forward for this limited offshore extent.

- Wave refraction: Incoming waves are refracted to the centre of the rip current. The rip current opposes wave propagation and therefore increases wave steepness; waves will break close to the rip channel. The corresponding radiation stress produces a water level set-up which deviates the rip flow such that it merges with the longshore current [977].
- Outer bar: The seaward excursion of the rip current is further reduced in situations where a second longshore bar is present

further seaward. The onshore drift that results from wave breaking at this outer bar counteracts the seaward momentum of the rip current [551].

Rip currents may take on a pulsating behaviour, in response to increased refraction and wave breaking when the rip current is strong, and decreased refraction and breaking when the rip current is weak [366]. It has been suggested that infragravity waves and shear instability may also contribute to the highly variable and pulsating character of rip flows [551]

Flash rips

The rip currents described above are related to bar-rip morphologies. However, rip currents may also occur in situations of alongshore uniform topography. These rip currents are typically stronger than the morphological rip currents; they occur episodically, their lifetime is short (a few minutes) and they appear randomly in the surf zone [147,624,831]. They are called 'transient rip currents' or 'flash rips'.

A possible mechanism is the instability of the cross-shore balance between radiation stress gradient and wave setup, caused by an incipient seaward current which locally attenuates offshore wave heights. This reduces the radiation stress gradient opposing the incipient current, which therefore will 'flash'. The flash is offset by fluctuations in the incident wave field [620].

Flash rips were observed in laboratory experiments in association with infragravity motions produced by intersecting wave trains with directional spreading [312]. Field measurements suggest that these rips are generated in the surf zone as a vortex pair within an overall rotational flow field co-existing with infragravity waves [466].

These different observations and theories illustrate that the dynamics of rip currents is not yet fully elucidated.

Multiple crescentic bars

In the previous paragraphs we have considered coasts with a single alongshore bar. However, many coasts have several parallel breaker

bars; this is especially the case for coasts which ar not swell dominated. Here we consider a coast with two longshore breaker bars, an
outer and an inner bar. Under suitable conditions — moderate waves
with small to moderate incidence angle — the outer bar will self-
organise into a crescentic pattern, by the processes described before.
This may happen for the inner bar too, but often the crescentic patterns
of the outer and inner bar are coupled. In fact, the pattern of the outer
bar tends to steer the pattern of the inner bar [144,697].

Observations show that the inner bar can respond to the outer
bar pattern in different ways. In the first place, it can self-organise
independently from the outer bar; this is often observed when the
crescentic pattern of the outer bar is not yet fully developed [739].
However, when the crescentic pattern of the outer bar is well developed, the inner bar loses its freedom.

Out-of-phase coupling

The inner bar may adopt a mirror crescentic pattern, similar to the
shoreline pattern in the case of a single breaker bar (out-of-phase
coupling). The rip channel of the inner bar opposes the inside horn
of the outer crescentic bar and the inner crescentic bar (in the form
of a bay) opposes the rip channel of the outer bar. This corresponds
to the situation shown in the Figs. 7.31 and 7.32, which is expected
to occur for strong wave breaking on the outer bar.

A similar out-of-phase inner bar pattern, mirroring the outer bar
pattern, can arise when the outer bar migrates onshore and the inner
horns of the outer bar weld to the inner bar [739]. During this process
the horn can separate from the outer bar and merge with the inner bar.
This was observed under storm conditions, when the outer crescentic
bar was straightened and migrating offshore [18].

In-phase coupling

The situation is different when waves are not high enough for substantial breaking on the outer bar. In this case, wave heights at the
inner bar are primarily influenced by refraction [144, 145]. Wave
refraction over the outer crescentic bar system directs waves away

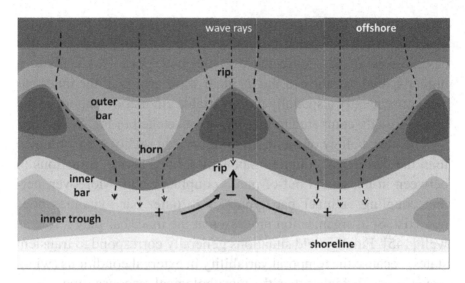

Fig. 7.34. Schematic representation of wave refraction over an outer bar (dashed lines are wave rays). Increasing darkness indicates increasing depth. The refraction focuses wave energy on the inner bar, away from the rip channel. The water level setup at the inner trough is highest (indicated by +) where wave breaking is most intense, and lowest (indicated by −) where wave breaking is weakest. The resulting circulation in the inner trough, which shapes the inner bar, is indicated by the arrows.

from the outer bar rip channel. The inner bar region experiencing strongest wave breaking is now shifted sideways. The circulation in the shoreward trough is towards the location opposite to the outer bar rip channel. The resulting inner bar crescentic pattern is in phase with the outer bar crescentic pattern, as shown in Fig. 7.34.

A rather similar, in-phase crescentic pattern of the inner bar may arise by a completely different process in a completely different situation. Energetic waves breaking on a crescentic outer bar under an angle, induce a meandering longshore current in the through between the outer and the inner bar. This meandering current erodes the inner bar, producing an undulating shape which is in phase with the shape of the outer bar [696].

When the inner bar couples in phase with the outer bar, under small or shore-normal wave incidence angles, the average spacing of rip channels at the inner bar is often smaller than the spacing of rip channels at the outer bar. A smaller spacing at the inner bar is observed when the spacing at the outer bar is large [141]. In this case, waves are less refracted over the outer bar. This leads to a higher spatial variation of the incident wave energy at the inner bar and consequently a spatially more variable circulation pattern in the inner through, compared to the situation sketched in Fig. 7.34.

The morphology of multi-bar systems is in reality more complex than the theoretical coupling modes described above. Situations in between in-phase and out-of-phase coupling occur. Moreover, perfect coupling of inner bar dynamics to outer bar dynamics does not exist. Self-organisation processes at the inner bar play a role as well [145]. Finally, field situations generally correspond to transient states, because the temporal variability in external conditions (wind, waves, tides) is higher than the morphological response time.

Importance of rip current systems

The greatest danger for sea swimmers are not sharks, but rip currents. Rip currents make each year many victims among bathers. The strongest rip currents occur when the wave energy dissipated by breaking at the longshore bar is maximum and when the rip channel depth is small; this corresponds to conditions of high waves and low tidal levels [110, 551]. Swimming against the rip current is not a good option, but only few swimmers know that the lateral extent of the current is limited. Swimming parallel to the shore gives the best chances to escape [584]. Knowledge when and where dangerous rip currents occur is important for lifeguards. Engineering interventions, both hard and soft, should be designed such that the development of dangerous rip currents be avoided.

Rip currents also carry sand offshore; however, most of the sand is deposited onshore of the closure depth. During calm periods of low swell waves this sand can be returned onshore by the dominating onshore orbital velocities of skewed waves.

A greater issue is the erosion of the shoreline facing the rip channels. Waves propagating onshore through rip channels do not break at the longshore bar, but break more closely to the shoreline. The ensuing water-level setup at the shore drives a circulation in the direction opposite to the rip cell circulation [128]. This counter circulation moves away the sediment suspended at the shore by the breaking waves, creating an inshore embayment, see Figs. 7.31 and 7.32. The shoreline cusps facing the crescentic bar, are often called 'megacusps', as distinct from the previously discussed beach cusps. However, they should not be confounded with the larger scale megacusps resulting from shoreline instability at high wave incidence angles (Sec. 7.5.1).

7.5.6. *Oblique Bars*

In Sec. 7.5.4 we have seen that shore-normal bars form under the influence of small waves approaching the shore under approximately right angles. In this section we will discuss the more frequent case of small/medium waves approaching the shore under an oblique angle which is sufficiently large for producing a significant longshore current (but smaller than the very oblique angles generating high-angle wave instability, see Sec. 7.5.1). Field observations show that in this case shoreline patterns may emerge as well, corresponding to oblique bars, see Fig. 7.35. For beaches with a longshore bar, oblique bars may occur together with crescentic bars. Oblique bars result from feedback of seabed irregularities to the longshore current and to the radiation stress; feedback to wave-refraction also plays a role. The bars run offshore from the beach, or offshore from the the first longshore breaker bar; in the latter case the longshore spacing of the bars is greater than in the former [493].

Generation mechanism

The generation mechanism of oblique bars is similar to the generation mechanism of shoreface-connected ridges, see Sec. 4.5.5. It is triggered by deflection of the longshore flow when crossing an oblique bar, see Fig. 7.36. Deflection of the longshore current $\vec{V}$ is

Fig. 7.35. Oblique, down-drift oriented bars along the shoreline of the Dutch Wadden island Terschelling. A system of crescentic bars with rip channels is visible further offshore. The tidal range is 2.5 m; the picture was taken at low water.

due to flow continuity and bottom friction torques induced by the oblique bars. Flow continuity requires $\vec{\nabla} \cdot ((D - z_b)\vec{V}) = 0$, D is the unperturbed water depth and z_b the emerging oblique bar height (we neglect the small water level perturbation η'). The shore-normal component of the deflected longshore current is either shoreward, in the case of an offshore down-drift oriented bar, or seaward, in the case of an offshore up-drift oriented bar. This shore-normal velocity component will transport sand shoreward in the first case and seaward in the second case.

Convergence of the shore-normal sand flux at the crest of the bar will produce bar growth and divergence will cause bar decay. Convergence or divergence of the sand flux depends on the cross-shore gradient of the suspended sand concentration. In the case of shoreward decreasing seabed stirring, convergence occurs for downdrift

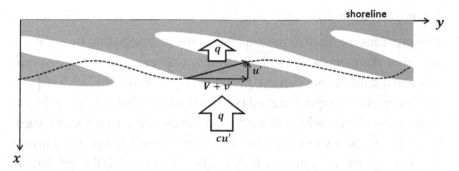

Fig. 7.36. Schematic representation of bar formation by a longshore current Bar is grey: Accretion; trough is white: Erosion. The longshore current (dotted line) is deflected landward over an offshore down-drift oriented bar; the onshore velocity component u' is the combined effect of flow continuity, bottom friction torques, vorticity conservation and radiation stress gradients. The bar will accrete if the landward flux of sediment caused by flow deflection is larger offshore than onshore, i.e., if the load of suspended sediment is larger offshore than onshore. This is normally the case in the nearshore surf zone, where a low oblique bar will consequently develop. The oblique bar is best developed in the zone where the radiation stress and the longshore current are strong due to intense wave breaking. Wave breaking at the bar is enhanced by wave refraction towards the bar.

oriented bars (onshore deflection of the longshore current). In the opposite case, convergence occurs for updrift bars (offshore deflection of the longshore current).

Feedback process for down-drift bars

The shoreward deflection of the longshore current over a down-drift oriented transverse bar causes growth of the bar if the ratio α_0/D (advected sediment load divided by depth, see (7.31)) decreases in shoreward direction. The growth of a down-drift oriented transverse bar is also stimulated by radiation stresses, resulting from wave breaking at the bar. Wave breaking at the bar is reinforced by refraction of waves towards the bar. Radiation stress gradients enhance the flow and the sand flux onto the bar. Bar accretion and onshore flow over the bar are mutually reinforcing processes. This implies that down-drift bars can emerge spontaneously by morphodynmic feedback, if hydrodynamic conditions are favourable for onshore decreasing transported sediment load.

Linear stability analysis

The previous qualitative description is illustrated below by a simple model of the morphodynamic feedback between a seabed perturbation and the longshore current, proposed by Ribas *et al.*[713,716]. An originally plane sloping seabed is perturbed by a low, offshore down-drift oriented bar, which deflects the longshore current shoreward. Semi-empirical expressions for the unperturbed longshore current $V(x)$ are given in Appendix D.3.2. The shoreward deflected flow is designated u' and the longshore perturbation of the longshore current v'. We assume that these flow velocities are much smaller than the wave-orbital velocity amplitude U (dominance of wave-induced currents over steady currents). In this case the wave-averaged sediment flux $\langle \vec{q}_s \rangle = \alpha_s \langle |\vec{u}|^3 \vec{u} \rangle$ can be approximated by

$$\langle \vec{q}_s \rangle = \alpha_0 (4u', V + v'), \quad \alpha_0 = (4/3\pi) \alpha_s U^3. \tag{7.34}$$

Here it is assumed that residual sand transport is mainly due to advection by steady currents of sand suspended from the seabed by waves.

The expression (7.34) is substituted in the sediment balance Eq. (7.30). The y-component of the longshore current perturbation, v', is eliminated from the expression of the sediment flux gradient by using the continuity equation, which to first order ($|z_b| \ll D$, $|v'|, |u'| \ll |V|$) reads

$$D(u'_x + v'_y) + u' D_x - V(z_b)_y = 0. \tag{7.35}$$

After a few manipulations we find

$$(z_b)_t + \frac{\alpha_0 V}{D}(z_b)_y = -3\alpha_0 u'_x - \alpha u' \left(\ln \left(\alpha_0^4 / D \right) \right)_x. \tag{7.36}$$

The first term on the left-hand side represents the time evolution of the bar and the second term stands for longshore migration. The terms at the right-hand side indicate growth (decay) of the bar, if they are positive (negative) for positive z_b.

At a down-current oriented bar the longshore current is deflected landwards ($u' < 0$, $u'_x < 0$ at the bar crest), due to flow continuity, frictional torque and radiation stress. If we assume that the wave stirring factor can be approximated by (7.31), then the x-derivative of the factor $\ln(\alpha_0^4/D)$ will be positive in the surf zone (in the case of saturated wave breaking it varies approximately according to D^5). Hence, the sediment balance equation (7.36) predicts growth at the bar crest $((z_b)_t > 0)$ in the surf zone. The bar will migrate downstream with a speed approximately given by $\alpha_0 V/h$. Just outside the surf zone, in the seaward extension of the bar, the flow is still shoreward, but increasing in landward direction ($u' < 0$, $u'_x > 0$). The wave-stirring factor α_0 now increases in landward direction, thus the gradient of $\ln(\alpha_0^4/D)$ is negative. This implies that $(z_b)_t < 0$: Just outside the surf zone in the extension of the bar a pool will develop. This mathematical analysis is equivalent to the earlier qualitative description of oblique bar growth.

Up-drift oblique bars

Shore-oblique bars may also develop in up-drift direction [493,715]. Up-drift oriented bars deflect the longshore current away from the coast ($u' > 0$ at the bar crest). In this situation positive feedback to bar growth may occur if the seabed stirring factor α_0/D increases shoreward in the surf zone. This may occur when spilling breakers dissipate much roller energy close to the shoreline [715]. In this case, the feedback mechanism for the development of up-drift bars mirrors the mechanism for down-drift bars.

Field measurements of the cross-shore distribution of the seabed stirring factor α_0 at the time of bar development are not available. Wave energy dissipation usually decreases onshore from the breaker line; however, the dissipation increases close to the shoreline by wave breaking on the beach face [274,298,761]. In this zone a shoreward increase of the seabed stirring factor may be expected. Other factors, such as infragravity waves and bottom ripples, also play a role in the cross-shore distribution of suspended sediment.

Observations show that under conditions with similar longshore currents transverse bars may either be present or absent and the migration of the bars is observed either in down-drift direction or in up-drift direction [493]. Simulations of oblique bar formation with numerical models exhibit a great sensitivity to the type of formulation used for sediment transport [337, 487]. Field evidence for the mechanisms underlying the formation of nearshore transverse bars is still scarce.

Tidal influence

The timescale for generation of longshore patterns, such as rip channels and shore-normal or oblique bars, is of the order of one or a few days [493]. The generation process may therefore strongly interfere with tidal motion. Especially at macrotidal beaches, the surf zone is swept forth and back in cross-shore direction. Observations show that bar-type longshore patterns develop both at microtidal and macrotidal beaches [131]; no systematic differences have been reported in literature [713]. This supports the idea that tidal motion does not produce a basic modification of the generation mechanism of longshore bar patterns. Instantaneous flow and sediment transport patterns do depend on the tidal phase, however.

7.5.7. *Beach Cusps*
Alongshore patterns in the swash zone

While the previous processes are acting mainly in the surf zone, a similar segregation between zones of dominating shoreward flow from zones with dominating seaward flow also occurs on the beach, in the swash zone. This process relates to swash, the uprush and downrush flow over the beach face produced by the final collapse of incident waves, see Sec. 7.2.2. The up- and downrush could follow identical paths, but this is often a morphodynamically unstable situation [745]. Swash circulation gives rise to the development of cuspate patterns, called beach cusps, with a longshore wavelength of typically ten to several tens of metres, see Fig. 7.37. These patterns exhibit a great regularity in form and spacing and may extend over coastal stretches of hundreds to thousands of metres.

Fig. 7.37. Sequence of beach cusps on a shingle beach situated in a small embayment of the exposed, mesotidal coast of Asturias (northern Spain). Top: Wave uprush. Bottom: After downrush. Photos by Maarten Michel.

Beach cusp morphology

The morphology of beach cusps is characterised by a sequence of steep, cuspate cross-shore bars (horns), separated by gently sloping embayments. In the developing phase of these structures the uprush

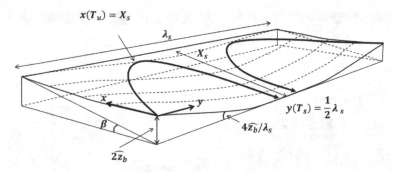

Fig. 7.38.   Schematic representation of a beach cusp and the flow pattern during the development phase. The backwash is diverted to the cusp bays and blocks the following swash towards the bay if the swash period T_s is longer than the period of the incoming wave. Uprush is then concentrated at the cusp horns. The height of the horn is $2\hat{z}_b$; β is the average beach slope, X_s the uprush length, T_u the uprush period, λ_s the cusp spacing and V the initial uprush velocity.

flow is concentrated at the cusps, while the downrush flow is concentrated in the embayments, see Fig. 7.38. This flow pattern stimulates the development of the beach-cusp pattern. The sediment-laden uprush flow nourishes the horn, especially with the coarsest sediment fraction; at the end of the uprush the swash flow has lost not only its kinetic energy, but also most of its (coarse) sediment. The downrush flow brings sediment in suspension from the embayment, which consequently erodes. The circulation from horn to embayment gets stronger while the cusp structure is building up, generating a positive feedback to beach-cusp development.

Field observations

Beach cusps are frequently observed on reflective beaches with a moderate to steep slope and coarse sediment. Beach cusp fields often develop in less than one day under conditions of calm weather with shore-normal incident waves. Beach cusp patterns even occur on beaches covered with boulders, see Fig. 2.1. Energetic swash flow and permeable substrate are clearly favourable for beach-cusp development.

The self-organising dynamics of beach-cusp formation was demonstrated in an experiment were an existing beach-cusp pattern was flattened by a bulldozer. The experiment showed that the beach-cusp pattern reappeared within less than half a day [173].

Beach-cusp circulation cells are strengthened when the swash period (period of uprush and downrush) is longer than the average period of the incoming waves. In that case the downrush flow counteracts the following swash flow uprush; the uprush is concentrated at the horn and almost entirely absent in the embayments. However, the beach-cusp pattern decays when the incident wave field becomes highly irregular and when the angle of incidence becomes oblique, generating longshore transport [573]. Beach cusps do not develop at weakly sloping, dissipative fine-grained beaches.

Cusp spacing

The observed spacing λ_s of beach cusps is variable and can be related to several swash and beach parameters: The swash excursion X_s, the wave period T, the wave height H, the beach slope β and the grain size d [171, 819]. These parameters are not independent; the observed cusp spacing can be represented by different relationships involving these parameters. The great regularity of beach cusp spacing and the strong correlation with the wave period has first directed search of the underlying physics to wave-forcing theories. As beach cusps are often observed at reflective beaches, it has been suggested that reflected waves might play a role. Incident waves which do not lose all their energy by breaking, but are partially reflected, may be trapped at the coast by refraction. Such waves are called edge waves (see section 7.3.2); their occurrence on reflective beaches has been established experimentally [433]. When such waves propagate along the coast in opposite directions they may form a standing wave. The standing wave modulates the swash flow of the broken incident wave; the alongshore wavelength of this modulation is constant and given by [363]

$$\lambda_s = (g/2m\pi)\beta T^2, \tag{7.37}$$

where $m = 1$ for synchronous edge waves and $m = 2$ for sub-harmonic edge waves. This relationship yields a fair representation of observed beach cusp spacing. Nevertheless, edge wave theories have become less credible to explain beach cusp generation, for several reasons: (1) both field observations and numerical models have shown that beach cusps may develop in the absence of edge waves [173,400,573,940] and (2) there is also no clear experimental support for the linear dependence of beach cusp spacing on beach slope [574,819].

Field observations and numerical simulations suggest a linear relationship between beach cusp spacing λ_s and swash excursion X_s [171,573,940],

$$\lambda_s = A X_s, \quad A = 1.5 - 1.7. \tag{7.38}$$

From other field analyses a somewhat larger range of cusp wavelengths was found [819], $A = 1.5 - 4$.

Swash-backwash asymmetry

Numerical investigations [172,573,940] provide evidence that beach cusps may develop as an instability inherent to the uprush-downrush process on a sloping beach. This theory is most widely accepted at present as the primary generation mechanism. The feedback mechanism was described earlier in this section. If uprush and downrush of swash flow takes place at exactly the same location, no structure will develop and sedimentation and erosion will be almost in balance. However, even a very small departure from this symmetric situation to an asymmetric situation where uprush and downrush are slightly shifted relative to each other, will produce a circulation which initiates the genesis of a cusp pattern. The cusp-pattern will further develop by positive feedback.

The model of Dean and Maurmeyer

It was first shown by Dean and Maurmeyer [214] that the cusp spacing (7.38) can be explained by the flow pattern produced by the cusp geometry. Based on observations, they assumed that for optimal morphodynamic feedback, the cusp spacing λ_s corresponds to twice the maximum lateral swash excursion $y(T_s)$, produced by deflection of the uprush from the lateral slope of the horn into the embayment during the swash period T_s.

We call $z_b(x, y)$ the cusp elevation relative to the unperturbed beach level and $2\hat{z}_b$ the height of the horn. The trajectory of a particle deflected by the cusp is shown in Fig. 7.38. Assuming constant swash depth D, the motion of the bore tip $x(t)$, $y(t)$ can be represented by the ballistic equations

$$d^2x/dt^2 + (r/D)dx/dt + 2g\beta = 0, \quad x(0) = 0,$$

$$dx/dt(0) = V,$$

$$d^2y/dt^2 + (r/D)dy/dt - 4g\hat{z}_b/\lambda_s = 0,$$

$$y(0) = dy/dt(0) = 0. \quad (7.39)$$

For simplicity we have assumed linear friction, $r \approx c_D u$, where u is a representative velocity during the swash motion. We further neglect infiltration and exfiltration during the swash motion. The swash excursion X_s follows from $dx/dt = 0$, the swash period T_s from $x(T_s) = 0$ and the wavelength λ_s from $\lambda_s/2 = y(T_s)$. For small friction the result is

$$\lambda_s = 4\sqrt{2\hat{z}_b X_s/\beta}. \quad (7.40)$$

If we assume that the average height of the horn $2\hat{z}_b$ can be expressed as a fraction ε of the average beach elevation over the uprush length, $2\hat{z}_b = \varepsilon\beta X_s$, we find

$$\lambda_s = 4\sqrt{\varepsilon}\, X_s. \quad (7.41)$$

For very strong friction the same relationship holds, with a factor 2 instead of 4. Field observations of medium-coarse sandy beaches indicate that well-developed beach cups correspond to $\lambda = (2.5\text{-}8)\sqrt{\varepsilon}\, X_s$ [573], in good agreement with the model of Dean and Maurmeyer.

Typical values of the prominence ε of full-grown beach cusps are in the range 0.1–0.3 [214, 573]. The cusp spacing predicted by the Dean-Maurmeyer model thus corresponds well with the empirical relationship (7.38).

The model of Dean and Maurmeyer suggests that the formation of beach cusps could be triggered by random initial beach features with a small prominence ε, that would progressively grow into a mature cusp morphology. There is some observational support for this scenario [573], but other observations indicate that beach cusps grow by merging of smaller cusp features [17]. At least in some cases, standing edge waves may play a role in triggering the regular spacing of beach cusps [167].

Although the model may not fully explain the initial emergence of beach cusps, it provides strong evidence for the development and stability of beach cusp fields through morphodynamic feedback between swash flow and beach morphology.

7.6. Coastal Profile Dynamics

Significance of cross-shore transport

In the previous section we have seen that wave-driven longshore transport plays a major role in the generation of shoreline patterns. However, coastal morphodynamics often depends as much, and sometimes even more, on wave-driven cross-shore transport. Although cross-shore sediment fluxes are several orders of magnitude smaller than longshore sediment fluxes, they contribute substantially to the large-scale coastal sediment balance, as they act along the full length of the coast. Cross-shore sediment transport influences not only the position of the shoreline; it determines also the average shape of the coastal profile at the shoreface. This is the result of mutual interaction, because the shape of the coastal profile influences cross-shore sediment transport.

It is generally accepted that, for an alongshore uniform coast, cross-shore sediment transport and shoreface profile mutually adjust, such that in the long term the profile fluctuates around an equilibrium state, where the net cross-shore transport is approximately constant along the profile. This constant is zero if we assume that (1) there is no cross-shore transport across the seaward boundary of the shoreface, (2) there is no net sand loss from the upper beach to the dunes, and (3) the mean sea level is constant.

7.6.1. *Shoreface Profile Models*

In Sec. 7.2.1 we have discussed coastal profile shapes observed in the field and we have presented different empirical expressions that fit large numbers of field data. In this section we will discuss in more detail several theories which may explain these profiles.

Dean's cross-shore profile model

Dean [212] was one of the first to propose a general criterion for cross-shore profile equilibrium in the surf zone, based on physical arguments. He assumed that the cross-shore profile is in equilibrium if wave energy dissipation per unit volume is everywhere the same.

Wave energy dissipation is given by the gradient dF/dx of the energy flux. We have $F = c_g E$ (see Appendix D.1.1), where c_g is the group velocity (D.14) and E the wave energy density (D.12). Assuming saturated wave breaking in the surf zone ($H = \gamma_{br} h$) and using the shallow-water approximation, it appears that F depends on the average water depth as $h^{5/2}$, according to linear wave theory.

Considering uniformity of wave energy dissipation per unit surface as equilibrium criterion, dF/dx = constant, we find $h^{5/2} \propto x$ or $h(x) \propto x^{2/5}$. However, the energy dissipation observed in laboratory experiments is not uniform across the surf zone; it is high just landward of the breaker line and decreases strongly in onshore direction [929].

The equilibrium criterion proposed by Dean is equivalent to $dF/dx \propto h(x)$, i.e., constant wave energy dissipation per unit volume. We then find the profile $h(x) \propto x^{2/3}$. Dean [212] showed that many observed surf zone profiles are fairly well represented by a function of this type.

Thermodynamic equilibrium hypothesis

Another criterion for profile equilibrium was derived by Jenkins and Inman [458], based on the assumption that equilibrium is characterised by maximum entropy production. This criterion is equivalent to finding the profile shape $h(x)$ for which wave energy dissipation is maximum. The profiles found following this criterion are different for the shoaling zone and the surf zone, because energy dissipation in the shoaling zone proceeds mainly through bottom friction and in the surf zone through wave breaking. Using linear wave theory, an analytical solution for the equilibrium profile can be found.

It appears that the shoaling profile depends critically on the closure depth h_{cl} and on the exponent n which relates the frictional energy dissipation ϵ_f to the wave-orbital velocity amplitude U, i.e., $\epsilon_f \propto U^n$. Assuming a profile shape of the form $A_{nb}(x - x_0)^{m_{nb}}$, thermodynamic equilibrium requires values of m_{nb} between 0.2 and 0.5.

The surf zone profile depends on assumptions made for the breakpoint condition and wave dissipation. Fitting the function $A_b x^{m_b}$ to the equilibrium solution yields a broad range of exponents, including $m_b=2/3$ and $m_b=2/5$. The latter value is obtained if breaking waves saturate at a constant value $H/h = \gamma_{br}$. Inman [442] showed that many profiles of swell dominated coasts can be described with an exponent $m_b = 0.365$. This exponent is consistent with thermodynamic equilibrium for the surf zone profile and the shoaling profile if the exponent n in the expression of the frictional energy dissipation is equal to 7/3.

It is not clear, however, whether the models of Dean and Jenkins/Inman are consistent with the general equilibrium requirement of uniform (x, y-independent) average sediment flux.

Cross-shore transport processes

The most usual criterion for profile equilibrium under the assumption of alongshore coastal uniformity is $\langle q(x) \rangle = 0$, where $\langle q(x) \rangle$ is the cross-shore sediment flux under average representative wave conditions.

Major contributions to cross-shore sediment transport are [729]:

- Seaward gravity-induced transport. Gravity-induced down-slope sediment transport is incorporated in the semi-empirical transport formulas (3.84 and 3.83) [45,46]. For fine sediment (ratio of settling velocity to wave-orbital velocity less than a few percent), down-slope suspension transport is relatively stronger than down-slope bedload transport [46].
- Seaward current. A net seaward flow exists in the lower part of the water column. It is a return flow compensating for the shoreward mass transport in the upper part of the water column between the wave-trough and wave-crest levels (Stokes transport). The near-bottom return flow is strongest in the breaker zone, where it is called undertow. An expression is given in the appendix (see D.26).
- Shoreward transport in the wave boundary layer. Wave propagation induces a net input of forward momentum to the viscous wave boundary layer, driving a residual forward drift in this very thin layer, called streaming. A theoretical expression is given in (3.30). However, in case of a rippled seabed, a counter drift exists at the top of the wave boundary layer, see Sec. 3.3.1.
- Shoreward transport due to wave skewness (D.19). The wave orbital motion becomes asymmetric when waves propagate into shallow water. The onshore orbital motion becomes stronger than the offshore orbital motion. This results under most circumstances in a net onshore sediment transport. However, as discussed in Secs. 3.3.1 and 3.8.2, the net sediment transport may also be directed offshore, due to settling lag effects. This happens in particular for fine sand, if the seabed is covered with steep ripples and/or if waves are steep and highly energetic.

• Shoreward transport due to acceleration asymmetry (Sec. 3.3.1). In the surf zone, wave asymmetry changes. Around the bar crest, where wave breaking is strongest, acceleration asymmetry (stronger onshore than offshore acceleration of the wave orbital motion) may overtake velocity asymmetry [275,397]. Some empirical sediment transport formulas which take these effects into account are given in Secs. 3.8.1 and 3.8.2.

Bowen's cross-shore profile model

In the following we will derive the shoreface equilibrium profile from a simple cross-shore transport model, following an approach first outlined by Bowen [104]. This model is based on expressions (3.83) and (3.84) for the wave-averaged cross-shore transport, in which different processes contributing to seaward and shoreward transport are introduced. The equilibrium requirement of zero net cross-shore transport then yields an equation from which the equilibrium profile $h(x)$ can be solved. The onshore incident wave field is simplified to a single-frequency wave, the height of which corresponds with the root mean square wave height H_{rms} and the frequency with the peak spectral frequency. Bowen considered a stronger bed-slope effect than in the suspended sediment formula (3.84), based on the auto-suspension criterion of Bagnold [45],

$$q_s = q_s^{wave}(1 - uh_x/w_s)^{-1}, \qquad (7.42)$$

where w_s is the sediment fall velocity and h_x the seabed slope. This criterion is physically realistic only if $|uh_x|/w_s \ll 1$. The analysis is restricted to situations where the amplitude U_1 of the first-order wave orbital velocity is much larger than other contributions u', for instance, from streaming or from higher-order wave contributions. With these assumptions, the transport formulas can be approximated by:

$$\langle q_s \rangle = \frac{\varepsilon_s \rho c_D}{g w_s \Delta \rho}\left(4\langle u'u_1{}^2|u_1|\rangle + \frac{h_x}{w_s}\langle u_1{}^4|u_1|\rangle\right), \qquad (7.43)$$

$$\langle q_b \rangle = \frac{\varepsilon_b \rho c_D}{g \Delta \rho \tan \varphi_r} \left(3 \langle u' u_1{}^2 \rangle + \frac{h_x}{\tan \varphi_r} \langle u_1{}^2 |u_1| \rangle \right). \quad (7.44)$$

The parameters ε_s, ε_b have order of magnitude 0.1 and 0.02, respectively. The friction coefficient c_D is of order 0.01 and $\delta\rho/\rho$ is the relative excess density of sand grains; $\tan \varphi_r \sim 0.13$ is the angle of repose for 0.3 mm sand grains (steepest slope for which no avalanching occurs).

Equilibrium for suspension transport in the surf zone

The following contributions to the cross-shore velocity in the surf zone are considered in the model:

$$u = u_1 + u', \quad u_1 = -U_1 \cos(kx + \omega t),$$
$$u' = -u_s + U_0 - U_2 \cos(2kx + 2\omega t), \quad (7.45)$$

where $u_s = 3kU_1^2/4\omega$ is the landward streaming velocity in the wave boundary layer (3.30), $U_0 = gkH^2/8h\omega$ is a seaward undertow current (D.26), $U_1 = \omega H/2 \sinh kh$ is the wave orbital velocity and $U_2 = 3kU_1^2/4\omega \sinh^2 kh$ is the second order Stokes contribution representing wave skewness (D.19). The wave number $k = \omega/c$ is given by the dispersion relation (D.10). Suspension transport is assumed to dominate over bedload transport in the surf zone. Substitution in (7.43) yields

$$\langle q_s \rangle = -\frac{\varepsilon_s c_D U_1^3}{g w_s \Delta\rho/\rho} \left[\frac{16}{3\pi}(u_s - U_0) + \frac{16}{5\pi} U_2 - \frac{h_x}{4w_s} U_1^2 \right]. \quad (7.46)$$

Equilibrium requires vanishing of the term between square brackets, which yields an expression for the seabed slope h_x. After substitution of the expressions for u_s, U_0, U_1, U_2 we find

$$h_x = \frac{16k w_s}{\pi \omega} \left[1 - \frac{2g \sinh^2 kh}{3\omega^2 h} + \frac{3}{5 \sinh^2 kh} \right]. \quad (7.47)$$

The first term between square brackets represents the contribution of undertow, the second term the contribution of streaming and the third term the contribution of wave skewness.

The equilibrium condition (7.47) depends on the fall velocity w_s. If we have equilibrium for sediment with a given median grain size, then we have no equilibrium for finer or coarser sediment (fall velocity w_s lower or higher than the average). According to (7.46) the coarse sediment fraction will be transported landward ($\langle q_s \rangle < 0$). The fine sediment fraction, by contrast, will be transported seaward. This shows that sediment sorting must take place on the shoreface, with coarser sediment higher in the profile and finer sediment at greater depth, as observed in the field. This result is due to the dependence of the suspended load on fall velocity in the transport formula (7.43).

The model also predicts an increase in profile steepness with increasing wave period (through the factor $1/\omega$); this feature is also in agreement with field observations [492].

The depth profile $h(x)$ can be obtained by numerical integration of (7.47). The result is shown in Fig. 7.39. Seaward from $x = 800\,\mathrm{m}$, the seaward current U_0 (undertow) becomes the dominating term in (7.47). The profile slope is then reversed; this might be interpreted as a breaker bar location. However, the slope does not become positive further offshore, illustrating the limitations of the model.

A local fit of a power law, $h \propto x^{m_b}$, yields the exponent m_b shown in the figure. Close to the shoreline, the value of m_b is close 2/5; it decreases further offshore in the surf zone.

Comparison with observed coastal profiles

The analytical model of Bowen has the merit of providing direct insight in the contribution of different processes to profile equilibrium. In this respect it may be helpful for interpreting observed equilibrium profiles in the field. However, as Bowen noted, it is probably too simple for a detailed comparison.

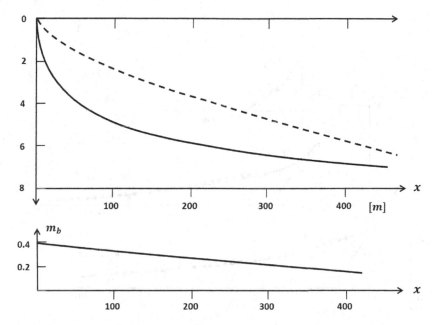

Fig. 7.39. Top panel: Solid line: Equilibrium surf zone profile $h(x)$ derived by integrating (7.47) for medium sand $w_s=0.03$ m/s and wave period $T=6$s. For $x > 800$ m the slope of the profile is reversed (not shown in the figure). The dashed line is the surf zone profile according to the Dean formula (7.1). Bottom panel. The exponent m_b, derived from a local fit $h \propto x^{m_b}$.

The coastal profiles investigated by Inman [442] can be represented by a power law with exponent $m_b = 2/5$. The analytical model has a lower exponent, except close to the shoreline. The shoaling zone is not represented; the model gives a negative slope in this zone.

Dean [212] and Bernabeu *et al.* [70] found a larger exponent, $m_b = 2/3$, for the beach profiles they examined. These profiles have a smaller steepness near the shoreline and a greater steepness going offshore.

Figure 7.40 shows a few typical coastal profiles for the micro/mesotidal Dutch coast. These profiles are averages over a period of several decades; this removes fluctuations related to migrating breaker bars. The surf zone slope exponent is similar to the exponent

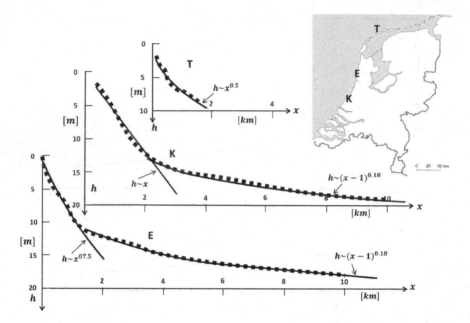

Fig. 7.40. Dotted lines: Coastal profiles averaged over a period of several decades along the Dutch coast at K = Katwijk, E = Egmond, T = Terschelling [905, 948]. At Katwijk and Egmond the profiles extend up to the closure depth (7.15). At any moment several breaker bars are present in these profiles at depths between 0 and 10 m. The bars are not visible in the long-term averaged profile due to their cross-shore migration. The solid lines represent parametric power-law fits, $A_b x^{m_b}$ in the breaker zone and $A_{nb}(x - x_0)^{m_{nb}}$ in the shoaling zone. For this micro-mesotidal dissipative coast ($\Omega > 6$) the exponent m_b is much larger than m_{nb}.

found by Dean and Bernabeu and much larger than the model exponent.

The model of Bowen does not include swash-induced transport near the shoreline. Infragravity motions are also absent from the model. The model may therefore be not representative for most field situations, especially near the shoreline.

Limitations of the model

The model of Bowen has several other shortcomings. Probably the most fundamental one is the representation of the wave field by a single monochromatic wave of constant amplitude and period. Even by choosing representative values for the wave amplitude and period, this simplified wave field does not interact with the coastal profile in the same way as a real wave field with time varying wave spectrum. The time varying wave spectrum smoothes gradients in the average cross-shore transport. Water-level variations due to tides and wind-setup have the same effect. The neglect of variability in waves and water levels leads to steeper seabed slopes than than occurring in nature. This holds in particular for coasts where tides are strong and where waves are mainly generated by local wind fields.

Other major shortcomings of wave-averaged transport models, such as (7.43) and (7.44), are the neglect of acceleration asymmetry in the wave-orbital motion and the assumption that sediment load is instantaneously coupled to the wave-orbital velocity. The influence of these processes to net sediment transport was discussed in Sec. 3.3.1, see also Fig. 3.5. The neglect of acceleration asymmetry can lead to underestimation of wave-induced shoreward transport. The coupling of sediment load to the instantaneous wave motion ignores phase lag effects, which can produce a reversal of the net sediment transport. Wave-averaged models, which are most often used in practice, can incorporate these effects only by introducing additional empirical terms in the transport formulas. The formulas (3.78), discussed in Sec. 3.8.2, are an example.

An alternative approach is the determination of wave asymmetry contributions from field data. Results show that contributions of acceleration asymmetry and phase lag between wave motion and suspended load cannot be ignored. Replacing the U_2 term in (7.46) by correlations between wave-orbital velocity and sediment concentration derived from field data, leads to a flatter profile near the shoreline [677], more in line with observations.

It should be concluded that practical applicability of simple analytical models is limited. More sophisticated process-based numerical models are needed for representing sediment transport in the surf zone. A comparison with observations remains difficult, however, because of the strong temporal and longshore variability of nearshore morphology. It is generally not clear what are conditions representative for equilibrium and when/if they are realised in the field.

Equilibrium situations can be realised in laboratory experiments. Much of our present insight in nearshore processes is based on such experiments. But translation of this insight to field situations is not obvious, due to scale, boundary effects and to natural variability in the field.

Shoaling zone

Like for the surf zone, Bowen's model can be applied for the shoaling zone. This is interesting, because it is difficult to analyse the equilibrium state of the shoaling zone with sophisticated numerical models. Main reasons are the weakness of cross-shore transport processes in the shoaling zone and the resulting long adaptation time scale [677].

Assuming that sand transport in the shoaling zone proceeds mainly through bedload, we use the formula (7.44). If we consider wave skewness (U_2) and streaming (u_s) in the wave boundary layer as major onshore transport mechanisms and gravity-driven down-slope transport as major offshore transport mechanism, we find, after substitution of these contributions in (7.44),

$$\langle q_b \rangle = -\frac{\varepsilon_b c_D U_1^2}{g \tan \varphi_r \Delta \rho / \rho} \left[3 \left(\frac{u_s}{2} + \frac{U_2}{4} \right) - \frac{h_x}{\tan \varphi_r} \frac{4 U_1}{3\pi} \right]. \quad (7.48)$$

At equilibrium the term between brackets vanishes. After substitution of the expressions for u_s, U_2, U_1 we find

$$h_x = \frac{27\pi \tan \varphi_r}{128} \frac{kH}{\sinh kh} \left(2 + \frac{1}{\sinh^2 kh} \right). \quad (7.49)$$

The numerically integrated profile $h(x)$ is shown in Fig. 7.41, where we have made the arbitrary choice of $x = 200$ m and $h = 6$ m as landward boundary of the shoaling zone. The profile slope is unrealistically steep. Possible causes of this discrepancy are similar to the

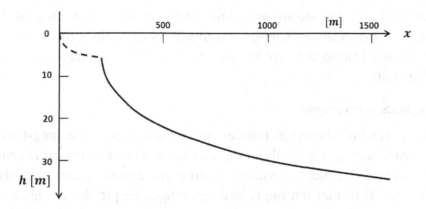

Fig. 7.41. Solid line: Equilibrium shoaling zone profile $h(x)$ derived by integrating (7.49) for fine sand $w_s=0.03$ m/s, wave period $T=6$ s and $H=1.5$ m. The landward boundary of the shoaling zone is chosen at $x=200$ m and $h=6$ m. The dashed line is the Bowen profile for the surf zone (from Fig. 7.39).

shortcomings discussed in the previous paragraph. We have neglected the smoothing affect of wave and water level variability on transport gradients. We have ignored phase lag effects that may reverse net wave-induced transport. This transport reversal has been observed in laboratory experiments for rippled seabeds and for sheet flow conditions [243, 356, 634, 717].

In Sec. 7.3 the role of infragravity waves bound to wave groups has been discussed. From theoretical arguments and from analysis of field data it appears that bound infragravity waves contribute to offshore transport [530, 571, 734, 772]. These effects, which reduce onshore transport in the shoaling zone, are not included in the model.

Sediment transport models at the intra-wave timescale can take these effects into account, see for instance, [449]. However, these models are still too demanding, from a computational point of view, for the simulation of long-term adaptation to a morphodynamic equilibrium.

It is not yet possible, for instance, to extend simulations to time intervals that include rare extreme conditions. Sediment resuspension at deeper water is probably raised by several orders of magnitude under exceptional storm conditions. Such conditions may have a substantial influence on the long-term equilibrium profile of the shoaling zone and lead to flat profiles, as occurring at the Dutch coast, see Fig. 7.40.

Instantaneous profiles

Instantaneous shoreface profiles are different from the long-term average profile, especially in the surf zone. The shoaling-zone profile is less variable; instantaneous and average profiles are generally similar. This does not imply that shoaling-zone profiles are always close to equilibrium; the adaptation to changes in external conditions (extreme storms, large-scale human interventions, sea-level rise) is slow.

The surf-zone profile can change substantially in a short time under varying external conditions. We may thus expect that the long-term average profile is approximately in equilibrium with long-term trends in external conditions, i.e., long-term average cross-shore sediment transport approximately constant over the profile. However, there is usually no correspondence between the external conditions (sea level, wave climate) prevailing at a given moment and the long-term average profile.

The response of the surf-zone profile to storms can be quite fast, of the order of days [741,893]. However, storm duration is generally too short for the realisation of an equilibrium profile. Fair weather periods have a much longer duration. However, morphologic activity is reduced under these conditions. Observed coastal profiles thus correspond in general to transient coastal states.

7.6.2. Longshore Breaker Bars

Field observations provide strong evidence that longshore bars arise when energetic waves break on the shoreface[894], as illustrated in Fig. 7.2. Longshore bars are typical features of the coastal profile in the storm season; under calm conditions longshore bars may be absent or hardly developed. The idea that wave breaking is the primary bar generation mechanism, is known as 'break-point hypothesis'. It appears that, once a bar is formed, it concentrates wave breaking at the bar. For this reason longshore bars are called breaker bars. Under calm conditions, longshore bars tend to break up in crescentic bars and to move slowly onshore. However, bar straightening may already occur under moderate wave conditions, in case of oblique wave incidence [184].

Break-point theory for bar formation

Seaward of the breaking point, the shape of the coastal profile is the result of a balance of wave-induced sediment transport and gravity-induced downslope transport. Net wave-induced transport in the shoaling zone is mainly onshore, due to velocity and acceleration skewness, as discussed earlier.

Landward of the break point, the net wave-induced transport may be reversed from onshore to offshore [233]. This happens because wave breaking modifies the velocity profile of the offshore return flow (the undertow), which compensates for the net onshore water transport higher in the vertical, between wave trough and wave crest. A large part of the energy dissipated by wave breaking is initially stored in a roller, near the surface (see Fig. 7.13). The turbulent decay of roller energy produces a shoreward shear stress on the underlying water column [729]. This shoreward stress pushes the undertow current down the water column. The undertow peaks close to the seabed [247, 450, 710] and therefore produces a strong off-shore directed shear stress at the seabed. Offshore sediment transport by undertow is further enhanced by increased suspended sediment concentrations in the zone where the energy of breaking waves is converted to turbulent motion. Observations show that plunging breakers produce a strong vortex motion at the plunging point, which creates a sand fountain when touching the seabed [983]. The undertow transport opposes wave-induced onshore transport; this results in convergence of sediment transport near the breaking point.

The transport convergence zone starts at the seaward limit of breaker-induced undertow. Bar growth will occur landward of this seaward limit, by stronger seabed stirring and stronger near-bottom return flow at the onshore side of the bar than at the offshore side [275, 842, 929]. Numerical models show that the simulation of the undertow current is crucial for predicting correctly the bar location and bar shape. The cross-shore location of the maximum of the undertow current is found to depend on the time scale at which roller energy is dissipated. Cross-shore diffusion of suspended sediment must be taken into account for the simulation of realistic bar profiles. In wave-averaged mathematical models, these effects are incorporated through empirical formulations [247, 256, 711]. It appears that these artefacts are not needed if the wave breaking process is modelled in detail at the intra-wave time scale [450].

The wave incidence angle also plays a role in breaker bar formation. The alongshore current induced by breaking of obliquely incident waves is strongest at the shoreward side of the bar. The turbulent stresses generated by the alongshore current decrease the near-bed undertow velocity, but enhance the resuspension of sediment. Model simulations indicate that the latter effect is more important than the former; the alongshore current strengthens the offshore sediment transport by undertow and the convergence of sediment transport at the bar crest [247, 924].

We have seen earlier (see Sec. 3.8.2) that under energetic skewed waves (stronger onshore than offshore near-bottom orbital motion), settling lag effects may reverse the direction of net sediment transport. Laboratory experiments revealed that this may occur at the landward flank of the bar [356]. In these experiments, the settling lag effect contributed to offshore directed transport and to bar growth, even more than undertow.

Bar migration, growth and decay

Although longshore bars do not emerge as free instabilities (according to the breakpoint mechanism), their behaviour depends strongly on morphodynamic feedback. The transformation of waves, when breaking at the bar, generates net sediment fluxes which determine bar migration and bar growth or decay. By analysing a large data set of surfzone profiles, Plant *et al.* [678] found that morphodynamic feedback depends crucially on the wave-breaking intensity parameter

$$\Gamma(X_{bar}) = H(X_{bar})/h(X_{bar}), \qquad (7.50)$$

where H is the root-mean-square wave height, h the mean water depth, both taken at the bar crest location X_{bar}. The generally observed behaviour of longshore bars is migration towards a surf zone location $x = X_{bar}^{eq}$ where waves start breaking. At this location $\Gamma(X_{bar}^{eq}) = \gamma_{br}$, see Fig. 7.42. The value of γ_{br} depends on the shoreface profile (the slope, in particular) and the wave characteristics (the wave steepness H/L, in particular), see Appendix section D.2.1. Observed values of γ_{br} range between 0.3 and 1.2 [138]. The equilibrium location X_{bar}^{eq} depends on the shoreface profile and on the wave height H. If the wave height increases, the equilibrium

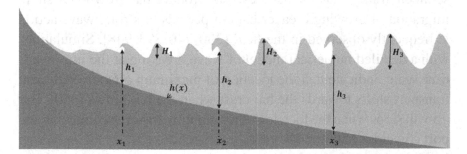

Fig. 7.42. Shoreface profile with incident waves of different heights. For larger waves the breakpoint is situated further offshore than for smaller waves. The breaker locations are $x_1 < x_2 < x_3$ for wave heights of resp. $H_1 < H_2 < H_3$. If the breaker criterion is independent of depth and wavelength, we have $\gamma_{br} = H_1/h_1 = H_2/h_2 = H_3/h_3$.

location X_{bar}^{eq} moves offshore and if wave height decreases, it moves onshore.

The migration direction of the bar depends on its location X_{bar} at a given time t. If the bar is close to its equilibrium position at time t, we get $\Gamma(X_{bar}) > \gamma_{br}$ when the wave height increases. In this case the bar moves offshore towards greater depths. However, if the bar at time t is situated seaward of its equilibrium position, we have $\Gamma(X_{bar}) < \gamma_{br}$, with onshore bar migration. When the wave height increases, this may still be the case; the bar will then continue to move onshore. In field situations, the equilibrium location may never been reached, because of continuously changing wave conditions [679].

Feedback processes

The physical interpretation follows from the dynamics according to the break-point hypothesis. We consider first the situation where the bar is located at time t landward of the equilibrium location X_{bar}^{eq}, i.e., $\Gamma(X_{bar}) > \gamma_{br}$. This may occur in periods of high waves (storms). The bar will be subject to intense wave breaking, which produces strong undertow (and eventually offshore transport due to settling lag under skewed waves). This will drive the bar offshore towards the equilibrium location X_{bar}^{eq} where $\Gamma(X_{bar}) = \gamma_{br}$. Convergence of sediment transport at the bar crest will produce bar growth. Offshore migration of growing breaker bars in periods of strong wave action is frequently observed in the field [330, 740, 893, 948]. Simulations with a detailed numerical model of wave breaking at the intra-wave time scale, indicate that the location of maximum offshore sediment transport shifts towards the bar crest when the bar grows [450]. Bar growth stops when the location of maximum offshore sediment transport coincides with the bar crest.

If the bar is located seaward of the equilibrium location X_{bar}^{eq}, we have $\Gamma(X_{bar}) < \gamma_{br}$. This may occur during calm post-storm periods. In this case the bar will move (slowly) onshore towards the equilibrium location. In situations where waves approach the coast under an angle and break on the bar, undertow and seabed stirring at the landward bar flank produce sediment transport convergence at the bar crest. This stimulates bar growth during onshore migration [256, 924]. However, in Sec. 7.5.5 we have seen that under small to moderate wave incident angles, longshore bars break up in crescentic bars, interrupted by rip channels. The crescentic bars also move onshore. This onshore migration is mainly due to net onshore flow over the bars, which is driven by radiation stresses associated with wave breaking on the bars [247, 679].

Wave breaking on the bar hardly occurs if the bar is located far seaward of the equilibrium location. In this case landward increase of wave skewness and asymmetry entail divergence of sediment transport at the bar crest. Observations show that during periods of low wave activity breaker bars generally decay while migrating shoreward [37]. Positive feedback to bar decay is due to increase of $h(X_{bar})$, decrease of Γ, stronger divergence of sediment transport and further decay.

Bar formation by standing waves

For a long time, cross-shore standing waves were considered major players in the formation of longshore bars [205, 386, 492]. This idea was supported by flume experiments, where bar patterns arise in response to standing reflected waves [649]. Observations of longshore bars at reflective beaches and observations of regularly spaced multiple bars also pointed to the standing wave formation mechanism. It was shown theoretically that standing wave patterns generate residual circulations capable of transporting sediment towards the antinodes. However, the spacing of longshore bars on natural beaches is generally too large to match the wavelength of reflected incident waves. Moreover, multiple bars also occur on dissipative beaches. It seemed therefore more plausible that bar formation is related to the standing wave pattern generated by reflected infragravity waves [774].

Under conditions of oblique wave incidence, reflected infragravity waves may become trapped in the nearshore zone and form a longshore pattern of standing edge waves. This longshore wave pattern was thought to be responsible for breaking up longshore bars into a pattern of crescentic bars [103].

When more field evidence on bar behaviour became available, in particular long-term records of video camera observations, it appeared that standing wave theories could not provide satisfactory explanations for the observed response of longshore bars to changing wave conditions. The observed behaviour follows more naturally from other theories, based on the break-point hypothesis. Standing or partially standing wave patterns do occur in the surf zone. It is probable that they influence the formation and behaviour of longshore bars; however, this influence is more subtle than thought in the past and not yet fully understood.

Bar formation by diffusion

Black *et al.* [83] showed that undertow is not the only mechanism for explaining bar formation. A bar-trough structure results in their model from diffusive redistribution of sand. The suspended sand concentration peaks at the outer surf zone boundary, where wave-orbital velocities are strongest. At this location a trough is scoured in the seabed. The suspended sand is diffused by oscillating infragravity currents and deposited at the edges of the trough. Wave-induced stirring strongly decreases in the interior of the surf zone. The gradient of the suspended sediment concentration is therefore stronger at the landward side of the trough than at the seaward side. Most diffusive transport thus takes place in onshore direction. This leads to onshore bar migration, without any contribution of wave skewness or asymmetry. For offshore transport, an undertow current has to be introduced in the model.

The diffusion model is not a comprehensive alternative for the break-point model. For instance, the diffusive bar is not sustained at long term without undertow current. The model illustrates, however, that diffusive processes can play a non-negligible role in bar formation and migration.

Multiple longshore bars

At many dissipative sandy coasts there is not just a single longshore surf zone bar, but a sequence of two, three or more shore-parallel bars. The cross-shore spacing of these bars is typically of the order of 50 to 300 metres. As indicated above, the widespread occurrence of parallel bar sequences first led to the hypothesis that their formation is related to standing reflected waves.

Today, a more common explanation is based on multiple wave breaking. According to this theory, the outer bar is associated with the initial breaking of incident waves. Waves are reformed afterwards in the trough behind the bar and break a second time, closer to the shore. If the coastal profile has a gentle slope, a third or fourth breaking may occur. In the case of strong tides, waves break on the outer bar only at low tide. At high tide, initial breaking occurs at the next inner bar. At storm dominated coasts, initial wave breaking occurs also at the first inner bar under moderate wave conditions and at the outer bar only under less frequent high wave conditions.

Outer bar control

A wealth of information on the behaviour of longshore bars can be derived from the images of video cameras, which have been installed at numerous beaches worldwide [406,893]. These observations show that the behaviour of inner bars is influenced by the behaviour of the outer bar [740]. According to the break-point theory, the outer bar develops in response to wave breaking and resulting near-bottom offshore undertow. Growth of the outer bar entails reduction of wave heights onshore. Wave breaking is reduced over some distance landward of the outer bar; in this zone bar formation is inhibited. Seaward migration of the outer bar, followed by its decay, leads to enhancement of wave breaking at the inner surf zone. This stimulates the development and seaward migration of inner bars [948]. This is

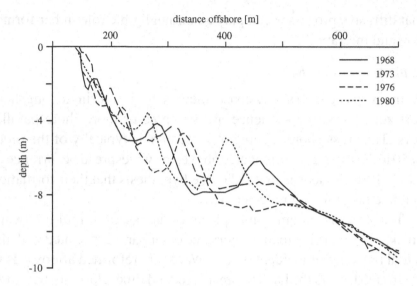

Fig. 7.43. Coastal profiles at Egmond (Netherlands, tidal range ≈2 m) in the period 1968–1980. In the surf zone (depths smaller than approximately 7 m) two or three subtidal bars are present in the coastal profile. A 15-year bar cycle can be observed, during which bars migrate on average offshore. The outer bar decays when it migrates beyond the 7 m depth contour; the 1968 outer bar has faded 1980. Offshore migration took the 1968 first inner bar in 1980 almost to the location of the 1968 outer bar. The second 1968 inner bar migrated in 1980 towards the 1968 location of the first inner bar. Redrawn from [905].

illustrated in Figs. 7.43 and 7.44 for the Dutch coast (Egmond) and for the Australian coast (Gold Coast).

7.7. Coastal Erosion

There are two types of coastal erosion. One type is erosion related to structural sand loss from the active coastal profile. The other type is erosion related to temporal shoreline fluctuations. Both types of erosion may occur simultaneously, which complicates the distinction in practice.

The first type refers to adaptation of the beach state to long-term trends and irreversible changes in external conditions. The case of relative sea-level rise is discussed before in Sec. 7.4.1. Other important causes of structural shoreline retreat are long-term trends in

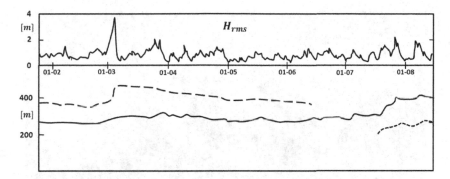

Fig. 7.44. Bar migration at the Gold Coast (Australia) in the period February–August 2006. The outer bar (bar crest = long-dashed line) moves offshore during a severe storm in the first week of March. During the calm weather period April–June, the outer bar moves onshore and decays. The decay of the outer bar triggers the offshore migration of the inner bar (solid line) in July, in a period of increasing wave action. The inner bar becomes outer bar, while a new inner bar (short-dashed line) is formed. Redrawn after [740].

wave climate, offshore transport of sand (especially the finer fraction) beyond the active coastal zone and trapping of sand in inlets and submarine canyons by littoral drift. The most serious impact on coastal stability is due to human interventions [376]. At many places the hydrodynamic conditions in the coastal zone have been changed and sediment supply has been altered, through sand mining, river damming and coastline fixation. One of the most frequent causes of structural coastal erosion due to human interventions is the interruption of longshore transport; this will be discussed later in this section.

The second type of coastal erosion is a temporary phenomenon with a reversible character. In the following paragraphs we will discuss this type of coastal erosion in more detail.

7.7.1. *Shoreline Fluctuations*

The shoreline position, defined here as the HW mark, is one of the most dynamic features of coastal systems. Shoreline fluctuations may have dramatic consequences at places where settlements are built close to the shore, see Fig. 7.45.

Fig. 7.45.   Shoreline erosion at Saint-Louis, Senegal, March 2015. High tide levels coinci-
dent with energetic ocean waves have eroded the beach and destroyed fishermen's houses on
the seafront. Settlements too close to the shoreline are ubiquitous in developing countries,
where the poor cannot afford buying land on safer places. Photo by the author.

Shoreline variability has usually a seasonal component, as shown
in Fig. 7.2. The upper beach is higher and the beach slope steeper in
summer than in winter. During winter, storm-related high water levels
and energetic breaking waves transport sand from the backshore to
the foreshore and even further offshore to the shoreface. The beach
profile is flattened and the shoreline (or HW line for tidal beaches) is
shifted onshore.

During summer, low-energy swell waves return the sand deposited
on the shoreface towards the beach. Offshore bars formed in the
winter season, break up into a series of crescentic bars when wave
action is reduced. Rip cell circulation stimulates onshore migration
of the crescentic bars, which may finally weld to the shore [144].
The bar pattern gradually disappears by infilling of the troughs [2].
Normally, beach erosion under high-energy wave conditions is a

much faster process (typically several days) than accretion under low-energy wave conditions (typically several months) [966].

The seasonal signal is generally less pronounced at storm-dominated coasts [680, 700]. Individual storms (high-energy waves and high water levels) cause offshore sand transport; the beach response depends on the pre-storm morphology. A sequence of moderate storms may cause as much erosion as an incidental extreme storm [166]. For tidal beaches, most sediment exchange takes place between the supratidal and lower-intertidal parts of the beach [700].

Complete post-storm recovery is not always achieved [531], especially at coasts subject to structural erosion. Beach recovery will be delayed if the interval between successive storms is too short. Besides, longshore transport can shift post-storm accretion to locations situated down-drift of the storm erosion sites.

Alongshore variability

Storm erosion is highly variable along the beach. Eroding coastal stretches ('erosion hotspots') alternate with coastal stretches which are hardly affected during storms [531].

Several factors that may contribute to longshore variability in shoreline erosion have been discussed before: Mega-cusps, related to high-angle wave-incidence instability, see Sec. 7.5.1; shoreline embayments, related to rip cells, see Sec. 7.5.5; sandwaves, related to intermittent shore-attachment of nearshore bars. Exfiltration of groundwater and wave focussing through refraction over nearshore sandbanks are other possible causes of local erosion hotspots. A general understanding of alongshore variability in storm response which allows prediction is still lacking [531].

Shoreline change prediction

Shoreline response to changing wave conditions can be very different depending on coastal typology, with even strong variations between nearby locations. Nevertheless, certain response characteristics are common to many beaches: The shoreline generally retreats under storm conditions and accretes under calm weather conditions. These shared characteristics can be modelled for each single location, providing a tool for local shoreline change prediction. The first ideas for such an approach were put forward by Wright, Short and Green [968] and later elaborated by Miller and Dean [605] and Yates, Guza and O'Reilly [975]. This shoreline response model is based on the following three assumptions: (1) the shoreline location X is at each time t representative for a particular beach state; (2) for each constant incident wave energy E, a unique equilibrium beach state exists, characterised by a shoreline position $X = X_{eq}$; (3) when the wave energy is kept constant, the shoreline position $X(t)$ tends asymptotically to the equilibrium position X_{eq}.

It is questionable whether these conditions are always fulfilled; one may expect that for a given constant wave energy, different shoreline positions are possible. The equilibrium shoreline position may depend not only on wave energy, but also on wave incidence angle, shoreface profile and longshore beach gradients. However, the approach based on the assumptions (1–3) is attractive due to its simplicity. The model can only be applied for stable beaches (fluctuating around a long-term equilibrium); for coasts with slow erosional or accretional trends, the model may still apply if the trend is subtracted from the data.

The evolution of the shoreline position towards the equilibrium position is assumed to follow the linear relaxation law

$$dX/dt = k(E)(X_{eq} - X), \quad X_{eq} = f(E). \qquad (7.51)$$

Because the equilibrium shoreline position retreats for increasing values of E, the function f should be a decreasing function of E. It can be parameterised by a linear relationship $f(E) = (b - E)/a$, as shown by Yates *et al.* [975]. The factor $k(E)$ is an increasing function of E, because shoreline retreat under high waves is a faster process than shoreline accretion under calm conditions; it can be modeled as $k(E) = aC^{\pm}E^{n}$, where C^{+} applies for retreat ($E > b - aX$) and C^{-} for accretion ($E < b - aX$). The exponent is taken as $n = 1/2$, but its precise value is not essential for the results. The parameters $a, b, C^{\pm}$ are determined by fitting Equation (7.51) to records of observed shoreline positions $X(t)$ and wave energies $E(t)$. Such records should have sufficient temporal resolution and length.

Fair estimates of observed shoreline change could be produced with this model for several field situations [975, 976]. It appears that the model is not very sensitive to the parametrisation of $k(E)$ and that $C^+ \approx C^-$. Besides, instead of E, other beach state indicators are possible. Davidson *et al.* [204] used the non-dimensional settling time Ω (7.3). In this latter more elaborate model, the equilibrium state indicator includes a memory component of previous beach states, which accounts for the evolution of the morphological beach state. This memory-dependent component alleviates the second condition (2) required for the simpler model.

Another elaboration of the model of Yates et al. was proposed by Jara et al. [453], who suggested that the number of free parameters of the shoreline prediction model can be reduced. Therefore they related the variation of the shoreline position $dX(t)/dt$ by a parabolic relaxation law to an equilibrium coastal profile constrained for each E by a given grainsize (fixing the profile-shape exponents m_a, m_b in (7.1), (7.6)), a given sediment volume in the active zone, a given berm height and a given position of the closure depth. In this way the model could be calibrated from a reduced dataset $(X(t), E(t))$.

For dissipative storm-dominated coasts (such as the Dutch coast), the shoreline position is not a good indicator for storm impact. The typical response under storm conditions is not shoreline retreat, but shoreline advance, related to a decrease of the beach slope [700]. However, the sand volume of the beach decreases systematically under storm conditions. In this case the model described above could be applied for predicting the beach sand volume, instead of the shoreline position. A drawback is the difficulty to obtain suitable time records for beach volume change, required for fitting the free model parameters.

Shoreline prediction as management tool

Shoreline change prediction models are important tools for coastal management and planning. Systematic comparison of predicted shoreline retreat with observations provides a means to distinguish temporary fluctuations from trends in shoreline position. This is necessary for planning shoreline maintenance interventions, such as beach or shoreface nourishments. Shoreline prediction models can also be used for establishing shoreline retreat statistics and for estimating the recurrence of extreme retreat events. This is crucial for defining hazard zones where construction is regulated (or simply forbidden). Reliable estimates require long time series of observed shoreline positions for tuning the model parameters.

7.7.2. Interruption of Littoral Drift

Interventions in the coastal zone, which influence hydrodynamic conditions, also influence the shoreline position. Most drastic impacts result from interventions that change the longshore transport (littoral drift). An illustration is shown in Fig. 7.46.

Littoral drift

Littoral drift is induced by the dissipation of obliquely incident waves in the surf zone, see Appendix D.3.2. Along coasts with persistent high-incidence wave angles, the net yearly littoral drift can be very high; in some cases the yearly transit of sand along the coast in the surf zone can exceed one million m^3. More usual is a net annual littoral drift in the order of one to a few hundreds of thousands m^3 [492].

It is generally difficult to make a precise estimate of littoral drift from field measurements. The best estimate is obtained by building a dam across the surf zone and by measuring the resulting sand accumulation at the up-drift side and erosion at the lee-side. The drawback of this method is obvious, however. Other methods are described by Komar [492], but uncertainty margins are substantial — of the order

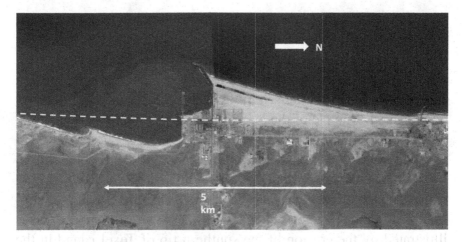

Fig. 7.46. Shoreline erosion at Nouakchott, Mauritania, February 2015. The construction of the 'Port de l'Amitié' in 1985 has blocked the southward littoral drift (estimated at 0.5–1 millon m³/year) and caused severe erosion at the lee-side of the harbour. Dotted line: Original coastline. Image Google Earth.

of a factor 2. The same holds for mathematical model simulations of littoral drift, which strongly rely on the transport formulas employed.

Many formulas have been proposed that relate the bulk littoral drift (integrated over the surf zone width) to easily measurable parameters. It appears that littoral drift depends primarily on wave height H and wave incidence angle θ_{br} at the breaker line. Other important parameters are the surf zone slope β, the sand grain-size d and the wave period T. A few recent formulas are given in Appendix D.3.2.

Lee-side erosion

Modification of littoral drift producing lee-side erosion occurs for interventions that block the littoral drift, such as jetties, harbour moles and groynes. It also occurs for constructions that modify the incident wave field in the surf zone, such as shore parallel ('detached') break-waters and shoreface nourishments. Shielding of incident waves leads to the formation of so-called salients. These are local sand protrusions with a lee-side erosional embayment; if there is no dominant wind direction, erosional embayments exists at both sides of the salient.

The only way to avoid lee-side erosion by structures that block the littoral drift, is sand bypassing. Extensive literature exists on the subject [99]. Many harbour entrances are equipped today with sand bypassing systems using stationary pumps. Lee-side erosion can be effectively neutralised in this way, as illustrated by the well documented example of the large Tweed river bypass system at the Australian Gold Coast [142].

Submarine canyons and river mouths are also obstacles to littoral drift. Shoals in the ebb-tidal deltas of coastal inlets act as 'stepping stones' which allow bypass of longshore transport. Erosion of ebb-tidal deltas entails increased erosion of the down-drift shore. This is illustrated by the erosion of the southern tip of Texel island in the Dutch Waddensea, where the volume of the ebb-tidal delta strongly decreased, due to the sediment demand of Texel inlet after closure of the Zuyderzee in 1930 [280], see Figs. 6.2 and 6.38. Another illustration is given in Fig. 7.47, which shows the impact of the artificial breach of the Senegal River through the sand spit 'Langue de Barbarie'. Lee-side erosion at coastal inlets due to interruption of littoral drift by the ebb-jet is sometimes termed "hydraulic groyne effect" [29].

7.7.3. *Coastal Erosion Control*

A global inventory of coastal erosion in 1982 led to the conclusion that at least 70% of the beaches worldwide had retreated over the past century [79]. One may expect that this figure will be higher over the next century, as a result of accelerating sea-level rise. For Europe, another study estimated that about 20% of the coastline is currently subject to strong erosion [246].

Erosion control is a hot topic at many places around the world. A vast literature exists on coastal protection practices; much information can be found on the Internet, see for example the Coastal Engineering Manual of the US Army Corps of Engineers, http://chl.erdc.usace.army.mil/cem, and the Coastal Wiki, http://www.coastalwiki.org.

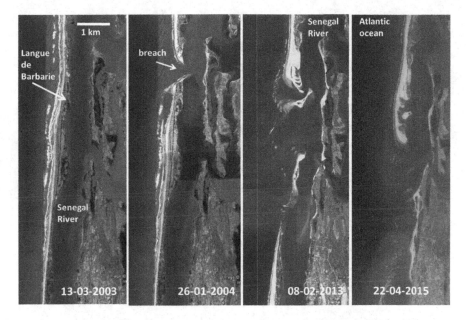

Fig. 7.47. Interrupted longshore sand transport at the Senegal River mouth. In November 2003 an artificial outlet of the Senegal river was created to protect the town of Saint-Louis from river flooding. The initial 4 m wide breach through the sand spit (Langue de Barbarie) interrupted the strong southward littoral drift (ca. $0.5 \ 10^6 \ m^3$). This deprived the southern part of the spit of sand supply, with rapid regression as a result. Wave-induced onshore flow pushes the remains of the eroded spit towards the former inner Senegal River bank. Images from Google Earth.

In this section we will discuss some elements for estimating the feasibility and effectiveness of different erosion control measures. The focus is on structural erosion, in contrast with the previously discussed temporal erosion related to shoreline fluctuations. Major causes of structural erosion are (see Secs. 7.4.1 and 7.7): Sea-level rise, trends in wave climate, sediment trapping in coastal inlets and submarine canyons, sand mining, sediment retention by river damming, sediment fixation by land reclamation, shoreline structures and interruption of longshore transport.

Coastal sediment cells and sediment management

Understanding in each situation the particular causes of coastal erosion is crucial for the development of effective response strategies. From the knowledge of sediment sinks, sources and transport pathways it is possible to identify sediment cells. A sediment cell is defined as a set of coastal subsystems which communicate through the exchange of sediment. Sediment cells may consist of coastal stretches limited alongshore by headlands or man-made structures, and limited offshore by the closure depth; they may contain river mouths as sediment sources and back-barrier basins and dune areas as sediment sinks. Erosion response strategies should be defined at the level of sediment cells, in order to appreciate the integral effects of interventions. Such strategies, based on the sediment cell concept, are called sediment management strategies.

Institutional measures

The most effective way of combating erosion consists of tackling the causes. Beach sand mining is an important cause of coastal erosion in many developing countries. A ban on this practice is an effective measure. This seems obvious, but enforcement is often a bottleneck.

Sediment retention by river dams and interruption of littoral drift by harbour jetties are other important causes of coastal erosion. In practice, the removal of these causes conflicts with other interests that generally prevail.

Accepting coastal erosion is often inevitable. Measures are then needed to mitigate coastal erosion impacts. The most obvious measure consist of imposing set-back lines: Regulations which allow the removal of constructions within a certain distance from the shoreline. Such regulations exist in several countries, where the coastal zone is not densely built-up. However, it is not a realistic option in densely built-up zones.

If strategies for tackling the causes of coastal erosion, or for accepting the consequences, are not feasible, one has to resort to coastal protection measures. In the following sections we will discuss 'hard' measures and 'soft' measures.

Hard protection measures

Along many coasts, seawalls or seadikes have been built to protect coastal settlements against storm damage or to protect low-lying hinterland from flooding. Construction costs range typically between 1,000 and 10,000 US$/m, depending on height and availability of materials [418]. Regular maintenance is required, with particular attention to scouring at the toe of the structure by wave reflection and wave breaking, which undermines the stability. The sub-aerial beach in front of a seawall erodes during storms; complete disappearance of the beach has been observed in many cases. The risk of beach loss is less for seadikes with a gentle slope and irregular surface [109], but inevitable for highly exposed coasts.

Groynes and shore parallel (detached) breakwaters are often used to prevent (or retard) local erosion of beach material. Typical costs are of the order of 1,000–4,000 US$/m for groyne construction and approximately twice as much for detached breakwaters. If erosion is caused by gradients in longshore sand transport, the construction of groynes will reduce the rate of erosion. However, if littoral drift is strong, groynes will shift (or even aggravate) the problem by inducing lee-side erosion. To minimise this problem, groyne spacing, length and height should be carefully chosen.

If beach erosion results from a combination of cross-shore and longshore sand fluxes, erosion rates can be reduced by the construction of detached breakwaters (eventually in combination with groynes). Sufficiently high detached breakwaters reduce beach erosion under storm condition, but prevent accretion under calm conditions. The coastal profile is modified, a new equilibrium profile will develop in response to the breakwater. In the case of strong littoral drift, a salient will develop behind the detached breakwater and erosion will occur at the down-drift side. The effectiveness of detached breakwaters depends strongly on the design; detailed design studies are required.

If beach erosion is caused by up-drift obstruction of longshore drift by harbour jetties, the problem can only be solved by the construction of sand bypassing systems, as discussed earlier.

Cliff erosion can be mitigated by toe protection measures, either hard or soft [818], see Sec. 7.4.2.

Soft protection measures

Hard structures — if properly designed — can slow down structural erosion, but they do not stop it. Coastal retreat is delayed, but continues. With the improvement of dredging techniques in the 1960ies and 1970ies, coastal nourishments have become an alternative for hard protection structures. By regular sand supplies, coastal erosion can be fully compensated. Fill sand is dredged from the sea bottom, sufficiently far offshore to avoid interference with the active coastal zone. The first experiments in Europe took place in 1950–1952, in Portugal, UK and Germany [370]. The Netherlands started later, but at present, the Dutch coastal nourishment programme is the largest in the world. Experience from the Dutch nourishment programme will be presented in the following. Coastal characteristics are indicated in Table 7.1.

Table 7.1.   Characteristics of the Holland coast [738, 801, 947]; all the figures represent long-term mean values. Longshore sediment transport is the annual mean of northward (positive) and southward (negative) wave-induced transports; at the central portion of the Holland coast northward and southward contributions are almost equal.

Wave height	1–1.5 m	Wave period	5–6 s
Highest wave heights	2.5–3 m	Highest wave period	$\approx$10 s
Tidal range	1.5–3 m	Tidal excursion	6–10 km
Net longshore drift velocity	3–6 cm/s	Net littoral drift	0–2 10^5 m^3/yr
Grain size	200 μm	Surf zone slope	0.01–0.005
Surf zone width	400–800 m	Longshore bars	1–4 (number)

Beach nourishment

Beach nourishment has several advantages over hard structures, but also some drawbacks. Advantages are: Increased beach width, no basic modification of coastal morphodynamics and no structural side effects. If the nourishment is not repeated, the beach returns after some time to its former state.

Major drawback is the relatively short lifetime of beach fill at places with strong structural erosion. Within one year (and sometimes after a single major storm) a large portion of the nourished sand is redistributed over the active zone. A first period of fast shoreline retreat is followed by a period of relative stability. After a few years the shoreline resumes the erosional trend from before the nourishment.

The positive effect of beach nourishment is typically dissipated after 2 to 5 years, depending on the structural erosion rate and the nourishment volume [909]. The amount of nourished sand is usually of the order of 50–200 m³/m. The costs depend primarily on the distance to the extraction location and range between 10 and 30 US$/m. The lifetime of the nourishment is increased when the nourished sand is slightly coarser than the original beach sand. The use of finer nourishment sand should be avoided. Another option for increasing the lifetime is a combination of beach nourishment with groynes. However, this supersedes some of the advantages of beach nourishments.

Shoreface nourishment

Because beach fill is redistributed over the active zone within few years, shoreface nourishment is an interesting alternative for beach nourishment. Nourishing the shoreface is easier than nourishing the beach (no pipelines needed) and therefore cheaper, typically 5–10 US$/m³, if sand extraction sites are not too far away. The nourished volume is larger than for beach nourishments; usual volumes are of the order of 300–500 m³/m.

Shoreface nourishments are most often used on coasts with a naturally barred profile. The nourishment is deposited as an elongated sand body at the seaward flank of the outer bar, or in the through between the outer end the inner bar. The nourished sand is naturally redistributed under the influence of cross-shore and longshore sand fluxes. Long-term nourishment programmes should therefore be based on the sediment cell concept.

The shoreface nourishment has a positive effect on the shoreline through two processes, the feeder process and the lee process [892]. The feeder process acts by sheltering the nearshore zone from the most energetic waves, that otherwise would cause beach erosion. The lee effect acts by shifting the littoral drift away from the nearshore zone; the resulting convergence of the littoral drift near the shoreline produces local accretion. This also implies that the nearshore littoral drift diverges at the down-drift side of the nourishment, resulting in shoreline erosion. The erosion disappears when the nourishment migrates and diffuses along the coast.

The shoreface nourishment itself can behave in a similar way as a crescentic bar, depending on length and position. Wave breaking on the crest of the nourishment creates an onshore flow which pushes the sand body onshore. At both sides of the nourishment seaward rip currents may develop, which carry sediment (and swimmers) offshore.

Risks of coastal nourishment

Preliminary model studies are required when planning coastal nourishments on coasts with frequent high-angle wave incidence. In the case of shoreface nourishment, initial lee erosion is a serious risk if littoral drift is strong. In Sec. 7.5.1 it was shown that shorelines can become unstable when waves approach the coast under a high angle (HAWI, [34]). Beach and shoreface nourishments entail the risk of creating shoreline undulations that could be amplified by HAWI [872]. Falques and Calvete [295] showed that amplification only occurs if undulated depth contours extend beyond the breaker zone. Numerical model studies further indicated that amplification is inhibited if high-angle wave incidence occurs less than 80% of the time [873], or if the incidence angle is reversed more than 30% of the time [475]. Amplification of shoreline undulations induced by coastal nourishments will thus occur only in rare cases.

Experience of coastal nourishment at the Dutch coast

Since 1990, the erosion control strategy of the Dutch coast is based on beach and shoreface nourishments. Sand for coastal nourishment is extracted from the offshore North Sea bottom at depths greater than 20 m.

Before 1990, several parts of the coast were structurally retreating. Erosion hotspots were located in particular near the Wadden Sea tidal inlets, see Fig. 7.48. There is strong evidence that the Wadden Sea is a major sediment sink for the Dutch coastal zone, see Figs. 6.38 and 6.48.

Coastal retreat was mitigated in the past by the construction of groynes; sea dikes were built at the most vulnerable places. The policy change in 1990 was stimulated by concerns about sea-level rise; nourishments were intended to create an additional sand buffer in front of the coast. Important support came from geological studies, which indicated that no major sand loss to deep water is occurring at the Dutch coast [64, 65, 170].

The 1990 coastal policy has fixed a coastal baseline, for the entire Dutch coast. When the actual coastline retreats beyond this baseline, the coast will be nourished. The position of the actual coastline depends on the nearshore sand volume, which is determined by annual monitoring of the bathymetric profile. The coastline definition is explained in Fig. 7.49. The incorporation of a sand volume in the definition of the coastline eliminates, at least partly, the influence of shoreline fluctuations and beach flattening due to individual storms or storm periods. The nourished volumes since 1975 are indicated in Table 7.2. An annual reservation is included in the long-term state

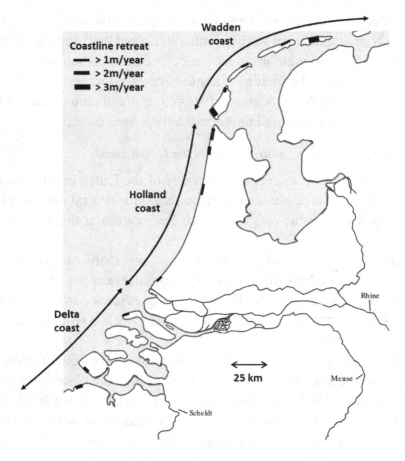

Fig. 7.48.   Dutch coast with major erosion hotspots before 1990.

budget for coastal nourishment, for a yearly nourishment volume of about 10 million m^3 (costs around 50 million euros). The nourishments have halted the structural erosion along the Dutch coast. At several places the dunefoot has advanced and beaches are widened.

Besides the Wadden Sea lagoons, the coastal dune belt is another important sediment sink along the Dutch coast. Profile measurements at two sites indicated dune accretion of the order of 10 m^3/m/year [885]; a similar figure was found for the Danish coast [1,164]. After beach nourishment, dune accretion increased by about 5 m^3/m/year [885].

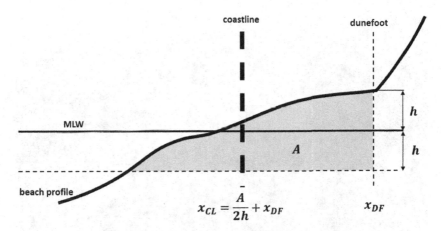

Fig. 7.49. Definition of the coastline. The position of the coastline x_{CL} is defined relative to the position of the dunefoot x_{DF} (in a fixed coordinate system) and is related to the sand volume A [m³/m] in grey, as indicated in the figure. This sand volume depends on the actual beach profile and the height h of the dunefoot relative to the mean low water level MLW. The volume A is computed from yearly bathymetric profiles, by extrapolation of the trend over the past 5 years. If the position x_{CL} is landward of the coastal baseline x_{CBL} fixed in 1990, the coast will be nourished.

Table 7.2. Sand volumes nourished at the Dutch coast [10^6 m³]. The three regions Delta, Holland and Wadden correspond to the Rhine–Meuse–Scheldt delta (110 km), the uninterrupted Holland coast (120 km) and the Wadden Sea islands (120 km), respectively, see Fig. 7.48. Beach and shoreface nourishments are indicated separately. Figures from [884].

	Nourishment type per coastal section						
	Wadden		Holland		Delta		
Period	beach	shoreface	beach	shoreface	beach	shoreface	Mean vol./year
1976–1990	14.8		13.8		32.6		4.1
1991–2000	18.6	4	18.5	6.3	24.8	1	7.3
2001–2012	7.9	33.6	15.3	42.9	21.4	12.7	11.2

Based on these findings, the Dutch coastal nourishment programme is not just targeted to erosion control, but also to reinforcement of the weakest links in the littoral dune belt. In 2011, a massive beach extension of 21.5 million m³ was realised along the coast south of the city of The Hague, see Fig. 7.50. This nourishment,

Fig. 7.50. The sand-engine, a massive coastal nourishment scheme of more than 20 million m³ south of The Hague, realised in 2011.

the so-called sand-engine, is primarily meant to strengthen the coastal defence line in this zone, which is crucial for protecting the most densely inhabited part of Holland from flooding by the sea. The name sand-engine refers to the objective of this nourishment: The sand body spreads along the coast by alternating longshore currents, widens the beach and nourishes the coastal dune belt [809].

Observed bar behaviour

The Dutch coast has typically 2 or 3 bars, with a characteristic cyclic behaviour [948]. The outer bar fades while migrating offshore; this offshore migration is triggered by storms [924].

When the outer bar dissipates, the next bar takes over and moves offshore towards the former position of the outer bar, while a new inner bar emerges at the shoreline. This cycle takes about 5 years at the southern part of the Holland coast and about 15 years at the northern part, see Fig. 7.43. The reason for this difference in timescale is not well understood; it may be related to the difference in shoreface profile (Fig. 7.40). Offshore bar migration does not imply a corresponding sand loss in the surf zone. A bar can migrate offshore while the net transport is onshore. The sand eroded from the decaying outer bar is mainly transported onshore and collected by the inner bar, which becomes the new outer bar [948].

Shoreface nourishments were placed at different positions relative to the bar pattern: Seaward of the outer bar location, at the furthest offshore location of the outer bar, at the seaward flank of the outer bar and in the trough between the outer and inner bar.

The morphodynamic response of the shoreface was different for each nourishment. The lifetime of the nourishment, as a distinct morphological feature, varied between 2 and 10 years. The most stable nourishments had the greatest volume and the greatest length. A few common characteristics emerge [226, 360, 650, 801, 884, 892, 909]:

- the nourishments interrupt the offshore migration cycle of the bar system, see Fig. 7.51;
- the nourishments develop a trough at the shoreward flank;
- the bar shoreward of the nourishment migrates onshore;
- after dissipation, the original bar pattern and migration cycle is restored.

Long-term sand budget of the Dutch coast

In all cases, shoreface nourishments have produced a net gain of sediment in the nearshore zone, landward of the nourishment. This sediment gain is smaller than the nourished volume — typically of the order of 50% after a few years. However, in general no significant effect of shoreface nourishments on the beach width is observed. The nourishment at Terschelling in the trough between the inner and outer bar is an exception: In this case the sand gain of the nearshore zone and the beach was much larger than the nourished volume. This was probably due to a lee-effect: Trapping of a longshore migrating sandwave behind the nourishment. For other shoreface nourishments the lee-effect was minor.

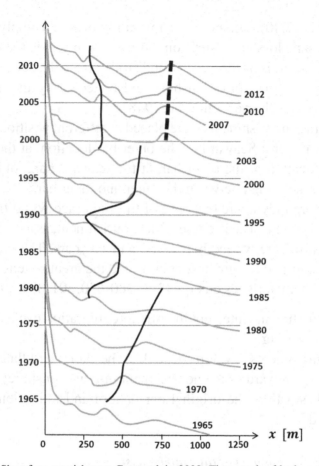

Fig. 7.51. Shoreface nourishment Camperduin 2002. Time stack of bathymetric profiles from 1965 to 2012 (in grey) — only some of the yearly measured profiles are shown in the figure. The vertical scale (distance of 5-year intervals) is 10 m. Bar locations are indicated by solid lines. The shoreface nourishment is indicated by the fat dotted line. The nourishment was deposited at the location where the outer bar had faded previously. The nourishment was stable for at least ten years. It interrupted the cycle of offshore bar migration; the offshore migration of the inner bar was halted. Redrawn from [884].

A sand budget calculation for the Holland coast over the period 1990–2005, based on bathymetric surveys, indicated a net sand volume increase of about 2–3 million m³/year in the nearshore zone (landward of the −8 m depth contour) [279, 909]. Accounting for the estimated northward sand export to the Wadden Sea of about

0.3–0.4 million m^3/year, the sand budget of the Holland coast was close to the average sand volume of about 3.5 million m^3/year nourished during this period.

The Dutch coastal zone (landward of the 20 m line, including back-barrier basins and littoral dune belt) is thus an approximately closed sand system, corresponding to a sediment cell. Only a minor part of the sand added to the coast is lost. Nourishments do not entail the risk of aggravating coastal erosion, except in rare cases. This is true, even if the design is mistakenly not well adapted to a local erosion problem. A sediment management strategy based on coastal nourishment is in the worst case a no-regret strategy. Such a conclusion does not hold for hard protection structures. In some cases hard structures can be more cost-effective than nourishments. However, the risk of irreversible collateral damage is non-negligible with the present state of knowledge.

Uncertainties and challenges

Numerical models provide guidance for the optimisation of nourishment designs. However, reliable predictions of the effectiveness of individual nourishments are still out of reach. The yearly bathymetric surveys of the Dutch coast, which started in the 50ies, were crucial for gaining insight in the response of the coastal system to interventions. This insight is further improved thanks to ARGUS cameras installed at several hot-spot locations for coastal erosion.

Important questions remain, related to the perspective of future sea-level rise. We are not sure of the feasibility and effectiveness of the sediment management strategy at long term. Great uncertainty exists regarding the sand budget of the active shoreface, beyond the breaker zone [279]. The co-evolution of this zone with sea-level rise will probably have a long-term impact on coastal stability. Sand transport on the outer shoreface is highly variable, depending on the combined influence of waves and currents driven by storms and tides. The lack of insight in this complex dynamics is a serious obstacle for answering the question to which extent the present sediment management strategy is a sufficient response to future sea-level rise. Other uncertainty factors are related to estimating and managing sediment exchange between the different compartments of the coastal sediment cell:

- net sediment transport towards tidal inlets and the related risk of increased sedimentation in the Wadden Sea;
- net sand exchange with the coastal dune belt, by aeolian accretion and storm erosion, and the related risk of dune breach.

For tackling these uncertainties, major steps in our knowledge of coastal processes have to be taken. A special challenge lies in the development of new observation techniques for the highly variable and intermittent sand transport processes at the interfaces of the different compartments of the coastal sediment cell.

7.8. Summary and Conclusions

In the past decade the knowledge of coastal morphodynamics has been enriched trough many field and model studies. A few examples:

- To the three best studied field sites in the world (Duck, USA; Gold Coast, Australia; Egmond, Netherlands), a fourth site has been added: Truc Vert, at the French Aquitaine coast, which is distinct from the other sites by its larger tidal range;
- The installation of ARGUS video cameras at many field sites around the world has provided invaluable insight in the great diversity of morphodynamic processes in the nearshore zone;
- In numerical model studies, the 2D-models are superseded by quasi-3D and 3D-models that enable a better representation of the highly complex 3D morphodynamics of the nearshore zone;
- Better insight has been gained in nearshore sediment transport processes. It has become clear, for example, that wave skewness is not synonymous with onshore transport; settling lag effects can produce the opposite. But it remains difficult to predict when this will be the case. The same holds for the effect of acceleration

asymmetry on the residual sediment flux. There is no doubt that acceleration asymmetry plays a role, but it is less clear in which situations and through which mechanism (transformation to shear-stress skewness in the wave-boundary layer or pressure gradient effect on particle suspension?).

The knowledge and understanding of nearshore morphodynamics remains largely based on empiricism. The spatial and temporal variability in the nearshore zone is a crucial factor for correct simulations and predictions of the evolution of the beach state. Simple models, which ignore this variability, do generally not give correct answers. For example, shoreface equilibrium profiles predicted by models based on a representative constant monochromatic wave field, are much steeper than observed in reality.

The issue of variability is a serious obstacle for simulating and predicting nearshore morphodynamics. It appears that nearshore morphodynamics is highly sensitive to even subtle changes in wave forcing related to wave height, angle of incidence and wave spectrum (for example, the proportion of sea waves compared to swell waves and the presence of infragravity waves). Changes in water levels due to tides and wind setup add to this variability. The morphological response time to different wave conditions is very different (short for high energetic waves, long for low swell waves); it depends also on the cross-shore location (short for the surf zone, long for the shoaling zone). The beach state is therefore never in equilibrium. Because of morphodynamic feedback, the response of the beach to changes in wave climate is highly dependent on its previous history.

Probably for this reason, only few simple rules seem to have general validity. The few rules that always 'work' have a qualitative character, for instance:

- beaches are eroded and flattened and bars move offshore in periods of high energetic waves;
- beaches accrete and steepen and bars move onshore in periods of low swell waves;

- oblique wave incidence induces straightening of longshore bars;
- a stable outer longshore bar stabilises the temporal variability of nearshore morphodynamics;
- tides stretch the extent of the surf zone and delay bar dynamics;
- the Dean parameter Ω distinguishes gently sloping dissipative beaches from steep reflective beaches.

However, more precise generic rules for specific transects cannot be given. For example, there is no good explanation for the alternation of stable coastal stretches and erosion hotspots along apparently homogeneous beaches. In general, the success of 3D numerical models for simulating the morphological response of the nearshore zone to interventions, such as coastal nourishments, is still far form guaranteed; coastal managers cannot yet safely rely on model predictions.

Interruption of the littoral drift is a major cause of human-induced coastal erosion. However, reliable methods for predicting the right magnitude of littoral drift do not yet exist. The most employed formulas for longshore sand transport differ in many cases by more than a factor 2. Comparison of simulations with observations is possible only in few cases, for instance at locations where jetties have been built across the whole surf zone. The uncertainty of other observation methods is of the same order as for existing empirical formulas.

Besides temporal and spatial variability, another crucial factor is the complexity due to the numerous morphodynamic feedback processes in the nearshore zone. Important feedback processes are related to:

- the interdependence of coastal bathymetry and wave refraction, through concentration of wave energy;
- the interdependence of wave breaking and bathymetry, through the generation of undertow currents and currents induced by radiation stresses;
- the interdependence of longshore gradients in longshore drift and bathymetry, especially for high-angle incident waves;

- the interdependence of longshore currents and bathymetry, through the generation of cross-shore sediment fluxes.

These feedbacks may induce morphodynamic instability and amplification of small perturbations of the nearshore bathymetry. In addition to these free morphodynamic instabilities one has to consider forced instabilities related to patterns in the wave field. Such patterns may arise from wave reflection at the shoreline (or at longshore bars), especially for infragravity waves, and from interferences of wave-fields approaching from (slightly) different angles.

The coastal tract approach has improved insight in large-scale sediment transport patterns, in particular insight in the mutual morphodynamic dependence of different compartments of the coastal system through sediment exchange. This has been translated in practice through the development of coastal management strategies based on sediment management within coastal cells. Knowledge of sediment fluxes between different compartments of the coastal cell is a prerequisite for effective sediment management strategies. However, little is known about some of these fluxes. This holds for sand transfer by aeolian processes across the beach-dune interface and for sand losses to deep water. Aeolian transport on the beach is extremely variable, spatially and temporally, depending in particular on beach moisture. Losses to deep water are generally believed to be small, if rocky headlands or submarine troughs are not nearby, but this is hardly underpinned with field observations. One may expect that the seasonal sediment exchange between nearshore and offshore zones of the active shoreface is influenced by tidal currents, which redistribute offshore sand deposits in alongshore direction.

Sediment management strategies are mainly based on coastal nourishments for maintaining beaches and for regulating the sand distribution within coastal cells. It appears that erosion hotspots are quite persistent and strongly reduce the lifetime of nourishments; frequent repetition of nourishments is therefore required. Shoreface nourishments are cheaper than beach nourishments; for the same price a much greater sand volume can be supplied to the coastal system, if borrow sites are not too far away. However, the influence of shoreface nourishments on beach width is often marginal. Depending on the purpose of the nourishment, this may raise questions about the cost-effectiveness of coastal nourishments. However, the effectiveness of hard protection measures is often not much better than for nourishments, and structural negative side effects are frequent, especially for coasts with strong littoral drift. The development of optimal management strategies for such coasts is still an important challenge.

The most promising way forward to overcome the numerous difficulties described above, is probably investing in further improvement of comprehensive physics-based numerical models. The detailed modelling of crucial processes, such as the wave-boundary layer dynamics and the wave-breaking dynamics, reduces the number of empirical parameters in such models, and therefore the general site-independent application. The required small size of spatial and temporal model scales is for the time being a major obstacle for application to field situations. As mentioned before, due to the dependence of nearshore morphodynamics on antecedent states, comparison with field situations requires simulation over long periods.

Laboratory experiments are helpful to improve the description of separate processes, for better modelling the wave-boundary layer and the wave breaking dynamics. Hydraulic models miss the three-dimensional complexity of field processes; their use for predictions directly applicable in the field is therefore limited.

The present predictive skills of morphodynamic models are still quite limited. Uncalibrated models may yield erroneous predictions. Field experiments remain indispensable. Results of field experiments are always site-specific. Generalisation of these results requires analysis of data from many different sites. Field-based research programmes should therefore include co-operation with research groups working at other sites. Data of field studies should be shared broadly within the scientific coastal community.

Appendix A

Basic Equations of Fluid Motion

A.1. General Nature of the Basic Equations

Newton's law

The equations describing water motion in seas and rivers are derived from Newton's law. Although at first sight these equations look quite straightforward, their solution for real situations is very complicated. The reason lies in the nonlinear nature of these equations. This non-linearity is related to the advection of momentum in a fluid and to boundary conditions that must be satisfied at the free surface. The nonlinearity of the equations of motion, together with viscous inter-action at the molecular scale, means that water motions at different spatial and temporal scales are linked to each other. The topography of river and sea basins is structured over an almost infinite range of scales, from the scale of the individual sediment grain to the scale of the entire basin; the hydrodynamic response to this topography interrelates all these scales. The equations can therefore only be solved after simplification. Such simplifications imply assumptions additional to Newton's law, which are valid only under certain conditions. Assumptions conflicting with Newton's law are avoided (as far as possible) by validation from observations.

Notations

Different notations will be used, to keep the appearance of some equations as simple as possible. Therefore we will use different notations.

Throughout the book, superscripts $(x), (y), (z)$ indicate the vector components along the x-axis, y-axis and z-axis. Subscripts x, y, z, t indicate partial derivatives with respect to x, y, z, t. With this notation, the x-derivative of the component u of the velocity vector $\vec{u}$ along the x-axis is written as $u_x^{(x)} \equiv \partial u / \partial x$, as an example. We have $u^{(x)} \equiv u,\ u^{(y)} \equiv v,\ u^{(z)} \equiv w$. The time derivative of the component v of the velocity vector $\vec{u}$ along the y-axis is written as $u_t^{(y)} \equiv \partial v / \partial t$.

In this part of the appendix we will also use another notation, where x, y, z are replaced by indices i, j with values 1, 2 or 3: $u^{(1)} \equiv u, u^{(2)} \equiv v, u^{(3)} \equiv w$. With this notation $x^{(1)} \equiv x, x^{(2)} \equiv y$, $x^{(3)} \equiv z, u_1^{(1)} \equiv \partial u / \partial x$ and $u_t^{(2)} \equiv \partial v / \partial t$. We use this other notation because it allows simplification of equations in 3 dimensions, by using the convention: If an index i, j, etc. appears twice in a product term, then summation over the repeated index is implied. For instance, $u_i^{(i)} \equiv \vec{\nabla} . \vec{u} \equiv u_x + v_y + w_z$.

Small-scale limit

The equations describing water motion and transport of substances are balance equations based on conservation properties of the balance variable f; this variable may represent, for instance, mass, momentum, vorticity, energy, suspended matter or sediment, etc. In practice it is not possible to describe the spatial structure for the full range of scales down to the molecular scale; the variable f is therefore defined as an average over a given spatial and temporal domain (the model scale) which should be taken smaller than the scales at which we want to resolve its variation. The difference between f and its real counterpart in nature is designated f', with $\langle f' \rangle = 0$, where the brackets stand for averaging over the temporal model scale. The model scale cannot be chosen arbitrarily, but depends on the scales of the physical processes which are not represented explicitly in the model. If these processes have a periodic character the model scale should be a multiple of the spatial or temporal periodicity; if these processes are aperiodic the model scale should be larger than the spatial or temporal correlation scales. The model scale is the lower limit at which the dynamics of the system is explicitly resolved. In practice there is also an upper limit; the validity of the balance equations does not extend to the global scale but is restricted to given boundaries at which the behaviour of the system need to be specified by boundary conditions.

Balance equations

In 3D space the balance equation has the following form

$$f_t + \Phi_i^{(i)} = -\Psi_i^{(i)} + P. \tag{A.1}$$

This equation describes the change of a variable f as a result of input, output, production and dissipation. Implicit within this equation is the scale at which f is described and at which the different balance terms are modelled. The various terms have the following meaning:

- First term on the left-hand side: Change of f in a unit volume per unit of time, averaged over temporal and spatial scales of small-scale fluctuations.
- Second term on the left-hand side: Gradient of advection of f, i.e., difference between input and output of f for a unit cell per unit of time. Often $\Phi^{(i)} = u^{(i)} f$, where $u^{(i)}$ is also defined as an average over the model scale. (But if, for instance, f is the energy of a propagating wave, then $u^{(i)}$ has to be replaced by the wave-group velocity.)
- First term on the right-hand side: Gradient of transport by fluctuations of the velocity field related to processes on a smaller scale than the model scale. If $\Phi^{(i)} = u^{(i)} f$ then $\Psi^{(i)}$ is given by $\Psi^{(i)} = \langle \overline{u'^{(i)} f'} \rangle$. If the small-scale processes fluctuate on time scales which are much smaller than the model scale and if these fluctuations are uncorrelated at the model time scale (random walk), then $\Psi^{(i)}$ can be described as a diffusion process, $\Psi^{(1)} = -K^{(1)} f_1$ etc. The diffusion coefficients $K^{(i)}$ cannot be obtained from (A.1) and should be specified by an additional 'closure' relationship.

- Second term on the right-hand side: Local production or destruction of f.

As the balance equation (A.1) applies within a limited domain, a solution can only be obtained if boundary conditions are specified. These are, on the one hand, conditions specifying f in the entire domain at an initial time t_0, and on the other hand, conditions which either specify f at the boundary or specify the flux of f through the boundary at any time.

Nonlinearity

The balance equation is often a nonlinear equation. The transport term $\Phi^{(i)} = u^{(i)} f$ is nonlinear if $u^{(i)}$ and f are mutually dependent variables. This is the case, for instance, if f represents the momentum $\rho u^{(i)}$, the salinity $S(x, y, z, t)$ or the seabed topography $z_b(x, y, t)$. Other nonlinearities may arise from the diffusion term and the production-destruction terms.

Reduction of dimensions

Balance equations may also be formulated in a 2D space (by averaging over depth or width) or in a 1D space (averaging over a cross-section A). In the latter case the balance equation takes the form

$$(A\overline{\overline{f}})_t + (A\overline{\overline{u}} \cdot \overline{\overline{f}})_x = A\overline{\overline{P}} - \Psi_x. \tag{A.2}$$

In this equation Ψ is the transport due to processes in the 'hidden' (transverse) dimensions; the average value over these hidden dimensions is indicated by the overbars. This is called dispersive transport,

$$\Psi = A\overline{\overline{(u - \overline{\overline{u}}) \cdot (f - \overline{\overline{f}})}}.$$

Under certain conditions this dispersive transport can be approximated by a gradient-type transport, with dispersion coefficient D,

$$\Psi \approx -AD\overline{\overline{f}}_x, \tag{A.3}$$

see Sec. 6.6.3.

A.2. Water Motion in Three Dimensions

Momentum balance

Application of Newton's law to fluids yields the momentum balance equation

$$(\rho u^{(i)})_t + (\rho u^{(j)} u^{(i)})_j = -p_i + \tau_{visc\,j}^{(i,j)} + \rho F^{(i)}. \qquad \text{(A.4)}$$

In the right-hand side of this equation the first term represents the pressure gradient, the second term the viscous stresses between fluid parcels and the third term external forces acting on the water mass, other than forces transmitted through shear stress. This last term corresponds, for instance, to the attraction force of the earth, $-g\rho$, and gravitational forces exerted by celestial bodies. These latter forces are only significant in very deep oceanic basins and will be left out of consideration from now. We are not interested in resolving water motion at the spatial and temporal scale of turbulence. Therefore the equations are averaged over temporal scales characteristic for turbulence; this yields

$$\rho u_t^{(i)} + \rho (u^{(j)} u^{(i)})_j = -p_i + \tau_j^{(i,j)} - g\rho \delta^{(i,3)}, \qquad \text{(A.5)}$$

where $\delta^{(i,j)} = 0$ if $i \neq j$ and $\delta^{(i,j)} = 1$ if $i = j$. Variations in density have been ignored in this equation; they are generally much smaller than variations in velocity (Boussinesq approximation). The stress terms $\tau^{(i,j)}$ now incorporate stresses produced by viscosity and stresses produced by turbulence (Reynolds stresses); the latter are much larger than the former (except in a very thin layer at the bottom), so we may write

$$\tau^{(i,j)} = -\rho \langle \overline{u'^{(i)} u'^{(j)}} \rangle, \quad u'^{(i)} = u^{(i)} - \langle \overline{u^{(i)}} \rangle, \qquad \text{(A.6)}$$

where the overbar stands for averaging over the spatial turbulent scale and $\langle \ldots \rangle$ for averaging over the temporal turbulent scale.

Boussinesq hypothesis

Temporal and spatial scales of turbulence are not particularly small; in shallow coastal waters they may span the whole water column and attain periods exceeding 10 minutes. Turbulent stresses are often related to local velocity gradients, although such an assumption is a rough simplification. Following an hypothesis of Boussinesq, we assume that the Reynolds stresses for three-dimensional turbulence can be represented by an eddy viscosity parameter N defined as

$$\tau^{(i,j)} = \rho N \left(u_j^{(i)} + u_i^{(j)} \right) - \frac{2}{3} \delta^{(i,j)} E^{turb},\qquad(\text{A}.7)$$

where E^{turb} is the turbulent energy. The mass balance (continuity equation) reads

$$u_i^{(i)} = 0.\qquad(\text{A}.8)$$

The Eqs. (A.5), (A.7) and (A.8) form together the so-called Boussinesq equations for water motion.

A.3. Horizontal Flow Equations

Hydrostatic approximation

If the horizontal scale of water motion is much larger than the vertical scale we may ignore horizontal momentum diffusion as it is much smaller than vertical momentum diffusion and we may ignore vertical accelerations in the vertical momentum balance (hydrostatic pressure assumption). The vertical momentum balance can then be integrated and we find for the hydrostatic pressure

$$p = p_s + \int_z^\eta g\rho dz,\qquad(\text{A}.9)$$

where η is the surface elevation relative to a horizontal reference level and p_s is the atmospheric pressure. The horizontal momentum balance equations read

$$u_t^{(i)} + u^{(j)} u_j^{(i)} + p_i/\rho = (N u_z^{(i)})_z, \quad i = 1, 2.\qquad(\text{A}.10)$$

These equations can be integrated to yield balance equations for the depth averaged velocity momentum. The result is (summation over repeated indices, $i, j = 1, 2$)

$$D\overline{u}_t^{(i)} + D\overline{u}^{(j)}\overline{u}_j^{(i)} + p_i/\rho + \tau_b^{(i)} - \tau_s^{(i)} = 0, \qquad (A.11)$$

$$\eta_t + (D\overline{u}^{(j)})_j = 0. \qquad (A.12)$$

In these equations $D = h + \eta$ is the total water depth, the overbar indicates the depth-averaged value, $\tau_b^{(i)}$ is the x_i-component of the bottom shear stress and $\tau_s^{(i)}$ the x_i-component of the shear stress at the water surface. A term $\left[D(\overline{u^2} - \overline{u}^2)\right]_x + D(\overline{uv} - \overline{u}\,\overline{v})_y$ has been neglected in the x-momentum equation and a similar term in the y-momentum equation.

A.4. Earth's Rotation

Coriolis acceleration

Fluid parcels moving at the earth's surface experience an acceleration due to the eastward rotation of the earth around the N-S axis, the so-called Coriolis acceleration. This acceleration can be considered a purely kinematic effect by noting that time derivation in a rotating frame introduces a term related to rotation of the axes of reference:

$$d\vec{r}/dt|_{fixed\,frame} = \vec{u} + \vec{\Omega} \times \vec{r},$$

where $\vec{r}$ indicates the position of a fluid parcel on the rotating earth ($\vec{r} = 0$ at the earth's centre), $\vec{u} \equiv d\vec{r}/dt$ the fluid velocity on the rotating earth, $\vec{\Omega}$ the rotation vector along the N-S rotation axis. The radial earth rotation frequency $\Omega \approx 7.3\,10^{-5}s^{-1}$. The fixed frame does not rotate. The acceleration then follows from

$$d^2\vec{r}/dt^2|_{fixed\,frame} = d\vec{u}/dt + \vec{\Omega} \times \vec{u} + \vec{\Omega} \times \vec{u} + \vec{\Omega} \times (\vec{\Omega} \times \vec{r}).$$

The term $d\vec{u}/dt$ is the acceleration on the rotating earth, $2\vec{\Omega} \times \vec{u}$ is the Coriolis acceleration and the term $\vec{\Omega} \times (\vec{\Omega} \times \vec{r})$ is the component of the centrifugal force compensated by earth's attraction and adjustment of the equilibrium sea surface slope. If no forces are acting on the fluid and in the absence of friction, $d^2\vec{r}/dt^2 = 0$ in the fixed frame.

In order to find the more usual formulas for the Coriolis acceleration we consider a system of axes on the rotating earth where θ is the azimuthal angle indicating longitude, ϕ the elevation angle indicating latitude and R is the earth radius. The x-direction is West-East, the y-direction South-North and the z-direction outward from the earth centre. In this coordinate system the vector components are

$$\vec{u} = (u = R\cos\phi \, d\theta/dt, \quad v = Rd\phi/dt, 0),$$

$$\vec{\Omega} = (0, \Omega\cos\phi, \Omega\sin\phi).$$

The Coriolis acceleration is then given by:

$$du/dt = fv, \; dv/dt = -fu, \; f = 2\Omega\sin\phi. \qquad\text{(A.13)}$$

Earth's rotation has a significant influence on stationary currents or on slowly varying currents, such as tidal motion. For currents varying on time scales much shorter than the Coriolis time scale $1/f$, such as wind waves or swell, the effect of earth's rotation can be ignored.

Momentum balance for large spatial and temporal scales

The influence of earth's rotation can also often be ignored in certain tidal rivers, estuaries or other narrow coastal systems. The Coriolis acceleration induced by a decrease or increase of the current velocity is counteracted by a cross-flow inclination of the water surface. The surface inclination in a channel along the x-axis induced by earth rotation $\eta_y = -fu/g$ is often small and dominated by the effects of river or channel meandering. However, Coriolis acceleration has to be taken into account in wide bays, estuaries and lagoons.

Earth's rotation needs to be included in the momentum balance equations (A.11). These equations can be rewritten in a simpler form by using the mass balance equation to eliminate the depth from the acceleration terms. After substitution of (A.13) and dropping the overbars of the depth-averaged velocities, the equations (Navier-Stokes equations) finally read

$$u_t + uu_x + vu_y - fv + g\eta_x + \frac{\tau_b^{(x)} - \tau_s^{(x)}}{\rho D} = 0, \quad \text{(A.14)}$$

$$v_t + uv_x + vv_y + fu + g\eta_y + \frac{\tau_b^{(y)} - \tau_s^{(y)}}{\rho D} = 0. \quad \text{(A.15)}$$

In these equations the usual notation of the partial derivatives has been used; $\tau_b^{(x)}$, $\tau_b^{(y)}$ are the x, y-components of the bottom shear stress and $\tau_s^{(x)}$, $\tau_s^{(y)}$ the x, y-components of the surface shear stress.

A.5. Vorticity Balance

For studying residual circulation or other spatial structures of the flow velocity field, it is often more practical to use the vorticity balance instead of the momentum balance. The vorticity balance describes angular momentum conservation in a fluid and can be derived from the momentum balance.

The horizontal depth-integrated vorticity and potential vorticity are defined as

$$\zeta = v_x - u_y, \quad \zeta^{pot} = \frac{\zeta + f}{D}, \quad \text{(A.16)}$$

where u, v are depth-integrated velocities. For potential flow ($u = \Phi_x$, $v = \Phi_y$) the vorticity ζ is zero, for circular rotational currents the vorticity equals twice the angular velocity. The potential vorticity ζ^{pot} is the sum of flow vorticity and planetary vorticity (Coriolis acceleration) divided by depth. The vorticity balance can be obtained from the depth-averaged momentum and mass balance equations using the curl operator. In this way the pressure gradients

p_x, p_y are eliminated and the result reads

$$\zeta_t^{pot} + u\zeta_x^{pot} + v\zeta_y^{pot} + \frac{1}{D}\left(\left(\frac{\tau_b^{(y)}}{\rho D}\right)_x - \left(\frac{\tau_b^{(x)}}{\rho D}\right)_y\right) = 0. \quad (A.17)$$

In the absence of friction, the left-hand side of Eq. (A.17) can be written as a time differential d/dt along the flow trajectory,

$$d\zeta^{pot}/dt = 0. \qquad (A.18)$$

Equation (A.18) states that, in absence of bottom friction, potential vorticity is conserved along a flow trajectory.

Appendix B

Tidal Propagation in One Dimension

B.1. 1D Tidal Equations

Channelised tidal flow

Flood and ebb flow in tidal inlet systems, such as tidal basins, estuaries and tidal rivers, is generally concentrated in a main channel. Tidal propagation also follows this main channel. This implies that tidal propagation in inlet systems can be described approximately by a set of one-dimensional equations where the longitudinal coordinate x follows the channel axis. The balance equations for mass and momentum describe how a water level change, imposed by external conditions at the inlet (tide, wave, wind) propagates landward and how a reflected wave propagates seaward. The balance equations are established for each infinitely short segment of the inlet system. The momentum balance equation is the equivalent of Newton's Law and describes flow acceleration or deceleration due to a pressure gradient produced by a water level difference over the segment and due to frictional momentum transfer to the bottom. The mass balance equation states that a difference between inflow and outflow produces an equivalent change of water volume in a segment through raising or lowering the water level.

One-dimensional flow schematisation

For the derivation of these equations, assumptions have to be made concerning the flow distribution over the basin cross-section $A(x, t)$.

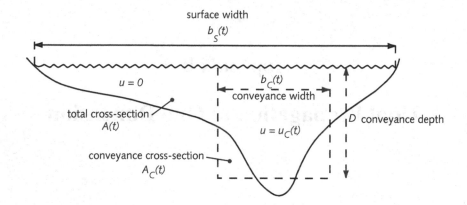

Fig. B.1.   Schematisation of the cross-section and definition of notations.

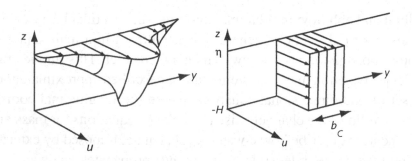

Fig. B.2.   1D-schematisation (right) of the flow velocity distribution (left).

The basic assumption is that for each cross-section a distinction can be made between two parts: (1) A momentum-conveying part (channel section) of the cross-section, where almost all along-channel flow is concentrated, and (2) a storage part, corresponding to banks and tidal flats, where water hardly flows in the along-channel direction, see Figs. B.1 and B.2. This is not an accurate reflection of the actual flow pattern, but it is generally sufficient for describing the major characteristics of tidal propagation [248]. The conveyance section has a cross-section $A_C(x, t)$, which is the product of a representative conveyance width $b_C(x, t)$ and a representative conveyance depth $D(x, t)$ at each time t. Instead of conveyance section, we will more often use the term channel cross-section. From the one-dimensional equations alone one cannot derive which part of the cross-section should be considered as conveying and which part as storage area. For practical computations estimates are made based on measuring campaigns or 2D and 3D numerical flow models. As a rule of thumb, the conveyance section can be taken as the part of the cross-section which is deeper than 1 to 2 metres [248].

1D flow variables

We consider a cyclic tide with period $T = 2\pi/\omega$. For describing water level and flow velocity we introduce the following quantities:

$Z_s(x, t)$ is the instantaneous water level relative to a horizontal reference level,

$Z_b(x)$ is the seabed level relative to the same reference level,

$\eta(x, t)$ is the tidal component of the instantaneous water level Z_s; the tidal average of this quantity is by definition equal to zero,

$h(x)$ is the tidal average of the instantaneous water depth D,

I is the tidally averaged water surface slope,

$b_S(x, t)$ is the total cross-sectional width at the water surface,

$Q(x, t)$ is the instantaneous discharge,

$u_C(x, t)$ is the average flow velocity in the conveyance section A_C.

These quantities are related to each other as follows

$$D = Z_s - Z_b = h + \eta, \quad h = \langle D \rangle = \langle Z_s \rangle - Z_b, \quad I = \langle Z_{sx} \rangle,$$
$$D_t = Z_{st} = \eta_t, \quad A_t = b_S \eta_t,$$
$$Q = A\overline{\overline{u}} = A_C u_C = D b_C u_C. \tag{B.1}$$

Angle brackets $\langle .. \rangle$ designate tidal averages and the double line over the flow velocity u stands for averaging over the total cross-section.

1D mass balance equation

The flow is characterised by two variables, η and u_C. They can be derived from the two equations for mass balance and momentum balance. The mass balance (continuity) equation reads

$$(A_C u_C)_x + b_S \eta_t = 0. \tag{B.2}$$

The first term describes the volume change due to inflow and out-flow, the second term describes the volume change due to water level change. The mass balance equation may also be written as

$$D_S(u_C)_x + D_S u_C (\ln(A_C))_x + \eta_t = 0 \tag{B.3}$$

with

$$D_S = A_C/b_S = D b_C/b_S. \tag{B.4}$$

The quantity D_S will be called 'propagation depth', because of its important influence on the propagation speed of the tidal wave. In tidal basins with large intertidal areas, the propagation depth is substantially smaller than the channel depth D, especially around high water. In that case the tidal variation of the propagation depth is stronger than the tidal variation of channel depth; tidal variation of the propagation depth is responsible for the high nonlinearity of the mass balance equation and for distortion of the tidal wave when propagating through the basin.

1D momentum balance equation

The momentum balance equation reads

$$(A_C u_C)_t + (A_C u_C^2)_x + Q_t^{trans} + g A_C Z_{sx} + \frac{b_C}{\rho} \tau_b = 0. \tag{B.5}$$

The first term describes the momentum change, the fourth term is the surface slope pressure gradient and the last term represents momentum transfer to the channel bed. The second and third terms describe the inflow-outflow balance of along-flow and cross-flow momentum respectively,

$$(A_C u_C^2)_x + Q_t^{trans} = (A\overline{\overline{u^2}})_x.$$

The surface stress is assumed zero. The bottom stress τ_b will be assumed proportional to the square of the velocity u_C as discussed in Chapter 3 Sec. 3.2.2,

$$\tau_b = \rho c_D |u_C| u_C.$$

Momentum exchange with storage areas

The term Q_t^{trans} describes the momentum carried to and from the storage zone

$$Q_t^{trans} = (A_t - A_{Ct})u^{ex}, \tag{B.6}$$

where ρu^{ex} is the momentum carried from the channel to the storage zone when $\eta_t > 0$ and the momentum carried from the storage zone to the channel when $\eta_t < 0$. Especially in the second case this momentum is smaller than the average momentum ρu_C carried by the flow in the channel.

The momentum balance can be written in a more convenient way by developing the product terms and using the mass balance equation; this yields

$$u_{Ct} + u_C u_{Cx} + g Z_{sx} + c_D \frac{|u_C| u_C}{D}$$

$$= \frac{A_t u_C}{A_C} \left(1 - \frac{A_{Ct}}{A_t} \right) \left(1 - \frac{u^{ex}}{u_C} \right). \tag{B.7}$$

We will ignore the term at the right-hand side of this equation. This is a reasonable approximation only if the tidal amplitude is very small compared to the depth or if the momentum exchanged between the channel and the storage zone is not very different from the momentum conveyed in the channel. This last assumption is more justified for flood than for ebb; it implies that during ebb the bottom friction term does not fully represent the total momentum loss. In tidal basins with a great width-to-depth ratio (>100), transverse mixing is a slow process relative to the tidal timescale. In these basins momentum loss on the tidal flats affects primarily the flow along the channel bank and to a lesser degree the flow at the centre of the channel.

In the following we will consider the momentum balance equation

$$u_t + u u_x + g(I + \eta_x) + c_D \frac{|u|\, u}{D} = 0, \tag{B.8}$$

instead of (B.7). This means that the cross-sectional geometry only plays a role in the mass balance equation and not in the momentum balance equation. From now on we leave out the index C from the flow velocity.

B.2. Simplification of Tidal Equations

The one-dimensional tidal Eqs. (B.2) and (B.8) have no simple analytical solution. They are usually solved by numerical methods. However, solutions in the form of analytical expressions give often better insight in tidal propagation dynamics. In some particular cases approximate analytical solutions exist. In this section we will examen in which particular situations the tidal equations can be simplified, such that analytical solutions can be obtained.

Long basins

In our definition, long tidal basins have a length comparable to or longer than the inverse tidal wave number $1/k = c/\omega = L/2\pi$. Here, c is the tidal wave propagation speed of magnitude $c \sim \sqrt{gh}$, $\omega = 2\pi/T$ the tidal angular frequency and L the tidal wavelength. In long basins, the magnitude of the x-derivatives of the tide elevation η and the tidal current velocity u is of order $k\eta$ and ku. The current velocity has order of magnitude ac/h, if friction is not very strong. The time derivatives of η and u have order of magnitude $\omega\eta$ and ωu, irrespective of the basin length. For the relative magnitude of the different terms in the momentum balance equation (B.8) we then find

$$O\left[\frac{u u_x}{u_t}\right] = \frac{a}{h}, \quad O\left[\frac{g\eta_x}{u_t}\right] = 1, \quad O\left[\frac{c_D u^2}{D u_t}\right] = \frac{c_D u}{h\omega}.$$

The momentum advection term uu_x in the tidal equations can be neglected if $a/h \ll 1$. We have $\omega = 1.4\,10^{-4}$, $c_D \approx 3\,10^{-3}$. The friction term dominates over the inertial term in long, shallow basins (depth of a few metres or less), where $c_D u \gg h\omega$. In such basins, the tidal current velocities are much smaller than $a\omega/kh$ and the term u_t in the momentum balance equation can be neglected.

The mass balance equation (B.2) can generally not be simplified; the different terms, $b_S\eta_t$, $u(A_C)_x$, $A_C u_x$, are all of comparable magnitude.

Short basins

If the basin length l is shorter than the inverse tidal wave number, the tidal current velocity is of order $u \sim al\omega/h$ and the x-derivative is of order $u_x \sim a\omega/h$. The relative magnitude of the terms in the momentum balance equation (B.8) is

$$O\left[\frac{uu_x}{u_t}\right] = \frac{a}{h}, \quad O\left[\frac{g\eta_x}{u_t}\right] = \frac{1}{kl}, \quad O\left[\frac{c_D u^2}{Du_t}\right] = \frac{c_D al}{h^2}.$$

The momentum advection term uu_x in the tidal equations can be neglected if $a/h \ll 1$. The inertial term u_t can be neglected if $kl \ll 1$. The friction term is important only in very shallow basins; otherwise the tide level η is approximately constant along the basin. For short basins, the mass balance equation (B.2) cannot be simplified.

Strongly converging basins

In strongly converging basins, the mass balance equation can be simplified. The x-derivative of the cross-section area is of order $A_x \sim A/l$. If $kl \ll 1$, the term $(u_C)_x$ in (B.3) is much smaller than $u_C(\ln(A_C))_x$ and can be neglected.

Linear friction

The quadratic form of the friction term greatly complicates the solution of the tidal equations. Therefore we replace the quadratic expression for frictional momentum dissipation by a linear expression

$$\tau_b = \rho c_D |u| u \approx \rho r u, \tag{B.9}$$

where r is the linearised friction coefficient. This linear expression represents a reasonable approximation of the friction experienced by the main tidal component if tidal flow is much stronger than the residual flow (river flow, Stokes drift), $\langle |u| \rangle \gg \langle u \rangle$. If river flow is significant compared to tidal flow, the quadratic expression generates additional tidal components which are absent in the linear expression [248]. In the absence of river flow, the coefficient r can be related to the friction coefficient c_D by requiring that the quadratic and linear expressions yield an identical energy dissipation, on average over the tidal period

$$r \approx c_D \frac{\langle |u|^3 \rangle}{\langle |u|^2 \rangle} = \frac{8}{3\pi} c_D u_{max}, \tag{B.10}$$

where u_{max} is the tidal velocity amplitude.

B.3. Nonlinear Tides in a Uniform Channel

The linear tides which are solution of the first-order tidal equations represent a very rough approximation of tidal propagation. In reality the tidal wave is distorted in shallow water due to the nonlinear dynamics of tidal wave propagation. Several terms in the mass- and momentum-balance equations contribute to this nonlinearity. It induces tidal asymmetry; flood and ebb flow are not mirror images of each other, but are different in strength an duration, see Chapter 6. Tidal asymmetry has important consequences for sediment transport in tidal inlet systems, see Fig. B.3. In this section we will investigate the influence of nonlinear terms in the tidal equations on distortion of the tidal wave. We will first consider the frictionless case and a uniform channel of infinite length.

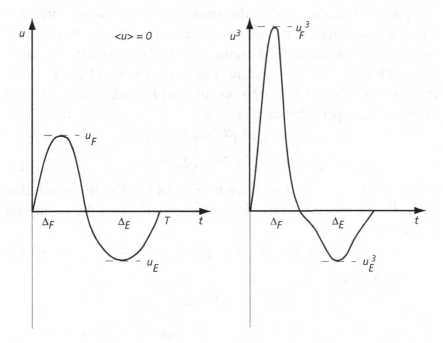

Fig. B.3. Sediment fluxes depend strongly on the tidal current velocity (often as a power $n \geq 3$, see Sec. 3.8). A small ebb-flood asymmetry can therefore produce a strong net tidally induced sediment transport.

Frictionless tidal propagation

The one-dimensional nonlinear tidal equations for frictionless flow read

$$b_S \eta_t + (b_C D u)_x = 0, \quad u_t + u u_x + g \eta_x = 0. \tag{B.11}$$

We will assume that the storage width b_S depends linearly on the tide level, $b_S = b_S^0 (1 + p \eta / h)$. The hypsometric parameter p is a measure for the relative tidal flat area,

$$p = \frac{h}{a} \frac{b_S^+ - b_S^-}{b_S^+ + b_S^-}, \tag{B.12}$$

where $b_S^\pm$ is the total wetted basin width at HW and LW, respectively. The instantaneous water depth is given by $D = h(1 + \eta/h)$ and we ignore for simplicity the tidal variation of the channel width b_C. Analytical solutions of the nonlinear tidal equations can be obtained by assuming that the nonlinear terms are small compared to the linear terms. This is the case if the tidal amplitude a is small compared to the depth h. We call $\epsilon = a/h$ and expand the tide level η and the velocity u as a perturbations series in ϵ,

$$\eta = \epsilon\eta^{(1)} + \epsilon^2\eta^{(2)} + \epsilon^3\eta^{(3)} + \cdots,$$

$$u = \epsilon u^{(1)} + \epsilon^2 u^{(2)} + \epsilon^3 u^{(3)} + \cdots. \tag{B.13}$$

At the boundary $x = 0$ we have $\eta = a\cos\omega t$. The linear solution $\eta^{(1)}, u^{(1)}$ corresponds to the first order terms in ϵ of the expansion of (B.11),

$$\eta_t^{(1)} + h_S u_x^{(1)} = 0, \quad u_t^{(1)} + g\eta_x^{(1)} = 0. \tag{B.14}$$

The solution is

$$\eta^{(1)} = a\cos\theta, \quad u^{(1)} = \frac{a\omega}{kh_S}\cos\theta,$$

$$\theta = kx - \omega t, \quad k = \frac{\omega}{\sqrt{gh_S}}, \tag{B.15}$$

where $h_S = hb_C/b_S^0$. The second order equations are linear in $\eta^{(2)}, u^{(2)}$,

$$\eta_{tt}^{(2)} - gh_S\eta_{xx}^{(2)} = -\frac{p}{h}\left(\eta^{(1)}\eta_t^{(1)}\right)_t$$

$$- \frac{h_S}{h}\left(\eta^{(1)}u^{(1)}\right)_{xt} + h_S(u^{(1)}u_x^{(1)})_x. \tag{B.16}$$

At the boundary $x = 0$ we have $\eta^{(2)} = 0$. The solution for the first and second order approximations is

$$\eta = a\cos\theta + \frac{a^2}{4h}\left(3 - p\frac{h_S}{h}\right)kx\sin 2\theta,$$

$$u = \frac{a\omega}{h_S k}\left[cos\theta + \frac{a}{4h_S}\left(\cos 2\theta + \left(3 - p\frac{h_S}{h}\right)\right.\right.$$

$$\left.\left. \times\left(-\frac{1}{2}\cos 2\theta + kx\sin 2\theta\right)\right)\right]. \tag{B.17}$$

These formulas clearly display tidal asymmetry. For small values of p (small intertidal area), tidal rise becomes faster and tidal fall slower as the tide propagates. The same applies to the flow velocity: When the tide turns from ebb to flood, the low-velocity period is shorter than for the turning of the tide from flood to ebb (for small p). Tidal asymmetry is related to differences in wave propagation speed during the tidal period, as will be shown below. From the function (B.17) an expression can be derived for the propagation speed at high water (HW, designated by superscript $+$) and low water (LW, designated by superscript $-$). Therefore we determine the location $x^+(t)$ of the wave crest by solving the equation

$$\frac{d}{dt}\eta_x(x^+(t), t) = 0 \tag{B.18}$$

in a small zone around the wave crest where $kx^+(t) - \omega t \ll 1$. The location $x^-(t)$ of the wave trough is determined in a similar way. Up to order a/h we find

$$c^{(\pm)} = \frac{dx^{(\pm)}}{dt} = \sqrt{gh_S}\left(1 \pm \frac{a}{h} + \frac{a}{2h_S} \mp \frac{ap}{2h}\right). \tag{B.19}$$

This expression is equal to (6.33) to first order in the small quantities a/h, $(b_S^+ - b_S^-)/(b_S^+ + b_S^-)$. The propagation speed at HW is higher than the propagation speed at LW, if the relative tidal flat area ap/h is small. This explains why the period of rising tide in this case becomes shorter and the period of falling tide longer as the tide propagates through the channel, see Fig. B.4.

Tidal propagation for strong friction

The one-dimensional nonlinear equations for friction-dominated tidal flow read

$$b_S\eta_t + (b_C Du)_x = 0, \quad g\eta_x + \frac{ru}{D} = 0. \tag{B.20}$$

Elimination of u from these equations gives

$$\eta_t = \frac{b_C}{b_S}\left(\frac{gD^2}{r}\eta_x\right)_x. \tag{B.21}$$

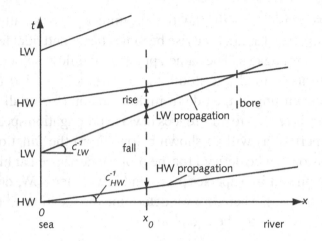

Fig. B.4. Schematic representation of the propagation paths $x^+(t)$ and $x^-(t)$ for nonlinear tidal wave propagation. The HW propagation path corresponds to $x^+(t)$ (or equivalently $t^+(x)$); $x^-(t)$ (or $t^-(x)$) represent the LW propagation path. In this example, LW-propagation is slower than HW-propagation. A tidal bore is generated when the HW wave crest catches up with the LW wave trough.

We expand η as a perturbation series in the small parameter $\epsilon = a/h$ and proceed in the same way as for the frictionless case. We consider a uniform channel of infinite length, with boundary condition $\eta = a \cos \omega t$ at $x = 0$. The first order terms in ϵ of the expansion of (B.21) are

$$\eta_t^{(1)} = \frac{ghhs}{r}\eta_{xx}^{(1)}. \tag{B.22}$$

The solution is, in complex notation,

$$\eta^{(1)} = \frac{a}{2}e^{i\theta} + cc, \quad \theta = (k + i\mu)x - \omega t, \quad k = \mu = \sqrt{\frac{\omega r}{2ghhs}}, \tag{B.23}$$

where cc indicates the complex conjugate. The linear second order equation is given by

$$\eta_t^{(2)} - \frac{\omega}{2k^2}\eta_{xx}^{(2)} = \frac{\omega}{2ak^2}\left(-p\eta^{(1)}\eta_{xx}^{(1)} + (\eta^{(1)^2})_{xx}\right)$$

$$= \frac{a\omega}{4}\left[i(p-4)e^{2i\theta} + 2e^{-2\mu x} + cc\right]. \tag{B.24}$$

At the boundary $x = 0$ we have $\eta^{(2)} = 0$. The solution for the first and second order approximations is

$$\eta = \frac{1}{2}e^{i\theta} + \frac{a}{8h}(p-4)(e^{2i\theta} - e^{2i\theta'}) + \frac{a}{4h}(1 - e^{-2\mu x}) + cc,$$

(B.25)

where $\theta' = (k + i\mu)x/\sqrt{2} - \omega t$. From the solution (B.25) and the condition (B.18) we can derive the propagation speed at high water and low water. Up to order a/h we have

$$c^{(\pm)} = \frac{dx^{(\pm)}}{dt} = \sqrt{\frac{2g\omega h h_S}{r}}\left(1 \pm \frac{a}{2h}(4 - 2\sqrt{2} - p)\right). \quad (B.26)$$

The propagation speed at HW is higher than the propagation speed at LW, as for the frictionless case.

Wave-propagation method

Approximate estimates of the tide propagation speed can be found by a simple method without solving the nonlinear equations. We call this approximate estimate the 'wave-propagation method'. For short periods around the wave crest and the wave trough the tide level η hardly varies with x. The frictionless tidal equations (B.11) are therefore approximately linear in short x and t intervals around the wave crest and the wave trough, apart from the small nonlinear terms $u_x u$ and $b_C D_x u$. The expressions for the HW and LW wave speed can therefore be obtained by substituting the HW and LW values of the tide level in the linear solution. According to this method, the HW and LW tide propagation speeds $c^{(\pm)}$ are given by

$$c^{(\pm)} = \sqrt{gD_S^{(\pm)}} = \sqrt{gh_S\frac{1 \pm a/h}{1 \pm pa/h}} \approx \sqrt{gh_S}(1 \pm \frac{a}{2h}(1 - p)).$$

In order to recover (B.19), we have to add the current velocity $u \approx (a/h_S)\sqrt{gD_S^{(\pm)}}$ and a small correction term $(a/2h - a/2h_S)\sqrt{gD_S^{(\pm)}}$.

For the situation with strongly dominating friction, the same method can be used, but it is less accurate. For friction-dominated tidal flow, the time derivative of the tide level η is not close to zero around the wave crest (and wave trough). The tidal equations are therefore not linear in a short time interval around the wave crest and the wave trough. If we ignore this fact, the propagation speed of the wave crest (resp. wave trough) is given by

$$c^{(\pm)} = \sqrt{\frac{2g\omega D^{(\pm)} D_S^{(\pm)}}{r}} \approx \sqrt{\frac{2g\omega h h_S}{r}} \left[1 \pm \frac{a}{2h}(2 - p) \right].$$

This differs from the second-order solution (B.26) of the nonlinear tidal equations; the HW propagation speed is overestimated and the LW speed underestimated. However, in most practical cases the difference is smaller than the inaccuracies inherent to the simple one-dimensional representation of tidal propagation.

The influence of hypsometry (the parameter p) on the tidal propagation speed yields the same factor for strong and weak friction. This is not surprising, because slowing down of tide propagation by tidal flat filling is a kinematic effect, which, to first order, is independent of friction.

In practical situations, tidal propagation is neither frictionless nor strongly friction-dominated, but somewhere in between. Therefore the wave-propagation method yields usually reasonable estimates for the HW and LW propagation speeds. We will use the wave-propagation method, because it yields simple expressions for wave asymmetry as a function of easily measurable bathymetric parameters.

Local excitation and damping of higher harmonics

Equations (B.16) and (B.24) illustrate how the first harmonic overtide M4 is generated and how it propagates. The r.h.s. of these equation represents the generation terms. For weak friction, Eq. (B.16) shows that the M4-tide (corresponding $\eta^{(2)}$, $u^{(2)}$) is generated by the positive correlation between water depth and tidal velocity in the mass-balance, by filling and emptying of tidal flats and by advection in the momentum balance. For strong friction, Eq. (B.24) shows that the M4-tide is generated by the ebb-flood asymmetry in the friction term of the momentum balance and by filling and emptying of tidal flats.

The l.h.s. of Eqs. (B.16) and (B.24) represents the propagation of the M4-tide. For weak friction, the M4-tide propagates as a wave. This wave is reflected at the basin boundaries and may become resonant if the length scale of the basin is close to 1/8 times the M2 tidal wavelength.

In the case of strong friction, the locally generated M4-tide may still be reflected at the basin boundaries, but it will not resonate because the M4 tidal wave is dissipated and diffused when propagating away from the location of generation. The frequently used representation of tidal asymmetry by a M4 tidal wave can therefore be misleading.

Linearisation of the friction term

In all the analytical models the bottom shear stress is linearised. However, the friction coefficient is in reality time-dependent and it equals zero at slack water. This may cast some doubts on the validity of the propagation method, where tidal asymmetry is related to different propagation speeds of the tidal wave in the periods around HW and LW. In fact, high and low slack water (HSW and LSW) almost coincide with HW and LW in friction-dominated, strongly convergent basins and in short, weakly converging basins. Tidal propagation is therefore almost frictionless at HW and LW. The friction-dependent propagation velocity may be considered as a reasonable approximation only some time before and after slack water, but not during slack water.

Numerical simulations show that quadratic friction does not strongly modify the overall asymmetry of the tidal curve, compared to linear friction. However, it does affect the shape of the tidal curve around slack water. This point has been investigated by

Lanzoni and Seminara [505], who showed that the tidal equation (B.8) with nonlinear friction produces physically unrealistic peaking of the tidal wave around HW and LW. Removal of this anomaly requires introduction of other nonlinear terms. The term $D_S u_x$ in the mass balance equation (B.3), in particular, contributes to smoothing the tidal curve around slack water. This indicates that models with linearised friction should be interpreted with care. The simple linear models or quasi-linear models with idealised basin topography do not aim to reproduce tidal curves in real field situations with any precision; the models are intended only to reproduce the gross characteristics of the tidal wave, and, in particular, the degree of tidal asymmetry.

B.4. Tidal Wave in a Uniform Basin of Finite Length

No friction — standing tidal wave

Tidal lagoons have a finite length. The channel cross-section is assumed uniform; the tidal wave will therefore be reflected at the channel end $x = l$. The tide is then a superposition of an incoming and a reflected wave; together these waves form a standing wave (see Fig. B.5)

$$\eta(x, t) = \frac{1}{2} a_l \left[\cos(k(x - l) - \omega t) + \cos(k(x - l) + \omega t) \right]$$

$$= a \frac{\cos(k(x - l))}{\cos kl} \cos \omega t,$$

$$u(x, t) = \frac{1}{2} \frac{ac}{h_S \cos kl} \left[\cos(k(x - l) - \omega t) - \cos(k(x - l) + \omega t) \right]$$

$$= u_0 \frac{\sin(k(x - l))}{\sin kl} \sin \omega t, \tag{B.27}$$

where

$$h_S = \frac{A_C}{b_S}, \quad a_l = \frac{a}{\cos kl}, \quad u_0 = u(0, 0) = -\frac{ac \tan kl}{h_S}, \quad c = \sqrt{g h_S}.$$

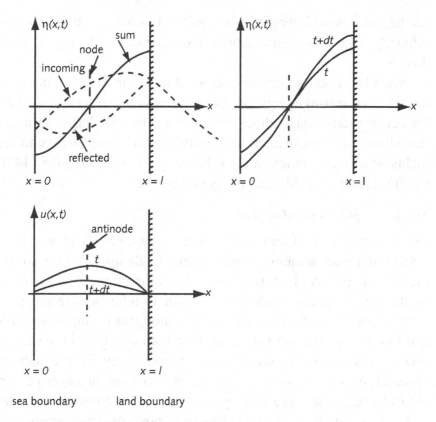

Fig. B.5.   Representation of a standing wave as the sum of an incoming wave at $x = 0$ and a reflected wave at $x = l$ in the absence of friction.

The velocity at the landward boundary $x = l$ is zero. Reflection of the tidal wave produces an increase of the tidal amplitude when approaching the landward boundary. The tidal variation of flow velocity is 90 degrees in advance of the tidal water level variation; flood flow corresponds to rising water and ebb flow to falling water. The mean water level during flood flow is the same as during ebb flow.

Resonance

Resonance will occur if the basin length is equal to a quarter of the tidal wave length,

$$l = L/4 = \pi/2k. \tag{B.28}$$

As the factor $\cos(kl)$ then becomes infinitely small, the tide becomes infinitely large. It is clear that bottom friction cannot be neglected in this case.

Apart from bottom friction, outward radiation of tidal energy at the inlet also causes energy loss from the basin to the open sea [612]. This energy radiation modifies the ocean tide near the basin entrance; modelling tidal propagation in a basin close to resonance requires inclusion of tidal motion in the offshore vicinity of the inlet [443]. Far from resonance this effect may be ignored.

Friction — partial standing wave

The tide in a basin of finite length with friction has the character of a damped, partial standing wave. If frictional damping is very strong the tidal amplitude decreases in landward direction; in the case of weak friction the amplitude increases due to reflection. In practice, the tidal amplitude does not strongly vary along the basin; around HW and LW the x- and t-derivatives of η are both small. In this case, the tidal equations can be considered as approximately linear equations around HW and LW. We therefore use the wave propagation method for estimating the wave propagation speed. We consider the tidal Eqs. (B.2) and (B.8), in which only the most important linear and nonlinear terms are retained and where the friction is linearised,

$$D_S u_x + \eta_t = 0, \quad u_t + g\eta_x + ru/D = 0. \tag{B.29}$$

We consider short time intervals around HW and LW where the depth D and the propagation depth D_S are replaced by constants, $D^\pm$, $D_S^\pm$. The solution involves a superposition of incoming and reflected tidal waves, similar to the frictionless case, which now includes exponential damping functions,

$$\eta = \frac{1}{2}a_l \left[e^{-\mu(x-l)} \cos(k(x-l) - \omega t) \right.$$

$$\left. + e^{\mu(x-l)} \cos(k(x-l) + \omega t) \right], \tag{B.30}$$

$$u = \frac{1}{2}\frac{a_l}{D_S}\frac{\omega}{\sqrt{k^2 + \mu^2}}\left[e^{-\mu(x-l)}\cos(k(x-l) - \omega t - \varphi)\right.$$

$$\left. - e^{\mu(x-l)}\cos(k(x-l) + \omega t + \varphi)\right], \quad \text{(B.31)}$$

where a_l is the tidal amplitude at the landward boundary,

$$a_l = \frac{a}{\sqrt{\cos^2 kl \cosh^2 \mu l + \sin^2 kl \sinh^2 \mu l}}. \quad \text{(B.32)}$$

For a resonant tidal basin length, $kl = \pi/2$, the tidal amplitude a_l and flow velocity u remain finite. The expressions (B.30) are approximately valid only around the HW and LW tidal phases. The complex wave number $k + i\mu$ and phase φ are given by

$$\binom{k}{\mu} = \frac{\omega}{\sqrt{2gD_S}}\sqrt{\pm 1 + \sqrt{1 + R^2}},$$

$$\cos\varphi = \frac{k}{\sqrt{k^2 + \mu^2}}, \quad R = \frac{r}{D\omega}. \quad \text{(B.33)}$$

The coefficient R is a measure of the strength of frictional damping relative to the inertial acceleration in the momentum balance equation. We have $k = k^{\pm}$, $c = c^{\pm}$, $\varphi = \varphi^{\pm}$, $R = R^{(\pm)}$, $D = D^{(\pm)}$, $D_S = D_S^{(\pm)}$ for HW and LW respectively. The time origin is chosen such that $t = 0$ coincides with HW at the landward basin boundary $x = l$.

Propagation speed of HW and LW

The propagation velocities of HW and LW throughout the basin are determined from the expression for $\eta(x, t)$ along the HW and LW trajectories. The difference in propagation speed between HW and LW causes a difference between periods of rising and falling tide. The difference between the ebb and flood flow durations can be determined from the expression of $u(x, t)$. This asymmetry between ebb and flood yields an expression for the difference between maximum flow velocities for ebb and flood, which can be related to a difference of sediment transport during the ebb and flood periods. The derivation is given below.

The times of HW and LW at any location are determined from the times of HW and LW at the basin end $x = l$, by solving $\eta_t = 0$. From the expression for η it follows that a solution should satisfy

$$e^{-\mu(x-l)}\sin(k(x-l) - \omega t) = e^{\mu(x-l)}\sin(k(x-l) + \omega t),$$

or

$$\tan(\omega t) = -\tan(k(x-l))\tanh(\mu(x-l)).$$

We consider basins which are short compared to the tidal wave length, i.e., $(\mu l)^2 \ll 1$, $(kl)^2 \ll 1$. In many tidal basins kl is close to 1; yet we approximate

$$\omega t \approx -k\mu(x-l)^2, \tag{B.34}$$

implying

$$t^{(\pm)}(x) - t^{(\pm)}(0) \approx \frac{k^{(\pm)}\mu^{(\pm)}}{\omega}\left[l^2 - (x-l)^2\right]. \tag{B.35}$$

The HW and LW trajectories $t^+(x)$, $t^-(x)$ are shown in Fig. B.6 for fictitious values of $k^\pm$, $\mu^\pm$.

Time of high water slack tide and low water slack tide

To determine the periods of ebb and flood, we have to determine the times HWS (HW slack tide) and LWS (LW slack tide). For this we solve the equation $u(x,t) = 0$,

$$e^{-\mu(x-l)}\cos(k(x-l) - \omega t - \varphi) = e^{\mu(x-l)}\cos(k(x-l) + \omega t + \varphi),$$

or

$$\tan(\omega t + \varphi) = \tanh(\mu(x-l))/\tan(k(x-l)).$$

For a tidal basin much shorter than the tidal wavelength ($kl < 1$) we find

$$\tan(\omega t + \varphi) \approx \frac{\mu}{k}\left(1 - \frac{1}{3}(k^2 + \mu^2)(x-l)^2\right).$$

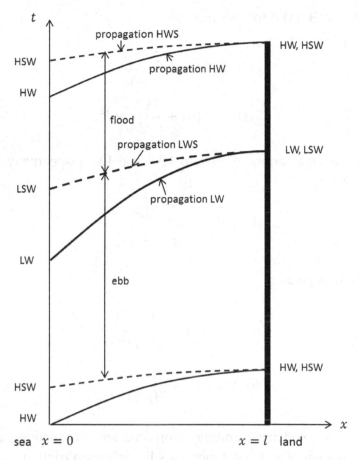

Fig. B.6. Location of HW, LW, and slack waters HWS and LWS as a function of time for a partially reflected damped wave in a tidal basin of length l.

Assuming $|\mu| < k$ and $\tan\varphi \approx \mu/k$, we have

$$t \approx -\frac{\mu k}{3\omega}(x - l)^2.$$

The times of high and low water slack are therefore given by

$$t_S^{(\pm)}(x) \approx t_S^{(\pm)}(0) + \frac{k^{(\pm)}\mu^{(\pm)}}{3\omega}(l^2 - (x - l)^2). \qquad \text{(B.36)}$$

Since high water and high water slack tide coincide when approaching the landward basin boundary ($x \to l$), we have

$$t_S^+(l) = t^+(l), \quad t_S^-(l) = t^-(l).$$

Then from (B.35) it follows that

$$t_S^+(l) \approx t^+(0) + \frac{2k^+\mu^+}{3\omega}l^2,$$

$$t_S^-(l) \approx t^-(0) + \frac{2k^-\mu^-}{3\omega}l^2, \tag{B.37}$$

where the superscripts $\pm$ refer to HW and LW respectively. From (B.33) we have

$$k^{(\pm)}\mu^{(\pm)} = \frac{\omega r}{2g D^{(\pm)} D_S^{(\pm)}}.$$

Substitution gives

$$t_S^+(0) \approx t^+(0) + \frac{rl^2}{3g}\frac{1}{D^+ D_S^+},$$

$$t_S^-(0) \approx t^-(0) + \frac{rl^2}{3g}\frac{1}{D^- D_S^-}. \tag{B.38}$$

It should be noted that the propagation time depends quadratically on the basin length l and that it increases linearly with friction. Without friction we have a standing tidal wave with zero time lag between inlet and basin head; the propagation time lag is entirely due to frictional influence on wave propagation. The quadratic dependence of propagation time on distance illustrates that friction induced wave propagation is basically a diffusion process.

Substitution of (B.38) in (B.36) finally gives for the difference $\Delta_{EF} = 2(t_S^-(x) - t_S^+(x)) - T$ between ebb and flood duration,

$$\Delta_{EF} \approx \Delta_{FR}^{inlet} + \frac{r}{3gh^2h_S^2}(3l^2 - (x-l)^2)(D^+ D_S^+ - D^- D_S^-), \tag{B.39}$$

where $\Delta_{FR}^{inlet} = 2(t^-(0) - t^+(0)) - T$ is the difference between the durations of falling and rising tide at the sea boundary.

The average time delay Δ_S between high (low) slack water and high (low) water at the sea boundary $x = 0$ is given by

$$\Delta_S = \frac{1}{2}\left(t_S^+(0) + t_S^-(0) - t^+(0) - t^-(0)\right) \approx \frac{rl^2}{3g}\frac{1}{hh_S}. \qquad \text{(B.40)}$$

Appendix C

Ocean and Shelf Tides

C.1. Tide Generation

Moon and sun

The regular daily upward and downward motion of the water surface along the coastline is the most visible expression of tidal forcing. In ancient Greece it was recognized that tides are in some way related to sun and moon, but this relationship could not be explained. The explanation of tidal motion as a consequence of gravitational forces was given by Newton. The lunar and solar tide generating forces acting on the oceans are extremely small; they are a factor $\approx 5.10^{-7}$ smaller than the gravitational force of the earth. The tide-generating force of the moon is about twice as large as that of the sun.

It may seem surprising that the small attraction forces of moon and sun are capable of producing water level elevations of several metres and tidal currents exceeding 1 m/s. The momentum dissipation associated with such strong tidal currents is much larger than the tide-generating forces, at least in continental shelf seas. This is precisely the reason why no substantial tidal motion is generated in shallow enclosed seas or lakes.

Ocean basin resonance

Tides are generated in ocean basins with water depths so large and tidal currents so weak that momentum dissipation is much smaller than the tide-generating forces. Ocean resonance plays an important role; tidal motion is amplified if the frequency of tidal forcing equals the frequency of free oscillations in the ocean basins. Tidal waves in wide ocean basins (width typically larger than 1000 km) rotate around points of zero amplitude, the amphidromic points; the rotation is counterclockwise in the Northern Hemisphere and clockwise in the Southern Hemisphere. Figure 6.15 shows the semi-diurnal tidal wave in the world's oceans; it comprises a large number of amphidromic systems, where the tide rotates in resonance with the 12 h 25 min periodic constituent of the gravitational force.

The tides generated in the ocean propagate to the continental shelf, where the tidal range often increases further. Different phenomena contribute to this amplification: resonant dimensions of the shelf sea, slowing down of wave-energy propagation ('shoaling') or concentration of the tidal energy flux in areas of reduced width ('funneling'). Tidal amplification in shelf seas is counteracted by frictional momentum dissipation. The strong tidal motion occurring in many shelf seas is thus not locally produced by tide generating forces, but results from co-oscillation with ocean tides and from local topographic amplification. Numerical tidal studies for the northwest European shelf [675] and for the East China shelf [477] show that the local tide generating force influences the co-oscillating semidiurnal tide in these shelf seas by no more than about 1%.

Semidiurnal periodicity

In order to explain tidal motion, it is not sufficient to consider only the gravitational forces exerted by moon and sun on the earth's water masses. Tide generation results from the local imbalance at the earth's surface of two opposing forces: The gravitational forces acting between the earth and the moon (and between the earth and the sun), and the centrifugal acceleration related to the orbital motions of these celestial bodies. In fact, only the tangential component of the resulting force is relevant, pointing towards the equator. Because the gravitational force and the centrifugal force cancel at any location on the earths surface just twice during each diurnal rotation, the major periodicity is approximately semidiurnal.

Diurnal tide and spring-neap cycle

Because the solar and lunar orbits do not coincide with the equatorial plane (the angle between orbital plane and equatorial plane is called declination), a daily inequality arises in the semidiurnal cycle. The daily inequality is strongest in shelf seas that resonate at diurnal frequency and which are situated close to amphodromic points of the semidiurnal tide. In such regions the tide is mainly diurnal.

Due to the ≈ 30-day orbital motion of the moon, the moon-earth and sun-earth axes approximately coincide every 15 days (syzygy). This causes a 15-day cycle of neap tide and spring tide; spring tide follows (with a short delay) full moon and new moon and neap tide follows half moon. In fact, the 15-day period corresponds to the frequency difference of the semidiurnal lunar component (M_2) and the semidiurnal solar component (S_2); the interference of these components produces the neap-spring variation of the tidal amplitude.

Other tidal components

The different cycles in the relative motions of moon, sun and earth surface generate tidal waves with corresponding periods. The lunar semidiurnal tidal component (M_2, period $\approx 12\,\text{h}\,25\,\text{min}$) is generally the largest tidal constituent (see Fig. C.1), followed by the semidiurnal solar tide S_2. Other important tidal constituents are:

- the diurnal component K_1, which is related to the declination of the lunar and solar orbits relative to the equatorial plane (period $\approx$ 24 h),
- the diurnal lunar component O_1 (period $\approx$ 26 h) and
- the diurnal solar component P_1 (period $\approx$ 24 h).

Tides with higher periodicity are generated due to the nonlinearity of tidal propagation. The periods of these higher harmonic components are multiples of the basic astronomical tidal periods.

Fig. C.1.   Coastal zones with dominant semi-diurnal tide.

These locally generated higher harmonic components are associated with tidal-wave distortion or tidal asymmetry. They play a crucial role in tide-topography interaction.

Asymmetric ocean tides

Symmetry of ocean tides applies to the different astronomic tidal components separately. If the frequencies ω_i of the major tidal components are not linearly related with integer coefficients (ω_i is different from any combination of sums and differences of ω_j, $j \neq i$), the tide is symmetric when averaged over a sufficiently long period. In this case the long-term average period of rising tide equals the average period of falling tide. However, certain combinations of astronomic components yield asymmetric tides, irrespective of the averaging period. The reason is that most astronomic tidal constituents result from a superposition of a limited number of basic periods. The periods of these tidal constituents correspond to sums and differences of the basic periods [245]. Therefore, different tidal constituents may interfere in such a way that flood and ebb are modulated in a systematic, asymmetric way. Such asymmetries may become significant in the case of strong diurnal tides; for dominant semidiurnal tides this effect is small [398].

A particular example is the combination of the M_2, K_1 and O_1 tides. The sum of the frequencies of the K_1 and O_1 tidal consituents equals the frequency of the M_2 tidal constituent. In regions where the amplitudes of the K_1 and O_1 tides are not small compared to the amplitude of the M_2 tide, the sum of these three tidal constituents yields an asymmetric tide, with an asymmetry depending on the respective phases of these tidal constituents.

C.2. Linear Tide

Simplifying assumptions

In this appendix we examine tidal propagation from ocean to coastal zone and in particular the transformation of the tide due to interaction with shelf topography. On wide semi-enclosed shelves the tide undergoes considerable modification. The tidal boundary conditions for coastal systems situated on such shelves are not well represented by the sinusoidal ocean tides; the distortion of the tidal wave in the shelf sea has to be considered. The discussion of tidal propagation in coastal seas is made more transparent by introducing a number of simplifications, listed below.

- Neglect of tide-generating forces by celestial bodies. The tide-generating forces are extremely small compared to the tidal momentum dissipation in shallow coastal seas.
- Neglect of vertical fluid accelerations. The horizontal scale of motion exceeds the vertical scale by several orders of magnitude.
- Neglect of the interaction of tidal flow with density stratification. Density stratification decreases energy dissipation and diminishes the interaction with seabed topography. In regions with strong tidal currents there is generally no strong density stratification.
- Neglect of velocity veering in the water column. The cross-flow acceleration due to earth rotation varies along the water column.

The resulting velocity veering affects momentum dissipation and invalidates the quadratic relationship between momentum dissipation and depth-averaged current velocity. This effect is of minor importance in shallow water (water depths of approximately 20 m or less).

- Neglect of horizontal momentum exchange. The horizontal scale of velocity gradients is much larger than the vertical scale; momentum exchange by horizontal velocity fluctuations is therefore much smaller than momentum exchange by vertical velocity fluctuations.
- Neglect of small-scale topographic structures. We average the tidal equations over spatial intervals which are small compared to the tidal wavelength but large compared to topographic features without significant influence on tidal propagation characteristics.
- Linearisation of frictional momentum dissipation. The quadratic nature of frictional momentum dissipation generates higher harmonic tides, in particular a tidal component with a frequency three times the frequency of the main tidal constituent. Ignoring this locally generated tidal component has only a minor influence on ebb-flood asymmetry.
- Neglect of other tidal constituents than the semidiurnal M_2 tide. Neglect of nonlinear terms terms in the tidal equations is equivalent to ignoring tidal distortion and excludes ebb-flood asymmetry. The magnitude of the nonlinear terms relative to the linear terms is of the order of the amplitude-to-depth ratio. If this ratio is small, tidal propagation can be described by linearised tidal equations. Analysis of tidal asymmetry requires consideration of the nonlinear terms.

Linear tidal equations

Considering these simplifications we may formulate tidal equations for the depth-averaged flow components u, v (we have dropped overbars), which are temporally averaged over turbulent fluctuations and spatially averaged over small scale topographic features. Pressure gradients are related to the water surface slope and frictional momentum dissipation is linearly related to the depth-averaged velocity. Momentum advection terms are neglected. The tidal equations then read

$$u_t - fv + g\eta_x + ru/h = 0, \tag{C.1}$$

$$v_t + fu + g\eta_y + rv/h = 0, \tag{C.2}$$

$$\eta_t + (hu)_x + (hv)_y = 0. \tag{C.3}$$

These equations describe several types of wave motion, depending on topography and boundary conditions. Two-dimensional tide propagation is strongly influenced by earth rotation, as the tidal radial frequency ω and the Coriolis parameter f have similar order of magnitude.

Kelvin wave along an infinite uniform coastline

In the vicinity of a coastline (at short distance compared to the tidal wavelength) the cross-shore flow component must vanish. The tidal wave that follows the shoreline with zero cross-shore flow is called Kelvin wave. We assume a straight infinite coastline and choose the x-axis shore-parallel and the y-axis shore-perpendicular. The along-shore tidal velocity u of the Kelvin wave is geostrophically balanced by a cross-shore water level slope

$$fu = -g\eta_y, \quad v = 0. \tag{C.4}$$

The tidal amplitude is therefore largest at the coast. For uniform depth h and negligible friction the Kelvin wave takes the form

$$\eta = ae^{-fy/c} \cos(kx - \omega t), \quad u = c\eta/h, \tag{C.5}$$

where $c = \sqrt{gh}$. The assumption of uniform depth implies that depth variations on the scale of the tidal wave length should be small.

Because of the exponential decrease of the tidal amplitude in cross-shore direction, the Kelvin wave can be considered as a coastally trapped wave. An important property of the Kelvin tidal wave is its propagation direction: It follows the coast at the right-hand-side on the northern hemisphere and at the left-hand side on the southern hemisphere.

Kelvin wave in a semi-enclosed coastal sea

We consider a rectangular bay or an approximately rectangular coastal sea of uniform depth, with a single open ocean boundary and characteristic length scale comparable to the tidal wavelength. The North Sea, the English Channel, the Yellow Sea and the Bering Sea are examples of coastal seas that can be schematised to a first approximation as semi-enclosed coastal seas. The incoming tide propagates from the ocean boundary ($x = 0$) into the basin as a damped Kelvin wave, following the right-hand side coastline (Northern Hemisphere) and, after reflection from the landward boundary ($x = l$), returns along the left-hand side coastline back towards the ocean

$$\eta = \frac{1}{2}a_l \, e^{-fy/c_1} e^{-\mu(x-l)} \cos(k(x - l) + \frac{fy}{c_2} - \omega t)$$
$$+\frac{1}{2}a_l \, e^{fy/c_1} e^{\mu(x-l)} \cos(k(x - l) + \frac{fy}{c_2} + \omega t), \quad \text{(C.6)}$$

where $y = 0$ is chosen as basin axis ($y = \pm b/2$ are the lateral basin boundaries) and where a_l is the tidal amplitude at the landward end of the basin axis ($x = l$, $y = 0$). Furthermore we have

$$c_1 = \frac{\omega}{k}(1 + R^2)^{1/2}, \quad c_2 = \frac{\omega}{\mu}(1 + R^2)^{1/2}, \quad R = r/\omega h,$$

$$k = \frac{\omega}{\sqrt{2gh}}(1 + (1 + R^2)^{1/2})^{1/2},$$

$$\mu = \frac{\omega}{\sqrt{2gh}}(-1 + (1 + R^2)^{1/2})^{1/2}.$$

The Kelvin wave satisfies the boundary condition $u = 0$ at the landward boundary $x = l$ only for $y = 0$. Hence, other solutions of the tidal equations (C.1) have to be considered as well (Poincaré waves). However, at sufficient distance (order b) from the landward boundary the Kelvin wave is the major tidal wave component.

The expression (C.6) shows that friction induces (1) a cyclonic rotation of the tidal wave crest and (2) a phase shift between the tide level η and the current velocity u. For strong friction the phase of u advances the phase of η by $T/8$; the phase advance is greater near the bottom than near the surface. Friction also modifies the strength and orientation of the tidal current vector down the vertical; near the bottom the current strength is reduced and each tide the current vector describes cyclonically an ellipse with a slightly anti-cyclonically rotated major axis [693].

Amphidromic point

In the absence of friction $(R = 0)$, the Kelvin wave rotates around an amphidromic point $\eta = 0$ situated at $x = l - T\sqrt{gh}/4$, $y = 0$. In the case of weak friction $(R^2 \ll 1)$, the amphidromic point is shifted in cyclonic direction relative to the propagation direction. From (C.6) we find the amphidromic point at $y = \pi R\sqrt{gh}/4f$.

Figure 6.27 shows the phase contours of the semidiurnal tidal wave in the North Sea. The tidal wave in the central North Sea is mainly driven by the Atlantic ocean tide at the northern boundary; the amphidromic point is shifted to the eastern part of the basin. The tidal wave in the Southern Bight is mainly driven by the tidal wave in the central North Sea. However, it is also influenced by the tidal wave in the English Channel, which causes a westward displacement of the amphidromic point.

C.3. Tidal Distortion Along the Coast

Distortion Kelvin wave is similar to 1D wave

If the tidal amplitude a is not very small compared to the mean water depth h, the nonlinear terms in the tidal equations cannot be neglected. While propagating along the coast the tidal wave will be distorted. For a Kelvin wave $(v = 0)$ this distortion follows from the equations of motion

$$u_t + uu_x + g\eta_x + c_D \frac{|u|u}{h+\eta} = 0,$$

$$\eta_t + [(h+\eta)u]_x = 0. \tag{C.7}$$

The form of these equations is identical to the 1D tidal equations discussed in Appendix B.3. A longshore propagating Kelvin wave is distorted in the same way as a tidal wave in a tidal basin without tidal flats: During propagation, tidal rise becomes faster, while tidal fall slows down. The distortion of the tidal wave with faster rise and slower fall can be represented by a M4 tidal component. If the M4 tidal component is generated by nonlinear propagation of the M2-tide, one should expect a relationship between the phases of the M2 and M4 tidal components. A numerical analysis of the tides in the southern North Sea shows that this is indeed the case for the tide traveling along the Dutch coast [692].

However, longshore propagating tidal waves are not always Kelvin waves. At the southern coast of the UK, for example, the M2 tidal wave has an amphidromic point. At the northern tip of the Holland coast tidal asymmetry changes due to interference of the amphidromic systems of the southern and central North Sea.

Field observations

Figure 6.27 shows the distortion of the tidal wave during its northward propagation along the Dutch coast. The strongest tidal asymmetry in the southern North Sea occurs along the coast of central and northern Holland. This is also the shallowest part of the Southern Bight, with relatively strong influence of friction on tidal propagation. Along the north-western coast of France the tidal wave is distorted in a similar way, see Fig. C.2.

Consequences of tidal distortion in coastal seas

In Sec. 6.5 the influence of offshore tidal asymmetry on the morphology of tidal basins is discussed in detail. This asymmetry corresponds

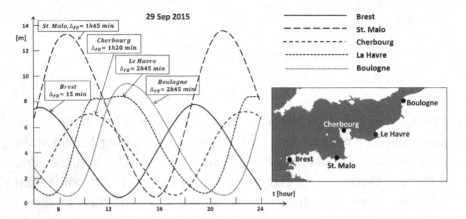

Fig. C.2. Distortion of the tidal wave while propagating along the north-western French coast at spring tide, 29 September 2015. Tidal asymmetry is characterised by the difference Δ_{FR} in duration between tidal fall and tidal rise. Brest is close to the ocean boundary, where the tide is almost sinusoidal. The tidal wave is amplified and distorted when propagating into the shallow Bay of Mont St. Michel (tidal curve at St. Malo). Propagating along the coast in north-eastern direction from Brest toward Cherbourg, Le Havre and Boulogne, the tidal asymmetry Δ_{FR} increases.

generally to a faster tidal rise and slower tidal fall and contributes to flood-dominant sediment transport at the inlet. Morphologic equilibrium requires neutralisation of this flood dominance; this can be achieved by return flow compensating for Stokes drift and by large intertidal flats slowing down HW propagation. The morphologic characteristics of tidal basins on wide-shelf coasts are therefore different from the characteristics of basins forced by symmetric ocean tides, see Sec. 6.5.1.

Distortion of the tidal wave in coastal seas also influences longshore residual sediment transport. A shorter tidal rise and longer tidal fall imply generally a shorter flood and longer ebb period, in particular in the case of important friction. Observations and numerical tidal models show, for instance, that along the Dutch coast net sand transport is directed to the north [674].

Appendix D

Wind Waves

D.1. Wave Theory

D.1.1. *Linear Theory*

In this appendix we will reproduce some results of classical wave theory, which are relevant for the chapter on wave-topography interaction. A more extensive treatment of wind wave theory can be found in many standard textbooks, for example [482, 812].

Monochromatic approximation

Waves incident on the coast are usually part of a wave field consisting of a broad spectrum of waves with different periods and wavelengths and with different shapes and propagation directions. This wave field exerts a force on the seabed and brings sediment in motion. Changes of near-shore coastal morphology results mainly from wave action averaged over long periods.

To avoid the difficulty of determining the action of many different individual waves it is assumed that the average sediment transport produced by the real wavefield is identical to the sediment transport produced by a hypothetical wavefield consisting of single-frequency (monochromatic) waves with the same energy as the average of the real wavefield and with a frequency and direction corresponding to the peak of the energy spectrum. We will adopt this approximation (although it is not always valid, see Chapter 7) and concentrate on single-frequency waves.

Based on this approximation, we will formulate equations for the description of wind wave propagation. The solution of these equations yields estimates of wave-orbital motion and wave-induced currents, which are important for wave-topography interaction.

Potential flow

In Sec. 3.3.1 we have demonstrated that almost no turbulence is generated by wave propagation; turbulent flow is restricted to a very thin layer near the bottom. This implies that above this turbulent layer, wave dynamics can be described by potential flow. The wave-orbital motion in this non-turbulent region is usually called 'free-stream wave-orbital flow'.

The turbulent flow in the wave-boundary layer is driven by the free-stream wave-orbital flow, which itself is hardly influenced by the flow in the wave-boundary layer. The flow in the wave-boundary layer is crucial for sediment transport.

We consider a single-frequency (monochromatic) wave, propagating in the negative x-direction, towards the shore. In the case of potential flow, the horizontal and vertical wave-orbital velocities $u(x, z, t)$, $w(x, z, t)$ in the plane x, z can be related to a flow potential $\Phi(x, z, t)$ by

$$u = \Phi_x, \quad w = \Phi_z. \tag{D.1}$$

The requirement of mass conservation implies that the function Φ has to satisfy the Laplace equation

$$\Phi_{xx} + \Phi_{zz} = 0. \tag{D.2}$$

Boundary conditions

To solve the Laplace equation we need to specify boundary conditions. At the offshore boundary x_∞ of the shoreface, the surface fluctuation $\eta(x_\infty, t)$ produced by the incoming wave, is assumed to be a sinusoidal function with given frequency $\omega = 2\pi/T$,

$$\eta(x_\infty, t) = a \cos \omega t. \tag{D.3}$$

Onshore propagating (in negative x-direction) in water of depth h relative to still water is described by

$$\eta(x, t) = \frac{H}{2}\cos(kx + \omega t), \tag{D.4}$$

where H is the wave height and $k = 2\pi/L$ the wave number. It is assumed that the wave will break on the shoreface and dissipate all its energy; the corresponding condition at the inshore boundary is the absence of a reflected wave.

The boundary condition at the seabed, $z = -h$, states that the velocity component normal to the seabed equals zero. If we assume that the seabed is almost horizontal over distances comparable to the wavelength, this condition is equivalent to zero vertical velocity at the bottom.

Finally two conditions need to be specified at the surface $z = \eta(x, t)$. One condition states that the velocity normal to the surface equals the surface displacement per unit time. The second condition specifies that the pressure at the surface equals the atmospheric pressure (taken as zero); this implies that the wave surface corresponds to a streamline for which the Bernoulli equation holds. The two surface conditions are nonlinear and thus produce a deformation of the initial sinusoidal wave shape. This deformation depends on the curvature of the wave surface; the degree of curvature can be characterised by the average magnitude of the surface slope ak. The boundary conditions at the surface $z = \eta(x, t)$ can be written as

$$w = \Phi_z = \frac{d\eta}{dt} = \eta_t + u\eta_x \tag{D.5}$$

and

$$\Phi_t + \frac{1}{2}(u^2 + w^2) + g\eta = 0. \tag{D.6}$$

At the top of the turbulent boundary layer (taken as horizontal, $z = 0$) we have the boundary condition

$$w = \Phi_z = 0. \tag{D.7}$$

Wave equations

We will first consider the linear approximation, and neglect the non-linear terms in (D.5) and (D.6). This is equivalent to neglecting terms of the order of $ka = 2\pi a/L$ relative to terms of order 1, as can be shown by substitution of the linear solution. The solution of the linearised wave equations reads

$$\Phi(x, z, t) = -\frac{H\omega}{2k} \frac{\cosh[k(z+h)]}{\sinh(kh)} \sin(kx + \omega t). \qquad \text{(D.8)}$$

For the horizontal wave-orbital velocity we have

$$u = -H\omega \frac{\cosh[k(z+h)]}{2\sinh(kh)} \cos(kx + \omega t). \qquad \text{(D.9)}$$

In the convention used here, forward wave-orbital velocities (in the onshore wave propagation direction) are negative. It should be noted that the wave-orbital velocity $u(x, z, y)$ decreases from the surface $z = \eta(x, t)$ to the bottom $z = -h$. This decrease is not caused by bottom friction, as for the tidal case, but by the requirement of potential flow. According to the potential flow hypothesis, the vertical gradient of the orbital velocity $u_z = w_x$. At the bottom $w = w_x = 0$; upward from the bottom $|u_z|$ increases as $|w_x|$ increases.

From the boundary condition (D.6) follows a relation between k and ω, the so-called dispersion relation. Therefore we substitute (D.8) and neglect the quadratic terms (which are of order ka) and find

$$c = \frac{\omega}{k} = \sqrt{\frac{g}{k} \tanh(kh)}. \qquad \text{(D.10)}$$

Wind wave propagation in deep water ($kh \gg 1$) does not depend on depth: $c \approx \sqrt{g/k}$; this contrasts with tidal wave propagation. In shallow water ($kh \ll 1$) we have $c \approx \sqrt{gh}$, just as for tidal waves.

Wave energy density

The wave energy per unit length and width (wave energy density) is given by the sum of potential and kinetic energy, averaged over the wavelength,

$$E = \frac{1}{L} \int_0^L \left(\frac{1}{2} g \rho \eta^2 + \int_{-h}^{\eta} \frac{1}{2} \rho (u^2 + w^2) dz \right) dx. \quad \text{(D.11)}$$

For linear waves the result is

$$E = \frac{1}{8} \rho \, g \, H^2. \quad \text{(D.12)}$$

Wave energy propagates with a different speed than the wave speed, as will be explained below.

Wave groups and energy propagation

Within the real irregular wave field, interference may take place between waves with different periods and wavelengths [530]. This interference produces modulations of the wave amplitude, known as wave groups (see Fig. 7.16). These wave groups propagate in the same direction as the constituent waves, but with a lower wave speed. This can be demonstrated by considering the superposition of two waves with close wavenumbers k and $k + \Delta k$. The sum of these two waves can be written

$$\eta_{sum} = \eta(k, \omega) + \eta(k + \Delta k, \omega + \Delta \omega)$$

$$= a_{env} \sin \left[k(x + ct) \right], \quad \text{(D.13)}$$

where a_{env} is the wave envelope given by

$$a_{env} \approx H \cos \left[\Delta k (x + c_g t) \right],$$

where Δk, $\Delta \omega$ have been assumed very small. The wave group velocity c_g is thus given by

$$c_g \approx d\omega / dk = nc. \quad \text{(D.14)}$$

The factor n follows from (D.10),

$$n = \frac{1}{2} + \frac{kh}{\sinh(2kh)}. \quad \text{(D.15)}$$

Wave energy also propagates with the group velocity c_g. If the wave energy propagated at a different speed it would have to travel through the nodes of the wave group, where the wave energy is nil; obviously this is not possible. The wave energy propagates in the same direction as the individual waves, which is the direction of the wave vector $\vec{k} = k^{(x)}, k^{(y)}$ ($2\pi/k^{(x)}, 2\pi/k^{(y)}$ are the wavelengths in x- and y-direction, resp.). We write $\vec{c}_g = c_g\vec{k}/k$. The energy flux is given by

$$\vec{F} = \vec{c}_g E. \qquad (D.16)$$

In the shoaling zone (sufficiently deep water ($kh \geq 1$), such that waves do not break), waves dissipate so little energy that

$$\vec{\nabla}.\vec{F} \approx 0. \qquad (D.17)$$

The group velocity c_g decreases when waves propagate in a direction of decreasing depth. For shore normal wave incidence, continuity of the energy flux (D.17) then implies growth of wave energy and wave amplitude; the wave height varies in the shoaling zone according to $H(x) \propto c_g^{-1/2}$.

In the case of oblique wave incidence (angle θ between wave crest and depth contour lines), wave height is also influenced by refraction. Snell's refraction law, $\sin\theta/c = $ constant, implies a decreasing angle of wave incidence when waves propagate into shallower water (the wave celerity c decreases). According to (D.17), the wave energy flux $c_g E \cos\theta$ is constant in onshore x direction. The wave celerity c_g decreases is onshore direction, while the cosine of the wave incidence angle increases. The wave height in the shoaling zone may therefore decrease in onshore direction for waves that approach the shore from deep water at high angles.

D.1.2. *Wave Skewness*

Second order Stokes theory

The expression (D.10) for the wave propagation speed shows that the wavelength $L = 2\pi c/\omega$ decreases with decreasing depth. Therefore the surface curvature of waves propagating onto the shoreface will increase. At some point, ka becomes on the order of 1, invalidating the assumption of linear wave propagation. In shallow water, waves become skewed.

Wave skewness is generated through the nonlinear surface boundary conditions (D.5, see also Fig. 7.10) and (D.6). The second order correction to the linear (first order) approximation can be obtained by expanding the flow potential Φ and the wave equations as a power series in the small parameter $\epsilon \equiv ka$ and by retaining the linear terms in ϵ. The solution reads [501]

$$\eta = \frac{H}{2}\cos(kx + \omega t)$$

$$+ \frac{kH^2}{16}\left(3\coth^3(kh) - \coth(kh)\right)\cos\left[2(kx + \omega t)\right], \quad \text{(D.18)}$$

$$u = U_0 - U_1\cos(kx + \omega t) - U_2\cos\left[2(kx + \omega t)\right] \quad \text{(D.19)}$$

with

$$U_0 = \frac{gH^2}{8ch} \approx \frac{H^2}{8h}\sqrt{\frac{g}{h}}, \quad U_1 = H\omega\frac{\cosh\left[k(z + h)\right]}{2\sinh(kh)} \approx \frac{H}{2}\sqrt{\frac{g}{h}},$$

$$U_2 = \frac{3}{16}H^2k\omega\frac{\cosh\left[2k(z + h)\right]}{\sinh^4(kh)} \approx \frac{3gH^2}{16\omega^2h^2}\sqrt{\frac{g}{h}}. \quad \text{(D.20)}$$

The approximations are for shallow water where $kh \ll 1$. The first constant term in (D.19) ensures zero net mass transport (orbital velocity averaged over depth and wave period). The second order nonlinear wave contribution introduces constant terms and terms with double wave period. The cosine terms with single and double period are in phase; this implies that nonlinear terms make wave crests higher and steeper, whereas wave troughs are made more shallow and flat. Wave-orbital velocities in the propagation direction (underneath wave crests) are larger than wave-orbital velocities against the propagation direction (underneath wave troughs). This asymmetry is usually termed 'velocity skewness'. The ratio of the second-order to the first order velocity amplitude is a measure of the degree of skewness; this ratio is called 'Ursell number U_r'. In the shallow-water approximation

$$U_r = \frac{U_2}{U_1} = \frac{3gH}{8\omega^2 h^2}. \tag{D.21}$$

A more general expression for velocity skewness is the parameter S_u, see (3.70). For the wave-orbital velocity given by (D.19) we have $S_u = 3U_2/4U_1 = 3U_r/4$.

Acceleration skewness

It should be noticed that short-wave asymmetry and tidal-wave asymmetry are of a different nature. Nonlinear deformation of the tidal wave leads to a sawtooth shaped wave surface, while nonlinear deformation of wind waves leads to a peaked wave surface. Tidal wave deformation is mainly due to a different propagation velocity of high water relative to low water; this difference relates to the great sensitivity of the tidal propagation speed to water depth. By contrast, short-wave propagation is not strongly dependent on water depth, provided the wavelength L is smaller than 2π times depth, as shown before. However, in very shallow water, when waves start breaking, the propagation speed of wind waves also becomes depth-dependent. At this stage, the wave surface takes a sawtooth shape, similar to tidal wave asymmetry. In the surf zone, the saw-tooth shape produces stronger wave-orbital acceleration in the wave propagation direction than in the opposite direction. This is usually called 'wave asymmetry' or 'acceleration skewness'.

The 2nd-order Stokes theory does not represent saw-toothed waves. Acceleration skewness requires introduction of an additional term in Eq. (D.19),

$$u = U_0 - U_1 \cos(kx + \omega t) - U_2 \cos[2(kx + \omega t)]$$
$$+ V_2 \sin[2(kx + \omega t)], \tag{D.22}$$

where stronger forward acceleration (negative x-direction) is represented by $V_2 > 0$. The velocity skewness of this wave is the same as for (D.19); the acceleration skewness parameter A_u (defined in (3.70)) is given by $A_u = -3V_2/4U_1$.

Near-bed wave-orbital asymmetry

In Sec. 3.3.1 it is shown that acceleration skewness in the free stream produces additional velocity skewness in the wave-boundary layer. This additional velocity skewness is important for wave-induced bed-load transport in the surf zone [256, 397] and should be incorporated in bedload transport formulas. This can be achieved by modelling the vertical profile of the wave-orbital velocity at the intra-wave time scale. However, the computations are very time consuming and not yet suited for simulating morphologic evolution of the shoreface over times scales of months-years.

A practical way to take into account the effect of acceleration skewness in the wave-boundary layer has been proposed by Abreu *et al.* [5]. They developed an empirical formula for the near-bed wave-orbital velocity including both skewed and saw-toothed waves,

$$u^{bed}(t) = U f_1 \frac{f_2 + \sin(\omega t)}{1 - r\cos(\omega t + \phi)}, \quad f_1 = \sqrt{1 - r^2},$$

$$f_2 = \frac{r\sin\phi}{1 + f_1}. \tag{D.23}$$

This formula represents acceleration skewness without velocity skewness for $0 < r < 1$ and $\phi = 0$ and velocity skewness without acceleration skewness for $0 < r < 1$ and $\phi = -\pi/2$. For other values of ϕ we have mixed asymmetry, see Fig. D.1. Broadly applicable expressions for the parameters r and ϕ as functions of the velocity skewness parameter S_u and the acceleration skewness parameter A_u have been derived by Ruessink *et al.* [743], by fitting (D.23) to a large data set of measured wave-orbital velocities in the shoreface zone.

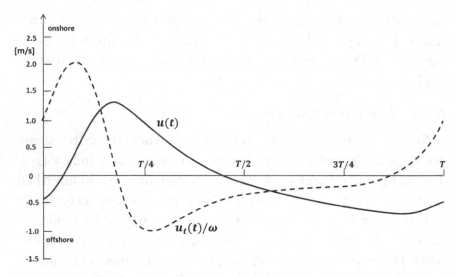

Fig. D.1.   Near-bottom wave-orbital velocity $u(t)$ (solid line) as a function of the wave phase and corresponding acceleration $u_t(t)$ multiplied by $T/2\pi$ (dotted line). In this example, corresponding to $\phi = -\pi/4, r = 0.75$ in Eq. (D.23), we have both velocity skewness (larger onshore than offshore velocities) and acceleration skewness (larger onshore than offshore acceleration).

Another empirical approach was suggested by Henderson *et al.* [388]. They noticed from field data that the phase advance in the frequency spectrum of u^{bed} (wave-orbital velocity in the wave-boundary layer) is almost frequency independent. This phase advance ϕ_{bed} with respect to the free stream velocity wave-orbital was estimated at 25°. If the free stream velocity is given by (D.22), the wave-orbital velocity in the wave-boundary layer can then be written

$$u^{bed} \propto - \cos(\theta + \phi_{bed}) - \frac{4}{3} [S_u \cos(2\theta + \phi_{bed})$$
$$- A_u \sin(2\theta + \phi_{bed})] , \tag{D.24}$$

where $\theta = kx + \omega t$. This yields a velocity skewness S_u^{bed} in the wave-boundary layer given by (assuming $S_u, A_u \ll 1$)

$$S_u^{bed} = \frac{\langle (u^{bed})^3 \rangle}{\langle |u^{bed}|^3 \rangle} = - \cos(\phi_{bed}) [S_u + \tan(\phi_{bed}) A_u] . \tag{D.25}$$

Using (D.25) in the bedload transport formula (3.68) gives a result similar to using the formula (3.71) proposed by Nielsen [635] for wave-induced bedload transport including acceleration skewness.

Undertow

Asymmetry of the wave orbital motion, associated with wave surface asymmetry, causes net sediment transport. More sediment will be suspended under wave crests than under wave troughs and therefore wave-induced sediment transport does not average to zero over a wave period. This leads normally to shoreward displacement of sediment. However, another first-order nonlinear velocity component influences residual sediment transport in the opposite way. This is the first term in (D.19), which represents the return flow compensating for the excess shoreward mass transport due to a greater water depth during shoreward orbital motion than during seaward orbital motion (Stokes drift). This return flow is usually called undertow.

Although derived from linear wave theory, the expression (D.19) gives a reasonable first order estimate for the depth averaged return current $\overline{U}_0$, even at the stage of wave breaking. Laboratory and field observations indicate [190, 845]

$$\overline{U}_0 = \frac{C_u}{8} \frac{gH^2}{ch}. \tag{D.26}$$

The coefficient C_u accounts for the effect of the surface roller in the breaking wave; for irregular waves C_u is found to be of order 1. Below the wave trough the average current is seaward, while shoreward mass transport is concentrated in the upper part of the vertical, between wave trough and wave crest, see Fig. D.2. Field observations indicate for the maximum depth-averaged undertow under breaking waves $\overline{U}_0 \sim (0.2 - 0.4)\sqrt{H}$, where the lower values hold for gently sloping beaches ($\beta \sim 0.02$) and the higher values for steeper beaches ($\beta \sim 0.04$) [572]. In the turbulent wave-boundary layer, the return flow can be represented by a logarithmic function of depth (see Sec. 3.3.2). Above the boundary layer, up to the wave-trough level, it

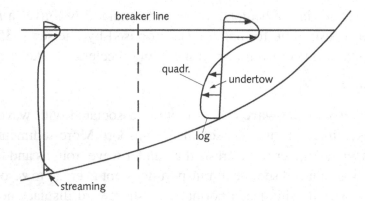

Fig. D.2. Flow profiles under non-breaking and breaking waves, averaged over the wave period. Shoreward mass transport near the surface is due to the covarying phases of wave-orbital motion and surface elevation. This near-surface shoreward transport is compensated by seaward transport below the wave trough level. In a thin layer at the bottom, shoreward drift is due to streaming (see Sec. 3.3.1). In the breaker zone, the shoreward mass transport is enhanced, as well as the seaward transport (undertow) lower in the vertical. In the roughness layer close to the bottom the profile is logarithmic; the undertow profile is quadratic.

can be represented by a quadratic function of depth [845]. In the wave breaking zone, the turbulent dissipation of roller energy induces an onshore shear stress on the fluid below [729]. The maximum return flow velocity therefore occurs close the bottom; it attains a value on the order of $2\overline{U}_0$ [190].

D.2. Wave Transformation in the Surf Zone

D.2.1. *Wave Energy Dissipation*

When waves travel shoreward into the shallow coastal zone, they lose energy due to bed friction and wave breaking. The energy loss ϵ_f due to bed friction for a monochromatic linear wave follows from (3.27),

$$\epsilon_f = \langle u\tau_b \rangle = \frac{1}{2}\rho f_w \langle |u|^3 \rangle = \frac{1}{12\pi}\rho f_w \left[\frac{H\omega}{\sinh kh}\right]^3, \qquad (\text{D.27})$$

where $H = 2a$ is the wave height and f_w is the friction coefficient given by (3.28). The energy dissipation due to breaking of spilling waves, ϵ_b can be derived from the theory of bore energy dissipation. This yields an expression [58,517,841]

$$\epsilon_b = \frac{1}{4}\rho g \frac{H^3}{hT},$$ (D.28)

where T is the wave period and where H the height of the breaking wave. We follow the usual assumption that the wave height H of the incident wave field follows a Rayleigh probability distribution $p(H)$,

$$p(H) = \frac{2H}{H_{rms}^2} \exp\left[-\left(\frac{H}{H_{rms}}\right)^2\right].$$

Thornton and Guza [841] derived a model for wave transformation based on the assumption that all the waves break if $H_{rms} > \gamma h$ and that the proportion of breaking waves with $H_{rms} < \gamma h$ is given by the weighting function $(H_{rms}/\gamma h)^n$. Multiplying the expression (D.28) for ϵ_b with the weighting function and averaging with the Rayleigh distribution $p(H)$ gives the result

$$\epsilon_b = \frac{3\sqrt{\pi}}{16}\rho g \frac{H_{rms}^3}{hT}\left(\frac{H_{rms}}{\gamma h}\right)^n.$$ (D.29)

Damping of the wave height when it travels into the surf zone can be estimated from the equation

$$\frac{dF}{dx} = \frac{d}{dx}(c_g E) = \frac{1}{8}\rho g \frac{d}{dx}(c_g H_{rms}^2) = \epsilon_f + \epsilon_b.$$ (D.30)

Dissipation of wave energy through wave breaking is much stronger than dissipation through friction in the surf zone [840]; the term ϵ_f can be neglected. The group velocity c_g in the surf zone is almost equal to the shallow water wave speed $c_g \approx c \approx \sqrt{gh}$. The depth is almost linearly related to the cross-shore distance x; we may thus substitute $x \approx h/\beta$, where β is the average bed slope. Equation (D.30) can now be written as a nonlinear differential equation in the variables h and $y = H_{rms}^2\sqrt{h}$. This equation is solved by separation of variables; for small depths ($h \ll L/20$, where L is the wavelength), the result is

$$H_{rms} = \gamma_{br} h, \quad \gamma_{br} = \left(\frac{2\sqrt{\pi}}{3} \frac{3+5n}{1+n} \frac{\beta}{kh} \gamma^n \right)^{1/(1+n)}. \quad \text{(D.31)}$$

For large values of n we find $\gamma_{br} \approx \gamma$. Experiments with spilling breakers on gently sloping beaches indicate γ values of 0.4–0.5 [409, 841]. Other estimates from field and laboratory observations suggest that γ increases linearly with $\tanh(H/L)$ ($H/L=$ deep-water wave steepness) or with kh, from a minimum value of 0.2–0.4 [30]. For swell waves on steep sloping beaches the value of γ_{br} was found to depend linearly on β/kh [704], $\gamma_{br} \approx 0.2 + \beta/kh$. Many other formulas for wave breaking can be found in the literature (see [535] for a review); in most formulas the breaker index γ_{br} depends on the wave steepness H/L and on the seabed slope β. Wave breaking is called 'saturated' if the wave height in the surf zone decreases according to (D.31) with a constant (depth independent) value of γ_{br}.

D.2.2. *Radiation Stress*

Wave breaking in the surf zone does not only affect the amplitude and the shape of the incident waves. It also generates water surface slopes and residual flow in the surf zone. This becomes apparent when the Reynolds equations (A.5) are integrated over the water column and the wave period. The integral of the nonlinear terms were called radiation stresses by Longuet-Higgins [539], who was first to realise that gradients in the radiation stresses transfer momentum from wave motion to depth-averaged residual flow.

After a few algebraic manipulations, using the mass balance equation and the boundary conditions at the bottom and surface, the depth-integrated momentum balance equations averaged over the wave period become

$$\frac{\partial}{\partial t} \int_{-h}^{\eta} \rho u \, dz + \frac{\partial}{\partial x} \int_{-h}^{\eta} (\rho u^2 + p) dz$$

$$+ \frac{\partial}{\partial y} \int_{-h}^{\eta} \rho u v \, dz - p_b \frac{\partial h}{\partial x} + \tau_b^{(x)} = 0,$$

$$\frac{\partial}{\partial t} \int_{-h}^{\eta} \rho v \, dz + \frac{\partial}{\partial y} \int_{-h}^{\eta} (\rho v^2 + p) dz$$

$$+ \frac{\partial}{\partial x} \int_{-h}^{\eta} \rho u v \, dz - p_b \frac{\partial h}{\partial y} + \tau_b^{(y)} = 0, \qquad (D.32)$$

where $D = h + \eta_0$, $\eta_0 = \langle \eta \rangle$ and $p_b \approx g\rho D$ is the pressure at the bottom. These equations describe the depth-averaged residual flow field $(u_0(x, y), v_0(x, y))$ and mean water surface inclination $(\eta_{0x}(x, y), \eta_{0y}(x, y))$ induced by gradients in the incident wave field. The wave-averaged bottom stress terms $\tau_{0h}^{(x)} = \langle \tau_b^{(x)} \rangle$ and $\tau_{0b}^{(y)} - \langle \tau_b^{(y)} \rangle$ take into account the residual flow momentum dissipated by bottom friction.

The radiation stresses S^{xx}, S^{xy}, S^{yy} correspond to the three non-linear terms in the wave equations:

$$S^{(xx)} = \left\langle \int_{-h}^{\eta} (p + \rho u^2) dz \right\rangle - \frac{1}{2} \rho g D^2,$$

$$S^{(yy)} = \left\langle \int_{-h}^{\eta} (p + \rho v^2) dz \right\rangle - \frac{1}{2} \rho g D^2,$$

$$S^{(xy)} = \left\langle \int_{-h}^{\eta} \rho u v \, dz \right\rangle. \qquad (D.33)$$

After averaging the equations over the wave period and introducing the radiation stresses (D.33) we find

$$(u_0{}^2)_x + (v_0 u_0)_y + g\eta_{0x} + \left[\tau^{(x)}{}_{0b}\right.$$

$$\left. + S^{(xx)}{}_x + S^{(xy)}{}_y\right]/\rho D = 0, \tag{D.34}$$

$$(v_0{}^2)_y + (u_0 v_0)_x + g\eta_{0y} + \left[\tau^{(y)}{}_{0b}\right.$$

$$\left. + S^{(yy)}{}_y + S^{(xy)}{}_x\right]/\rho D = 0, \tag{D.35}$$

$$(Du_0)_x + (Dv_0)_y = 0, \tag{D.36}$$

where $D(x) = h(x) + \eta_0(x)$ is the total water depth after averaging over the wave period. The variation of the residual flow over the water column has been ignored. This is a reasonable approximation, considering the small water depth in the surf zone.

Approximate analytical expressions of the radiation stresses can be obtained from linear wave theory. Upon substitution of the linear wave solution and assuming a uniform wave field propagating in a horizontal $x-y$-plain with an angle θ to the cross-shore x-axis, we find

$$S^{(xx)} \approx \left(n(1 + \cos^2\theta) - \frac{1}{2}\right) E,$$

$$S^{(yy)} \approx \left(n(1 + \sin^2\theta) - \frac{1}{2}\right) E,$$

$$S^{(xy)} \approx nE \sin\theta \cos\theta, \tag{D.37}$$

where E is the wave energy density, $E = \rho g H^2/8$.

D.3. Wave-induced Flow

The longshore uniform case

Here we consider the case where coastal morphology and the incident wave field are uniform in alongshore direction. In this case simple equations are obtained for the average cross-shore surface slope η_{0x},

$$g\rho D\eta_{0x} + S^{(xx)}{}_x = 0, \tag{D.38}$$

and for the average bottom shear stress $\tau^{(y)}{}_{0b}$ by a longshore current,

$$\tau^{(y)}{}_{0b} + S^{(xy)}{}_x = 0. \tag{D.39}$$

For small depth ($kD \ll 1$) the radiation stresses can be approximated by the following first order expressions:

$$S^{(xx)} \approx \left(\frac{1}{2} + \cos^2\theta\right) E, \quad S^{(yy)} \approx \left(\frac{1}{2} + \sin^2\theta\right) E,$$

$$S^{(xy)} \approx E\sin\theta\cos\theta, \tag{D.40}$$

Under the assumption of wave breaking saturation, the energy E in the surf zone can be related to the average water depth D, (D.31):

$$E = \frac{1}{8}\rho g\gamma_{br}^2 D^2(x). \tag{D.41}$$

D.3.1. *Shore-normal Wave Incidence*

Shoaling and set-down

We consider waves propagating in shore-normal direction ($\theta = 0$), seaward of the zone where waves are breaking, and alongshore uniform bathymetry. We assume that outside the surf zone energy dissipation is so small that the energy flux $F = ncE$ is approximately constant (not dependent on x or h). Substitution in (D.37) yields an expression for the radiation stress which is used in Eq. (D.38). We then find the following expression for the mean water level inclination

$$\frac{d\eta_0}{dh} = -\frac{F}{g\rho h}\frac{d}{dh}\left(\frac{2n - \frac{1}{2}}{nc}\right). \tag{D.42}$$

Here the depth D has been approximated by h and differentiation by x has been replaced by equivalent differentiation by h (we assume that h is a monotonously increasing function of x). Integration of the right hand side yields (far off-shore $\eta_0 = 0$)

$$\eta_0 = -\frac{1}{8}\frac{kH^2}{\sinh(2kh)} \approx -\frac{1}{16}\frac{H^2}{h}. \qquad (\text{D.43})$$

The last approximation is only valid for shallow water where $kh \ll 1$. When waves propagate into shallow water — but before wave breaking — the wave amplitude increases and a radiation stress gradient develops. The radiation stress gradient produces a decrease of mean water level, called 'set-down'. The increase of wave amplitude a and the decrease of depth h both contribute to wave set-down, see Fig. D.3.

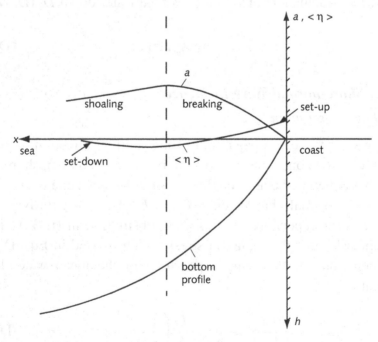

Fig. D.3.   Set-down and set-up of mean water level produced by shoaling and wave breaking.

Setup in the surf zone

Wave breaking in the surf zone results in a decrease of wave amplitude, according to (D.31). Here we assume that the proportionality factor γ_{br} between the wave height $H(x)$ in the surf zone and the local depth $D = h(x) + \eta_0(x)$ is constant. Substitution in the radiation stress formula (D.37), together with the approximation $kD \ll 1$, yields

$$E = \frac{1}{8} g \rho \gamma_{br}^2 D^2, \quad S^{(xx)} = \frac{3}{16} g \rho \gamma_{br}^2 D^2. \tag{D.44}$$

We use Eq. (D.38) for evaluating the influence of radiation stress on the water level inclination, assuming shore-normal wave-incidence ($\theta = 0$). Integration gives the setup generated η_0 at the shoreline generated by wave breaking,

$$\eta_0 = \frac{\frac{3}{8} \gamma_{br}^2}{1 + \frac{3}{8} \gamma_{br}^2} h_{br} \approx \frac{\frac{3}{8} \gamma_{br}}{1 + \frac{3}{8} \gamma_{br}^2} H_0, \tag{D.45}$$

where h_{br} is the depth at the breaker line and H_0 the deep water wave height (seaward of the surf zone). This set-up can be close to one metre under storm conditions. For $\gamma_{br} = 0.6$ we find a setup $\eta_0 / H_0 = 0.2$.

Although the application of linear wave theory in the surf zone may be questioned, the wave setup determined in this way is consistent with field observations. However, the scatter in the observations is considerable. Some field experiments indicate $\eta_0 / H_0 = 0.15 - 0.4$; other observations are best represented by fitting $\eta_0 / H_0 = 0.02 + 0.003 / \beta$ [705].

Due to the presence of wave groups the set-up is not constant, but varies with the period of the wave groups (Fig. 7.16). The highest waves in the wave-group center start breaking at a greater depth than the lowest waves at the wave group nodes. The set-up at the wave group center is larger and the waves propagate further onto the beach. This periodic feature is called surf beat.

D.3.2. *Oblique Wave Incidence*

Longshore current

Now we consider the case of a wave field approaching at an angle θ an alongshore uniform coast (all derivatives with respect to y equal zero). In addition to a gradient of the radiation stress component $S^{(xx)}$ in x-direction, causing a shoreward slope of the mean surface level, we also have a gradient in x-direction of the radiation stress component $S^{(xy)}$. This cannot be balanced by a longshore surface level inclination, since the shoreline is assumed to be infinitely long and uniform. Instead, the cross-shore gradient $S_x^{(xy)}$ drives a mean current along the coast [540]. This current, called the 'longshore current' or 'breaker current', is strongest where the cross-shore variation of wave amplitude is largest, i.e., in the surf zone. Its magnitude results from a balance between radiation stress gradient and momentum dissipation of the longshore current through turbulent bottom friction (Eq. (D.39)). For momentum dissipation through bottom friction we will assume a quadratic friction law,

$$\vec{\tau}_b = \rho c_D \sqrt{(u_0 + u)^2 + (v_0 + v)^2}(u_0 + u, v_0 + v). \qquad (D.46)$$

We use the symbol c_D for the friction coefficient of a steady flow with superposed wave field; this coefficient is larger than the friction coefficient for steady flow, see Sec. 3.3.2. Empirical estimates are given by the expressions (3.34) or (3.35).

For a wave field with incidence angle θ we write

$$u = U \cos\theta \cos\omega t, \quad v = U \sin\theta \cos\omega t.$$

Wave orbital velocities are in general substantially larger in the surf zone than the steady flow components; in this case we may neglect u_0^2, v_0^2 relative to U^2. Wave propagation in the surf zone is generally close to shore-normal, due to refraction. To a first approximation, $\sin\theta$ can be neglected compared to $\cos\theta$. With these approximations, and after averaging over the wave period we find

$$\tau^{(x)}{}_{0b} = 2ru_0, \quad \tau^{(y)}{}_{0b} = rv_0,$$

$$r \approx 2\rho c_D U/\pi \approx \rho c_D H \sqrt{g/D}/\pi. \tag{D.47}$$

The magnitude of the longshore current V_{br} follows from (D.39)

$$V_{br} = v_0 = \tau^{(y)}{}_{0b}/r = -S^{(xy)}{}_x/r. \tag{D.48}$$

The radiation stress $S^{(xy)}$ may be written as

$$S^{(xy)} = nE \cos\theta \sin\theta = F^{(x)} \frac{\sin\theta}{c}, \tag{D.49}$$

where $F^{(x)} = ncE \cos\theta$ is the energy flux perpendicular to the coast.

Outside the breaker zone the shoreward energy flux is approximately constant, $F_x^{(x)} \approx 0$. According to Snells' wave refraction law $\sin\theta/c = $ constant. Hence, outside the breaker zone $S_x^{(xy)} \approx 0$ and no longshore current is generated.

In the surf zone a simplified expression for $S^{(xy)}$ is obtained if we use for the propagation speed the approximation $c_g \approx c \approx \sqrt{gD}$. Then, Snell's law, $(\sin\theta/c)_x = -(\cos\theta/c)_y \approx 0$, reduces to $\sin\theta/c \approx \sin\theta_{br}/\sqrt{gh_{br}}$. For the wave height we use the breaking saturation approximation $H \approx \gamma_{br} D$, according to (D.31). After substitution in (D.48), using (D.47) and (D.49), we find for the longshore current

$$V_{br} = \frac{\tau^{(y)}{}_{0b}}{r} = -\frac{S^{(xy)}{}_x}{r}$$

$$\approx -\frac{5\pi}{16} \frac{\gamma_{br}}{1 + \frac{3}{8}\gamma_{br}^2} \frac{\beta}{c_D} g^{1/2} h_{br}^{-1/2} D \cos\theta \sin\theta_{br}, \tag{D.50}$$

where $\beta = h_x$ is the average surf zone slope. The x-derivation introduces the factor $(1 + 3\gamma_{br}^2/8)$, corresponding to the width increase of the surf zone due to wave setup, see (D.45). The index br designates values evaluated at the breaker line; X_{br} corresponds to the width of the surf zone. In this expression we have neglected a factor $(1 - \tan\theta/5)$, following from the x-derivative of $\cos\theta$.

The longshore velocity is proportional to the depth; the velocity distribution is thus triangular in the case of constant slope β: Zero velocity outside the surf zone, maximum velocity at the edge of the surf zone, and a gradual velocity decrease to zero at the coast, see Fig. D.4. In reality, momentum diffusion perpendicular to the shore makes the velocity distribution smoother and less triangular. Observations show that in the case of a barred shoreface the maximum longshore current occurs between the bar and the shoreline [116]. The longshore breaker current is localised in a narrow strip along the coast; it can exceed 1 m/s, with an important lateral velocity shear. This lateral shear may generate instability (meandering) of the longshore current [105, 292, 652].

The average longshore current velocity $\bar{V}_{br}$ in the surf zone is obtained by averaging the depth-weighted velocity (D.50) over the surf zone width X_{br}. The result is

$$\bar{V}_{br} = -V_1\sqrt{gH_{br}}\sin\theta_{br}\cos\theta_{br}, \quad V_1 = \frac{5\pi\beta}{24c_D}\frac{\gamma_{br}^{1/2}}{1+\frac{3}{8}\gamma_{br}^2}.$$

(D.51)

In the simpler empirical formula of the longshore current given by Komar [492] we have $V_1 = 1$.

If the surf zone bathymetry is not uniform in alongshore direction, the wave height H will vary with the alongshore coordinate y. This influences the setup η_0 along the coast, according to (D.34), and introduces a term $\rho g D\eta_{0y}$ in the in the momentum balance (D.48), according to (D.35). The result is [48]

$$\bar{V}_{br} = \sqrt{gH_{br}}(-V_1\sin\theta_{br}\cos\theta_{br} - V_2(H_{br})_y),$$

$$V_2 = \frac{\pi}{4c_D\gamma_{br}^{3/2}}\frac{1+\frac{1}{8}\gamma_{br}^2}{1+\frac{3}{8}\gamma_{br}^2}.$$

(D.52)

Considering a case where $\beta = 0.02, c_D = 0.01, \gamma_{br} = 0.7, \theta_{br} = 10°, H_{br} = 2m$, (D.51) gives V_{br}=-0.7 m/s. For a longitudinal

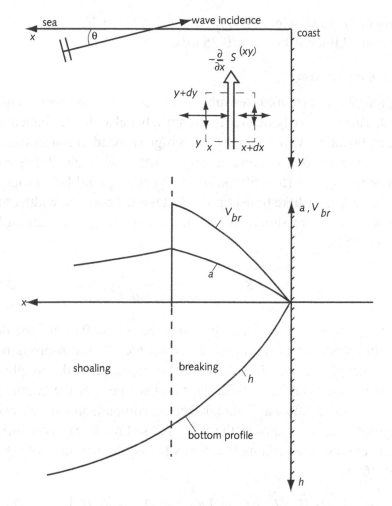

Fig. D.4. Longshore current V_{br} in the surf zone produced by breaking of obliquely incident waves. The black arrows in the upper panel illustrate the shoreward decrease of wave-orbital velocity and the resulting longshore gradient in the radiation stress (open arrow). The shoreward decrease of wave-orbital velocity is related to the shoreward decrease of wave amplitude a, according to the wave-saturation hypothesis, shown in the lower panel. The strength of the longshore current depends locally on the cross-shore gradient of the radiation stress, which is largest shoreward of the breaker line and zero at the shoreline. Outside the breaker zone the wave amplitude increases shoreward due to shoaling, but this increase produces no alongshore radiation-stress gradient.

gradient in wave height $(H_{br})_y = \pm 5.10^{-4}$, the V_2-term in (D.52) gives an additional $V_{br} = \mp 0.25$ m/s.

Longshore transport

The longshore current causes substantial sediment transport along the coast. This is not only due to its strength, but also due to seabed stirring by breaking waves, which causes high suspended concentrations in the surf zone. A common range of the net wave-induced longshore transport is $q_{br} \approx (5 - 50) \times 10^4 \, \text{m}^3/\text{year}$. Bagnold [45] proposed for the bulk longshore transport q_{br} (integrated over the width of the surf zone) an expression related to the average longshore current V_{br} (D.51, D.52),

$$q_{br} = K_{Bag}\frac{F_{br}}{(1 - p_b)g\Delta\rho}\frac{V_{br}}{U}, \qquad (D.53)$$

where $p_b \approx 0.4$ is the seabed porosity, $\Delta\rho \approx 1650 \, \text{kg/m}^3$ the density difference water-sediment, $F = ncE\cos\theta$ the shore-normal wave energy flux and $U \approx Hc/2h \approx \gamma_{br}c_{br}/2$ is the amplitude of the wave-orbital velocity at the breaker line. For the coefficient K_{Bag} the value 0.28 was found, based on comparison with observed longshore transport rates in the field. All sediment transport formulas are expressed as volumetric fluxes [m³/s]. Substitution of (D.52) gives [657]

$$q_{br} = \sqrt{gH_{br}}\,H_{br}^2\cos\theta_{br}\left[K_1\beta\sin\theta_{br} - K_2H_y\right]. \qquad (D.54)$$

The coefficients K_1, K_2 are related to V_1, V_2 by a factor $K_{Bag}/(4\gamma_{br}(1 - p_b)\Delta\rho/\rho) \approx 0.1$. Typical values are $K_1 \sim 1$, $K_2 \sim 2.5$, based on (D.51, D.52), K_{Bag} and assuming for the breaker index $\gamma_{br} \approx 0.7$. For oblique wave incidence, the first (K_1-) term in (D.54) is generally much larger than the second (K_2-) term, except in situations with pronounced gradients in coastal bathymetry, such as transverse shoreline bars and rip cell morphology. The K_2-term counteracts local shoreline advance/retreat related to wave refraction [439].

The frequently used CERC formula [152] for wave-induced long-shore transport corresponds to the first term of the Bagnold formula, with the exception of the seabed slope factor β,

$$q_{br} = K_{CERC} \frac{\sqrt{g H_{br}} H_{br}^2}{(1 - p_b)\, \gamma_{br}^{0.5}\, \Delta\rho/\rho} \sin(2\theta_{br}). \tag{D.55}$$

The coefficient K_{CERC} can be related to the Bagnold coefficient K_{Bag} for different values of the slope parameter β. The CERC manual recommends $K_{CERC} \approx 0.05$.

Another popular empirical formula for q_{br} was derived by Kamphuis [476],

$$q_{br} = 2.3 \frac{\beta^{0.75} T_p^{1.5} H_{br}^2}{(1 - p)\, \Delta\rho\, d^{0.25}} \left[\sin(2\theta_{br}) \right]^{0.6}, \tag{D.56}$$

where H_{br} is the significant wave height at breaking, T_p is the peak wave period and d the grain size diameter; all quantities are expressed in standard units [kg, m, s].

A recent formula given by Van Rijn [914] reads

$$q_{br} = 2.2 \; 10^{-4} \; g^{0.5} \; \beta^{0.4} \; d^{-0.6} \; H_{br}^{3.1} \sin(2\theta_{br}). \tag{D.57}$$

The predictions of these formulas differ on average by a factor 2 for steep surf zones and differ otherwise even more; the largest values are generally given by the CERC formula and the smallest values by Kamphuis'es formula.

Longshore sediment transport is maximum at a wave incidence angle of 45° at the breaker line; at this angle the shoreward transport of longshore wave-orbital momentum is highest (corresponding to highest radiation stress $S^{(xy)}$). The longshore transport is smaller at angles of wave incidence which are either more shore-normal or more shore-parallel. The angle of wave incidence is smaller at the breaker line than in deep water, due to refraction. Refraction is related to the depth dependency of the wave propagation speed; it is strongest for waves with a great wavelength compared to depth ($kh \ll 1$), in which case $c \approx \sqrt{gh}$. In the case of a straight coastline and shore-parallel depth contours, it can be shown that the angle of wave incidence at the breaker line, θ_{br}, is practically always smaller than 45°, even for very short waves and almost shore-parallel deep water waves [294].

D.4. Infragravity Waves

The radiation stress $S = S^{(xx)}$ associated with a group of shore-normal incident waves with wavenumber bandwith Δk can be approximated by the expression [539]

$$S \approx S_0 + S_g \sin(\Delta k(x + c_g t)),$$

where c_g is the group velocity (D.14). The gradient of the radiation stress induces a velocity u_g and a variation of the mean water level, η_g, according to the momentum balance,

$$(u_g)_t + S_x/D\rho + g(\eta_g)_x = 0$$

in the absence of frictional dissipation. If we neglect the last term at the l.h.s., which is smaller than the other two, we find that the wave group induces an additional velocity u_g which is directed offshore where the wave group envelope is largest and onshore where the wave group envelope is smallest. The gradient in the velocity u_g produces a water level variation given by $(\eta_g)_t = -D(u_g)_x$, according to the continuity equation. The mean water level variation η_g is in phase with the velocity u_g induced by the wave group. It thus appears that the wave group induces an infragravity wave, bound to the wave group, with wavelength equal to the group length, see Fig. 7.16.

Bibliography

1. Aagaard, T., Davidson-Arnott, R., Greenwood, B. and Nielsen, J. (2004) Sediment supply from shoreface to dunes: Linking sediment transport measurements and long-term morphological evolution. *Geomorphology* **60**: 205–224.
2. Aagaard, T., Hughes, M. and Andersen, R.S. (2006) Hydrodynamics and sediment fluxes across an onshore migrating intertidal bar. *J. Coast. Res.* **22**: 247–259.
3. Aagaard, T. and Greenwood, B. (2008) Infragravity wave contribution to surf zone sediment transport. The role of advection. *Mar. Geol.* **251**: 1–14.
4. Aagaard, T. and Jensen, S.G. (2013) Sediment concentration and vertical mixing under breaking waves. *Mar. Geol.* **336**: 146–159.
5. Abreu, T., Silva, P.A., Sancho, F. and Temperville, A. (2010) Analytical approximate wave form for asymmetric waves. *Coastal Eng.* **57**: 656–667.
6. Aleman, N., Robin, N., Certain, R., Anthony, E.J. and Barusseau, J.-P. (2015) Longshore variability of beach states and bar types in a microtidal, storm-influenced, low-energy environment. *Geomorphology* **241**: 175–191.
7. Allan, E., Pontee, N., Pye, K. and Parsons, A. North West Estuaries Processes Reports — Ribble Estuary. Halcrow Report, Sefton council.
8. Allard, J., Chaumillon, J. and Fenies, H. (2009) A synthesis of morphological evolutions and Holocene stratigraphy of a wave-dominated estuary: The Arcachon lagoon, SW France. *Cont. Shelf Res.* **29**: 957–969.
9. Alldridge, A.L. (1979) The chemical composition of macroscopic aggregation in two neretic seas. *Limnol. Oceanogr.* **24**: 855–866.
10. Allen, J.L.R. (1968) *Current Ripples.* N. Holland Publ. Comp., Amsterdam, 433 pp.
11. Allen, J.R.L. (1980) Sandwaves: A model of origin and internal structure. *Sed. Geol.* **26**: 281–328.
12. Allen, J.R.L. (1982) Sedimentary structures — Their characteristics and physical basis. Vol. 1 In: *Developments in Sedimentology*, Vol. 30A, Elsevier, New York, 593 p.

13. Allen, G.P., Sauzay, G., Castaing, S.P. and Jouanneau, J.M. (1977) Transport and deposition of suspended sediment in the Gironde estuary, France. In: *Estuarine Processes*, Vol. II. Ed., M.Wiley, Acdemic Press, New York, pp. 63–81.

14. Allen, G.P., Castaing, P., Frodefond, J.M. and C. Migniot (1979) Quelques effets a long terme des amenagements sur la sedimentation dans l'estuaire de la Gironde. *Publications du CNEXO: Actes de Colloques* 9: 115–138.

15. Allen, G.P., Salomon, J.C., Bassoulet, P., Du Penhoat, Y. and De Grandpre, C. (1980) Effects of tides on mixing and suspended sediment transport in macrotidal estuaries. *Sediment. Geol.* 26: 69–90.

16. Allen, J.R.L. (2000) Morphodynamics of Holocene salt marshes: A review sketch from the Atlantic and Southern North Sea coasts of Europe. *Quaternary Science Revs.* 19: 1155–1231.

17. Almar, R., Coco, G., Bryan, K.R., Huntley, D.A., Short, A.D. and Senechal, N. (2008) Video observations of beach cusp morphodynamics. *Mar. Geology* 254: 216–223.

18. Almar, R., Castelle, B., Ruessink, B.G., Senechal, N., Bonneton, P. and Marieu, V. (2010) Two- and three-dimensional double sandbar system behaviour under intense wave forcing and a meso-macro tidal range. *Cont. Shelf Res.* 30: 781–792.

19. Amos, C.L., Daborn, G.L., Christian, H.A., Atkinson, A. and Robertson, A. (1992) In situ erosion measurements on fine-grained sediments from the Bay of Fundy. *Mar. Geol.* 108: 175–196.

20. Amos, C.L. (1995) Siliciclastic tidal flats. In: *Geomorphology and sedimentology of estuaries*. Ed., G.M.E. Perillo, Elsevier, New York, pp. 273–306.

21. Amos, C.L., Bergamasco, A., Umgiesser, G., Cappucci, S., Cloutier, D., DeNat, L., Flindt, M., Bonardi, M. and Cristante, S. (2004) The stability of tidal flats in Venice Lagoon — The results of *in-situ* measurements using two benthic, annular flumes. *Journal of Marine Systems*, 51: 211–241.

22. Andersen A.G. (1967) On the development of stream meanders. *Proc. 12th Congr. IAHR.* 1: 370–378.

23. Andersen, T.J. and Pejrup, M. (2001) Suspended sediment transport on a temperate, microtidal mudflat, the Danish Wadden *Sea. Mar. Geol.* 173: 69–85.

24. Andersen, T.J., Pejrup, M. and Nielsen, A.A. (2005) Long-term and high-resolution measurements of bed level changes in a temperate, microtidal coastal lagoon. *Marine Geol.* 226: 115–125.

25. Anderson, T.J. (2001) Seasonal variability in erodibility of two temperate microtidal mudflats. *Est. Coast. Shelf Sci.* 45: 507–524.

26. Antia, E.E. (1996) Rates and patterns of migration of shoreface-connected sandy ridges along the Southern North Sea Coast. *J. Coast. Res.* 12: 38–46.

27. Antia, E.E. (1996) Shoreface-connected ridges in German and US Middle Atlantic Bights: Similarities and contrasts. *J. Coast. Res.* **12**: 141–146.

28. Anthony, E.J. (2013) Storms, shoreface morphodynamics, sand supply, and the accretion and erosion of coastal dune barriers in the southern North Sea. *Geomorphology* **199**: 8–21.

29. Anthony, E.J. (2013) Wave influence in the construction, shaping and destruction of river deltas: A review. *Mar. Geol.* **361**: 53–78.

30. Apotsos, A. (2007) *Setup in the Surf Zone.* PhD Thesis, MIT and WHOI.

31. Arens, S.M. (1996) Patterns of sand transport on vegetated foredunes. *Geomorphology* **17**: 339–350.

32. Ashley, G.M. (1990) Classification of large-scale subaqeous bedforms: A new look at an old problem. *J. Sed. Petrol.* **60**: 160–172.

33. Ashton, A., Murray, A.B. and Arnault, O. (2001) Formation of coastline features by large-scale instabilities induced by high-angle waves. *Nature* **414**: 296–300.

34. Ashton, A. and Murray, A.B. (2006) High-angle wave instability and emergent shoreline shapes: 1. Modeling of sand waves, flying spits, and capes. *J. Geophys. Res.* **111**: F04011.

35. Ashton, A. and Murray, A.B. (2006) High-angle wave instability and emergent shoreline shapes: 2. Wave climate analysis and comparisons to nature. *J. Geophys. Res.* **111**: F04012.

36. Ashton, A.D., Walkden, M.J.A. and Dickson, M.E. (2011) Equilibrium responses of cliffed coasts to changes in the rate of sea level rise. *Marine Geol.* **284**: 217–229.

37. Aubrey, D. (1979) Seasonal patterns of onshore/offshore sediment movement. *J. Geophys. Res.* **84**: 6,347–6,354.

38. Aubrey, D.G. and Speer, P.E. (1985) A study of non-linear tidal propagation in shallow inlet/estuarine systems. Part I: Observations. *Est. Coast. Shelf Sci.* **21**: 185–205.

39. Aubrey, D.G. and Friedrichs, C.T. (1988) Seasonal climatology of tidal non-linearities in a shallow estuary. In: *Hydrodynamics and Sediment Dynamics of Tidal Inlets.* Eds., D.G. Aubrey and L. Weishar, Springer Verlag, p. 103–124.

40. Austin, J.A. (2004) Estimating effective longitudinal dispersion in the Chesapeake Bay. *Est. Coast. Shelf Sci.* **60**: 359–368.

41. Austin, M.J., Masselink, G., Hare, T.J. and Russell, P.E. (2007) Relaxation time effects of wave ripples on tidal beaches. *J. Geophys. Res.* **34**: L16606.

42. Austin. M.J., Scott, T.M., Brown, J.W., Brown, J.A., MacMahan, J.H., Masselink, G. and Russell, P.E. (2010) Temporal observations of rip current circulation on a macro-tidal beach. *Cont. Shelf Res.* **30**: 1149–1165.

43. Avoine, J., Allen, G.P., Nichols, M., Salomon, J.C. and Larsonneur, C. (1981) Suspended-sediment transport in the Seine estuary, France: Effect of man-made modifications on estuary-shelf sedimentology. *Mar. Geol.* **40**: 119–137.
44. Bagnold, R.A. (1946) Motion of waves in shallow water. Interactions of waves and sand bottom. *Proc. Royal Soc. (A)* **187**: 1–15.
45. Bagnold, R.A. (1963) Mechanics of marine sedimentation. In: *The Sea,* Vol. 3. Ed., M.N. Hill, Wiley-Interscience, pp. 507–528.
46. Bailard, J.A. (1981) An energetics total load sediment transport model for a plane sloping beach. *J. Geophys. Res.* **86**: 10,938–10,954.
47. Bakker, W.T. (1969) The dynamics of a coast with a groyne system. *Proc. 11th Int. Coastal Eng. Conf., ASCE,* pp. 492–517.
48. Bakker, W.T. (1971) The influence of longshore variation of the wave height on the littoral drift. Rijkswaterstaat, Directorate for Water Management and Hydraulic Research, Dept for Coastal Research, The Hague. Study Report WWK 71–19.
49. Baldock, T.E. and Huntley, D.A. (2002) Long-wave forcing by the breaking of random gravity waves on a beach. *Proceedings of the Royal Society of London, A.* **458**: 2177–2201.
50. Baldock, T.E., Tomkins, M.R., Nielsen, P. and Hughes M.G. (2004) Settling velocity of sediments at high concentrations. *Coastal Eng.* **51**: 91–100.
51. Baldock, T.E., Alsina, J.A., Caceres, I., Vicinanza, D., Contestabile, P., Power, H. and Sanchez-Arcilla, A. (2011) Large-scale experiments on beach profile evolution and surf and swash zone sediment transport induced by long waves, wave groups and random waves. *Coastal Eng.* **58**: 214–227.
52. Baquerizo, A., Losada, M.A. and Smith, J.M. (1998) Wave reflection from beaches: a predictive model. *J. Coast. Res.* **14**: 291–298.
53. Barwis, J.H. (1978) Sedimentology of some South Carolina tidal-creek point bars and a comparison with their fluvial counterparts. In: *Fluvial Sedimentology.* Ed., A.D. Miall. *Can. Soc. Petr. Geol. Mem.* **5**: 129–160.
54. Bartholomae, A., Kubicki, A., Badewien, T.H. and Flemming, B.W. (2009) Suspended sediment transport in the German Wadden Seaseasonal variations and extreme events. *Ocean Dynamics.*
55. Basco, D. R. (1985) A qualitative description of wave breaking. *Journal of Waterway, Port, Coastal and Ocean Engineering* **11**: 171–188.
56. Bass, S.J., Aldridge, J.N., McCave, I.N. and Vincent, C.E. (2002) Phase relationship between fine sediment suspensions and tidal currents in coastal seas. *J. Geophys. Res.* **107**: 10-1–10-6.
57. Battjes, J.A. (1974) Surf similarity. *Procs. 14th Int. Conf. Coastal Engineering, ASCE,* pp. 466–480.
58. Battjes, J.A. and Janssen, J.P.F.M. (1978) Energy loss and set-up due to breaking of random waves. *Pro. 16th Int. Conf. Coast. Eng., ASCE:* pp. 569–587.

59. Battjes, J.A. (1988) Surf-zone dynamics. *Ann. Rev. Fluid Mech.* **20**: 257–293.
60. Bauer, B.O., Davidson-Arnott, R.G.D., Nordstrom, K.E., Ollerhead, J. and Jackson, N.L. (1996) Indeterminacy in aeolian sediment transport across beaches. *J. Coast. Res.* **12**: 641–653.
61. Bauer, B.O., and Davidson-Arnott, R.G.D. (2002) A general framework for modeling sediment supply to coastal dunes including wind angle, beach geometry, and fetch effects. *Geomorphology* **49**: 89–108.
62. Bearman, G. (Ed.) (1991) *Waves, Tides and Shallow-Water Processes*. The Open University, Pergamon Press, Oxford.
63. Bearman, J.A., Friedrichs, C.T., Jaffe, B.E. and Foxgrover, A.C. (2010) Spatial trends in tidal flat shape and associated environmental parameters in South San Francisco Bay. *Journal of Coastal Research* **26**: 342–349.
64. Beets, D.J., Roep, Th.B. and de Jong, J. (1981) Sedimentary sequences of the sub-recent North Sea coast of the Western Netherlands near Alkmaar. In: *Holocec Marine Sedimentation in the North Sea Basin*. Eds., S.D. Nio, R.T.E. Schuttenhelm and Tj.C.E. van Weering. *I. A. S. Spec. Publ.* **5**: 133–145.
65. Beets, J.D., van der Valk, L. and Stive, M.J.F. (1992) Holocene evolution of the coast of Holland. *Mar. Geol.* **103**: 423–443.
66. Beets, J.D. and van der Spek, A.J.F. (2000) The holocene evolution of the barrier and the back-barrier basins of Belgium and the Netherlands as a function of late Weichselian morphology, relative sea level rise and sediment supply. *Neth. J. Geosci.* **79**: 3–16.
67. Bell, P.S. (1999) Shallow water bathymetry derived from an analysis of X-band marine radar images of waves. *Coast. Eng.* **37**: 513–527.
68. Bendixen, M., Clemmensen, L.B. and Kroon, A. (2013). Sandy berm and beach-ridge formation in relation to extreme sea-levels: A Danish example in a micro-tidal environment. *Marine Geol.* **344**: 53–64.
69. Berendsen, H.J.A. (1998) Birds-eye view of the Rhine–Meuse Delta (The Netherlands). *J. Coast. Res.* **14**: 740–752.
70. Bernabeu, A.M., Medina, R. and Vidal, C. (2003) A morphological model of the beach profile integrating wave and tidal influences. *Mar. Geol.* **197**: 95–116.
71. Berné, S., Lericolais, G., Marsset, T., Bourillet, J.F. and De Batist, M. (1998) Erosional offshore sand ridges and lowstand shorefaces: examples from tide- and wave-dominated environments in France. *J. Sed. Res.* **68**: 540–555.
72. Berné, S., Vagner, P., Guichard, F., Lericolais, G., Liu, Z., Trentesaux, A., Yin, P. and Yi, H.I. (2002) Pleistocene forced regressions and tidal sand ridges in the East China Sea. *Mar. Geol.* **188**: 293–315.
73. Berni, C., Barthlemy, E. and Michallet, H. (2013) Surf zone cross-shore boundary layer velocity asymmetry and skewness: An experimental study on a mobile bed. *J. Geophys. Res.* **118**.

74. Bertier, C. (2011) Dynamique et suivi du bouchon vaseux dans l'estuaire de la Loire. Sminaire Technique. Le transport sdimentaire: Principes et expriences sur le bassin ligrien, Vierzon 24 Novembre 2011.
75. Besio, G., Blondeaux, P. and Vittori, G. (2006) On the formation of sand waves and sand banks. *J. Fluid Mech.* **557**: 1–27.
76. Best, J. (2005) The fluid dynamics of river dunes: A review and some future research directions. *J. Geophys. Res.* **110**: F04S02.
77. Bijker, E.W. (1967). Some considerations about scales for coastal models with moveable bed. Thesis, Delft Technological University.
78. Bijker, E.W., Kalkwijk, J.P.Th. and Pieters, T. (1974) Mass transport in gravity waves on a sloping bottom. In: *Procs. 14th Int. Coast. Eng. Conf., ASCE*, pp. 447–465.
79. Bird, E.C.F. (1985) Coastline Changes. Wiley, Chichester.
80. Bird, E.C.F. (1996) Coastal erosion and rising sea-level. In: *Sea-Level Rise and Coastal Subsidence.* Ed., J.D. Milliman and B.U. Haq. Kluwer Ac. Publ., Dordrecht.
81. Bird, E.C.F. (2008) *Coastal Geomorphology: An Introduction*, Second Edition. JohnWiley and Sons, Ltd.
82. Black, K.P., Gorman, R.M. and Symonds, G. (1995) Sediment transport near the breakpoint associated with cross-shore gradients in vertical eddy diffusivity. *Coastal Eng.* **26**: 153–175.
83. Black, K.P., Gorman, R.M. and Byran, K.R. (2002) Bars formed by horizontal diffusion of suspended sediment. *Coastal Eng.* **47**: 53–75.
84. Blanton, J.O., Lin, G. and Elston, S.A. (2002) Tidal current asymmetry in shallow estuaries and tidal creeks. *Cont. Shelf Res.* **22**: 1731–1743.
85. Blanton, J.O., Seim, H., Alexander, C., Amft, J. and Kineke, G. (2003) Transport of salt and suspended sediments in a curving channel of a coastal plain estuary: Satilla River, GA. *Est. Coast. Shelf Sci.* **57**: 993–1006.
86. Blondeaux, P. and Seminara, G. (1985) A unified bar-bend theory of rivers. *J. Fluid Mech.* **157**: 449–470.
87. Blondeaux, P. (1990) Sand ripples under sea waves. Part 1. Ripple formation. *J. Fluid Mech.* **218**: 1–17.
88. Blondeaux, P. (2001) Mechanics of coastal forms. *Ann. Rev. Fluid Mech.* **33**: 339–369.
89. Bokuniewicz, H. (1995) Sediementary systems of coastal-plain estuaries. In: *Geomorphology and sedimentology of estuaries.* Ed., G.M.E. Perillo, Elsevier, Amsterdam: pp. 49–67.
90. Bolla Pittaluga, M., Repetto, R. and Tubino, M. (2001) Channel bifurcation in one-dimensional models: A physically based nodal point condition. In: *2nd IAHR Symp. Riv. Coas. Est. Morph.* Ed., S. Ikada. Obihiro, Japan, pp. 305–314.

91. Bolla Pittaluga, M., Repetto, R. and Tubino M. (2003) Channel bifurcation in braided rivers: Equilibrium configurations and stability. Water Resources Res. **39**(3): 1046.
92. Boothroyd, J.C. (1985) Tidal inlets and tidal deltas. In: *Coastal Sedimentary Environments*. Ed., R.A.Davis, Springer-Verlag, New York, pp. 445–532.
93. Borsje, B.W. (2006) *Biological Influence on Sediment Transport and Bed Composition for the Western Wadden Sea*. MSc. Thesis, Twente University.
94. Borsje, B.W., De Vries, M.B., Bouma, T.J., Besio, G., Hulscher, S.J.M.H. and Herman, P.M.J. (2009) Modeling bio-geomorphological influences for offshore sand waves. *Continental Shelf Res.* **29**: 1289–1301.
95. Borsje, B.W., Hulscher, S.J.M.H., Herman, P.M.J. and De Vries, M.B. (2009) On the paramaterization of biological influences on offshore sand wave dynamics. *Ocean Dynamics* 1–13.
96. Borsje, B.W. (2012) *Biogeomorphology of coastal seas*. PhD Thesis, Twente University, Netherlands.
97. Borsje, B.W., Kranenburg, W.M., Roos, P.C., Matthieu, J. and. Hulscher S.J.M.H. (2014) The role of suspended load transport in the occurrence of tidal sand waves. *J. Geophys. Res. Earth Surf.* **119**.
98. Borsje, B.W., Bouma, T.J., Rabaut, M., Herman, P.M.J. and Hulscher, S.J.M.H. (2014) Formation and erosion of biogeomorphological structures: A model study on the tube-building polychaete Lanice conchilega. *Limnol. Oceanogr.* **59**: 1297–1309.
99. Boswood, P.K. and Murray, R.J. (2001) World-wide Sand Bypassing Systems: Data Report. Conservation technical report No. 15, Queensland Government, Environmental Protection Agency, ISSN 1037-4701.
100. Bouchet, J-M., Deltreil, J-M., Manaud, F., Maurer, D. and Trut, G. (1997) Etude Integree du Bassin dArcachon. Report Ifremer DEL/AR/RDN/1997-09.
101. Bouma, T.J., De Vries, M.B., Low, E., Peralta, G., Tanczos, I.C., Van de Koppel, J. and Herman, P.M.J. (2005) Trade-offs related to ecosystem-engineering: A case study on stiffness of emerging macrophytes. *Ecology* **86**: 2187–2199.
102. Boehlich, M.J. (2003) Tidedynamik der Elbe. Mitteilungsblatt der Bundesanstalt fr Wasserbau Nr. **86**: 55–60.
103. Bowen, A.J. and Inman, D.L. (1971) Edge waves and crescentic bars. *J. Geophys. Res.* **76**: 8662–8671.
104. Bowen, A.J. (1980) Simple models of nearshore sedimentation; beach profiles and longshore bars. In: *The coastline of Canada*, Ed., S.B. McCann, Geological Survey of Canada, Ottawa, pp. 1–11.
105. Bowen, A.J. and Holman, R.A. (1989) Shear instabilities of the mean longshore current, 1. Theory. *J. Geophys. Res.* **94**: 18023–18030.

106. Bowen, M.M. and Geyer, W.R. (2003) Salt transport and the time-dependent salt balance of a partially stratified estuary. *J. Geophys. Res.* **108**: 27.1–27.15.
107. Boyd, R., Forbes, D.L. and Heffler, D.E. (1988) Time-sequence observations of wave-formed sand ripples on an ocean shoreface. *Sedimentol.* **35**: 449–464.
108. Boyd, R., Dalrymple, R.W. and Zaitlin, B.A. (1992) Classification of clastic coastal depositional environments. *Sed. Geol.* **80**: 139–150.
109. Bradbury, A., Rogers, J. and Thomas, D. (2012) Toe structures management manual, Report SC070056/R, UK Environment Agency.
110. Brander, R.W. (1999) Field observations on the morphodynamic evolution of a low-energy rip current system. *Mar. Geol.* **157**: 199–217.
111. Brenon, I. and Le Hir, P. (1999) Modelling the turbidity maximum in the Seine estuary: Identification of formation processes. *Est. Coast Shelf Sci.* **49**: 525–544.
112. Brown, J.M. and Davies, A.G. (2010) Flood/ebb tidal asymmetry in a shallow sandy estuary and the impact on net sand transport. *Geomorphology* **114**: 431–439.
113. Bruun, P. (1954) Coast erosion and the development of beach profiles. Beach Erosion Board, US Army Corps of Eng., *Tech. Mem.* **44**: 1–79.
114. Bruun, P. (1954) Migrating sand waves or sand humps, with special reference to investigations carried out on the Danish North Sea Coast. *Coastal Eng., Am. Soc. of Civ. Eng.* **1954**: 269–295.
115. Bruun, P. (1962) Sea-level rise as a cause of shore erosion. *Proc. Am. Soc. Civ. Eng., J. Water Harbors Div.* **88**: 117–130.
116. Bruun, P. (1963) Longshore currents and longshore throughs. *J. Geophys. Res.* **68**: 1065–1078.
117. Bruun. P., Mehta, A.J. and Johnson, I.J. (1978) Stability of tidal inlets: Theory and engineering. In: *Developments in Geotechnical Engineering.* Elsevier Science, Amsterdam, p. 510.
118. Bruun, P. (1988) The Bruun rule of erosion by sea-level rise: A discussion on large-scale two- and three-dimensional usages. *J. Coast. Res.* **4**: 627–648.
119. Burchard, H. and Baumert, H. (1998) The formation of estuarine turbidity maxima due to density effects in the salt wedge. A hydrodynamic process study. *J. Phys. Oceanography* **28**: 309–321.
120. Burchard, H., Floeser, G., Stanev, J.V., Badewien, T.H. and Riethmueller, R. (2007) Impact of Density Gradients on Net Sediment Transport into the Wadden Sea. *J. Phys. Oceanography* **38**: 566–587.
121. Butt, T., Russell, P. and Turner, I. (2001) The influence of swash infiltration-exfiltration on beach face sediment transport: Onshore or offshore? *Coastal Eng.* **42**: 35–52.

122. Byrne, R.J., Gammisch, R.A. and Thomas, G.R. (1981) Tidal prism-inlet area relationships for small tidal inlets. In: *Procs. 17th Int. Coast. Eng. Conf.,* *ASCE,* New York, pp. 2517–2533.

123. Caballeria, M., Coco, G., Falqués, A. and Huntley, A.D. (2002) Selforganization mechanisms for the formation of nearshore crescentic and transverse sand bars. *J. Fluid Mech.* **465**: 379–410.

124. Cadee, G. (1979) Sediment reworking by the polychaete heteromastus filiformis on a tidal flat in the Dutch Wadden Sea. Netherlands *J. Sea Res.* **13**: 441–456.

125. Callender, R.A. (1969) Instability and river channels. *J. Fluid Mech.* **36**: 465–480.

126. Calvete, D., Falqués, A., De Swart, H.E. and Walgreen, M. (2001) Modelling the formation of shoreface-connected sand ridges on storm-dominated inner shelves. *J. Fluid Mech.* **441**: 169–193.

127. Calvete, D., De Swart, H.E. and Falqués, A. (2002) Effect of depth-dependent wave stirring on the final amplitude of shoreface-connected ridges. *Cont. Shelf Res.* **22**: 2763–2776.

128. Calvete, D., Dodd, N., Falqués, A. and van Leeuwen, S.M., (2005) Morphological development of rip channel systems: Normal and near normal wave incidence. *J. Geophys. Res.* **110**: C10006.

129. Calvete, D., Coco, G., Falques, A. and Dodd, N. (2007) (Un)predictability in rip channel systems. *Geophys. Res. Lett.* **34**: (L05605).

130. Callaghan, D.P., Bouma, T.J., Klaassen, P., van der Wal, D., Stive, M.J.F. and Herman, P.M.J. (2010) Hydrodynamic forcing on salt-marsh development: Distinguishing the relative importance of waves and tidal flows. *Estuarine. Coastal and Shelf Sci.* **89**: 73–88.

131. Camenen, B. and Larroude, P. (1999) Nearshore and transport modelling: application to Trucvert beach. In: *Proc. IAHR Symp. on River, coastal and Estuarine Morphodynamics,* Vol. II: 31–40.

132. Camenen, B. and Larroude, P. (2003) Comparison of sediment transport formulae for the coastal environment. *Coastal Engineering* **48**: 111–132.

133. Camenen, B. (2009) Estimation of the wave-related ripple characteristics and induced bed shear stress. *Estuarine, Coastal and Shelf Sci.* **84**: 553–564.

134. Camenen, B. and Van Bang, D.P. (2011) Modelling the settling of suspended sediments for concentrations close to the gelling concentration. *Continental Shelf Res.* **31**: 106–111.

135. Canestrelli, A., Fagherazzi, S., Defina, A. and Lanzoni, S. (2010) Tidal hydrodynamics and erosional power in the Fly River delta, Papua New Guinea. *J. Geophys. Res.* **115**: F04033.

136. Carbajal, N. and Montano, Y. (2001) Comparison between predicted and observed physical features of sandbanks. *Est. Coast. Shelf Sci.* **52**: 435–443.

137. Carrier, G. F. and Greenspan, H. P. (1958) Water waves of finite amplitude on a sloping beach, *J. Fluid Mech.* **4**: 97–109.
138. Carter, R.W.G. (1988) *Coastal Environments*. Academic Press, London, 617 pp.
139. Cartwright, D.E. (1999) *Tides, A Scientific History*. Cambridge University Press, UK, 292 pp.
140. Castaing, P. (1989) Co-oscillating tide controls long-term sedimentation on the Gironde estuary, France. *Mar. Geol.* **89**: 1–9.
141. Castelle, B., Bonneton, P., Dupuis, H. and Senechal, N. (2007) Double bar beach dynamics on the high-energy meso-macrotidal French Aquitanian Coast: A review. *Marine Geol.* **245**: 141–159.
142. Castelle, B., Turner, I.K., Bertin, X. and Tomlinson, R. (2009) Beach nourishments at Coolangatta Bay over the period 1987–2005: Impacts and lessons. *Coastal Eng.* **56**: 940–950.
143. Castelle, B., Michallet, H., Marieu, V., Leckler, F., Dubardier, B., Lambert, A., Berni, C., Bonneton, P., Bartelemy, E. and Bouchette, F. (2010) Laboratory experiment on rip current circulations over a moveable bed: Drifter measurements. *J. Geophys. Res.* **115**: C12008.
144. Castelle, B., Ruessink, B. G., Bonneton, P., Marieu, V., Bruneau, N. and Price, T. D. (2010) Coupling mechanisms in double sandbar systems, Part 1: Patterns and physical explanation. *Earth Surf. Proc. Landforms* **35**: 476–486.
145. Castelle, B., Ruessink, B. G., Bonneton, P., Marieu, V., Bruneau, N. and Price, T. D. (2010) Coupling mechanisms in double sandbar systems, Part 2: Impact on alongshore variability of inner-bar rip channels. *Earth Surf. Proc. Landforms* **35**: 771–781.
146. Castelle, B. and Coco, G. (2012) The morphodynamics of ripchannels on embayed beaches. *Cont. Shelf Res.* **43**: 10–23.
147. Castelle, B., Almar, R., Dorel, M., Lefebvre, J.P., Senechal, N., Anthony, E.J., Laibi, R., Chuchla, R. and du Penhoat, Y. (2014) Rip currents and circulation on a high-energy low-tide-terraced beach (Grand Popo, Benin, West Africa). In: *Procs. 13th Int. Coastal Symp.* Eds., A.N. Green and J.A.G. Cooper. Durban, S-Africa, *J. Coastal Res.* Special Issue 66, ISSN 0749-0208.
148. Castro-Orgaz, O., Girldez, V., Mateos, L. and Dey, S. (2012) Is the von Karman constant affected by sediment suspension? *J. Geophys.Res.* **117**: F04002.
149. Cayocca, F. (2001) Long-term morphological modeling of a tidal inlet: the Arcachon Basin, France. *Coastal Eng.* **42**: 115–142.
150. Charru, F. and Hinch, E.J. (2006) Ripple formation on a particle bed sheared by a viscous liquid. Part 1. Steady flow. *J. Fluid Mech.* **550**: 111–121.
151. Cauchat, J., Guillou, S., Van Bang, D.P. and Nguyen K. (2013) Modelling sedimentation-consolidation in the framework of a one-dimensional two-phase flow model. *J. Hydraulic Research* **51**: 293–305.

152. CERC Shore Protection manual, I-III. Army Corps of Engineers, US Govt. Printing Office.

153. Chang, H., Simons, D.B. and Woolisher, D.A. (1971) Flume experiments on alternate bar formation. *Proc. ASCE, Waterways, Harbors Coast. Eng. Div.* **97**: 155–165.

154. Chang, J.H. and Choi. J.Y. (2001) Tidal-flat sequence controlled by holocene sea-level rise in Gomso Bay, West Coast of Korea. *Est. Coast. Shelf Science* **52**: 391–399.

155. Chang, T.S., Joerdel, O., Flemming, B.W. and Bartholomae, A. (2006) The role of particle aggregation/disaggregation in muddy sediment dynamics and seasonal sediment turnover in a back-barrier tidal basin, East Frisian Wadden Sea, southern North Sea. *Mar. Geology* **235**: 49–61.

156. Chantler, A.G. (1971) The applicability of regime theory to tidal water courses. *J. Hydraul. Res.* **12**: 181–191.

157. Chen, J., Liu, C., Zhang, C. and Walker, H.J. (1990) Geomorphological development and sedimentation in Qiantang estuary and Hangzou bay. *J. Coast. Res.* **6**: 559–572.

158. Chen, S. and Doolen, G.D. (1998) Lattice Boltzmann method for fluid flows. *Annu. Rev. Fluid Mech.* **30**: 329–364.

159. Cherlet, J., Besio, G., Blondeaux, P., Van Lancker, V., Verfaillie, E. and Vittori, G. (2007) Modeling sand wave characteristics on the Belgian Continental Shelf and in the Calais–Dover Strait. *J. Geophys. Res.* **112**: C06002.

160. Chernertsky, A.S., Schuttelaars, H.M. and Talke, S.A. (2010) The effect of tidal asymmetry and temporal settling lag on sediment trapping in tidal estuaries. *Ocean Dynamics* **60**: 1219–1241.

161. Choi, K. (2011) Tidal rhythmites in a mixed-energy, macrotidal estuarine channel, Gomso Bay, west coast of Korea. *Mar. Geol.* **280**: 105–115.

162. Choi, K., Hong, C.M., Kimb, M.H., Oh, C.R. and Jung, J.H. (2013) Morphologic evolution of macrotidal estuarine channels in Gomso Bay, west coast of Korea. Implications for the architectural development of inclined heterolithic stratification. *Mar. Geol.* **346**: 343–354.

163. Christiansen, T., Wiberg, P.L. and Milligan, T.G. (2000) Flow and sediment transport on a tidal salt marsh surface. *Estuarine Coastal Shelf Sci.* **50**: 315–331.

164. Christiansen, M.B. and Davidson-Arnott, R. (2004) Rates of landward sand transport over the foredune at Skallingen, Denmark and the role of dune ramps. *Danish Journal of Geography* **104**: 31–43.

165. Chung, D.H. and Van Rijn, L.C. (2003) Diffusion approach for suspended sediment transport. *J. Coast. Res.* **19**: 1–11.

166. Ciavola, P. and Stive, M.J.F. (2012) Thresholds for storm impacts along European coastlines: Introduction. *Geomorphology* **143-144**: 1–2.

167. Ciriano, Y., G. Coco, K.R. Bryan, and Elgar S. (2005) Field observations of swash zone infragravity motions and beach cusp evolution. *J. Geophys. Res.* **110**: C02018, doi:10.1029/2004JC002485.

168. Clarke, L.B. and Werner, B.T. (2004) Tidally modulated occurrence of megaripples in a saturated surf zone. *J. Geophys. Res.* **109**: C01012: 1–15.

169. Claudin, P., Charu, F. and Andreotti, B. (2011) Transport relaxation time and length scales in turbulent suspensions. *J. Fluid Mech.* 1–16.

170. Cleveringa, J. (2000) Reconstruction and modelling of Holocene coastal evolution of the western Netherlands. Thesis, Utrecht University, Geologica Ultraiectina **200**.

171. Coco, G., O'Hare, T.J. and Huntley, D.A. (1999) Beach cusps: A comparison of data and theories for their formation. *J. Coast. Res.* **15**: 741–749.

172. Coco, G., Huntley, D.A. and O'Hare, T.J. (2000) Investigation of a self-organization model for beach cusp formation and development. *J. Geophys. Res.* **105**: 21,9991–22,002.

173. Coco, G., Burnet, T.K. and Werner, B.T. (2003) Test of self-organisation in beach cusp formation. *J. Geophys. Res.* **108**(C3): 46.

174. Coco, G., Murray, B. and Green, O.G. (2007) Sorted bed forms as self-organized patterns: 1. Model development. *J. Geophys. Res.* **112**: F03015.

175. Coco, G., Murray, A. B., Green, M. O., Thieler, E. R. and Hume, T. M. (2007) Sorted bed forms as self-organized patterns: 2. Complex forcing scenarios, *J. Geophys. Res.* **112**: F03016.

176. Coco, G., Zhou, Z., Van Maanen, B., Olabarrieta, M., Tinocoa, R. and Townend, I. (2013) Morphodynamics of tidal networks: Advances and challenges. *Mar. Geology* **346**: 1–16.

177. Coleman, N.L. (1970) Flume studies of the sediment transfer coefficient. *Water Resources Res.* **6**: 801–809.

178. Coleman, N.L. (1981) Velocity profile with suspended sediment. *J. Hydraul. Res.* **19**: 211–229.

179. Coleman, S.E. and Melville, B.W. (1996) Initiation of bed forms on a flat sand bed. *J. Hydr. Eng.* **122**: 301–310.

180. Collins, M.B., Amos, C.L. and Evans, G. (1981) Observations of some sediment-transport processes over intertidal flats, the Wash, UK. In: *Int. Ass. Sediment Spec. Publ.* **5**: 81–98.

181. Collins, M.B. (1983) Supply, distribution and transport of suspended sediment in a macrotidal environment: Bristol Channel, UK *Can. J. Fish. Aquat. Sci* **40**(suppl.): 44–59.

182. Colombini, M. and Stocchino, A. (2011) Ripple and dune formation in rivers. *J. Fluid Mech.* **673**: 121–131.

183. Comoy, M. (1881) *Etude pratique sur les marées fluviales*. Gauthiers-Villars, Paris.

184. Contardo, S. and Symonds, G. (2015) Sandbar straightening under wind-sea and swell forcing. *Marine Geology* **368**: 25–41.
185. Cooper, J.A. and Pilkey, O.H. (2004) Sea-level rise and shoreline retreat: time to abandon the Bruun Rule. *Global Planetary Change* **43**: 157–171.
186. Cornaglia, P. (1889) Delle Spiaggie. Accademia Nazionale dei Lincei, *Atti. Cl. Sci. Fis., Mat. e Nat. Mem.* **5**: 284–304.
187. Cornish, V. (1898) On sea beaches and sandbanks. *Geograph. J.* **11**: 528–559, 628–647.
188. Cowell, P.J., Stive, M.J.F., Niedoroda, A.W., de Vriend, H.J., Swift, D.P.J., Kaminsky, G.M. and Capobianco, M. (2003) The coastal-tract: A conceptual approach to aggregate modelling of low-order coastal change. *J. Coast. Res.* **19**: 812–827.
189. Cowell, P.J., Stive, M.J.F., Niedoroda, A.W., Swift, D.P.J., de Vriend, H.J., Buijsman, M.C., Nicholls, R.J., Roy, P.S., Kaminsky, G.M., Cleveringa, J., Reed, C.W. and De Boer, P.L. (2003) The coastal-tract (Part 2): Applications of aggregated modeling of lower-order coastal change. *J. Coast. Res.* **19**: 828–848.
190. Cox, D.T. and Kobayashi, N. (1998) Application of an undertow model to irregular waves on plane and barred beaches. *J. Coast. Res.* **14**: 1314–1324.
191. Cuadrado, D.G., Perillo, G.M.E and Vitale, A.J. (2014) Modern microbial mats in siliciclastic tidal flats: Evolution, structure and the role of hydrodynamics. *Mar. Geol.* **352**: 367–380.
192. CUR (1992) Artificial sand fills in water. Centre for civil engineering and codes (CUR), Report 152, Gouda, The Netherlands.
193. Dagniaux, M.F. (2013) *Analysis of the Morphological Behaviour along the Luanda Coast.* M.Sc. Thesis, Delft University.
194. D'Alpaos, A., Lanzoni, S., Marani, M. and Rinaldo, A. (2010) On the tidal prism-channel area relations. *J. Geophys. Res.* **115**: F01003.
195. Dalrymple, R.W. (1984) Morphology and internal structure of sand waves in the Bay of Fundy. *Sedimentol.* **31**: 365–382.
196. Dalrymple, R.W., Zaitlin, B.A. and Boyd, R. (1992) Estuarine facies models: conceptual basis and stratigraphic implications. *J. Sed. Petrol.* **62**: 1130–1146.
197. Dalrymple, R.W. and Rhodes, R.N. (1995) Estuarine Dunes and Bars. In: *Geomorphology and Sedimentology of Estuaries.* Ed., G.M.E. Perillo. *Developments in Sedimentology* **53**, Elsevier, Amsterdam, pp. 359–422.
198. Dalrymple R.W., Mackay, D.A., Ichaso A.A. and Choi, K.S. (2012) Processes, morphodynamics, and facies of tide-dominated estuaries. In: *Principles of Tidal Sedimentology.* Eds., R.A. Davis and R.W. Dalrymple. Springer.
199. Damgaard, J.S., Van Rijn, L.C., Hall, L.J. ans Soulsby, R. (2001) Intercomparison of engineering methods for sand transport. In: *Sediment Transport Modelling in Marine Coastal Environments.* Eds., L.C. van Rijn, A.G.

Davies, J. Van de Graaf and J.S. Ribberink. Aqua Publications, Amsterdam, CJ1–CJ12.

200. Dankers, P.J.T. and Winterwerp, J.C. (2007) Hindered settling of mud flocs: Theory and validation. *Continental Shelf Res.* **27**: 1893–1907.

201. Dastgheib, A., Roelvink, J.A. and Wang, Z.B. (2008) Long-term process-based morphological modeling of the Marsdiep Tidal Basin. *Mar. Geology* **256**: 90–100.

202. Davidson-Arnott, R.G.D. and Van Heyningen, A. (2003) Migration and sedimentology of longshore sandwaves, Long Point, Lake Erie, Canada. *Sedimentology* **50**: 1123–1137.

203. Davidson-Arnott, R.G.D., Yang, Y., Ollerhead, J., Hesp, P.A. and Walker, I.J. (2008) The effects of surface moisture on aeolian sediment transport threshold and mass flux on a beach. *Earth Surface Processes and Landforms* **33**: 55–74.

204. Davidson, M.A., Splinter, K.D. and Turner, I.L. (2013) A simple equilibrium model for predicting shoreline change. *Coastal Eng.* **73**: 191–202.

205. Davies, A.G. (1982) On the interaction between surface waves and undulations of the sea bed. *J. Mar. Res.* **40**: 331–368.

206. Davies, A.G., Ribberink, J.S., Temperville, A. and Zyserman, J.A. (1997) Comparisons between sediment transport models and observations made in wave and current flows above plain beds. *Coastal Eng.* **31**: 163–169.

207. Davies, J.L. (1980) *Geographical Variation in Coastal Development*. Longman, New York, 212 pp.

208. Davis, R.A. (1985) Beach and nearshore zone. In: *Coastal Sedimentary Environments*. Ed., R.A. Davis, Springer-Verlag, pp. 379–444.

209. Davies, A.G. and Thorne, P.D. (2005) Modeling and measurement of sediment transport by waves in the vortex ripple regime. *J. Geophys. Res.* **110**: C05017.

210. Davies, A.G. and Thorne, P.D. (2008) Advances in the study of moving sediments and evolving seabeds. *Surv. Geophys.*

211. Dean, R.G. (1973) Heuristic models of sand transport in the surf zone. *Proc. Conf. Eng. Dynamics in the Surf Zone*, Sydney, pp. 208–214.

212. Dean, R.G. (1977) Equilibrium beach profiles: US Atlantic coast and Gulf coasts. *Ocean Eng. Tech. Rep.* **12**, Univ. of Delaware, Newark, 45 pp.

213. Dean, R.G. (1987) Coastal sediment processes: Towards engineering solutions. *Proc. Coastal Sediments, ASCE*, pp. 1–24.

214. Dean, R.G., and Maurmeyer, E.M. (1980) Beach cusps at Point Reyes and Drakes Bay beaches, California. In: *Procs. Int. Conf. Coast. Eng. ASCE*, New York, 863–884.

215. Dean, R.G. (1991) Equilibrium beach profiles: Characteristics and applications. *J. Coast. Res.* **7**: 53–84.

216. Dean, R.G. and Dalrymple, R.A. (2002) *Coastal Processes with Engineering Applications*. Cambridge Univ. Press, 475 pp.
217. De Bok, C. and Stam, J.M. (2002) *Long-term Morphology of the Eastern Scheldt*. Report Rijkswatertstaat, RIKZ 2002/108x.
218. De Bakker, A.T.M., Tissier, M.F.S. and Ruessink, B.G. (2014) Shoreline dissipation of infragravity waves. *Cont. Shelf Res.* **72**: 73–82.
219. De Haas, H. and Eisma, D. (1993) Suspended-sediment transport in the Dollard estuary. *Neth. J. Sea Res.* **31**: 37–42.
220. Dehouck, A., Dupuis, H. and Senechal, N. (2009) Pocket beach hydrodynamics: The example of four macrotidal beaches, Brittany, France. *Mar. Geol.* **266**: 1–17.
221. Deigaard, R., Drønen, N., Fredsøe, J., Jensen, J.H. and Jørgensen, M.P. (1999) A morphology stability analysis for a long straight barred coast. *Coast. Eng.* **36**: 171–195.
222. De Jong, H. and Gerritsen, F. (1985) Stability parameters of the Western Scheldt estuary. In: *Procs. 19th Int. Coast. Eng. Conf. ASCE*, New York, pp. 3079–3093.
223. De Jonge, V.N. and Van Beusekom, J.E.E. (1995) Wind- and tide-induced resuspension of sediment and microphytobenthos from tidal flats in the Ems estuary. *Limnol. Oceanogr.* **40**: 788–778.
224. De Jonge, V.N., Schuttelaars, H.M., Van Beusekom, J.E.E., Talke S.A. and De Swart, H.E. (2014) The influence of channel deepening on estuarine turbidity levels and dynamics, as exemplified by the Ems estuary. *Estuarine, Coastal and Shelf Science* **139**: 46–59.
225. J. Deloffre J., Verney, R., Lafite, R., Lesueur, P., Lesourd, S. and Cundy, A.B. (2007) Sedimentation on intertidal mudflats in the lower part of macrotidal estuaries: Sedimentation rhythms and their preservation. *Mar. Geol.* **241**: 19–31.
226. De Sonneville, B. and Van der Spek, A. (2012) Sediment- and morphodynamics of shoreface nourishments along the North-Holland coast. *Procs. 33rd Int. Conf. Coastal Engineering, Santander*, Spain.
227. De Swart, H.E. and Zimmerman J.T.F. (2009) Morphodynamics of tidal inlet systems. *Ann. Rev. Fluid Mech.* **41**: 203–29.
228. Dette, H.H. (2002) Sandbewegung im Küstenbereich. *Die Küste* **65**: 215–256.
229. De Vriend H.J., Bakker, W.T. and Bilse, D.P. (1994) A morphological behaviour model for the outer delta of mixed-energy tidal inlets. *Coast. Eng.* **23**: 305–327.
230. De Vriend, H. (2001) Long-term morphological prediction. In: *River, Coastal and Estuarine Morphodynamics*. Eds., G.Seminara and P. Blondeaux. Springer, Berlin, pp. 163–190.

231. De Vriend, H.J., Wang, Z.B., Ysebaert, T., Herman, P.M.J. and Ding, P.X. (2011) *Eco-Morphological Problems in the Yangtze Estuary and the Western Scheldt. Wetlands*, Springer.

232. De Vries, S., Van Thiel de Vries, J.S.M. and Ruessink, G. (2013) Modelling aeolian sediment accumulations on a beach. In: *7th International Conference on Coastal Dynamics*, Arcachon, France, 24–28 June.

233. Ghyr-Nielsen, M and Sorensen, T. (1970) Sand transport phenomena on coasts with bars. *Procs. 12th Coastal Eng. Conf., ASCE*, pp. 855–866.

234. Dibbits, H.A.M.C. (1950) Nederland Waterland, a historical-technical perspective. Oosthoek, Utrecht, 286 pp. (In Dutch)

235. Dibajnia, M. and Watanabe, A. (1996) A transport rate formula for mixed-size sands. In: *Proc. Int. Conf. Coast. Eng.*, Orlando, Florida. ASCE: 3791–3804.

236. Dieckmann, R.M., Osterthun, M., Partenscky, H.W. (1987) Influence of water-level elevation and tidal range on the sedimentation in a German tidal flat area. *Progress in Oceanography* **18**: 151–166.

237. DiLorenzo, J.L. (1988) The overtide and filtering response of small inlet/bay systems. In: *Hydrodynamics and Sediment Dynamics of Tidal Inlets*. Eds., D.G. Aubrey and L. Weishar. Springer, Verlag, New York, pp. 24–53.

238. Di Silvio, G. (1989) Modelling the morphological evolution of tidal lagoons and their equilibrium configurations. In: *22nd IAHR Congress, IAHR*, pp. C169–175.

239. Di Silvio, G. (1991) Averaging operations in sediment transport modelling: Short-step versus long-step morphological simulations. In: *Int. Symp. Transp. Susp. Sed. Mod.* Ed., L. Montefusco. Univ. Florence, pp. 723–739.

240. Dissanayake, D.M.P.K., Roelvink, J.A. and Van der Wegen, M. (2009) Modelled channel patterns in a schematized tidal inlet. *Coastal Eng.* **56**: 1069–1083.

241. Dissanayake, D.M.P.K., Ranasinghe, R. and Roelvink, J.A. (2012) The morphological response of large tidal inlet/basin systems to relative sea level rise. *Climatic Change* **113**: 253–276.

242. Dohmen-Janssen, C.M. and Hanes, D.M. (2002) Sheet flow dynamics under monochromatic nonbreaking waves. *J. Geophys. Res.* **107**(C10): 3149.

243. Dohmen-Janssen, C.M., Kroekenstoel, D.F., Hassan, W.N. and Ribberink J.S. (2002) Phase lags in oscillatory sheet flow: Experiments and bed load modelling. *Coastal Eng.* **46**: 61–87.

244. Dolan, R. (1971) Coastal landforms: crescentic and rhythmic. *Geol. Soc. Am. Bul.* **82**: 177–180.

245. Doodson, A.T. (1921) The harmonic development of the tide-generating potential. *Proc. R. Soc. London, Ser. A* **100**: 305–329.

246. Doody, P., Ferreira, M., Lombardo, S., Lucius, I., Misdorp, R., Niesing, H., Salman, A. and Smallegange, M. (2004) Living with coastal erosion

in Europe: Sediment and space for sustainability: Results from the EURO-SION study. European Commission: Luxembourg. ISBN 92-894-7496-3. 38 pp.

247. Droenen, N. and Deigaard, R. (2007) Quasi-three-dimensional modelling of the morphology of longshore bars. *Coastal Eng.* **54**: 197–215.

248. Dronkers, J.J. (1964) *Tidal Computations in Rivers and Coastal Waters.* North-Holland Publ. Co., Amsterdam, 518 pp.

249. Dronkers, J.J. (1970) Research for the coastal area of the delta region of the Netherlands. *Proc. 12th Int. Coast. Eng. Conf. Washington, ASCE,* Ch. 108.

250. Dronkers, J. and Zimmerman, J.T.F. (1982) Some principles of mixing in coastal lagoons. Oceanologica Acta SP, pp. 107–117.

251. Dronkers, J. (1982) Conditions for gradient-type dispersive transport in one-dimensional tidally averaged transport models. *Est. Coast. Shelf Sci.* **14**: 599–621.

252. Dronkers, J. (1984) Import of fine marine sediment in tidal basins. In: *Procs. Int. Wadden Sea Symp, Neth. Inst. for Sea Res. Publ. Series* **10**: 83–105.

253. Dronkers, J. (1986) Tidal asymmetry and estuarine morphology. *Neth. J. Sea Res.* **20**: 117–131.

254. Dronkers, J. (1998) Morphodynamics of the Dutch Delta. In: *Physics of Estuaries and Coastal Seas.* Ed., J. Dronkers and M.B.A.M. Scheffers, Balkema, Rotterdam, pp. 297–304.

255. Duan, J.G. (2005) Analytical approach to calculate the rate of bank erosion. *J. Hydraul. Eng., ASCE* **131**: 980–990.

256. Dubarbier, B., Castelle, B., Marieu, V. and Ruessink, G. (2014) Process-based modeling of cross-shore sandbar behaviour. *Coastal Eng.* **95**: 35–50.

257. Dupont, J-P., Lafite, R., Huault, F., Hommeril, P. and Meyer, R. (1994) Continental/marine ratio changes in suspended and settled matter across macrotidal estuary (the Seine estuary, northwestern France). *Mar. Geol.* **120**: 27–40.

258. Dyer, K.R. (1986) *Coastal and Estuarine Sediment Dynamics.* John Wiley, Chichester. p. 342.

259. Dyer, K.R. and Huntley, D.A. (1999) The origin, classification and modelling of sand banks and ridges. *Cont. Shelf Res.* **19**: 1285–1330. *Int. Coast. Eng. Conf. Washington, ASCE,* Ch. 108.

260. Dyer, K.R. and Manning, A.J. (1999) Observation of the size, settling velocity and effective density of flocs and their fractal dimension. *J. Sea Res.* **41**: 87–95.

261. Dyer, K.R., Christie, M.C. and Wright, E.W. (2000) The classification of intertidal mudflats. *Cont. Shelf Res.* **20**: 1039–1060.

262. Dyer, K.R., Christie, M.C. and Manning, A.J. (2004) The effect of suspended sediment on turbulence within an estuarine turbidity maximum. *Est. Coast. Shelf Sci.* **59**: 237–248.

263. Dyer, K.R. Personal communication.

264. Eidsvik, K.J. (2004) Some contributions to the uncertainty of sediment transport predictions. *Cont. Shelf Res.* **24**: 739–754.
265. Edmonds, D.A. and Slingerland, R.L. (2008) Stability of delta distributary networks and their bifurcations. *Water Resources Res.* **44**: w09426.
266. Edmonds, D.A. (2012) Stability of backwater influenced bifurcations: A study of the Mississippi–Atchafalaya bifurcation. Geophysical Research Letters, **39**: L08402.
267. Einstein, H.A. (1950) The bed-load function for sediment transportation in open channel flows. *US Dept. Agri. Techn. Bull.* 1026.
268. Einstein, H.A. and Krone, R.B. (1962) Experiments to determine modes of cohesive sediment transport in salt water. *J. Geophys. Res.* **67**: 1451–1461.
269. Eisma, D., Bernard, P., Cadee, G.C., Ittekot, V., Kalf, J., Laane R., Martin, J.M., Mook, W.G., Van Put, A. and Schuhmacher, T. (1983) Suspended-matter particle size in some Wets-European estuaries; Part I: *Particle Size Distribution. Neth. J. Sea Res.* **28**: 193–214.
270. Eitner, V. (1996) Morphological and sedimentological development of a tidal inlet and its catchment area (Otzumer Balje, Southern North Sea). *J. Coast. Res.* **12**: 271–293.
271. Ehlers, J. (1988) The morphodynamics of the Waddensea. Balkema, Rotterdam.
272. Elfrink, B. and Baldock, T. (2002) Hydrodynamics and sediment transport in the swash zone: A review and perspectives. *Coastal Eng.* **45**: 149–167.
273. El Ganaoui, O., Schaaff, E., Boyer, P., Amielh, M., Anselmet, F. and Grenz, C. (2004) The deposition and erosion of cohesive sediment determined by a multi-class model. *Est. Coast. Shelf Sci.* **60**: 457–475.
274. Elgar, S., Guza, R.T., Raubenheimer, B., Herbers, T.H.C. and Gallagher, E. (1997) Spectral evolution of shoaling and breaking waves on a barred beach. *J. Geophys. Res.* **102**: 15 797–15 805.
275. Elgar, S., Gallagher, E.L. and Guza, R.T. (2001) Nearshore sandbar migration. *J. Geophys. Res.* **106**: 11,623–11,727.
276. Elias, E., Stive, M., Bonekamp, H. and Cleveringa, J. (2003) Tidal inlet dynamics in response to human intervention. *J. Coast. Eng.* **45**: 629–658.
277. Elias, E.P.L. and Van der Spek, A.J.F. (2006) Long-term morphodynamic evolution of Texel Inlet and its ebb-tidal delta (The Netherlands). *Mar. Geology* **225**: 5–21.
278. Elias, E.P.L., Cleveringa, J., Buijsman, M.C., Roelvink, J.A. and Stive M.J.F. (2006) Field and model data analysis of sand transport patterns in Texel Tidal inlet (the Netherlands). *Coastal Eng.* **53**: 505–529.
279. Elias, E., Van Koningsveld, M., Tonnon, P.K. and Wang, Z.B. (2007) Sediment budget analysis and testing hypotheses for the Dutch coastal system. Report WL/Delft Hydraulics.

280. Elias, E.P.L., Van der Spek, A.J.F., Wang, Z.B. and De Ronde, J. (2012) Morphodynamic development and sediment budget of the Dutch Wadden Sea over the last century. *Neth. J. of Geosciences Geologie en Mijnbouw* **91**: 293–310.

281. Elias, E. and Van der Spek, A. (2014) Grootschalige morfologische veranderingen in de Voordelta. Deltares report 1207724 (in Dutch).

282. Elliott, A.J. (1987) Observations of meteorologically induced circulation in the Potomac estuary. *Est. Coast. Mar. Sci.* **6**: 285–299.

283. Engelund, F. and Hansen, E. (1972) A monograph on sediment transport in alluvial streams. Technical Press, Copenhagen, 62 pp.

284. Engelund, F. (1970) Instability of erodible beds. *J. Fluid Mech.* **42**: 225–244.

285. Engelund, F. and Hansen, E. (1972) *A Monograph on Sediment Transport in Alluvial Streams*, 3rd Edn. Technical Press, Copenhagen.

286. Erikson, L.H., Larson, M. and Hanson, H. (2002) Laboratory investigation of beach scarp and dune recession due to notching and subsequent failure. *Mar. Geol.* **245**: 1–19.

287. Escoffier, F.F. (1940) The stability of tidal inlets. *Shore and Beach* **8**: 111–114.

288. Eysink, W. D. (1990) Morphologic response of tidal basins to changes. *Proc. 22nd Coastal Engineering Conference, ASCE*, Delft, Vol. 2, pp. 1948–1961.

289. Fagherazzi, S., Carniello, L., D'Alpaos, L. and Defina, A. (2006) Critical bifurcation of shallow microtidal landforms in tidal flats and salt marshes. *Proc. Natl. Acad. Sci. U. S. A.* **103**(22): 8337–8341.

290. Fagherazzi, S., Kirwan, M.L., Mudd, S.M., Guntenspergen, G.R., Temmerman, S., DAlpaos, A., Van de Koppel, J., Rybczyk, J.M., Reyes, E., Craft, C. and Clough, J. (2011) Numerical models of salt marsh evolution: Ecological, geomorphic, and climatic factors. *Rev. Geophys.* **50**: RG1002.

291. Falqués, A., Calvete, D. and Montoto, A. (1998) Bed-flow instabilities of coastal currents. In: *Physics of Estuaries and Coastal Seas*. Eds., J. Dronkers and M.B.A.M. Scheffers. Balkema, Rotterdam, pp. 417–424.

292. Falqués, A. and Iranzo, I. (1994) Numerical simulation of vorticity waves in the nearshore. *J. Geophys. Res.* **99**: 835–841.

293. Falqués, A., Coco, G. and Huntley, D.A. (2000) A mechanism for the generation of wave-driven rhythmic patterns in the surf zone. *J. Geophys. Res.* **105**(C10): 24071–24087.

294. Falqués, A. (2003) On the diffusivity in coastline dynamics. *Geophys. Res. Letters* **30**(21): OCE 4.

295. Falqués, A., Calvete, D. (2005) Large-scale dynamics of sandy coastlines. Diffusivity and instability. *J. Geophys. Res.* **110**: C03007.

296. Falqués, A., Dodd, N., Garnier, R., Ribas, F., MacHardy, L.C., Larroude, P., Calvete, D. and Sancho, F. (2008) Rhythmic surf zone bars and morphodynamic self-organization. *Coastal Eng.* **55**: 622–641.

297. Feddersen, F. (2004) Effect of wave directional spread on the radiation stress: comparing theory and observations. *Coast. Eng.* **51**: 473–481.

298. Feddersen, F. and Trowbridge, J.H. (2005) The effect of wave breaking on surf-zone turbulence and alongshore currents: A modeling study. *J. Phys. Oceanography* **35**: 2187–2203.

299. Field, M.E., Nelson, C.H., Cacchione, D.A. and Drake, D.E. (1981) Sand waves on an epicontinental shelf: Northern Bering Sea. *Mar. Geol.* **42**: 233–258.

300. Figueiredo, A.G., Swift, D.J.P., Stubblefield, W.L. and Clarke, T.L. (1981) Sand ridges on the inner Atlantic shelf of North America: Morphometric comparisons with Huthnance stability model. *Geomarine Letters* **1**: 187–191.

301. Figueiredo, A.G., Sanders, J.E. and Swift, D.J.P. (1982) Storm-graded layers on inner continental shelves: Examples from Southern Brazil and the Atlantic coast of the central United States. *Sed. Geol.* **31**: 171–190.

302. Fischer, H.B., List, E.J., Koh, R.C.Y., Imberger, J. and Brooks, N.H. (1979) *Mixing in Inland and Coastal Waters*. Academic Press, New York.

303. FitzGerald, D.M. (1977) *Hydraulics, Morphology and Sediment Transport at Price Inlet, South Carolina*. Ph.D. dissertation, Geol. Dep., Univ. South Carolina, 84 pp.

304. FitzGerald, D.M. (1988) Shoreline erosional-depositional processes associated with tidal inlets. In: *Hydrodynamics and Sediment Dynamics of Tidal Inlets*. Eds., D.G. Aubrey and L. Weishar. Springer-Verlag, New York, pp. 186–225.

305. Flemming, B.W. (1988) Zur klassifikation subaquatischer, strömung stransversaler transportkörper. *Bochumer Geologische und Geotechnische Arbeiten*, **29**: 44–47.

306. Flemming, B.W., Delafontaine, M.T. (2000) Mass physical properties of muddy intertidal sediments: Some applications, misapplications and non-applications. *Cont. Shelf Res.* **20**: 1179–1197.

307. Foda, M.A. (2003) Role of wave pressure in bedload sediment transport. *J. Waterway, Port, Coastal and Ocean Eng.* **129**: 243–249.

308. Foster, D. L., Bowen, A.J., Holman, R.A. and Natoo, P. (2006) Field evidence of pressure gradient induced incipient motion. *J. Geophys. Res.* **111**: C05004.

309. Foti, E. and Boldeaux, P. (1995) Sea ripple formation: The turbulent boundary case. *Coastal Eng.* **25**: 227–236.

310. Fourrière, A., Claudin, P. and Andreotti, B. (2010) Bedforms in a turbulent stream: Formation of ripples by primary linear instability and of dunes by non-linear pattern coarsening. *J. Fluid Mech.* **649**: 287–328.

311. Foussard, V. (2009) Evolution hydro-geomorphologique de lestuaire de la Seine, au regard des usages passes et presents. Rapport de Synthese, GIP Seine Aval.

312. Fowler, R. and Dalrymple, R. (1991) Wave group forced nearshore circulation. In *Proc. 22nd Int. Conf. on Coast. Eng.* **I**: 729–742.
313. Francalanci, S., Bendoni, M, Rinaldi, M. and Solari, L. (2013) Ecomorphodynamic evolution of saltmarshes: Experimental observations of bank retreat processes. *Geomorphology* **195**: 53–65.
314. Fredsøe, J. (1974) On the development of dunes in erodible channels. *J. Fluid Mech.* **64**: 1–16.
315. Fredsøe, J. (1978) Meandering and braiding of rivers. *J. Fluid Mech.* **84**: 609–624.
316. Fredsøe, J. (1982) Shape and dimensions of stationary dunes in rivers. *J. Hydr. Div., ASCE* **111**: 1041–1059.
317. Fredsoe, J., Andersen, K.H. and Sumer, M.B. (1999) Wave plus current over a ripple-covered bed. *Coastal Eng.* **38**: 177–221.
318. Fredsøe J. and Deigaard R. (1992) *Mechanics of Coastal Sediment Transport.* World Scientific Publishing, Singapore.
319. Friedrichs, C.T., and Aubrey, D.G. (1988) Non-linear tidal distortion in shallow wellmixed estuaries: A synthesis. *Estuarine, Coastal and Shelf Science* **27**: 521–545.
320. Friedrichs, C.T., Aubrey. D.G. and Speer, P.E. (1990) Impacts of relative sea-level rise on evolution of shallow estuaries. In: *Coastal and Estuarine Studies 38, Residual Currents and Long-Term Transport.* Ed., R.T. Cheng, Springer-Verlag, New York, pp. 105–122.
321. Friedrichs and Madsen. O.S. (1992) Non-linear diffusion of the tidal signal in frictionally dominated embayments. *J. Geophys. Res.* **97**: 5637–5650.
322. Friedrichs, C.T., Lynch, D.R. and Aubrey. D.G. (1992) Velocity asymmetries in frictionally-dominated tidal embayments: Longitudinal and lateral variability. In: *Dynamics and Exchanges in Estuaries and the Coastal Zone.* Ed., D. Prandle. Springer-Verlag, New York, pp. 277–312.
323. Friedrichs C.T. and Aubrey, D.G. (1994) Tidal propagation in strongly convergent channels. *J. Geophys. Res.* **99**: 3321–3336.
324. Friedrichs C.T. (1995) Stability shear stress and equilibrium cros-sectional geometry of sheltered tidal channels. *J. Coast. Res.* **11**: 1062–1074.
325. Friedrichs, C.T. and Aubrey, D.G. (1996) Uniform bottom shear stress and equilibrium hypsometry of intertidal flats. In: *Mixing in Estuaries and Coastal Seas, Coastal Estuarine Stud.* 50. Ed., C. Pattiaratchi. AGU, Washington D.C., pp. 405–429.
326. Friedrichs, C.T., Armbrust, B.D. and De Swart, H.E. (1998) Hydrodynamics and equilibrium sediment dynamics of shallow funnel-shaped tidal estuaries. In: *Physics of Estuaries and Coastal Seas.* Ed., J. Dronkers and M.B.A.M. Scheffers, Balkema, Rotterdam, pp. 315–328.
327. Friedrichs, C.T. (2011) Tidal flat morphodynamics: A synthesis. In: *Treatise on Estuarine and Coastal Science*, Vol. 3, Estuarine and Coastal Geology and

Geomorphology. Ed., J. D. Hansom and B.W. Fleming, Elsevier, Amsterdam, pp. 137–170.

328. Fruergaard, M., Andersen, T.J., Nielsen, L.H., Madsen, A.T., Johannessen, P.N., Murray, A.S., Kirkegaard, L. and Pejrup, M. (2011) Punctuated sediment record resulting from channel migration in a shallow sand-dominated micro-tidal lagoon, Northern Wadden Sea, Denmark. *Mar. Geology* **280**: 91–104.

329. Gallagher, B. (1971) Generation of surfbeat by nonlinear wave interactions. *J. Fluid Mech.* **49**: 1–20.

330. Gallagher, E.L., Guza, T. and Elgar, S. (1998) Observations of sandbar evolution on a natural beach. *J. Geophys. Res.* **103**: 3203–3215.

331. Gallagher, E.L., Elgar, S. and Thornton, E.B. (1998) Observations and predictions of megaripple migration in a natural surf zone. *Nature* **394**: 165–168.

332. Gallagher, E.D. (2008) Bioturbation. *Biol. Ocean. Processes, EEOS* 630.

333. Gallagher, E.L. (2011) Computer simulations of self-organized megaripples in the nearshore. *J. Geophys. Res.* **116**: F01004.

334. Gallop, S.L., Bryan, K.R., Coco, G. and Stephens, S.A. (2011) Storm-driven changes in rip channel patterns on an embayed beach. *Geomorphology* **127**: 179–188.

335. Gao, S. and Collins, M. (1994) Tidal inlet equilibrium in relation to cross-sectional area and sediment transport patterns. *Est. Coast. Shelf Science* **38**: 157–172.

336. Gao, P. (2008) Transition between two bed-load transport regimes: Saltation and sheet flow. *J. Hydraulic Res.* **134**: 340–349.

337. Garnier, R., Calvete, D., Falques, A. and Caballeria, M. (2006) Generation and nonlinear evolution of shore-oblique/transverse sand bars. *J. Fluid Mech.* **567**: 327–360.

338. Gensac, E., Gardel, A., Lesourd, S. and Brutier, L. (2015) Morphodynamic evolution of an intertidal mudflat under the influence of Amazon sediment supply — Kourou mud bank, French Guiana, South America. *Estuarine, Coastal and Shelf Science* **158**: 53–62.

339. Gerkema, T. (2000) A linear analysis of tidally generated sand waves. *J. Fluid Mech.* **417**: 303–322.

340. Gerritsen, F., Dunsbergen, D.W. and Israel, C.G. (2003) A rational stability approach for tidal inlets, including analysis of the effect of wave action. *J. Coast. Res.* **19**: 1066–1081.

341. Geyer, W.R. and Farmer, D.M. (1989) Tide-induced variation of the dynamics of a salt wedge estuary. *J. Phys. Ocean.* **19**: 1060–1072.

342. Geyer, W.R. (1993) The importance of suppression of turbulence by stratification on the estuarine. *Estuaries* **16**: 113–125.

343. Gibson, R.E., England, G.L., Hussey, M.J.L. (1967). The theory of one-dimensional consolidation of saturated clays. *Geotechnique* **17**: 261–273.
344. Glenn, S.M. and Grant, W.D. (1987) A suspended sediment stratification correction for combined wave and current flows. *J. Geophys. Res.* **92**: 8244–8264.
345. Godin, G. (1991) Compact approximations to the bottom friction term, for the study of tides propagating in channels. *Cont. Shelf Res.* **11**: 579–589.
346. Goff, J.A., Swift, D.J.P., Duncan, C.S., Mayer, L.A. and Hughes-Clarke, J. (1999) High-resolution swath sonar investigation of sand ridge, dune and ribbon morphology in the offshore environment of the New Jersey margin. *Mar. Geol.* **161**: 307–337.
347. Goldstein, E.B., Murray, A.B. and Coco, G. (2011) Sorted bedform pattern evolution: Persistence, destruction and self-organized intermittency. *Geophys. Res. Lett.* **38**: L24402.
348. Gourlay, M.R. (1968) Beach and dune erosion tests, Rep. m935/m936, Delft Hydraul. Lab., Delft.
349. Graas, S. and Savenije, H.H.G. (2008) Salt intrusion in the Pungue estuary, Mozambique: Effect of sand banks as a natural temporary salt intrusion barrier. *Hydrol. Earth Syst. Sci. Discuss.* **5**: 2523–2542.
350. Grabemann, I., Uncles, R.J., Krause, G. and Stephens, J.A. (1997) Behaviour of turbidity maxima in the Tamar and Weser estuaries. *Est. Coast. Shelf Science* **45**: 235–246.
351. Grabowski, R.C., Droppo, I.G. and Wharton, G. (2011) Erodibility of cohesive sediment: The importance of sediment properties. *Earth Science Reviews* **105**(3–4): 101–120.
352. Graf, W.H. (1971) *Hydraulics of Sediment Transport*. McGraw-Hill, NY, 513 pp.
353. Grant, W.D. and Madsen, O.S. (1979) Combined wave and current interaction with a rough bottom. *J. Geophys. Res.* **84**: 1797–1808.
354. Grant, W.D. and Madsen, O.S. (1982) Movable bed roughness in unsteady oscillatory flow. *J. Geophys. Res.* **87**: 469–481.
355. Grasmeijer, B.T., Kleinhans, M.G. (2004) Observed and predicted bed forms and their effect on suspended sand concentrations. *Coastal Eng.* **51**: 351–371.
356. Grasso, F., Michallet, H. and Barthlemy, E. (2011) Sediment transport associated with morphological beach changes forced by irregular asymmetric, skewed waves, *J. Geophys. Res.* **116**: C03020.
357. Green, M.O., Black, K.P. and Amos, C.L. (1997) Control of estuarine sediment dynamics by interactions between currents and waves at several scales. *Mar. Geol.* **144**: 97–116.
358. Green, M.O. (2011) Very small waves and associated sediment resuspension on an estuarine intertidal flat. *Est. Coastal Shelf Sci.* **93**: 449–459.

359. Groen, P. (1967) On the residual transport of suspended matter by an alternating tidal current. *Neth. J. Sea Res.* **3**: 564–574.
360. Grunnet, N.M., Walstra, D-J.,R. and Ruessink, B.G. (2004) Process-based modelling of a shoreface nourishment. *Coast. Eng.* **51**: 581–607.
361. Guézennec, L., Lafite, R., Dupont, J-P., Meyer, R. and Boust, D. (1999) Hydrodynamics of suspended particulate matter in the tidal freshwater zone of a microtidal estuary. *Estuaries* **22**: 717–727.
362. Gust, G. and Walger, E. (1976) The influence of suspended cohesive sediments on boundary-layer structure and erosive activity of turbulent seawater flow. *Mar. Geol.* **22**: 189–206.
363. Guza, R.T. and Inman, D.L. (1975) Edge waves and beach cusps. *J. Geophys. Res,* **80**: 2997–3012.
364. Guza, R.T., Thornton and Holman, R.A. (1984) Swash on steep and shallow beaches. Procs. 19th Coastal Engineering Conference, *Am. Soc. of Civ. Eng.* **1984**: 708–723.
365. Gyr, A. and Hoyer, K. (2006) Sediment transport. A geophysical phenomenon. Springer, 208 pp.
366. Haas, K.A., Svendse, I.A., Haller, M.C. and Zhao, Q. (2003) Quasi-three-dimensional modeling of rip current systems. *J. Gephys. Res.* **108**: 10-1–10-21.
367. Haller, M.C., Dalrymple, R.A. and Svendsen, I.A. (2002) Experimental study of nearshore dynamics on a barred beach with rip channels. *J. Geophys. Res.* **107**(C6): 14–21.
368. Haller, M.C., Honegger, D. and Catalan, P.A. (2014) Current Observations via Marine Radar. *J. Waterway, Port, Coastal, Ocean Eng.* **140**: 115–124.
369. Hallermeyer, R.J. (1981) A profile zonation for seasonal sand beaches from wave climate. *Coast. Eng.* **4**: 253–277.
370. Hamm, L., Capobianco, M., Dette, H.H., Lechugad, A., Spanhoff, R. and Stive, M.J.F. (2002) A summary of European experience with shore nourishment. *Coastal Eng.* **47**: 237–264.
371. Hands, E.W. and Shepsis, V. (1999) Cyclic movement at the entrance to Willapy Bay, Washington, USA. In: *Coastal Sediments*. Ed., N.C. Kraus and W.G. McDougal. ASCE, pp. 1522–1536.
372. Hanes, D. and Huntley, D. (1986) Continuous measurements of suspended sand concentration in a wave dominated nearshore environment. *Cont. Shelf Res.* **6**: 585–596.
373. Hanes, D.M, Alymov, V. and Chang, Y.S. (2001) Wave-formed sand ripples at Duck, North Carolina. *J. Geophys. Res.* **106**: 22,575–22,592.
374. Hansen, D.V. (1965) Currents and mixing in the Columbia river estuary. In: *Ocean Science and Ocean Engineering*, Vol. 2. The Marine Technology Society, Washington D.C., pp. 943–955.

375. Hansen E. (1967) The formation of meanders as a stability problem. *Hydr. Lab. Techn. Univ. Denmark, Res.* Progress Paper 13.

376. Hapke, C.J., Kratzmann, M.G. and Himmelstoss, E.A. (2013) Geomorphic and human influence on large-scale coastal change. *Geomorphology* **199**: 160–170.

377. Haring, J. (1970) Historische ontwikkeling in het Noordelijk Deltabekken 1879–1966 Nota W-70.060, Deltadienst, Rijkswaterstaat (in Dutch).

378. Harms, J.C. (1969) Hydraulic significance of some sand ripples. *Geol. Soc. Amer. Bull.* **80**: 363–396.

379. Harris, P.T. (1988) Large-scale bedforms as indicators of mutually evasive sand transport and the sequential infilling of wide-mouthed estuaries. *Sediment. Geol.* **57**: 273–298.

380. Harris, P.T., Baker, E.K., Cole, A.R. and Short, S.A. (1993) A preliminary study of sedimentation in the tidally dominated Fly River delta, Gulf of Papua. *Cont. Shelf. Res.* **13**: 441–472.

381. Harris, P.T., Hughes, M.G., Baker, E.K., Dalrymple, R.W. and Keene, J.B. (2004) Sediment transport in distributary channels and its export to the pro-deltaic environment in a tidally dominated delta: Fly River, Papua New Guinea. *Cont. Shelf Res.* **24**: 2431–2454.

382. Haslett, S.K., Cundy, A.B., Davies, C.F.C., Powell, E.S. and Croudace, I.W. (2003) Salt marsh sedimentation over the past c. 120 years along the West Cotentin coast of Normandy (France): Relationship to sea-level rise and sediment supply. *J. Coast. Res.* **19**: 609–620.

383. Hasselmann, K. On the nonlinear energy transfer in a gravity-wave spectrum, part 1. General theory. *J. Fluid Mech.* **12**: 481–500.

384. Hattori, M., Kawamata, R. (1980) Onshore-offshore transport and beach profile change. *Proc. Coast. Eng. Conf.* II, pp. 1175–1193.

385. Hayes, M.O. (1980) General morphology and sediment patterns in tidal inlets. *Sediment. Geol.* **26**: 139–156.

386. Heathershaw, A.D. and Davies, A.G. (1985) Resonant wave reflection by transverse bedforms and its relation to beaches and offshore bars. *Mar. Geol.* **62**: 321–338.

387. Hedegaard, I.B., Deigaard, R. and Fredsøe, J. (1991) Onshore/offshore sediment transport and morphological modelling of coastal profiles. *Procs. Coast. Sed.* **91**: 643–654.

388. Henderson, S.M., Allen, J. S. and Newberger, P. A. (2004) Nearshore sand-bar migration predicted by an eddy-diffusive boundary layer model. *J. Geophys. Res.* **109**: C06024.

389. Hennings, I., Lurin, B., Vernemmen, C. and Vanhessche, U. (2000) On the behaviour of tidal currents due to the presence of submarine sand waves. *Mar. Geol.* **169**: 57–68.

390. Hequette, A., Hemdane, Y. and Anthony, E.J. (2008) Sediment transport under wave and current combined flows on a tide-dominated shoreface, northern coast of France. *Mar. Geol.* **249**: 226–242.
391. Hesp, P.A. (2012) Surfzone-beach-dune interactions. Jubilee Conference Proceedings, NCK-Days 2012.
392. Hibma, A., de Vriend, H.J. and Stive, M.J.F. (2003) Numerical modelling of shoal pattern formation in well-mixed elongated estuaries. *Est. Coast. Shelf Sci.* **57**: 981–991.
393. A. Hibma, A., Schuttelaars, H.M. and De Vriend, H.J. (2004) Initial formation and long-term evolution of channel-shoal patterns. *Cont. Shelf Res.* **24**: 1637–1650.
394. Hino, M. (1963) Turbulent flow with suspended particles. *J. Hydraul. Div.*, ASCE, **89**: 161–185.
395. Hino, M. (1974) Theory on formation of rip current and cuspidal coast. *Procs. 14th Int. Coast. Eng. Conf., ASCE*, pp. 901–919.
396. Hitching, E. and Lewis, A.W. (1999) Bed roughness over vortex ripples. In: *Proc. 4th Int. Symp. Coast. Eng. and Coast. Sed. Processes*. Long Island, ASCE, pp. 18–30.
397. Hoefel, F, and Elgar, S. (2003) Wave-induced sediment transport and sandbar migration. *Science* **299**: 1885–1887.
398. Hoitink, A.F.J., Hoekstra, P. and van Mare, D.S. (2003) Flow asymmetry associated with astronomical tides: Implications for residual transport of sediment. *J. Geophys. Res.* **108**: 13-1–13-8.
399. Hollebrandse, F.A.P. (2005) *Temporal Development of the Tidal Range in the Southern North Sea.* Master Sci. Thesis, Techn. Univ. Delft.
400. Holland, K.T. and Holman R.A. (1996) Field observations of beach cusps and swash motions. *Mar. Geol.* **134**: 77–93.
401. Holland, K.T. and Puleo, J.A. (2001) Variable swash motions associated with foreshore profile change. *J. Geophys. Res.* **106**: 4613–4623.
402. Holloway, P.E. (1981) Longitudinal mixing in the upper reaches of the Bay of Fundy. *Est. Coast. Shelf Sci.* **13**: 495–515.
403. Holman, R.A., Lippmann, T.C., O'Neill, P.V. and Haines, J.W. (1993) The application of video image processing to the study of nearshore processes. *Oceanography* **6**: 78–85.
404. Holman, R.A. (2001) Pattern formation in the nearshore. In: *River, Coastal and Estuarine Morphodynamics*. Eds., G. Seminara and P. Blondeaux, Springer-Verlag, Berlin, pp. 141–162.
405. Holman, R.A., Symonds, G., Thornton, E.B. and Ranasinghe, R. (2006) Rip spacing and persistence on an embayed beach. *J. Geophys. Res.* **111**: (C06006).
406. Holman, R.A. and Stanley, J. (2007) The history and technical capabilities of Argus, *Coast. Eng.* **54**: 477–491.

407. Holman, R.A., Lalejini, D.M., Edwards, K. and Veeramony, J. (2014) A parametric model for barred equilibrium beach profiles. *Coastal Eng.* **90**: 85–94.
408. Horel, J.D. (1984) Complex principal component analysis: theory and examples. *J. Clim. Appl. Meteorol.* **23**: 1660–1673.
409. Horikawa, K. and Kuo, C.T. (1966) A study on wave transformation in the surf zone. *Proc. 10th Int. Coast. Eng. Conf.*, Tokyo, ASCE, pp. 217–233.
410. Horn, D.P. (2002) Beach groundwater dynamics. *Geomorphology* **48**: 121–146.
411. Horton, R.E. (1945) Erosional development of streams and their drainage basins; hydrophysical approach to quantitative morphology. *Geol. Soc. Am. Bull.* **56**: 275–370.
412. Houbolt, J.J.H.C. (1968) Recent sediments in the Southern Bight of the North Sea. *Geologie en Mijnbouw* **47**: 245–273.
413. Houwing, E.J. (1999) Determination of the critical erosion threshold of cohesive sediments on intertidal mudflats along the Dutch Wadden Sea Coast. *Estuarine Coastal and Shelf Science*, **49**: 545–555.
414. Howarth, M.J. and Huthnance, J.M. (1984) Tidal and residual currents around a Norfolk sandbank. *Est. Coast. Shelf Sci.* **19**: 105–117.
415. Hoyt, J.H. (1967) Barrier island formation. *Geological Society of America Bulletin* **78**: 1125–1136.
416. Hsu, J.R.C., Silvester, R. and Xia, Y.M. (1989) Static equilibrium bays — New relationships. *J. Waterway, Port, Coastal and Ocean Eng.*, ASCE **115**: 285–298.
417. Hsu, T-J., Hanes, D.M. (2004) Effects of wave shape on sheet flow sediment transport. *J. Geophys. Res.* C05025.
418. Hudson, T., Keating, K. and Pettit, A. (2015) Cost estimation for coastal protection — Summary of evidence. Report-SC080039/R7, UK Environment Agency.
419. Huettel, M., Roy, H., Precht, E. and Ehrenhauss, S. (2003) Hydrodynamical impact on biogeochemical processes in aquatic sediments. *Hydrobiologia* **494**: 231–236.
420. Hughes, F.W. and Rattray, M. (1980) Salt flux and mixing in the Columbia river estuary. *Est. Coast. Shelf Sci.* **10**: 470–493.
421. Hughes, M. G. (1995) Friction factors for wave uprush, *J. Coastal Res.* **11**: 1089–1098.
422. Hughes, M.G. and Baldock, T.E. (2004) Eulerian flow velocities in the swash zone: Field data and model predictions. *J. Geophys. Res.* **109**: C08009.
423. Hulscher, S.J.M.H., de Swart, H.E. and de Vriend, H.J. (1993) The generation of offshore tial sand banks and sand waves. *Cont. Shelf Res.* **13**: 1183–1204.

424. Hulscher, S.J.M.H. (1996) Tidal-induced large-scale regular bed form patterns in a three-dimensional shallow water model. *J. Geophys. Res.* **101**: 20727–20744.

425. Hulscher, S.J.M.H. (2001) Comparison between predicted and observed sandwaves and sandbanks in the North Sea. *J. Geophys. Res.* **106**: 9327–9338.

426. Hulscher, J.M.H. and Dohmen-Janssen, C.M. (2005) Introduction to special section on marine sand wave and river dune dynamics. *J. Geophys. Res.* **110**: F04S01.

427. Hume, T.M. and Herdendorf, C.E. (1992) Factors controlling tidal inlet characteristics on low drift coasts. *J. Coast. Res.* **8**: 355–375.

428. Hunkins, K. (1981) Salt dispersion in the Hudson estuary. *J. Phys. Ocean.* **11**: 729–738.

429. Hunt, I.A. (1959) Design of seawalls and breakwaters. *J. Waterw. Harb. Div.*, *ASCE* **85**: 123–152.

430. Hunt, J.R. (1986) Particle aggregate break-up by fluid shear. In: *Estuarine Cohesive Sediment Dynamics*. Ed., A.J.Mehta. *Lecture Notes on Coastal and Estuarine Studies*, Vol. 14. Springer-Verlag, Berlin, pp. 85–109.

431. Hunt, S., Bryan, K.R. and Mullarney, J.C. (2015) The influence of wind and waves on the existence of stable intertidal morphology in meso-tidal estuaries. *Geomorphology* **228**: 158–174.

432. Hunter, K.A. and Liss, P.S. (1982) Organic matter and the surface charge of suspended particles in estuarine waters. *Limnol. Oceanogr.* **27**: 322–335.

433. Huntley, D.A. and Bowen, A.J. (1973) Field observations of edge waves. *Nature* **243**: 160–161.

434. Huntley, J.R., Nicholls, R.J., Liu, C. and Dyer, K.R. (1994) Measurements of the semi-diurnal drag coefficient over sand waves. *Cont. Shelf Res.* **14**: 437–456.

435. Huntley, D.A., Coco, G., Bryan, K.R. and Murray, A.B. (2008) Influence of defects on sorted bedform dynamics. *Geophys. Res. Letters* **35**: L02601.

436. Huthnance, J.M. (1973) Tidal current asymmetries over the Norfolk sandbanks. *Est. Coast. Mar. Sci.* **1**: 89–99.

437. Huthnance, J.M. (1982) On one mechanism forming linear sandbanks. *Est. Coast. Mar. Sci.* **14**: 79–99.

438. Idier, D., Astruc, D. and Hulscher, S.J.M.H. (2004) Influence of bed roughness on dune and megaripple generation. *Geophys. Res. Lett.* **31**: L13214.

439. Idier, D., Falques, A., Ruessink, B.G. and Garnier, R. (2011) Shoreline instability under low-angle wave incidence. *J. Geophys. Res. Earth-surface.* **116**: F04031.

440. Ikeda, S., Parker, G. and Sawai, K. (1981) Bend theory of river meanders. Part 1. Linear development. *J. Fluid Mech.* **112**: 363–377.

441. Inch, K., Masselink, G., Puleo, J.A., Russell, P. and Conley, D.C. (2015) Vertical structure of near-bed cross-shore flow velocities in the swash zone of a dissipative beach. *Continental Shelf Res.* **101**: 98–108.

442. Inman, D.L., Elwany, M.H. and Jenkins, S.A. (1993) Shorerise and bar-berm profiles on ocean beaches. *J. Geophy. Res.* **98**: 18181–18199.

443. Ippen, A.T. and Goda, Y. (1963) Wave-induced oscillations in harbors: the solution for a rectangular harbor connected to the open sea. *Rep. Hydr. Lab MIT*, p. 59.

444. Ippen, A.T. (1966) *Estuary and Coastline Hydrodynamics*. McGraw-Hill, New York, 744 pp.

445. Israel, C.G. and Dunsbergen, D.W. (1999) Cyclic morphological development of the Ameland Inlet, The Netherlands. In: *River, Coastal and Estuarine Morphodynamics. Proc. Conf. IAHR*, pp. 705–715.

446. Izumi, N. and Parker, G. (1995) Inception of channelization and drainage basin formation: upstream-driven theory. *J. Fluid Mech.* **283**: 341–363.

447. Jackson, P.S. and Hunt, J.C.R. (1975) Turbulent flow over a low hill. *Quart. J. R. Met. Soc.* **101**: 929–955.

448. Jackson, R.G. (1976) Sedimentological and fluid-dynamic implications of the turbulent bursting phenomenon in geophysical flows. *J. Fluid Mech.* **77**: 531–560.

449. Jacobsen, N.G., Fredsoe, J. and Jensen, J.H. (2014) Formation and development of a breaker bar under regular waves. Part 1: Model description and hydrodynamics. *Coastal Eng.* **88**: 182–193.

450. Jacobsen, N.G. and Fredsoe, J. (2014) Formation and development of a breaker bar under regular waves. Part 2: Sediment transport and morphology. *Coastal Eng.* **88**: 55–68.

451. Jaffee, B. and Rubin, D. (1996) Using non-linear forecasting to determine the magnitude and phasing of time-varying sediment suspension in the surf zone. *J. Geophys. Res.* **101**: 14,238–14,296.

452. Janssen-Stelder, B. (2000) The effect of different hydrodynamic conditions on the morphodynamics of a tidal mudflat in the Dutch Wadden Sea. *Cont. Shelf Res.* **20**: 1461–1478.

453. Jara, M.S., González, M. and Medina R. (2015) Shoreline evolution model from a dynamic equilibrium beach profile. *Coastal Engineering* **99**: 1–14.

454. Jarret J.T. (1976) *Tidal Prism-Inlet Area Relationships*. GITI, Rep.3, US Army Eng. Waterw. Exp. Station, Vicksburg.

455. Jay, D.A. and Smith. J.D. (1990) Residual circulation in shallow estuaries. 1. Highly stratified, narrow estuaries. *J. Geophys. Res.* **95**: 711–731.

456. Jay, D.A. (1991) Green's law revisited: Tidal long-wave propagation in channels with strong topography. *J. Geophys. Res.* **96**: 20,585–20,598.

457. Jay, D.A. and Musiak, J.D. (1994) Particle trapping in estuarine tidal flows. *J. Geophys. Res.* **99**: 20,445–20,461.
458. Jenkins, S.A. and Inman, D.L. (2006) Thermodynamic solutions for equilibrium profiles. *J. Geophys. Res.* **111**: C02003.
459. Jensen, B.L., Sumer, B.M. and Fredsoe, J. (1989) Turbulent oscillatory boundary layers at high Reynolds numbers. *J. Fluid Mech.* **206**: 265–297.
460. Jeuken, M.C.J.L. (2000) *On the Morphologic Behaviour of Tidal Channels in the Westerschelde Estuary.* PhD thesis, Utrecht University.
461. Jewell, S.A., Walker, D.J and Fortunato, A.B. (2012) Tidal asymmetry in a coastal lagoon subject to a mixed tidal regime. *Geomorphology* **138**: 171–180.
462. Ji, Z.G. and Mendoza, C. (1997) Weakly nonlinear stability analysis for dune formation. *J. Hydr. Eng.* **123**: 979–985.
463. Jimenez, J.A. and Madsen, O.S. (2003) A simple formula to estimate settling velocity of natural sediments. *J. of Waterway, Port, Coastal, and Ocean Eng.* **129**: 70–78.
464. Jiyu, C., Cangzi, L., Chongle, Z. and Walker, H.J. (1990) Geomorphological development and sedimentation in Qiantang estuary and Hangzou Bay. *J. Coast. Res.* **6**: 559–572.
465. Johnson, D.W. (1919) Shore Processes and Shoreline Development. Prentice Hall, NY, 584 pp.
466. Johnson, D. (2004) *The Spatial and Temporal Variability of Nearshore Currents.* Thesis, University of Western Australia.
467. Jones, N.V. and Elliot, M. (2000) Coastal zone topics: Process, ecology and management, 4. The Humber estuary and adjoining Yorkshire and Lincolnshire coasts. *Est. Coast. Sci. Ass.*, Hull, UK.
468. Jonsson, I.G. (1966) Wave boundary layers and friction factors. In: *Proc. Int. Conf. Coast. Eng.*, Tokyo, Japan. ASCES, pp. 127–148.
469. Jonsson, I.G. and Carlsen, N.A. (1976) Experimental and theoretical investigations in an oscillatory rough turbulent boundary layer. *J. Hydraul. Res.* **14**: 45–60.
470. Jouanneau, J.M. and Latouche, C. (1981) The Gironde estuary. In: *Contributions to Sedimentology 10 E. Schweizerbartsche Verlagsbuchhandlung, Nagele und Obermiller*, Stuttgart, 115 pp.
471. Julien, P.Y. and Wargadalam, J. (1995) Alluvial channel geometry: Theory and applications. *J. Hydr. Eng.* **121**: 312–325.
472. Jumars, P.A. and Nowell, A.R.M. (1984) Fluid and sediment dynamic effects on marine benthic community structure. *American Zoologist*, **24**: 45–55.
473. Kaczmarek, L.M., Biegowski, J. and Ostrowski, R. (2004) Modelling crossshore intensive sand transport and changes of bed grain size distributions versus field data. *Coast. Eng.* **51**: 501–529.

474. Kaergaard, K. and Fredsoe, J. (2013) Numerical modeling of shoreline undulations part 1: Constant wave climate. *Coastal Eng.* **75**: 64–76.

475. Kaergaard, K. and Fredsoe, J. (2013) Numerical modeling of shoreline undulations part 2: Varying wave climate and comparison with observations. *Coastal Eng.* **75**: 77–90.

476. Kamphuis, J.W. (1991) Alongshore sediment transport rate. *J Waterway, Port, Coastal and Ocean Eng. Div, ASCE* **117**: 624–640.

477. Kang, S.K., Lee, S.R. and Lie, H.J. (1998) Fine-grid tidal modelling of the Yellow and East China seas. *Cont. Shelf Res.* **18**: 739–772.

478. Kappenberg, J. and Fanger, H.-U. (2007) Sedimenttransportgeschehen in der tidebeeinflussten Elbe, der Deutschen Bucht und in der Nordsee. Report GKSS 2007/20.

479. Kennedy, J.F. (1969) The formation of sediment ripples, dunes and antidunes. *Ann. Rev. Fluid Mech.* **1**: 147–168.

480. Kerssens, P.J.M., Prins, A. and van Rijn, L.C. (1979) Model for suspended sediment transport. *J. Hydraul. Div., ASCE*, **105**: 461–476.

481. Kikkert, G.A., Pokrajac, D., O'Donoghue, T. and Steenhauer, K. (2013) Experimental study of bore-driven swash hydrodynamics on permeable rough slopes. *Coastal Eng.* **79**: 42–56.

482. Kinsman, B. (1965) *Wind Waves*. Prentice-Hall, Englewood Cliffs, N.J.

483. Kirby, R. and Parker, W.R. (1983) Distribution and behaviour of fine sediment in the Severn Estuary and Inner Bristol Channel. *UK. Can. J. Fish. Aquat. Sci.* **40**(suppl.): 83–95.

484. Kirby, R. (1992) Effects of sea-level on muddy coastal margins. In: *Dynamics and Exchanges in Estuaries and the Coastal Zone.* Eds., D. Prandle. Springer-Verlag, New York, pp. 313–334.

485. Kirby, R. (2000) Practical implications of tidal flat shape. *Cont. Shelf Research* **20**: 1061–1077.

486. Kitinades, P.K. and Kennedy, J.F. (1984) Secondary currents and river-meander formation. *J. Fluid Mech.* **144**: 217–229.

487. Klein, M.D. and Schuttelaars, H.M. (2005) Morphodynamic instabilities of planar beaches: Sensitivity to parameter values and process formulations. *J. Geophys. Res.* **110**: F04S18.

488. Kleinhans, M., Jagers, B., Mosselman, E. and Sloff, K. (2006) Effect of upstream meanders on bifurcation stability and sediment division in 1D, 2D and 3D models. In: *Procs. Int. Conf. Fluvial Hydr.*, Lisbon, Portugal. Eds., R.M.L. Ferreira, E.C.T.L. Alves, J.G.A.B. Leal and A.H. Cardoso. Taylor and Francis/Balkema, London, UK, pp. 1355–1362.

489. Kleinhans, M.G., Weerts, H.J.T., Cohen, K.M. (2010) Avulsion in action: Reconstruction and modelling sedimentation pace and upstream flood water levels following a Medieval tidal-river diversion catastrophe (Biesbosch, The Netherlands, 1421–1750 AD). *Geomorphology* **118**: 65–79.

490. Kohsiek, L.H.M. and Terwindt, J.H.J. (1981) Characteristics of foreset and topset bedding in megaripples related to hydrodynamic conditions on an intertidal shoal. In: *Holocene Marine Sedimentation in the North Sea Basin.* Eds: S.D. Nio. R.T.E. Schuttenhelm and Tj.C.E. van Weering. *Int. Ass. Sed. Soc. Publ.* **5**: 27–37.

491. Kohsiek, L.H.M., Buist, H.J., Bloks, P., Misdorp, R., van der Berg, J.H. and Visser, J. (1988) Sedimentary processes on a sandy shoal in a mesotidal estuary (Oosterschelde, The Netherlands). In: *Tide-influenced Sedimentary Environments and Facies.* Eds., P.L. de Boer *et al.* Reidel Publ. Co., pp. 210–214.

492. Komar, P.D. (1998) *Beach Processes and Sedimentation.* Prentice Hall, London, p. 544.

493. Konicki, K.M. and Holman, R.A. (2000) The statistics and kinematics of of transverse bars on an open coast. *Mar. Geol.* **169**: 69–101.

494. Kraak, A., Balfoort, H.M., Vroon, J. and Hallie, F. (2002) Tradition, trends and tomorrow. *The 3rd Coastal Policy Document of The Netherlands.* RIKZ/Rijkswaterstaat, The Hague.

495. Kranenburg, W. M., Ribberink, J. S., Uittenbogaard, R. E. and Hulscher, S. J. M. H. (2012) Net currents in the wave bottom boundary layer: on wave shape streaming and progressive wave streaming. *J. Geophys. Res.* **117**: (F03005).

496. Krone, R.B. (1986) The significance of aggregate properties to transport processes. In: *Estuarine Cohesive Sediment Dynamics.* Eds., A.J. Mehta. *Lecture Notes on Coastal and Estuarine Studies*, Vol. 14. Springer-Verlag, Berlin, pp. 66–84.

497. Kroon, A. Personal communication

498. Kroon, A. and Masselink, G. (2002) Morphodynamics of intertidal bar morphology on a macrotidal beach under low-energy wave conditions, North Lincolnshire, England. *Mar. Geology* **190**: 591–608.

499. Kroon, A., Larson, M., Mller, I., Yokoki, H., Rozynski, G. Cox, J. and Larroude, P. (2008) Statistical analysis of coastal morphological data sets over seasonal to decadal time scales. *Coastal Engineering* **55**: 581–600.

500. Lacey, G. (1929) Stable channels in alluvium. *Proc. Inst. Civ. Eng.*, London, **229**: 259–290.

501. Lamb, H. (1932) *Hydrodynamics.* Cambridge Univ. Press.

502. Langhorne, D.N. (1973) A sand wave field in the outer Thames Estuary, G.B. *Mar. Geol.* **14**: 129–143.

503. Lanckneus, J., De Moor, G. and Stolk, A. (1994) Environmental setting, morphology and volumetric evolution of the Middelkerke Bank (southern North Sea). *Mar. Geol.* **121**: 1–21.

504. Lanuru, M. (2004) The spatial and temporal patterns of erodibility of an intertidal flat in the East Frisian Wadden Sea, Germany. Thesis, Kiel University.

505. Lanzoni, S. and Seminara, G. (1998) On tide propagation in convergent estuaries. *J. Geophys. Res.* **103**: 30,793–30,812.
506. Lanzoni, S. (2000) Experiments on bar formation in a straight flume 1. Uniform sediment. *Water Resources Res.* **36**: 3337–3349.
507. Lanzoni, S. and Seminara, G. (2002) Long-term evolution and morphodynamic equilibrium of tidal channels. *J. Geophys. Res.* **107**: 1-1–1-13.
508. Langlois, V. and Valance A. (2007) Initiation and evolution of current ripples on a flat sand bed under turbulent water flow. *Eur. Phys. J. E* **22**: 201–208.
509. Larsen, S.M., Greenwood, B. and Aagaard, T. (2015) Observations of megaripples in the surf zone. *Marine Geology* **364**: 1–11.
510. Larras, J. (1963) *Embouchures, Estuaires, Lagunes et Deltas*. Collection Centre de Chatou, Eyrolles, France, 171pp.
511. LeBlond, P.H. (1978) On tidal propagation in shallow rivers. *J. Geophys. Res.* **83**: 4717–4721.
512. Lavoine, E. and Lechalas, M.C. (1885) La Seine maritime et son estuaire. *Encyclopdie des travaux publics*, 312p.
513. Lee, B.J., Toorman, E., Molz, F., Wang, J. (2011) A two-class population balance equation yielding bimodal flocculation of marine or estuarine sediments. *Water Research* **45**: 2131–2145.
514. Le Hir, P., Roberts, W., Cazaillet, O., Christie, M., Bassoullet, P. and Bacher, C. (2000) Characterization of intertidal flat hydrodynamics. *Cont. Shelf Res.* **20**: 1433–1459.
515. Le Hir, P., Ficht, A., Silva Jacinto, R., Lesueur, P., Dupont, J-P, Lafitte, R., Brenon, I., Thouvenin, B. and Cugier, P. (2001) Fine sediment transport and accumulations at the mouth of the Seine estuary (France). *Estuaries* **24**: 950–963.
516. LeHir, P., Cayocca, F. and Waeles, B. (2011) Dynamics of sand and mud mixtures: A multiprocess-based modelling strategy. *Continental Shelf Res.* **31**: 135–149.
517. Le Mehaute, B. (1962) On non-saturated breakers and the wave run-up. *Proc. 8th Int. Conf. on Coast. Eng. ASCE*, pp. 77–92.
518. Leopold L.B. and Wolman M.G. (1957) River channel patterns: Braided, meandering and straight. *Geol. Survey Prof. Paper* 282-B.
519. Leopold, L.B., Wolman, M.G. and Miller, J.P. (1964) *Fluvial Processes in Geomorphology*. Freeman, San Francisco.
520. Lessa, G. (1996) Tidal dynamics and sediment transport in a shallow macrotidal estuary. In: *Mixing in Estuaries and Coastal Seas, Coastal and Estuarine Studies, Am. Geophys. Un.* **50**: 338–360.
521. Lesser, G.R. (2009) *An Approach to Medium-term Coastal Morphological Modelling*. PhD Thesis, Delft University of Technology.

522. Levoy, F., Anthony, E.J., Monfort, O. and Larsonneur, C. (2000) The morphodynamics of megatidal beaches in Normandy, France. *Mar. Geol.* **171**: 39–59.

523. Li, M.Z. and Amos, C.L. (1998) Predicting ripple geometry and bed roughness under combined waves and currents in a continental shelf environment. *Cont. Shelf Res.* **18**: 941–947.

524. Li, M.Z. and Amos, C.L. (1999) Field observations of bedforms and sediment transport thresholds of fine sand under combined waves and currents. *Mar. Geol.* **158**: 147–160.

525. Li, M.Z. and Gust, G. (2000) Boundary layer dynamics and drag reduction in flows of high cohesive sediment suspensions. *Sedimentol.* **47**: 71–86.

526. Liang, G. and Seymour, R.J. (1991) Complex principle component analysis of wave-like sand motions. In: *Proc. Coastal Sediments*, New York, ASCE, pp. 2175–2186.

527. Lincoln, J.M. and Fitzgerald, D.M. (1988) Tidal distortions and flood dominance at five small tidal inlets in Southern Maine. *Mar. Geol.* **82**: 133–148.

528. Linley, E.A.S. and Field, J.G. (1982) The nature and significance of bacterial aggregation in a nearshore upwelling ecosystem. *Est. Coast. Shelf Sci.* **14**: 1–11.

529. Lippmann, T.C. and Holman, R.A. (1990) The spatial and temporal variability of sandbar morphology. *J. Geophys. Res.* **95**: 11,575–11,590.

530. List, J.H. (1986) Wave groupiness as a source for nearshore long waves. *Proc. Int. Conf. Coast. Eng. ASCE*, New York, pp. 497–511.

531. List, J.H., Farris, A.S. and Sullivan, C. (2006) Reversing storm hotspots on sandy beaches: Spatial and temporal characteristics. *Mar. Geology* **226**: 261–279.

532. Liu, Z. (1985) A preliminary study of tidal current ridges. *Chin. J. Ocean. Limnol.* **3**: 118–133.

533. Liu, Z., Huang, Y. and Zhang, Q. (1989) Tidal current ridges in the southwestern Yellow Sea. *J. Sed. Petr.* **59**: 432–437.

534. Liu, Z., Berné, S., Saito, Y., Yu, H., Trentesaux, A., Uehara, K., Yin, P., Liu, J.P., Li, C., Hu, G. and Wang, X. (2007) Internal architecture and mobility of tidal sand ridges in the East China Sea. *Cont. Shelf Res.* **27**: 1820–1834.

535. Liu, Y., Niu, X and Yu, X. (2011) A new predictive formula for inception of regular wave breaking. *Coastal Eng.* **58**: 877–889.

536. Lohrer, A., Thrush, S. and Gibbs, M. (2004) Bioturbators enhance ecosystem function through complex biogeochemical interactions. *Nature* **431**: 1092–1095.

537. Longuet-Higgins, M.S. (1953) Mass transport in water waves. Royal Soc. London, *Phil. Trans.* **245A**: 535–581.

538. Longuet-Higgins, M.S. and Stewart, R.W. (1962) Radiation stress and mass transport in gravity waves, with application to 'surf beats'. *J. Fluid Mech.* **8**: 563–583.

539. Longuet-Higgins, M.S. and Stewart, R.W. (1964) Radiation stresses in water waves: A physical discussion with applications. *Deap-Sea Res.* **11**: 529–562.

540. Longuet-Higgins, M.S. (1970) Longshore currents generated by obliquely incident sea waves. *J. Geohys. Res.* **75**: 6778–6801.

541. Louisse, C.J. and Kuik, T.J. (1990) Coastal defence alternatives in The Netherlands. *Int. Conf. Coast. Eng. ASCE*, 1862–1875.

542. Louda, J.W., Loitz, J.W., Melisiotis, A. and Orem, W.H. (2004) Potential sources of hydrogel stabilisation of Florida Bay lime mud sediments and implications for organic matter preservation. *J. Coast. Res.* **20**: 448–463.

543. Loureiro, C., Ferreira, O. and Cooper, A.G. (2013) Applicability of parametric beach morphodynamic state classification on embayed beaches. *Mar. Geol.* **346**: 153–164.

544. Louters, T. and Gerritsen, F. (1994) The riddle of the sands. *Min. Publ. Works*, The Netherlands, RIKZ-90.040.

545. Louters, T., van den Berg, J.H. and Mulder, J.P.M. (1998) Geomorphological changes of the Oosterchelde tidal system during and after the implementation of the Delta project. *J. Coast. Res.* **14**: 1134–1151.

546. Lueck, R.G. and Lu, Y. (1997) The logarithmic layer in a tidal channel. *Cont. Shelf Res.* **17**: 1785–1801.

547. Lumborg, U. and Pejrup, M. (2005) Modelling of cohesive sediment transport in a tidal lagoon an annual budget. *Mar. Geol.* **218**: 1–16.

548. Lundberg, L. (2005) Development of the Pungwe river basin Joint Integrated Water Resources Management Strategy. Technical report, Government of the Republic of Mozambique. Swedish International Development Cooperation Agency (Sida).

549. Lyard, F., Lefevre, F., Letellier, T. and Francis, O. (2006) Modelling the global ocean tides: Modern insights from FES2004. *Ocean Dynamics* **56**: 394–415.

550. Lynch, D.K. (1982) Tidal Bores *Scientific American* **247**: 134–143.

551. MacMahan, J. H., Thornton, E. B. and Reniers, A. J. H. M. (2006) Rip current review. *Coastal Eng.* **53**: 191–208.

552. Madsen, O.S., Wright, L.D., Boon, J.D. and Chisholm, T.A. (1993) Wind stress, bed roughness and sediment suspension on the inner shelf during an extreme storm event. *Cont. Shelf Res.* **13**: 1303–1324.

553. Madsen, O. S. (1994) Spectral wave-current bottom boundary layer flows. In: *Procs. 24th International Conference Coas Eng.* 384–398.

554. Makaske, B. (2001) Anastomosing rivers: A review of their classification, origin and sedimentary products. *Earth-Science Reviews* **53**: 149–196.

555. Malarkey, J., Davies, A.G. (2002) Discrete vortex modelling of oscillatory flow over ripples. *Appl. Ocean Res.* **24**(3): 127–145.
556. Malarkey, J., Magar, V. and Davies, A.G. (2015) Mixing efficiency of sediment and momentum above rippled beds under oscillatory flows. *Cont. Shelf Res.* **108**: 76–88.
557. Malarkey, J., Davies, A.G. and Li, Z . (2003) A simple model of unsteady sheet-flow sediment transport. *Coastal Eng.* **48**: 171–188.
558. Malarkey, J., Pan, S., Li, M., ODonoghue, T., Davies, A.G. and OConnor, B.A. (2009) Modelling and observation of oscillatory sheet-flow sediment transport. *Ocean Engineering* **36**: 873–890.
559. Malikides, M., Harris, P.T. and Tate, P.M. (1989) Sediment transport and flow over sand waves in a non-rectilinear tidal environment. *Cont. Shelf Res.* **9**: 203–221.
560. Mallet, C., Howa, H.L., Garlan, T., Sottolichio, A. and Le Hir, P. (2000) Residual transport model in correlation with sedimentary dynamics over an elongate tidal sandbar in the Gironde estuary (Southwestern France). *J. Sed. Res.* **70**: 1005–1016.
561. Manning, A.J., Martens, C., De Mulder, T., Vanlede, J., Winterwerp, J. C., Ganderton, P. and Graham G. W. (2007) Mud Floc Observations in the Turbidity Maximum Zone of the Scheldt Estuary During Neap Tides. *J. Coastal Research*, SI 50 (Proceedings of the 9th International Coastal Symposium), pp. 832–836.
562. Manning, A.J., Spearman, J.R., Whitehouse, R.J.S., Pidduck, E.L., Baugh, J.V. and Spencer, K.L. (2013) Flocculation Dynamics of Mud: Sand Mixed Suspensions In: *Sediment Transport Processes and Their Modelling Applications*. Ed., A.J. Manning. *Earth Planetary Science.*
563. Marani, M., Lanzoni S., Zandolin D., Seminara G., and Rinaldo A. (2002) Tidal meanders. *Water Resour. Res.* **38**(11): 1225.
564. Marciano, R., Wang, Z.B., Hibma, A. and De Vriend, H. (2005) Modeling of channel patterns in short tidal basins. *J. Geophys. Res.* **110**: F01001.
565. Marieu, V., Bonneton, P., Foster, D.L. and Ardhuin, F. (2008) Modeling of vortex ripple morphodynamics. *J. Geophys. Res.* **113**: C09007.
566. Mariotti, G. and Fagherazzi, S. (2011) Asymmetric fluxes of water and sediments in a mesotidal mudflat channel. *Cont. Shelf Res.* **31**: 23–36.
567. Mariotti, G., and Fagherazzi, S. (2013) Wind waves on a mudflat: The influence of fetch and depth on bed shear stresses, *Cont. Shelf Res.* **60S**: 99–110.
568. Marsset, T., Tessier, B., Reynaud, J-Y., De Batist, M. and Plagnol, C. (1999) The Celtic Sea banks: An example of sand body analysis from very high-resolution seismic data. *Mar. Geol.* **158**: 89–109.
569. Masetti R., Fagherazzi, S. and Montanari, A. (2008) Application of a barrier island translation model to the millennial-scale evolution of Sand Key, Florida. *Continental Shelf Res.* **28**: 1116–1126.

570. Masselink, G., and Short, A.D. (1993) The effect of tide range on beach morphodynamics and morphology: A conceptual beach model. *J. Coast. Res.* **9**: 785–800.

571. Masselink, G. (1995) Group bound long waves as a source of infragravity waves in the surf zone. *Cont. Shelf Res.* **15**: 1525–1547.

572. Masselink, G. and Black, K.P. (1995) Magnitude and cross-shore distribution of bed return flow measured on natural beaches. *Coastal Eng.* **25**: 165–190.

573. Masselink, G. and Pattiaratchi, C.B. (1998) Morphological evolution of beach cusps and associated swash circulation patterns. *Mar. Geol.* **146**: 93–113.

574. Masselink, G. (1999) Alongshore variation in beach cusp morphology in a coastal embayment. *Earth Surface Processes and Landforms* **24**: 335–347.

575. Masselink, G. and Hughes, M. (2003) *Introduction to Coastal Processes and Geomorphology*. Oxford University Press.

576. Masselink, G. and Puleo, J.A. (2006) Swash-zone morphodynamics. *Cont. Shelf Res.* **26**: 661–680.

577. Masselink, G., Kroon, A. and Davidson-Arnott, R.G.D. (2006) Morpho-dynamics of intertidal bars in wave-dominated coastal settings A review. *Geomorphology* **73**: 33–49.

578. Masselink, G., Russell, P., Turner, I. and Blenkinsopp, C. (2009) Net sediment transport and morphological change in the swash zone of a high-energy sandy beach from swash event to tidal cycle time scales. *Marine Geol.* **267**: 18–35.

579. Masselink, G., Aagaard, T. and Kroon, A. (2011) Destruction of inter-tidal bar morphology during a summer storm surge event: Example of positive morphodynamic feedback. *Proc. ICS 2011, J. Coastal Res.* **64**: 105–109.

580. Masselink, G. and Van Heteren, S. (2014) Response of wave-dominated and mixed-energy barriers to storms. *Mar. Geology* **352**: 321–347.

581. Mazires, A., Gillet, H., Idier, D., Mulder, T., Garlan, T., Mallet, C., Marieu, V. and Hanquiez, V. (2015) Dynamics of inner-shelf, multiscale bedforms off the south Aquitaine coast over three decades (South east Bay of Biscay, France). *Continental Shelf Res.* **92**: 23–36.

582. McAnally, W.H., Friedrichs, C., Hamilton, D., Hayter, E., Shrestha, P., Rodriguez, H., Sheremet, A. and Teeter, A. (2007) Management of Fluid Mud in Estuaries, Bays, and Lakes. I: Present State of Understanding on Character and Behavior. *J. Hydr. Eng.* **133**: 9–22.

583. McBride, R.A. and Moslow, T.F. (1991) Origin, evolution and distribution of shoreface sand ridges, Atlantic inner shelf. U.S.A. *Marine Geol.* **97**: 57–85.

584. McCarroll, R.J., Castelle, B., Brander, R.W. and Scott, T. (2015) Modelling rip current flow and bather escape strategies across a transverse bar and rip channel morphology. *Geomorphology* **246**: 502–518.

585. McCave, I.N. (1971) Sand waves in the North Sea off the coast of Holland. *Mar. Geol.* **10**: 199–225.
586. McCave, I.N. and Langhorne, D.N. (1982) Sand waves and sediment transport around the end of a tidal sand bank. *Sedimentol.* **29**: 95–110.
587. McDowell, D.M. and O'Connor, B.A. (1977) *Hydraulic Behaviour of Estuaries.* MacMillan Press, London, 292 pp.
588. McLean, S.R., Wolfe, S.R. and Nelson, J.M. (1999) Predicting boundary shear stress and sediment transport over bed forms. *J. Hydr. Eng.* **125**: 725–736.
589. Meene, J.W.H. van de, Boersma, J.R. and Terwindt, J.H.J. (1996) Sedimentary structures of combined flow deposits from the shoreface-connected ridges along the central Dutch coast. *Mar. Geol.* **131**: 151–175.
590. Mehta, A.J. and Partheniades, E. (1975) An investigation of the depositional properties of flocculated fine sediments. *J. Hydr. Res.* **13**: 361–381.
591. Mehta, A.J. and Ozsoy, E. (1978) Inlet Hydraulics. In: *Stability of Tidal Inlets.* Ed., P. Bruun, Elsevier Scientific Publishing Company, p. 510
592. Mehta, A.J. (1986) Characteristics of cohesive sediment properties and transport processes in estuaries. In: Estuarine *Cohesive Sediment Dynamics.* Ed., A.J. Mehta. *Lecture Notes Coastal and Estuarine Studies* 14, Springer-Verlag, Berlin, pp. 427–445.
593. Mehta, A.J. (1996) Interaction between fluid mud and water waves. In: *Environmental hydraulics.* Eds., V.P. Singh and W.H. Hager. Kluwer Ac. Publ., Dordrecht, pp. 153–187.
594. Mehta, A., J. (2013) An Introduction to Hydraulics of Fine Sediment Transport World Scientific, *Advanced Series on Ocean Engineering*, Vol. 38, 1066 p.
595. Mei, C.N., Fan, S. and Jin, K. (1997) Resuspension and transport of fine sediments by waves. *J. Geophys. Res.* **102**: 15,807–15,821.
596. Meyer-Peter, E. and Mueller. R. (1948) Formulas for Bed-Load Transport. In: *Procs. 2nd Conference IAHR Congress, Int. Assoc. for Hydraul. Res.*, Stockholm, pp. 39–64.
597. Miche, R. (1944) Mouvements ondulatoires des mers en profondeur constante on descroisante. *Annales des Ponts et Chaussees*, pp. 25–78, 131–164, 270–292, 369–406.
598. Mietta, F., Chassagne, C., Manning, A. and Winterwerp, J. (2009) Influence of shear rate, organic matter content, pH and salinity on mud flocculation. *Ocean Dynamics* **59**: 751–76.
599. Migniot, C. (1968) Etude des propriétés physiques de différents sédiments très fins et de leur comportement sous des actions hydrodynamiques. *La Houille Blanche* **7**: 591–620.
600. Migniot, C. (1998) Mission Mont Saint Michel — Synthèse des connaissances hydro-sédimentaires. *Direction Départementale de l'Equipement de la Manche*, 111 pp.

601. Migniot, C. (1993) Bilan de l'hydrologie et de l'hydrosdimentaire de l'estuaire de la Loire au cours des deux dernières décennies. *Association pour la Protection de l'Environnement de l'Estuaire de la Loire*, 51 pp.

602. Mikes D. and Manning A.J. (2010) Assessment of flocculation kinetics of cohesive sediments from the Seine and Gironde estuaries, France, through laboratory and field studies. *J. of Waterway, Port, Coastal, and Ocean Engineering* **136**: 306–318.

603. Miles, J., Butt, T. and Russell, P. (2006) Swash zone sediment dynamics: A comparison of a dissipative and an intermediate beach. *Mar. Geol.* **231**: 181–200.

604. Miles, J., Thorpe, A., Russell, P. and Masselink, G. (2014) Observations of bedforms on a dissipative macrotidal beach. *Ocean Dynamics* **64**: 225–239.

605. Miller, J. K. and Dean, R. G. (2004) A simple new shoreline change model. *Coastal Eng.* **51**: 531–556.

606. Miller, J.K and Dean, R.G. (2007) Shoreline variability via empirical orthogonal function analysis: Part I temporal and spatial characteristics. *Coastal Eng* . **54**: 111–131.

607. Mitchener, H., Torfs, H. (1995) Erosion of mud-sand mixtures. *Coastal Eng.* **29**: 1–25.

608. Mitrovica, J.X. and Milne, G.A. (2002) On the origin of late Holocene sea-level highstands within equatorial ocean basins. *Quaternary Sci. Rev.* **21**: 2179–2190.

609. Monin, A.S. and Yaglom, A.M. (1971) Statistical fluid mechanics: Mechanics of turbulence. MIT Press, MA,. 769 pp.

610. Montserrat, F. (2011) *Estuarine Ecosystem Engineering Biogeomorphology in the Estuarine Intertidal*. Thesis, Technical Univ. Delft.

611. Morton, R.A., Clifton, H.E., Buster, N.A., Peterson, R.L. and Gelfenbaum, G. (2007) Forcing of large-scale cycles of coastal change at the entrance to Willapa Bay, Washington. *Mar. Geol.* **246**: 24–41.

612. Miles, J. and Munk, W. (1961) The harbor paradox. ASCE *J. Waterw. Harb. Div.* **87**: 111–130.

613. Mitchell, S.B. (2013) Turbidity maxima in four macrotidal estuaries. *Ocean and Coastal Management* **79**: 62–69.

614. Moerman, E. (2011) *Long-Term Morphological Modelling of the Mouth of the Columbia River*. Thesis, Delft Univ., p. 153

615. Moore, R.D., Wolf, J., Souza, A. and Flint, S. (2009) Morphological evolution of the Dee Estuary, Eastern Irish Sea, UK: A tidal asymmetry approach. *Geomorphology* **103**: 588–596.

616. Moulton, M., Elgar, S. and Raubenheimer, B. (2014) A surfzone morphological diffusivity estimated from the evolution of excavated holes. *Geophys. Res. Lett.* **41**: 4628–4636.

617. Mudersbach, C., Wahl, T., Haigh, I.D. and Jensen, J. (2013) Trends in high sea levels of German North Sea gauges compared to regional mean sea level changes. *Cont. Shelf Res.* **65**: 111–120.

618. Mukhopadhyay, S.K. (2007) The Hooghly Estuarine System, NE Coast of Bay of Bengal, India. Workshop on Indian Estuaries, NIO, Goa (2007)

619. Munk, W.H. and Anderson, E.R. (1948) Notes on the theory of the thermocline. *J. Mar. Res.* **7**: 276–295.

620. Murray, A.B., LeBars, M. and Guillon, C. (2003) Tests of a new hypothesis for non-bathymetrically driven rip currents. *J. Coast. Res.* **19**: 269–277.

621. Murray, A.B., Thieler, E.R. and Tighe, B. (2003) Sorted patterns on shallow shelves: Instability and finite-amplitude selforganization. Third International Symposium on River, Coastal and Estuarine Morphodynamics, Barcelona: pp. 365–376

622. Murray, A.B. and Thieler, E.R. (2004) A new hypothesis and exploratory model for the formation of large-scale inner-shelf sediment sorting and rippled scour depressions. *Continental Shelf Res.* **24**: 295–315.

623. Murray, A.B., Goldstein, E.B. and Coco, G. (2014) The shape of patterns to come: From initial formation to long-term evolution. *Earth Surf. Process. Landforms* **39**: 62–70.

624. Murray, T. , Cartwright, N., Tomlinson , R. (2013) Video-imaging of transient rip currents on the Gold coast open beaches. In: *Proc. 12th Int. Coastal Symp.* (Plymouth, England). Eds., D.C. Conley, G. Masselink, P.E. Russell, and T.J. OHare, *J. Coastal Res.*, Special Issue **65**: 1809–1814.

625. Nahon, A., Bertin, X., Fortunato, A.B. and Oliveira, A. (2012) Process-based 2DH morphodynamic modeling of tidal inlets: A comparison with empirical classifications and theories. *Mar. Geol.* **291–294**: 1–11.

626. Neumeier, U. and Ciavola, P. (2004) Flow resistance and associated sedimentary processes in a Spartina maritama salt marsh. *J. Coast. Res.* **20**: 435–447.

627. Nicholls, R.J., Birkemeyer, W.A. and Lee, G.H. (1998) Evaluation of depth of closure using data from Duck, NC, USA. *Mar. Geol.* **148**: 179–201.

628. Nichols, M.M. and Biggs, R.B. (1985) Estuaries. In: *Coastal Sedimentary Environments*. Eds., R.A. Davis, Springer-Verlag, New York, pp. 77–186.

629. Nichols, M.M. (1989) Sediment accumulation rates and relative sea-level rise in lagoons. *Mar. Geol.* **88**: 201–219.

630. Nichols, M.M. and Boon, J.D. (1994) Sediment transport processes in coastal lagoons. In: *Coastal Lagoon Processes*. Ed., B. Kjerfve, Elsevier, Amsterdam, pp. 157–219.

631. Nicolis, G. and Prigogine, I. (1989) *Exploring Complexity*. Freeman and Co., New York, 313 pp.

632. Nidzieko, N., J. (2010) Tidal asymmetry in estuaries with mixed semidiurnal/diurnal tides. *J. Geophys. Res.* **115**: C08006.

633. Niederoda, A. W. and Tanner, W. F. (1970) Preliminary study on transverse bars. *Marine Geol.* **9**: 41–62.

634. Nielsen, P. (1992) Coastal bottom boundary layers and sediment transport. In: *Advanced Series on Ocean Engineering*, IV. World Scientific.

635. Nielsen, P. and Callaghan, D.P. (2003) Shear stress and sediment transport calculations for sheet flow under waves. *Coastal Eng.* **47**: 347–354.

636. Nienhuis, J. H., Perron, J. T., Kao, J. C. T. and Myrow, P. M. (2014) Wavelength selection and symmetry breaking in orbital wave ripples. *J. Geophys. Res. Earth Surf.* **119**: 2239–2257.

637. Nikora, V., Goring, D., McEwan, I. and Griffiths, G. (2001) Spatially averaged open-channel flow over rough bed. *J. Hydr. Eng.* **127**: 123–133.

638. Nordstrom, C.E., Psuty, N. and Carter, R.W.G. (1990) Eds., *Coastal Dunes*: Processes and Morphology, John Wiley and Sons, Chichester, 392 pp.

639. North Solent Shoreline Management Plan (2008) New Forest District Council publication, Appendix C.

640. O'Brien, M.P. (1969) Equilibrium flow areas of inlets and sandy coasts. J. Waterw. Harbor *Coast. Eng. Div.* **95**: 43–52.

641. O'Connor, B.A., Nunes, C.R.,and Sarmento, A.J.N.A. (1996) Sand wave dimensions and statistics. In: *CSTAB Handbook and Final Report*. Ed., B.A. O'Connor. Univ. Liverpool, pp. 336–353.

642. Odd, N.V.M. and Owen, M.W. (1972) A two-layer model of mud transport in the Thames estuary. In: *Procs. Instn. Civ. Eng.*, Suppl. paper **75175**: 175–205.

643. O'Donoghue, T. and Wright, S. (2001) Experimental study of graded sediments in sinusoidal oscillatory flow. In: *Coastal Dynamics*, Lund, Sweden, pp. 918–927.

644. O'Donoghue, T., Doucette, J.S., Van der Werf, J.J. and Ribberink, J.S. (2006) The dimensions of sand ripples in full-scale oscillatory flows. *Coastal Eng.* **53**: 997–1012.

645. Oertel, G.F. (1977) Geomorphic cycles in ebb deltas and related patterns of shore erosion and accretion. *J. Sed. Petr.* **47**: 1121–1131.

646. Oertel, G.F. (1988) Processes of sediment exchange between tidal inlets, ebb deltas and barrier islands. In: *Hydrodynamics and Sediment Dynamics of Tidal Inlets*. Eds., D.G. Aubrey and L. Weishar. Springer-Verlag, New York: pp. 297–318.

647. Off, T. (1963) Rhythmic linear sand bodies caused by tidal currents. *AAPG Bull.* **47**: 324–341.

648. Officer, C.B. (1976) *Physical Oceanography of Estuaries and Associated Coastal Waters*. John Wiley, New York.

649. O'Hare, T.J. and Davies, A.G. (1990) A laboratory study of of sand bar evolution. *J. Coastal Res.* **6**: 531–544.
650. Ojeda, E., Ruessink, B.G. and Guillen, J. (2008) Morphodynamic response of a two-barred beach to a shoreface nourishment. *Coastal Eng.* **55**: 1185–1196.
651. Olabarrieta, M., Warner, J.C. and Kumar, N. (2011) Wave-current interaction in Willapa Bay. *J. Geophys. Res.* **116**: C12014.
652. Oltman-Shay, J., Howd, P.A. and Birkemeier, W.A. (1989) Shear instabilities in the longshore current, 2. Field observations. *J. Geophys. Res.* **94**: 18031–18042.
653. Oost, A.P., de Haas, H., Ijnsen, F., van den Boogert, J.M. and de Boer, P.L. (1993) The 18,6 yr nodal cycle and its impact on tidal sedimentation. *Sedimentary Geology* **87**: 1–11.
654. Osborne, P. and Greenwood, B. (1993) Sediment suspension under waves and currents: Time scales and vertical structure. *Sedimentology*, **40**: 599–622.
655. Osman, M.A. and Thorne, C.R. (1988) Riverbank stability analysis. I: *Theory. J. Hydraul. Eng.* **11**: 134–150.
656. Owen, M.W. (1971) The effect of turbulence on the settling velocities of silt flocs. In: *Proc. 14th Conf. IAHR*, Paris, D **4**: 1–5.
657. Ozasa, H. and Brampton, A.H. (1980) Mathematical modelling of beaches backed by seawalls. *Coastal Eng.* **4**: 47–63.
658. Parchure, T.M. and Mehta, A.J. (1985) Erosion of soft cohesive sediment deposits. *J. Hydraulic Engineering-ASCE*, **111**: 1308–1326.
659. Parker, B.B. (1991) The relative importance of the various nonlinear mechanisms in a wide range of tidal interactions. In: *Tidal Hydrodynamics*. Ed., B.B. Parker, Wiley, New York, 237–268.
660. Parker, G. (1976) On the cause and the characteristic scales of meandering and braiding in rivers. *J. Fluid Mech.* **76**: 457–480.
661. Parker, G., Lanfredi, N.W. and Swift, D.J.P. (1982) Seafloor response to flow in a southern hemisphere sand ridge field: Argentine inner shelf. *Sed. Geol.* **33**: 195–216.
662. Parker, W.R. and Kirby, R. (1982) Time dependent properties of cohesive sediment relevant to sedimentation management European experience. *Estuarine Comparisons*, Ed., V.S. Kennedy, Academic, New York.
663. Partheniades, E. (1965) Erosion and deposition of cohesive soils. ASCE *J. Hydr. Div.* **91**: 105–139.
664. Pattiaratchi, C.B. and Collins, M.B. (1987) Mechanisms for linear sandbank formation and maintenance in relation to dynamical oceanographic observations. *Progr. Oceanogr.* **19**: 117–176.
665. Pedersen, J.B.T. and Bartholdy J. (2006) Budgets for fine-grained sediment in the Danish Wadden Sea. *Mar. Geology* **235**: 101–117.

666. Pejrup, M., Mikkelsen, O.A. (2010) Factors controlling the field settling velocity of cohesive sediment in estuaries. *Estuarine, Coastal and Shelf Science* **87**: 177–185.

667. Pelnard-Considère, J.R. (1954) Essai de théorie de l'évolution des formes de rivages en plages de sable et de galets. Soc. Hydrotechnique de France, IVmes Journées de l'Hydraulique, Les Energies de la Mer, Paris, Question 3, 1953.

668. Perillo, G.M.E. (1995) Definitions and geomorphologic classifications of estuaries. In: *Geomorphology and Sedimentology of Estuaries*. Ed., G.M.E. Perillo. Elsevier, Amsterdam, pp. 17–47.

669. Perlin, A. and Kit, E. (2002) Apparent roughness in wave-current flow: Implication for coastal studies. *J. Hydraulic Eng.* **128**: 729–741.

670. Pethick, J.S. (1984) *An introduction to coastal geomorphology*. Arnold, London, 260 pp.

671. Pethick, J.S. (1992) Saltmarsh geomorphology. In: *Saltmarshes: Morphodynamics, Conservation and Engineering Significance*. Eds., J.R.L. Allen and K. Pye, Cambridge Univ. Press: pp. 41–63.

672. Pethick, J.S. (1996) The geomorphology of mudflats. In: *Estuarine Shores: Evolution, Environments and Human Alterations*. Eds., K.R. Nordstrom and C.T. Roman, Wiley, Chichester, pp. 185–211.

673. Pethick, J.S. (1996) The Dyfi estuary and the Aberdyfi coast. Report to the Countryside Council for Wales.

674. Pingree, R.D. and Griffiths, D.K. (1979) Sand transport paths around the British isles resulting from M2 and M4 tidal interactions. *J. Mar. Biol Ass. UK* **59**: 497–513.

675. Pingree, R.D. and Griffiths, D.K. (1987) Tidal friction for semidiurnal tides. *Cont. Shelf Res.* **7**: 1181–1209.

676. Pires, A.R., Freitas, M.C., Andrade, C., Taborda, R., Ramos, R., Pacheco, A., Ferreira, O., Bezerra, M. and Cruces, A. (2011) Morphodynamics of an ephemeral tidal inlet during a life cycle (Santo Andre Lagoon, SW Portugal). *J. Coastal Research SI* **64**: 1565–1569.

677. Plant, N.G., Ruessink, B.G. and Wijnberg, K.M. (2001) Morphologic properties derived from a simple cross-shore sediment transport model. *J. Geophys. Res.* **106**(C1): 945–958.

678. Plant, N.G., Freilich, M.H and Holman, R.A. (2001) Role of morphologic feedback in surf zone sandbar response. *J. Geophys. Res.* **106**: 973–989.

679. Plant, N.G., Holland, K.T. and Holman, R.A. (2006) A dynamical attractor governs beach response to storms. *Geophys. Res. Letters* **33**: L17607.

680. Poate, T., Masselink, G., Russell, P. and Austin, M. (2014) Morphodynamic variability of high-energy macrotidal beaches, Cornwall, UK. *Marine Geol.* **350**: 97–111.

Dynamics of Coastal Systems

681. Pontee, N.I., Whitehead, P.A. and Hayes, C.M. (2004) The effect of freshwater flow on siltation in the Humber estuary, north east UK. *Est. Coast. Shelf Sci.* **60**: 241–249.

682. Postma, H. (1954) Hydrography of the Dutch Wadden Sea. *Arch. Néerl. Zool.* **12**: 319–349.

683. Postma, H. (1957) Size frequency distributions of sands in the Dutch Wadden Sea. *Archs. Neerl. Zool.* **13**: 319–349.

684. Postma, H. (1961) Transport and accumulation of suspended matter in the Dutch Wadden Sea. *Neth. J. Sea Res.* **1**: 148–190.

685. Postma, H. (1967) Sediment transport and sedimentation in the estuarine environment. In: *Estuaries*. Ed., G.H. Lauff, *Am. Ass. Adv. Sci.* 83, Washington, D.C., pp. 158–179.

686. Powell, M.A., Thieke, R.J. and Mehta, A.J. (2004) Ebb and flood delta volumes at Florida's sandy tidal entrances. In: Proceedings Physics of Estuaries and Coastal Seas, 2004.

687. Prandle, D. and Rahman, M. (1980) Tidal response in estuaries. *J. Phys. Ocean.* **10**: 1552–1573.

688. Prandle, D. (2003) Relationships between tidal dynamics and bathymetry in strongly convergent estuaries. *J. Phys. Ocean.* **33**: 2738–2750.

689. Prandle, D. (2004) Salt intrusion in partially mixed estuaries. *Est. Coast. Shelf Sci.* **59**: 385–397.

690. Prandle, D. (2004) Sediment trapping, turbidity maximum and bathymetric stability in macrotidal estuaries. *J. Geophys. Res.* **109**: C09001.

691. Prandle D. (2009) Estuaries. Dynamics, Mixing, Sedimentation and Morphology Cambridge University Press, 248 p.

692. Prandle, D. (1980) Co-tidal Charts for the Southern North Sea. *Deutsche Hydrografische Zeitschrift* **33**: 68–81.

693. Prandle, D. (1982) The vertical structure of tidal currents. *Geophys. and Astrophys. Fluid Dyn.* **22**: 29–49.

694. Price, W.A. (1947) Equilibrium of form and forces in tidal basins on coasts of Texas and Louisiana. *Bul. Am. Ass. Petr. Geol.* **31**: 1619–1663.

695. Price, T.D. and Ruessink, B.G. (2008) Morphodynamic zone variability on a microtidal barred beach. *Mar. Geol.* **251**: 98–109.

696. Price, T. D., Castelle, B., Ranasinghe, R. and Ruessink, B.G. (2013) Coupled sandbar patterns and obliquely incident waves. *J. Geophys. Res.* **118**: 1677–1692.

697. Price, T.D., Ruessink, B.G. and Castelle, B. (2014) Morphological coupling in multiple sandbar systems — A review. *Earth Surf. Dynam.* **2**: 309–321.

698. Puleo, J.A., Holland, K.T., Plant, N.G., Slinn, D.N. and Hanes, D.M. (2003) Fluid acceleration effects on suspended sediment transport in the swash zone. *J. Geophys. Res.* **108**: C11.

699. Pullen, T. and She, K. (2002) A numerical study of breaking waves and a comparison of breaking criteria. *Proc. 28th Int. Conf. Coast. Eng.*, Cardiff, ASCE, pp. 293–305.

700. Quartel, S., Kroon, A. and Ruessink, B.G. (2008) Seasonal accretion and erosion patterns of a microtidal sandy beach. *Mar. Geol.* **250**: 19–33.

701. Ranasinghe, R. and Pattiaratchi, C. (2003) The seasonal closure of tidal inlets: Causes and effects. *Coastal Eng.* **45**: 601–627.

702. Ranasinghe, R., Symonds, G., Black, K. and Holman, R. (2004) Morphodynamics of intermediate beaches: a video imaging and numerical modelling study. *Coast. Eng.* **51**: 629–665.

703. Ranasinghe, R., Callaghan, D. and Stive, M.J.F. (2012) Estimating coastal recession due to sea level rise: Beyond the Bruun Rule. *Climatic Change* **110**: 561–574.

704. Raubenheimer, B., Guza, R.T. and Elgar, S. (1996) Wave transformation across the inner surf zone. *J. Geophys. Res.* **101**: 25,589–25,597.

705. Raubenheimer, B., Guza, R.T. and Elgar, S. (2001) Field observations of wave-driven setdown and setup. *J. Geophys. Res.* **106**: 4629–4638.

706. Raubenheimer, R., Elgar, S, and Guza, T. (2004) Observations of swash zone velocities: A note on friction coefficients. *J. Geophys. Res.* **109**: C01027, 1–8.

707. Raudkivi, A.J. and Witte, H.H. (1990) Development of bed features. *J. Hydr. Eng.* **116**: 1063–1079.

708. Raudkivi, A.J. (1997) Ripples on stream bed. *J. Hydr. Eng.* **123**: 58–64.

709. Redfield, A.C. (1965) Ontogeny of a salt marsh estuary. *Science* **147**: 50–55.

710. Reniers, A.J.H.M., Thornton, E.B., Stanton, T.P. and Roelvink, J.A. (2004) Vertical flow structure during Sandy Duck: observations and modelling. *Coast. Eng.* **51**: 237–260.

711. Reniers, A.J.H.M., Roelvink, J.A. and Thornton, E.B. (2004) Morphodynamic modeling of an embayed beach under wave group forcing. *J. Geophys. Res.* **109**: C01030.

712. Ribas, F., Falqués, A, Plant, N. and Hulscher, S. (2001) Self-organization in surf zone morphodynamics: Alongshore uniform instabilities. In: *Procs. 4th Int. Conf. Coastal Dynamics*. Eds., H. Hanson and M. Larson. ASCE, pp. 1068–1077.

713. Ribas, F., Falqués, A. and Montoto, A. (2003) Nearshore oblique sand bars. *J. Geophys. Res.* **108**: C4.

714. Ribas, F. and Kroon, A. (2007) Characteristics and dynamics of surfzone transverse finger bars. *J. Geophys. Res.* **112**: F03028.

715. Ribas, F., De Swart, H.E., Calvete, D. and Falques, A. (2012) Modeling and analyzing observed transverse sand bars in the surf zone. *J. Gephys. Res.* **117**: F02013.

716. Ribas, F., Falqués, A, De Swart, H.E., Dodd, N., Garnier, R. and Calvete, D. (2015) Understanding coastal morphodynamic patterns from depth-averaged sediment concentration. *Rev. Geophysics* **53**: 362–410.

717. Ribberink, J.S. and Al-Salem A.A. (1994) Sediment transport in oscillatory boundary layers in cases of rippled beds and sheet flow. *J. Geophys. Res.* **99**: 707–727.

718. Ribberink, J.S. and Al-Salem, A.A. (1995) Sheet flow and suspension of sand in oscillatory boundary layer. *Coastal Eng.* **25**: 205–225.

719. Ribberink, J.S. (1998) Bed-load transport for steady flows and unsteady oscillatory flows. *Coastal Eng.* **34**: 59–82.

720. Richards, K.J. (1980) The formation of ripples and dunes on an erodible bed. *J. Fluid Mech.* **99**: 597–618.

721. Richardson, J.F., Zaki, W.N. (1954) Sedimentation and fluidisation: Part 1. *Trans. Inst. Chem. Eng.* **32**: 35–53.

722. Ridderinkhof, H. (1990) *Residual Currents and Mixing in the Wadden Sea.* Thesis, Utrecht University, p. 91.

723. Ridderinkhof, H., van der Ham, R. and van der Lee, W. (2000) Temporal variations in concentration and transport of suspended sediments in a channel-flat system in the Ems-Dollard estuary. *Cont. Shelf Res.* **20**: 1479–1493.

724. Riethmüller, R., Fanger, H.U., Grabemann, I., Krasemann, H.L., Ohm, K., Böning, J., Neumann, L.J.R., Lang, G., Markofsky, M. and Schubert, R. (1988) Hydrographic measurements in the turbidity maximum of the Weser estuary. In: *Physical Processes in Estuaries.* Eds., J. Dronkers and W. van Leussen. Springer-Verlag, Berlin, pp. 332–344.

725. RIKZ National Institute for Coastal and Marine Management (1994) Average tidal curves for the Dutch tidal waters. 1991.0. (De gemiddelde getijkromme,in Dutch). RIKZ, The Netherlands, ISBN 90-369-0453-6.

726. Ritter, A. (1892) Die Fortpflanzung der Wasserwellen. *Zeitschrift des Vereines Deutscher Ingenieure* **36**: 947–954.

727. Robins, P. (2008) Present and future flooding scenarios in the Dyfi estuary, Wales. CAMS report 2008–2012. Bangor Univ.

728. Robinson, A.H.W. (1965) Residual currents in relation to shoreline evolution of the east Anglian coast. *Mar. Geol.* **4**: 57–84.

729. Roelvink, D.J.A. and Stive, M.J.F. (1989) Bar-generating cross-shore flow mechanisms on a beach. *J. Geophys. Rev.* **94**: 4785–4800.

730. Rosati, J.D., Dean, R.G. and Walton, T.L. (2013) The modified Bruun Rule extended for landward transport. *Mar. Geol.* **340**: 71–81.

731. Rosen, P.S. (1978) A regional test of the Bruun Rule on shoreline erosion. *Mar. Geology* **26**: 7–16.

732. Roy, P.S., Cowell, P.J., Ferland, M.A. and Thom, B.G. (1994) Wave dominated coasts. In: *Coastal Evolution: Late Quaternary Shoreline*

Morphodynamics. Eds. R.W.G. Carter and C.D. Woodroffe, Cambridge University Press, pp. 121–185.

733. Rubin, D.M. and Ikeda, H. (1990) Flume experiments on the alignment of transverse, oblique and longitudinal dunes in directionally varying flows. *Sedimentol.* **37**: 673–684.

734. Ruessink, B.G. (1998) Bound and free infragravity waves in the nearshore zone under breaking and nonbreaking conditions. *J. Geophys. Res.* **103**: 12,795–12,805.

735. Ruessink, B.G., van Enckevort, I.M.J., Kingston, K.S. and Davidson, M.A. (2000) Analysis of observed two- and three-dimensional nearshore bar behaviour. *Mar. Geol.* **169**: 161–183.

736. Ruessink, B.G., van Enckevort, I.M.J., Kingston, K.S. and Davidson, M.A. (2002) Analysis of 2- and 3-dimensional nearshore bar nehaviour. In: *Coast3D-Egmond; The Behaviour of a Straight Sandy Coast on the Time Scale of Storms and Seasons*. Eds., L.C. Van Rijn, B.G. Ruessink and J.P.M. Mulder. ISBN 90-800356-5-3, Aqua Publ., Amsterdam, L1–L23.

737. Ruessink, B.G. and Jeuken, M.C.J.L. (2002) Dunefoot dynamics along the Dutch coast. *Earth Surface Processes and Landforms* **27**, 1043–1056.

738. Ruessink, B.G., Wijnberg, K.M., Holman, R.A., Kuriyama, Y. and van Enckevort, I.M.J. (2003) Intersite comparison of interannual nearshore bar behaviour. *J. Geophys. Res.* **108**: C8 3249.

739. Ruessink, B.G., Coco, G., Ranasinghe, R. and Turner, I.L. (2007) Coupled and noncoupled behavior of three-dimensional morphological patterns in a double sandbar system, *J. Geophys. Res.* **112**: C07002.

740. Ruessink, B.G., van den Berg, T.J.J. and Van Rijn L.C. (2009) Modeling sediment transport beneath skewed asymmetric waves above a plane bed. *J. Geophys. Res.* **114**: C11021.

741. Ruessink, B.G., Pape, L. and Turner, I.L. (2009) Daily to interannual cross-shore sand bar migration: Observations from a multiple sandbar system. *Cont. Shelf Res.* **29**: 1663–1677.

742. Ruessink, B.G., Michallet, H., Abreu, T., Sancho, F., van der A, D.A., van der Werf, J.J. and Silva, P.A. (2011) Observations of velocities, sand concentrations, and fluxes under velocity-asymmetric oscillatory flows. *J. Geophys. Res.* **116**: C03004.

743. Ruessink, B.G., Ramaekers, G. and Van Rijn, L.C. (2012) On the parameterization of the free-stream non-linear wave orbital motion in nearshore morphodynamic models. *Coastal Eng.* **65**: 56–63.

744. Ruggiero, P., Holman, R.A. and Beach, R.A. (2004) Wave run-up on a high-energy dissipative beach. *J. Geophys. Res.* **109**: C06025.

745. Russel, R.J. and McIntire, W.G. (1965) Beach cusps. *Geol. Soc. Am. Bull.* **76**: 307–320.

746. Ryu, S.O. (2003) Seasonal variation of sedimentary processes in a semi-eclosed bay: Hampyong Bay, Korea. *Est. Coast. Shelf Sci.* **56**: 481–492.

747. Salomons, W. and Mook, W.G. (1981) Field observations of isotopic composition of particulate organic carbon in the Southern North Sea and adjacent estuaries. *Mar. Geol.* **41**: 11–20.

748. Sato, S. and Horikawa, K. (1986) Laboratory study on sand transport over ripples due to asymmetric oscillatory flows. *Proc. 20th Int. Coastal Eng. Conf.*, 1481–1495.

749. Savenije, H.H.G. (2003) The width of a bankfull channel; Lacey's formula explained. *J. Hydrol.* **276**: 176–183.

750. Savenije, H.H.G. (2005). Salinity and Tides in Alluvial Estuaries. Elsevier, Amsterdam, 197 pp.

751. Scharp, J.C. (1949). Hydrografie. In: *Handboek der Geografie van Nederland.* Eds., G.J.A. Mulder, J.J. De Erven, Z. Tijl, **I**: 378–529.

752. Schielen, R., Doelman, A. and De Swart, H.E. (1993) On the nonlinear dynamics of free bars in straight channels. *J. Fluid Mech.* **252**: 325–356.

753. Schijf, J.B. and Schönfeld, J.C. (1953) Theoretical considerations on the motion of salt and fresh water. In: *Proc. Minn. Int. Hydraul. Conv.*, Mineanopolis, p. 321.

754. Schramkowski, G.P., Schuttelaars, H.M. and de Swart, H.E. (2002) The effect of geometry and bottom friction on local bed forms in a tidal embayment. *Cont. Shelf Res.* **22**: 1821–1833.

755. Schröder, M. and Siedler, G. (1989) Turbulent momentum and salt transport in the mixing zone of the Elbe estuary. *Est. Coast. Shelf Sci.* **28**: 615–638.

756. Schuurman, F., Shimizu, Y., Iwasaki, Y. and Kleinhans M.G. (2016) Dynamic meandering in response to upstream perturbations and floodplain formation. *Geomorphology* **253**: 94–109.

757. Schuerch, M., Rapaglia, J., Liebetrau, V., Vafeidis, A. and Reise, K. (2012) Salt Marsh Accretion and Storm Tide Variation: An Example from a Barrier Island in the North Sea. *Estuaries and Coasts* **35**: 486–500.

758. Schumm, S.A. (1969) River metamorphosis. *J. Hydr. Div. Proc.* ASCE **96**: 201–222.

759. Schuttelaars, H.M. and De Swart, H.E. (1999) Formation of channels and shoals in a short tidal embayment. *J. Fluid Mech.* **386**: 15–42.

760. Schwab, W.C., Baldwin, W.E., Hapke, C.J., Lentz, E.E., Gayes, P.T., Denny, J.F., List, J.H., Warner, J.C. (2013). Geologic evidence for onshore sediment transport from the inner continental shelf. *J. Coast. Res.* **29**: 526–544.

761. Scott, C.P., Cox, D.T., Maddux, T.B. and Long, J.W. (2005) Large-scale laboratory observations of turbulence on a fixed barred beach. *Meas. Sci. Technol.* **16**: 1903–1912.

762. Scott, T., Masselink, G. and Russell, P. (2011) Morphodynamic characteristics and classification of beaches in England and Wales. *Marine Geol.* **286**: 1–20.

763. Scott, T., Masselink, G., Austin, M.J. and Russell, P. (2014) Controls on macrotidal rip current circulation and hazard. *Geomorphology* **214**: 198–215.

764. Scully, M.E. and Friedrichs, C.T. (2007) Sediment pumping by tidal asymetry in a partially mixed estuary. *J. Geophys. Res.* **112**: C07028.

765. Seminara, G. and Tubino, M. (1997) Bed formation in tidal channels: Analogy with fluvial bars. In: *Morphology of Rivers, Estuaries and Coasts*. Ed., DiSilvio, IAHR, London.

766. Seminara, G. and Blondeax, P. Eds., (2001) *River, Coastal and Estuarine Morphodynamics*. Springer, Berlin, p. 211.

767. Seminara, G. and Tubino, M. (2001) Sand bars in tidal channels. Part 1. Free bars. *J. Fluid Mech.* **440**: 49–74.

768. Seminara, G., Lanzoni, S., Bolla Pittaluga, M. and Solari, L. (2001) Estuarine Patterns: An introduction to their morphology and mechanics. In: *Geomorphological Fluid Mechanics*. Eds., M.J. Balmforth and E. Provenzale. Springer, pp. 455–499.

769. Senechal, N., Coco, G., Bryan, K.R. and Holman, R.A. (2011) Wave runup during extreme storm conditions. *J. Geophys. Res.* **116**.

770. Sha, L.P. (1998) Sand transport patters in the ebb-tidal delta off Texel Inlet, Wadden Sea, The Netherlands. *Mar. Geol.* **86**: 137–154.

771. Sha, L.P. and van den Berg, J.H. (1993) Variation in ebb-tidal delta geometry along the coast of the Netherlands and the German Bight. *J. Coast. Res.* **9**: 730–746.

772. Shi, N.C. and Larsen, L.H. (1984) Reverse transport induced by amplitude-modulated waves. *Mar. Geol.* **54**: 181–200.

773. Shepard, F.P. (1973) *Submarine Geology*. Harper and Row, New York, 517 pp.

774. Short, A.D. (1975) Multiple offshore bars and standing waves. *J. Geophys. Res.* **80**: 3838–3840.

775. Short, A.D. (1991) Macro-meso tidal beach morphodynamics — An overview. *J. Coast. Res.* **7**: 417–436.

776. Shri, R. S. and Chugh, M.A. (1961) Tides in Hooghly River. *Int. Ass. of Sci. Hydrology Bulletin*, **6**(2): 10–26.

777. Simpson, J.H., Crawford, W.R., Rippeth, T.P., Campbell, A.R. and Cheok, J.V.S. (1996) The vertical structure of turbulent dissipationn in shelf seas. *J. Phys. Ocean.* **26**: 1579–1590.

778. Singh Chauhan, P.P. (2009) Autocyclic erosion in tidal marshes. *Geomorphology* **110**: 45–57.

779. Sistermans, P.J.G., Van de Graaf, J. and Van Rijn, L.C. (2001) Vertical sorting of graded sediments by waves and currents. In: *Proc. Int. Conf. Coast. Eng.*, Sydney, Australia. ASCE, pp. 2780–2793.
780. Sleath, J.F.A. (1976) On rolling grain ripples. *J. Hydr. Res.* **14**: 69–80.
781. Sleath, J.F.A. (1984) *Sea Bed Mechanics*. Wiley, New York.
782. Sleath, J.F.A. (1987) Turbulent oscillatory flow over rough beds. *J. Fluid Mech.* **182**: 369–409.
783. Sleath, J.F.A. (1991) Velocities and shear stress in wave-current flows. *J. Geophys. Res.* **96**: 15237–15244.
784. Slingerland, R. and Smith, N.D. (2004) River avulsions and their deposits. *Annu. Rev. Earth Planet. Sci.* **32**: 257–285.
785. Small, C. and Nicholls, R.J. (2003) A global analysis of human settlement in coastal zones. *J. Coast. Res.* **19**: 584–599.
786. Smit, P., Zijlema, M. and Stelling, G. (2013) Depth-induced wave breaking in a non-hydrostatic, near-shore wave model. *Coastal Eng.* **76**: 1–16.
787. Smith, J.B. and FitzGerald, D.M. (1994) Sediment transport patterns at the Essex River Inlet ebb-tidal delta, Massachusetts, USA. *J. Coast. Res.* **10**: 752–774.
788. Smith, J.D. (1969) Geomorphology of a sand ridge. *J. Geol.* **77**: 39–55.
789. Smith, J.D. and McLean, S.R. (1977) Spatially averaged flow over a wavy surface. *J. Geophys. Res.* **82**: 1735–1746.
790. Solari, L., Seminara, G., Lanzoni, S., Marani, M. and Rinaldo, A. (2002) Sand bars in tidal channels. Part 2: Tidal meanders. *J. Fluid Mech.* **451**: 203–238.
791. Sonu, C.J. (1968) Collective movement of sediment in littoral environment. In: *Proc. Int. Conf. Coast. Eng.*, ASCE, pp. 373–400.
792. Sottolichio, A. and Castaing, P. (1999) A synthesis on seasonal dynamics of highly concentrated structures in the Gironde estuary. Comptes Rendus *Acad. Sci. Paris, Earth and Planetary Sciences* **329**: 795–800.
793. Sottolichio, A., Kervella, S and Bertier, C. (2014) Synoptic observations of the dynamics of intertidal flats in the Loire estuary (France). *Procs. 17th Physics of Estuaries and Coastal Seas (PECS) Conf.*, Brazil.
794. Soulsby, R.L., Atkins, R. and Salkfield, P. (1994) Observations of the turbulent structure of a suspension of sand in a tidal current. *Cont. Shelf Res.* **14**: 429–235.
795. Soulsby, R.L. and Damgaard, J.S. (2005) Bedload sediment transport in coastal waters. *Coastal Engineering* **52**: 673–689.
796. Soulsby, R.L. (1997) *Dynamics of Marine Sands*. Thomas Telford, pp. 249.
797. Soulsby, R.L. and Whitehouse, R.J.S. (1997) Threshold of sediment motion in coastal environments. In *Proc. Australasian Coastal Eng. and Ports Conf.*, Christchurch, pp. 149–154.

798. Soulsby, R.L., Whitehouse, R.J.S., Marten, K.V. (2012) Prediction of time-evolving sand ripples in shelf seas. *Cont. Shelf Res.* **38**: 47–62.

799. Soulsby, R.L., Manning, A.J., Spearman, J., Whitehouse, R.J.S. (2013) Settling velocity and mass settling flux of flocculated estuarine sediments. *Mar. Geol.* **339**: 1–12.

800. Southard, J.B. and Dingler, J.R. (1971) Flume study of ripple propagation behind mounds on flat sand beds. *Sedimentol.* **16**: 251–263.

801. Spanhoff, R., Biegel, E.J., Burger, M. and Dunsbergen, D.W. (2003) Shoreface nourishments in the Netherlands, Proceedings Coastal Sediments 2003, Fifth International Symposium on Coastal Engineering and Science of Coastal Sediment Processes.

802. Spanhoff, R. Private communication.

803. Speer, P.E. and Aubrey, D.G. (1985) A study of non-linear tidal propagation in shallow inlet/estuarine systems. Part II: Theory. *Estuarine, Coast. Shelf Sci.* **21**: 207–224.

804. Stal, L. J. (2010) Microphytobenthos as a biogeomorphological force in intertidal sediment stabilization. *Ecological Engineering* **36**: 236–245.

805. Stive, M.J.F., Roelvink, D.J.A. and De Vriend, H.J. (1991) Large-scale coastal evolution concept. *Procs. 22nd Int. Conf. Coast. Eng. ASCE*, pp. 1962–1974.

806. Stive, M.J.F., De Vriend, H.J., Nicholls, R. and Capobianco, M. (1992) Shore nourishment and the active zone: A time-scale dependent view. In: *Int. Conf. Coastal Eng.* Ed., B.L. Edge, ASCE, pp. 2464–2473.

807. Stive, M.J.F., Capobianco, M., Wang, Z.B., Ruol, P. and Buijsman, M.C. (1998) Morphodynamics of a tidal lagoon and adjacent coast. In: *Physics of Estuaries and Coastal Seas*. Eds., J. Dronkers and M.B.A.M. Scheffers, Balkema, Rotterdam, pp. 397–407.

808. Stive, M.J.F., Aarninkhof, S.G.J., Hamm, L., Hanson, H., Larson, M., Wijnberg, K.M., Nicholls, J. and Capobianco, M. (2002) Variability of shore and shoreline evolution. *Coastal Eng.* **47**: 211–235.

809. Stive M.J.F., de Schipper M.A., Luijendijk A.P., Aarninkhof S.G.J., van Gelder-Maas C., van Thiel de Vries J.S.M., de Vries S., Henriquez M., Marx S. and Ranasinghe R. (2013) A new alternative to saving our beaches from local sea-level rise: The sand engine. *J. Coastal Res.* **29**: 1001–1008.

810. Stive, M.J.F. (2004) How important is global warming for coastal regions? An editorial comment. *Climatic Change* **64**: 27–39.

811. Stockdon, H.F., Holman, R.A., Howd, P.A. and Sallenger, A.H. (2006) Empirical parameterization of setup, swash, and runup. *J. Coastal Eng.* **53**: 573–588.

812. Stoker, J.J. (1957) Water waves. *Interscience,* New York.

813. Stolper, D., List, J.H. and Thieler, E.R. (2005) Simulating the evolution of coastal morphology and stratigraphy with a new morphological-behaviour model (GEOMBEST). *Marine Geol.* **218**: 17–36.

814. Stouthamer, E., Berendsen, H.J.A., (2000) Factors controlling the Holocene avulsion history of the Rhine-Meuse delta (The Netherlands). *Journal of Sedimentary Research* **70**: 1051–1064.

815. Straaten, L.M.J.U. and Kuenen, P.H. (1957) Accumulation of fine-grained sediments in the Dutch Wadden Sea. *Geol. Mijnbouw* (N.S.) **19**: 329–354.

816. Sturm, O. (2011) Prefeasibility study on a barrier downstream of HCMC. MSc. Thesis Delft Univ.

817. Sumer, B.M., Kozakiewicz, A., Fredse, J. and Deigaard, R. (1996) Velocity and concentration profiles in the sheet flow layer of movable bed. *J. of Hydr. Eng., ASCE,* **122**: 549–558.

818. Sunamura, T. (1992) *Geomorphology of Rocky Coasts.* Wiley, Chichester.

819. Sunamura, T. (2004) A predictive relationship for the spacing of beach cusps in nature. *Coast. Eng.* **51**: 697–711.

820. Sutherland, A.J. (1967) Proposed mechanism for sediment entrainmment by turbulent flows. *J. Geophys. Res.* **72**: 6183–6194.

821. Swift, D.J.P., Duane, D.B. and McKinney, T.F. (1973) Ridge and swale topography of the middle Atlantic Bight, North America: Secular response to the Holocene hydraulic regime. *Marine Geology* **15**: 227–247.

822. Swift, D.J.P., Parker, G., Lanfredi, N.W., Perillo, G. and Figge, K. (1978) Shoreline-connected sand ridges on American and European shelves — A comparison. *Est. Coast. Mar. Sci.* **7**: 227–247.

823. Swift, D.J.P. and Field, M.E. (1981) Evolution of a classic sand ridge field: Maryland sector, North American inner shelf. *Sedimentol.* **28**: 461–482.

824. Swift, D.J.P. and Thorne, J.A. (1991) Sedimentation on continental margins, I: A general model for shelf sedimentation. In: *Shelf Sand and Sandstone Bodies.* Eds., D.J.P. Swift, G.F. Oertel, R.W. Tillman and J.A. Thorne. *Int. Ass. Sed.*, Blackwell, Oxford, pp. 3–31.

825. Swart, D.H. (1776) Coastal sediment transport. *Computation of Longshore Transport.* rep. R**968**: Part I. WL Delft Hydraulics, Delft.

826. Syvitski, J.P.M., Saito, Y. (2007) Morphodynamics of deltas under the influence of humans. *Global and Planetary Change* **57**: 261–282.

827. Syvitski, J.P.M., Overeem, I., Brakenridge, G.R., Hannon, M. (2012) Floods, floodplains, delta plains A satellite imaging approach. *Sedimentary Geology* **267–268**: 1–14.

828. Talmon, A.M., Van Mierlo, M.C.L.M. and Struiksma, N. (1995) Laboratory measurements of the direction of sediment transport on transverse alluvial-bed slopes. *J. Hydr. Res.* **33**: 495–517.

829. Talke, S.A. and De Swart, H.E. (2006) Hydrodynamics and Morphology in the Ems/Dollard Estuary: Review of Models, Measurements, Scientific

Literature, and the Effects of Changing Conditions. University of Utrecht Report IMAU R06-01.

830. Talukdar, S., Kumar, B., Dutta, S. (2012) Predictive capability of bedload equations using flume data. *J. Hydrol. Hydromech.* **60**: 45–56.

831. Tang, E.C.S. and Dalrymple, R.A. (1989) Nearshore Circulation: B. Rip currents and wave groups. In: *Nearshore Sediment Transport.* Ed., R.J. Seymour, Plenum Press, New York.

832. Temmerman, S., Govers, G., Meire, P. and Wartel, S. (2003) Modelling long-term marsh growth under changing tidal conditions and suspended sediment concentrations, Scheldt estuary, Belgium. *Mar. Geol.* **193**: 151–169.

833. Temmerman, S., Bouma, T.J., Govers, G. and Lauwaet, D. (2005) Flow paths of water and sediment in a tidal marsh: Relations with marsh developmental stage and tidal inundation height. *Estuaries* **28**: 338–352.

834. Temmerman, S., Bouma, T.J., Govers, G., Wang, Z.B., De Vries, M.B. and Herman, P.M.J. (2005) Impact of vegetation on flow routing and sedimentation patterns: Three-dimensional modeling or a tidal marsh. *J. Geophys. Res.* **110**: F04019.

835. Temmerman, S., Bouma, T.J., Van de Koppel, J., Van der Wal, D., De Vries, M.B. and Herman, P. M. J. (2007) Vegetation causes channel erosion in a tidal landscape. *Geology* **35**: 631–634.

836. Temmerman, S., De Vries, M.B. and Bouma, T.J. (2012) Coastal marsh die-off and reduced attenuation of coastal floods: A model analysis. *Global and Planetary Change* **92–93**: 267–274.

837. Ten Brinke, W.B.M., Dronkers, J. and Mulder, J.P.M. (1994) Fine sediments in the Eastern-Scheldt tidal basin before and after partial closure. *Hydrobiol.* **282/283**: 41–56.

838. Terwindt, J.H.J. (1971) Sand waves in the southern North Sea. *Mar. Geol.* **10**: 51–67.

839. Thompson, C.E.L., Amos, C.L., Lecouturier M. and Jones, T.E.R. (2004) Flow deceleration as a method of determining drag coefficient over roughened flat beds. *J. Geophys. Res.* **109**: C03001, 1–12.

840. Thornton, E.B. and Guza, R.T. (1982) Energy saturation and phase speeds measured on a natural beach. *J. Geophys. Res.* **87**: 9499–9508.

841. Thornton, E.B. and Guza, R.T. (1983) Transformation of wave height distribution. *J. Geophys. Res.* **88**: 5925–5938.

842. Thornton, E., Humiston, R. and Birkemeyer, W. (1996) Bar-trough generation on a natural beach. *J. Geophys. Res.* **101**: 12,097–12,110.

843. Tiessen, M.C.H., van Leeuwen, S.M., Calvete, D. and Dodd, N., (2010) A field test of a linear stability model for cresentic bars. *Coastal Eng.* **57**: 41–51.

844. Tissier, M., Bonneton, P., Michallet, H. and Ruessink, B.G. (2015) Infragravity-wave modulation of short-wave celerity in the surf zone. *J. Geophys. Res.* **120**: 6799–6814.
845. Tomasicchio, G.R. and Sancho, F. (2002) On wave induced undertow at a barred beach. *Proc. 28th Int. Conf. Coast. Eng.*, Cardiff, ASCE, pp. 557–569.
846. Toorman, E.A. (2003) Validation of macroscopic modelling of particle-laden turbulent flows. *Procs. 6th Belgian National Congress on Theoretical and Applied Mechanics*, Gent.
847. Toublanc, F., Brenon, I., Coulombier, T. and LeMoine O. (2015) Fortnightly tidal asymmetry inversions and perspectives on sediment dynamics in a macrotidal estuary (Charente, France). *Cont. Shelf Res.* **94**: 42–54.
848. Townend, I., Wang, Z.B., Spearman, J. and Wright, A. (2008) Volume and surface changes in estuaries and tidal inlets. *Procs. ICCE Hamburg*, 2008.
849. Traykovski, P., Hay, A.E., Irish, J.D. and Lynch, J.F. (1999) Geometry, migration and evolution of wave-orbital ripples at LEO-(15) *J. Geophys. Res. C* **104**: 1505–1524.
850. Traykovski, P., Geyer, W.R. and Sommerfield, C. (2004) Rapid sediment deposition and fine-scale strata formation in the Hudson estuary. *J. Geophys. Res.* **109**: F02004.
851. Traykovski, P., Trowbridge, J. and Kineke, G. (2015) Mechanisms of surface wave energy dissipation over a high-concentration sediment suspension. *J. Geophys. Res. Oceans* **120**: 1638–1681.
852. Trembanis, A.C., Wright, L.D., Friedrichs, C.T., Green, M.O. and Hume, T. (2004) The effects of spatially complex inner shelf roughness on boundary layer turbulence and current and wave friction: Tairua embayment, New Zealand. *Cont. Shelf Res.* **24**: 1549–1571.
853. Trenhaile, A.S. (2009) Modeling the erosion of cohesive clay coasts. *Coastal Eng.* **56**: 59–72.
854. Trowbridge, J. and Madsen, O.S. (1984) Turbulent wave boundary layers (1) Model formulation and first-order solution, *Journal of Geophysical Research*, **89**: 7989–7997.
855. Trowbridge, J.H. and Madsen, O.S. (1984) Turbulent wave boundary layers: 2. Second-order theory and mass transport. *J. Geophys. Res.* **89**: 7999–8007.
856. Trowbridge, J.H. (1995) A mechanism for the formation and maintenance of the shore-oblique sand ridges on storm-dominated shelves. *J. Geophys. Res.* **100**: 16071–16086.
857. Tsujimoto, T. (2010) Diffusion coefficient of suspended sediment and kinematic eddy viscosity of flow containing suspended load. In: *Procs. River Flow 2010* Eds., Dittrich, Koll, Aberle and Geisenhainer. Bundesanstalt fr Wasserbau ISBN 978-3-939230-00-7

858. Turner, I.L., Whyte, D., Ruessink, B.G. and Ranasinghe, R. (2007) Observations of rip spacing, persistence and mobility at a long, straight coastline. *Mar. Geol.* **236**: 201–211.

859. Turrell, W.R. and Simpson, J.H. (1988) The measurement and modelling of axial convergence in shallow well-mixed estuaries. In: *Physical Processes in Estuaries.* Eds., J. Dronkers and W. van Leussen, Springer-Verlag, Berlin, pp. 130–145.

860. Uncles, R.J. and Jordan, M.B. (1979) Residual fluxes of water and salt at two stations in the Severn Estuary. *Es. Coast. Mar. Sci.* **9**: 287–302.

861. Uncles, and Jordan, M.B. (1980) A one-dimensional representation of residual currents in the Severn estuary and associated observations. *Est. Coast. Shelf Sci.* **10**: 39–60.

862. Uncles, R.J. Elliot, R.C.A. and Weston, S.A. (1985) Observed fluxes of water, salt and suspended sediment in a partly mixed estuary. *Est. Coast. Shelf Sci.* **20**: 147–167.

863. Uncles, R.J. Elliot, R.C.A. and Weston, S.A. (1986) Observed and computed lateral circulation patterns in a partially mixed estuary. *Est. Coast. Shelf Sci.* **22**: 439–457.

864. Uncles, R.J. and Stephens, J.A. (1990) Salinity stratification and vertical shear transport in an estuary. In: *Coastal and Estuarine Studies* 38, Residual currents and long-term transport. Ed., R.T. Cheng, Springer-Verlag, New York, pp. 137–150.

865. Uncles, R.J. and Stephens, J.A. (1993) The nature of the turbidity maximum in the Tamar estuary. *Est. Coast. Shelf Sci.* **36**: 413–431.

866. Uncles, R.J. (2002) Estuarine physical processes research: Some recent studies and progress. *Est. Coast. Shelf Sci.* **55**: 829–856.

867. Uncles, R.J., Bale, A.J., Brinsley, M.D., Frickers, P.E., Harris, C., Lewis, R.E., Pope, N.D., Staff, F.J., Stephens, J.A., Turley, C.M. and Widdows, J. (2003) Intertidal mudflat properties, currents and sediment erosion in the partially mixed Tamar Estuary, UK. *Ocean Dynamics* **53**: 239–251.

868. Uncles, R.J., Stephens, J.A. and Law, D.J. (2006) Turbidity maximum in the macrotidal, highly turbid Humber Estuary, UK: Flocs, fluid mud, stationary suspensions and tidal bores. *Estuarine, Coastal and Shelf Science* **67**: 30–52.

869. Van der A, D.A., ODonoghue, T., Davies, A.G. and Ribberink, J.S. (2011) Experimental study of the turbulent boundary layer in acceleration-skewed oscillatory flow. *J. Fluid Mech.* **684**: 251–283.

870. Van der A, D.A., Ribberink, J.S., Van der Werf, J.J., O'Donoghue, T., Buijsrogge, R.H. and Kranenburg W.M. (2013) Practical sand transport formula for non-breaking waves and currents. *Coastal Engineering* **76**: 26–42.

871. Van den Berg, J.H. (1987) Bed form migration and bedload transport in some rivers and tidal environments. *Sedimentology* **34**: 681–698.

872. Van den Berg, N., Falques, A. and Ribas, F. (2011) Long-term evolution of nourished beaches under high angle wave conditions. *J. Marine Systems* **88**: 102–112.

873. Van den Berg, N., Falques, A. and Ribas, F. (2012) Modeling large scale shoreline sand waves under oblique wave incidence. *J. Geophys. Res.* **117**: F03019.

874. Vandenbruwaene, W., Meire, P. and Temmerman, S. (2012) Formation and evolution of a tidal channel network within a constructed tidal marsh. *Geomorphology* **151–152**: 114–125.

875. Vandenbruwaene, W., Plancke, Y., Verwaest, T. and Mostaert F. Interestuarine comparison: Hydro-geomorphology Hydro- and geomorphodynamics of the TIDE estuaries Scheldt, Elbe, Weser and Humber. TIDE Report Flanders Hydraulics Research WL 2013–770.

876. Van de Koppel, J., Van der Wal, D., Bakker, J.P. and Herman, P.M.J. (2005) Self-organization and vegetation collapse in salt marsh ecosystems. *Am. Nat.* **165**: E1–E12.

877. Van de Kreeke, J. Stability of tidal inlets; Escoffier's analysis. *Shore and Beach* **60**: 9–12.

878. Van de Kreeke, J. and Iannuzzi, R.A. (1998) Second-order solutions for damped cooscillating tide in narrow canal. *J. Hydr. Eng.* **124**: 1253–1260.

879. Van de Kreeke, J. (2004) Equilibrium and cross-sectional stability of tidal inlets: application to the Frisian Inlet before and after basin reduction. *Coastal Eng.* **51**: 337–350.

880. Van de Kreeke, J. and Hibma, A. (2005) Observations on silt and sand transport in the throat section of the Frisian Inlet. *Coastal Eng.* **52**: 159–175.

881. Van der Molen, J. (2002) The influence of tides, wind and waves on the net sand transport in the North Se *Cont. Shelf Res.* **22**: 2739–2762.

882. Van der Spek, A.F.J. (1994) *Large-scale Evolution of Holocene Tidal Basins in the Netherlands.* Thesis, Utrecht University, 191 pp.

883. Van der Spek, A.F.J. (1997) Tidal asymmetry and long-term evolution of Holocene tidal basins in The Netherlands: simulation of paleo-tides in the Schelde estuary. *Mar. Geol.* **141**: 71–90.

884. Van der Spek, A. and Elias, E. (2013) The effects of nourishments on autonomous coastal behaviour. *Procs. Coastal Dynamics Conf,* pp. 1753–1762.

885. Van der Wal, D. (1999) *Aeolian Transport of Nourishment Sand in Beach-Dune Environments.* Thesis, University of Amsterdam.

886. Van der Wal, D., Pye, K. and Neal, A. (2002) Long-term morphological change in the Ribble estuary, northwest England. *Mar. Geol.* **189**: 249–266.

887. Van der Wegen, M., Dastgheib, A. and Roelvink, J.A. (2010) Morphodynamic modeling of tidal channel evolution in comparison to empirical PA relationship. *Coastal Eng.* **57**: 827–837.
888. Van der Wegen, M. and Roelvink, J.A. (2012) Reproduction of estuarine bathymetry by means of a process-based model: Western Scheldt case study, the Netherlands. *Geomorphology* **179**: 152–167.
889. Van der Werf, J.J. (2006) *Sand Transport Over Rippled Beds in Oscillatory Flow*. Doctoral Thesis, Department of Civil Engineering, University of Twente, The Netherlands.
890. Van Dongeren, A.R. and De Vriend, H.J. (1994) A model of morphological behaviour of tidal basins. *Coast. Eng.* **22**: 287–310.
891. Van Doorn, T. (1981) Experimental investigation of near bottom velocities in water waves with and without a current. Rep. M1423, Delft Hydraul. Lab., Delft, Netherlands.
892. Van Duin, M.J.P., Wiersma, N.R., Walstra, D-J.R., Van Rijn, L.C. and Stive, M.J.F. (2004) Nourishing the shoreface: observations and hindcasting of the Egmond case, The Netherlands. *Coast. Eng.* **51**: 813–837.
893. Van Enckevort, I.M.J. and Ruessink, B.G. (2003) Video observations of nearshore bar behaviour. Part 1: Alongshore uniform variability. *Cont. Shelf Res.* **23**: 501–512.
894. Van Enckevort, I.M.J., Ruessink, B.G., Coco, G., Suzuki, K., Turner, I.L., Plant, N.G. and Holman, R.A. (2004) Observations of nearshore crescentic sandbars. *J. Geophys. Res.* **109** (C06028). doi:10.1029/2003JC002214
895. Van Goor, M.A., Zitman, T.J., Wang, Z.B. and Stive, M.J.F. (2003) Impact of sea-level rise on the morphological equilibrium state of tidal inlets. *Mar. Geol.* **202**: 211–227.
896. Van Heteren, S., Baptist, M.J., Van Bergen Henegouwen, V.N., Van Dalfsen, J.A., Van Dijk, T.A.G.P., Hulscher, N.H.B.M., Knaapen, M.A.F., Lewis, W.E., Morelissen, R., Passchier, S., Penning, W.E., Storbeck, F., Van der Spek, A.F.J., Van het Groenewoud, H. and Weber, A. (2003) *Eco-Morphodynamics of the Seafloor*. Delft Cluster Publ. 03.01.05-04, Univ. Delft, p. 52.
897. Van Katwijk, M.M., Bos, A.B., Hermus, D.C.R. and Suykerbuyk, W. (2010) Sediment modification by seagrass beds: Muddification and sandification induced by plant cover and environmental conditions. *Est. Coastal Shelf Sci.* **89**: 175–181.
898. Van Lancker, V.R.M. and Jacobs, P. (2000) The dynamical behaviour of shallow-marine dunes. In: *Marine Sandwave Dynamics*. Eds., A. Trentesaux and T. Garlan. *Procs. Int. Workshop*, Univ. Lille, pp. 213–220.
899. Van Leussen, W. (1994) *Estuarine Macro-Flocs and Their Role in Fine-Grained Sediment Transport*. Thesis, Utrecht University, p. 488.

900. Van Maanen, B., Coco, G. and Bryan, K.R. (2013) Modelling the effects of tidal range and initial bathymetry on the morphological evolution of tidal embayments. *Geomorphology* **191**: 23–34.

901. Van Maldegem, D.C., Mulder, H.P.J. and Langerak, A. (1991) A cohesive sediment balance for the Scheldt estuary. *Neth. J. Aq. Ecol.* **27**: 247–256.

902. Van Maren, D.S. and Winterwerp, J.C. (2013) The role of flow asymmetry and mud properties on tidal flat sedimentation. *Cont. Shelf Res.* **60S**: S71–S84.

903. Van Rijn, L.C. (1984) Sediment transport, part II: Suspended load transport. *J. Hydraul. Div. Proc. ASCE* **110**: 1613–1641.

904. Van Rijn, L.C. (1993) *Handbook Sediment Transport in Rivers, Estuaries and Coastal Seas*. Aqua Publ., Amsterdam.

905. Van Rijn, L.C., Ruessink, B.G. and Mulder, J.P.M. (2002) Summary of project results. In: *Coast3D-Egmond; The Behaviour of a Straight Sandy Coast on the Time Scale of Storms and Seasons*. ISBN 90-800356-5-3, Aqua Publ., Amsterdam.

906. Van Rijn, L.C., Caljauw, M. and Kleinhout, K. (2002) Basic features of morphodynamics at the Egmond site on the medium-term time scale of seasons. In: *Coast3D-Egmond; The Behaviour of a Straight Sandy Coast on the Time Scale of Storms and Seasons*. Eds., L.C. Van Rijn, B.G. Ruessink and J.P.M. Mulder. ISBN 90-800356-5-3, Aqua Publ., Amsterdam, pp. J1–J20.

907. Van Rijn, L.C. (2007) United view of sediment transport by currents and waves I: Initiation of motion, Bed roughness and Bed load transport. *J. Hydr. Eng., ASCE*, **133**: 649–667.

908. Van Rijn, L.C. (2007) United view of sediment transport by currents and waves II: Suspended transport. *J. Hydr. Eng., ASCE*, **133**: 668–689.

909. Van Rijn, L. (2010) Coastal erosion control based on the concept of sediment cells. Report EU project Concepts and Science for Coastal Erosion, Conscience, www.conscience-eu.net

910. Van Rijn, L.C. (2011) Analytical and numerical analysis of tides and salinities in estuaries; part I: tidal wave propagation in convergent estuaries. *Ocean Dynamics* **61**: 1719–1741.

911. Van Rijn, L.C. (2011) *Principles of Fluid Flow and Surface Waves in Rivers, Estuaries and Seas*. Aqua Publ., The Netherlands.

912. Van Rijn, L. (2011) Comparison Hydrodynamics and Salinity of Tide Estuaries: Elbe, Humber, Scheldt and Weser. TIDE Report, Deltares 1203583-000.

913. Van Rijn. L.C., Ribberink, J.S., Van Der Werf, J. and Walstra, D.J.R. (2013) Coastal sediment dynamics: recent advances and future research needs. *J. Hydraulic Res.* **51**: 475–493.

914. Van Rijn, L.C. (2014) A simple general expression for longshore transport of sand, gravel and shingle. *Coastal Eng.* **90**: 23–39.

915. Van Straaten, L.M.J.U. and Kuenen, P.H. (1957) Accumulation of fine grained sands in the Dutch Wadden sea. *Geol. en Mijnbouw* **19**: 406–413.

916. Van Veen, J. (1950) Eb-en vloedschaar systemen in de Nederlandse getijde-wateren. *Tijdschrift Kon. Ned. Aardrijkskundig Genootschap* **67**: 303–325.

917. Van der Veen, H.H. and Hulscher, S.M.J.H. (2009) Predicting the occurrence of sand banks in the North Sea. *Ocean Dynamics*.

918. Venier, C., Figueiredo da Silva, J., McLelland, S.J., Duck, R.W. and Lanzoni, D.S. (2012) Experimental investigation of the impact of macroalgal mats on flow dynamics and sediment stability in shallow tidal areas. *Est. Coastal Shelf Sci.* **112**: 52–60.

919. Verney R., Lafite R., Brun-Cottan J.C. and Le Hir P., (2010) Behaviour of a floc population during a tidal cycle: Laboratory experiments and numerical modelling. *Cont. Shelf Res.* **31**: S64–S83.

920. Vincent, C.E. and Green. M.O. (1990) Field measurements of the suspended sand concentration profiles and fluxes and of the resuspensionm coefficient over a rippled bed. *J. Geophys. Res.* **95**: 11591–11601.

921. Vincent, C.E., Stolk, A. and Porteer, C.F.C. (1998) Sand suspension and transport on the Middelkerke Bank (southern North Sea) by storms and tidal currents. *Mar. Geol.* **150**: 113–129.

922. Vittori, G. (2003) Sediment suspension due to waves. *J. Geophys. Res.* **108**(C6), **4**: 1–17.

923. Walcott, R.I. (1972) Past sea levels, eustasy and deformation of the Earth. *Quaternary Research* **2**: 1–14.

924. Walstra, D.J.R., Reniers, A.J.H.M., Ranasinghe, R., Roelvink, J.A. and Ruessink, B.G. (2012) On bar growth and decay during interannual net offshore migration. *Coastal Eng.* **60**: 190–200.

925. Walther, R. (2009) Developpement et exploitation d'un modele hydrosed-imentaire a trois dimensions sur l'estuaire de la Loire. Rapports Sogreah 1711822 R8 et R10, prepared for the GIP Loire Estuaire.

926. Walton, T.L. and Adams, W.D. (1976) Capacity of inlet outer bars to store sand. *Procs. 15th Int. Coast. Eng. Conf., ASCE*, New York, pp. 1919–1937.

927. Wang, Z.B. (1992) Fundamental considerations on morphodynamic modeling in tidal regions. Rep. Z331, part I, Delft Hydraul., Delft.

928. Wang, Z.B., Fokkink, R.J., de Vries, M. and Langerak, A. (1995) Stability of river bifurcations in 1D morphological models. *J. Hydraul. Res.* **33**: 739–750.

929. Wang, P., Ebersole, B.A. and Smith, E.R. (2003) Beach-profile evolution under spilling and plunging breakers. *J. Waterway, Port, Coast. Ocean Eng., ASCE* **129**: 41–46.

930. Wang, Y., Zhang, Y., Zou, X., Zhu, D. and Piper, D. (2012) The sand ridge field of the South Yellow Sea: Origin by river-sea interaction. *Mar. Geol.* **291–294**: 132–146.

931. Wang, Z.B., Hoekstra, P., Burchard, H., Ridderinkhof, H., De Swart, H.E. and Stive, M.J.F. (2012) Morphodynamics of the Wadden Sea and its barrier island system. *Ocean and Coastal Management* **68**: 39–57.

932. Wang, Z.B. and Townend, I.H. (2012) Influence of the nodal tide on the morphological response of estuaries. *Mar. Geology* **291–294**: 73–82.

933. Wargo, C.A. and Styles, R. (2007) Along channel flow and sediment dynamics at North Inlet, South Carolina. *Est. Coastal and Shelf Sci.* **71**: 669–682.

934. Watanabe, Y. and Tubino, M. (1992) Influence of bed load and suspended load on alternate bars. *Proc. Hydr. Eng. JSCE*, 36.

935. Wells, J.T. (1997) Tide-dominated estuaries and tidal rivers. In: *Geomorphology and Sedimentology of Estuaries*. Ed., G.M.E. Perillo, Elsevier, Amsterdam, pp. 179–205.

936. Warrick, R.A., Barrow, E.M. and Wigley, T.M.I. Eds. (1993) Climate and sea level change: Observations, projections and implications. Cambridge Univ. Press, 424 pp.

937. Weerman, E.J., Van de Koppel, J., Eppinga, M.B., Montserrat, F., Liu, Q-X and Herman, P.M.J. (2010) Spatial self-organization on intertidal mudflats through biophysical stress divergence. *The American Naturalist* 176(1).

938. Weiss, R. (2012) The mystery of boulders moved by tsunamis and storms. *Marine Geology* **295–298**: 28–33.

939. Wells, J.T., Adams, C.E., Park, Y.A. and Frankenberg, E.W. (1990) Morphology, sedimentology and tidal channel processes on a high-tide-range mudflat, west coast of South Korea. *Marine Geology* **95**, 111–130.

940. Werner, B.T. and Fink, T.M. (1993) Beach cusps as self-organized patterns. *Science*, **260**: 968–971.

941. Werner, B.T. and Kocurek, G. (1999) Bedform spacing from defect dynamics. *Geology* **27**: 727–730.

942. Werner, S.R., Beardsley, R.C. and Williams, A.J. (2003) Bottom friction and bed forms on the southern flank of Georges Bank. *J. Geophys. Res.* **108**(C11), GLO 5: 1–21.

943. West, J.R. and Mangat, J.S. (1986) The determination and prediction of longitudinal dispersion coefficients in a narrow, shallow estuary. *Est. Coast. Shelf Sci.* 161–181.

944. Whitehead, P. (2011) River Elbe River Engineering and Sediment Management Concept. ABP mer, R. 1805.

945. Whitehouse, R.J.S. and Mitchener, H.J. (1998) Observations of the morphodynamic behaviour of an intertidal mudflat at different timescales. In: *Sedimentary Processes in the Intertidal Zone*. Eds., K.S. Black, D.M. Paterson and A. Cramp, Geological Society, London, Special Publ. **139**: 255–271.

946. Wiberg, P.L. and Harris, C.K. (1994) Ripple geometry in wave-dominated environments. *J. Geophys. Res.* C1 **99**: 775–789.

947. Wijnberg, K. (1995) *Morphologic Behaviour of a Barred Coast Over a Period of Decades.* Thesis, Utrecht University, ISBN 90-6266-125-4, 245 pp.

948. Wijnberg, K. and Terwindt, J.H.J. (1995) Extracting decadal morphological behaviour from high-resolution long-term bathymetric surveys along the Holland coast using eigenfunction analysis. *Marine Geol.* **126**: 301–330.

949. Williams, P.B. and Kemp, P.H. (1971) Initiation of ripples on flat sediment beds. *J. Hydr. Div. ASCE* **97**: 505–522.

950. Williams, G.P. (1986) River meanders and channel size. *Journal of Hydrology*, **88**: 147–164.

951. Wilson, K.C. (1998) Mobile bed-friction at high shear stress. *J. Hydraul. Div.* **115**: 97–103.

952. Wind, H.G. and Vreugdenhil, C.B. (1986) Rip-current generation near structures. *J. Fluid Mech.* **171**: 459–476.

953. Winkelmolen, A.H. and Veenstra, H.J. (1980) The effect of a storm surge on nearshore sediments in the Ameland-Schiermonnikoog area. *Geologie Mijnbouw* **59**: 97–111.

954. Winterwerp, J.C. (2001) Stratification effects by cohesive and non-cohesive sediments. *J. Geophys. Res.* **106**: 22559–22574.

955. Winterwerp, J.C. (2002) On the flocculation and settling velocity of estuarine mud. *Cont. Shelf Res.* **22**: 1339–1360.

956. Winterwerp, J. and Van Kesteren, W. (2004). *Introduction to the Physics of Cohesive Sediment Dynamics in the Marine Environment.* Elsevier, Amsterdam.

957. Winterwerp, J.C., Manning, A.J., Martens, C., De Mulder, T. and Vanlede J. (2006) A heuristic formula for turbulence-induced flocculation of cohesive sediment. *Estuarine, Coastal and Shelf Science* **68**: 195–207.

958. Winterwerp, J.C. (2013) On the response of tidal rivers to deepening and narrowing; risks for a regime shift towards hyper-turbid conditions. Flemish-Dutch Scheldt Committee report, 83 pp.

959. Wolanski, E., King, B. and Galloway, D. (1995) Dynamics of the Turbidity Maximum in the Fly River Estuary, Papua New Guinea. *Estuarine, Coastal and Shelf Sci.* **40**: 321–337.

960. Wolanski, E., King, B. and Galloway, D. (1997) Salinity intrusion in the Fly River estuary, Papua New Guinea. *J. Coast. Res.* **13**: 983–994.

961. Wolanski, E., Moore, K., Spagnol, S., D'Adamo, N. and Pattiaratchi, C. (2001) Rapid, human-induced siltation of the Macro-Tidal Ord River Estuary, Western Australia. *Estuarine, Coastal and Shelf Science* **53**: 717–732.

962. Wood, R. and Widdows, J. (2002) A model of sediment transport over an intertidal transect, comparing the influence of biological and physical factors. *Limnol. Ocean.* **47**: 848–855.

963. Woodroffe, C.D. (2002) Coasts, form, processes and evolution. Cambridge Univ. Press, 623 pp.

964. Wright, L.D., Coleman, J.M. and Thom, B.G. (1973) Processes of channel development in a high tide-range environment: Cambridge Gulf-Ord River Delta, Western Australia. *J. Geol.* **81**: 15–41.

965. Wright, L.D., Chappel, J., Thom, B.G., Bradshow, M.P. and Cowell, P. (1979) Morphodynamics of reflective and dissipative beach and inshore systems: Southeastern Australia. *Mar. Geol.* **32**: 105–140.

966. Wright, L.D. and Short, A.D. (1984) Morphodynamic variability of surf zones and beaches: A synthesis. *Mar. Geol.* **56**: 93–118.

967. Wright, L.D. (1985) River Deltas. In: *Coastal Sedimentary Environments*. Ed., R.A. Davis. Springer-Verlag, New York, pp. 1–76.

968. Wright, L.D., Short, A.D. and Green, M.O. (1985) Short-term changes in the morphodynamic states of beaches and surf zones: An empirical predictive model. *Marine Geol.* **62**: 339–364.

969. Xing, F., Wang, Y.P. and Wang, H.Y. (2012) Tidal hydrodynamics and fine-grained sediment transport on the radial sand ridge system in the southern Yellow Sea. *Mar. Geol.* **291–294**: 192–210.

970. Xu, F., Wang, D.-P. and Riemer, N. (2010) An idealized model study of flocculation on sediment trapping in an estuarine turbidity maximum. *Cont. Shelf Res.* **30**: 1314–1323.

971. Xu, F., Tao, J., Zhou, Z., Coco, G. and Zhang, C. (2016) Mechanisms underlying the regional morphological differences between the northern and southern radial sand ridges along the Jiangsu Coast, China. *Marine Geol.* **371**: 1–17.

972. Yalin, M.S. (1964) Geometrical properties of sand waves. *Proc. Am. Soc. Civil Eng.* **90**: 105–119.

973. Yalin, M.S. and da Silva, A.M.F. (1992) Horizontal turbulence and alternating bars. *J. Hydrosci. Hydraul. Eng.* **9**: 47–58.

974. Yang, B.C., Dalrymple, R.W., Gingras, M.K., Chun, S.S. and Lee, H.J. (2007) Up-estuary variation of sedimentary facies and ichnocoenoses in an open-mouthed, macrotidal, mixed-energy estuary, Gomso Bay, Korea. *Journal of Sedimentary Research* **77**: 757–771.

975. Yates, M. L., Guza, R. T. and OReilly W. C. (2009) Equilibrium shoreline response: Observations and modelling. *J. Geophys. Res.* **114**: C09014.

976. Yates, M. L., Guza, R. T., OReilly W. C., Hansen, J.E. and Barnard, P.L. (2011) Equilibrium shoreline response of a high wave energy beach. *J. Geophys. Res.* **116**: C04014.

977. Yu, J. and Slinn, D.N. (2003) Effects of wave-current interaction on rip currents. *J. Geophys. Res.* **108**: 33-1–33-19.

978. Yu, Q., Flemming, B. and Gao, S. (2010) Scale-dependent Characteristics of Equilibrium Morphology of Tidal Basins along the Dutch-German North Sea

Coast. *Procs. 15th Conf. Physics of Estuaries and Coastal Seas*, Colombo, Sri Lanka, pp. 290–293.

979. Yu, Q., Wang, Y., Gao, S., Flemming, B. (2012) Modeling the formation of a sand bar within a large funnel-shaped, tide-dominated estuary: Qiantangjiang Estuary, China. *Marine Geology* **299–302**: 63–76.

980. Yu, Q., Wang, Y., Flemming, B. and Gao, S. (2012) Modelling the equilibrium hypsometry of back-barrier tidal flats in the German Wadden Sea (Southern North Sea). *Cont. Shelf Res.* **49**: 90–99.

981. Zagwijn, W.H. (1983) Sea level changes in the Netherlands during the Eemian. *Geol. Mijnb.* **62**: 437–450.

982. Zagwijn, W.H. (1986) Nederland in het Holoceen. Rijks Geologische Dienst, *Haarlem*, 46 pp.

983. Zang, D.P., Sunamura, T. (1994) Multiple bar formation by breaker induced vortices: A laboratory approach. In: *Proc. 24th, Int. Coast. Eng.* Kobe, Japan. ASCE, pp. 2856–2870.

984. Zhang, K., Douglas, B.C. and Leatherman, S.P. (2004) Global warming and coastal erosion. *Climatic Change* **64**: 41–58.

985. Zedler, E.A. and Street, R.L. (2001) Large-eddy simulation of sediment transport: Current over ripples. *J. Hydr. Eng.* **127**: 444–452.

986. Zhou, Z., Coco, G., Van der Wegen, M., Gong, Z., Zhang, C. and Townend, I. (2015) Modeling sorting dynamics of cohesive and non-cohesive sediments on intertidal flats under the effect of tides and wind waves. *Cont. Shelf Res.* **104**: 76–91.

987. Zimmerman, J.T.F. (1973) The influence of the subaqeous profile on wave-induced bottom stress. *Neth. J. Sea Res.* **6**: 542–549.

988. Zimmerman, J.T.F. (1976) Mixing and flushing of tidal embayments in the western Wadden Sea, I: Distribution of salinity and calculation of mixing time scales. *Neth. J. Sea Res.* **10**: 149–191.

989. Zimmerman, J.T.F. (1978) Topographic generation of residual circulation by oscillatory (tidal) currents. *Geophys. Astrophys. Fluid Dyn.* **11**: 35–47.

990. Zimmerman, J.T.F. (1981) Dynamics, diffusion and geomorphological significance of tidal residual eddies. *Nature* **290**: 549–555.

991. Zimmerman, J.T.F. (1986) The tidal whirlpool: A review of horizontal dispersion by tidal and residual currents. *Neth. Journal of Sea Research* **20**: 133–154.

Printed in the United States
By Bookmasters